Lecture Not

Computer S

Lecture Notes in Computer Science

Vol. 245: H.F. de Groote, Lectures on the Complexity of Bilinear Problems. V, 135 pages. 1987.

Vol. 246: Graph-Theoretic Concepts in Computer Science. Proceedings, 1986. Edited by G. Tinhofer and G. Schmidt. VII, 307 pages. 1987.

Vol. 247: STACS 87. Proceedings, 1987. Edited by F.J. Brandenburg, G. Vidal-Naquet and M. Wirsing. X, 484 pages. 1987.

Vol. 248: Networking in Open Systems. Proceedings, 1986. Edited by G. Müller and R.P. Blanc. VI, 441 pages. 1987.

Vol. 249: TAPSOFT '87. Volume 1. Proceedings, 1987. Edited by H. Ehrig, R. Kowalski, G. Levi and U. Montanari. XIV, 289 pages. 1987.

Vol. 250: TAPSOFT '87. Volume 2. Proceedings, 1987. Edited by H. Ehrig, R. Kowalski, G. Levi and U. Montanari. XIV, 336 pages. 1987.

Vol. 251: V. Akman, Unobstructed Shortest Paths in Polyhedral Environments. VII, 103 pages. 1987.

Vol. 252: VDM '87. VDM – A Formal Method at Work. Proceedings, 1987. Edited by D. Bjørner, C.B. Jones, M. Mac an Airchinnigh and E.J. Neuhold. IX, 422 pages. 1987.

Vol. 253: J.D. Becker, I. Eisele (Eds.), WOPPLOT 86. Parallel Processing: Logic, Organization, and Technology. Proceedings, 1986. V, 226 pages. 1987.

Vol. 254: Petri Nets: Central Models and Their Properties. Advances in Petri Nets 1986, Part I. Proceedings, 1986. Edited by W. Brauer, W. Reisig and G. Rozenberg. X, 480 pages. 1987.

Vol. 255: Petri Nets: Applications and Relationships to Other Models of Concurrency. Advances in Petri Nets 1986, Part II. Proceedings, 1986. Edited by W. Brauer, W. Reisig and G. Rozenberg. X, 516 pages. 1987.

Vol. 256: Rewriting Techniques and Applications. Proceedings, 1987. Edited by P. Lescanne. VI, 285 pages. 1987.

Vol. 257: Database Machine Performance: Modeling Methodologies and Evaluation Strategies. Edited by F. Cesarini and S. Salza. X, 250 pages. 1987.

Vol. 258: PARLE, Parallel Architectures and Languages Europe. Volume I. Proceedings, 1987. Edited by J.W. de Bakker, A.J. Nijman and P.C. Treleaven. XII, 480 pages. 1987.

Vol. 259: PARLE, Parallel Architectures and Languages Europe. Volume II. Proceedings, 1987. Edited by J.W. de Bakker, A.J. Nijman and P.C. Treleaven. XII, 464 pages. 1987.

Vol. 260: D.C. Luckham, F.W. von Henke, B. Krieg-Brückner, O. Owe, ANNA, A Language for Annotating Ada Programs. V, 143 pages. 1987.

Vol. 261: J. Ch. Freytag, Translating Relational Queries into Iterative Programs. XI, 131 pages. 1987.

Vol. 262: A. Burns, A.M. Lister, A.J. Wellings, A Review of Ada Tasking. VIII, 141 pages. 1987.

Vol. 263: A.M. Odlyzko (Ed.), Advances in Cryptology – CRYPTO '86. Proceedings. XI, 489 pages. 1987.

Vol. 264: E. Wada (Ed.), Logic Programming '86. Proceedings, 1986. VI, 179 pages. 1987.

Vol. 265: K.P. Jantke (Ed.), Analogical and Inductive Inference. Proceedings, 1986. VI, 227 pages. 1987.

Vol. 266: G. Rozenberg (Ed.), Advances in Petri Nets 1987. VI, 451 pages. 1987.

Vol. 267: Th. Ottmann (Ed.), Automata, Languages and Programming. Proceedings, 1987. X, 565 pages. 1987.

Vol. 268: P.M. Pardalos, J.B. Rosen, Constrained Global Optimization: Algorithms and Applications. VII, 143 pages. 1987.

Vol. 269: A. Albrecht, H. Jung, K. Mehlhorn (Eds.), Parallel Algorithms and Architectures. Proceedings, 1987. Approx. 205 pages. 1987.

Vol. 270: E. Börger (Ed.), Computation Theory and Logic. IX, 442 pages. 1987.

Vol. 271: D. Snyers, A. Thayse, From Logic Design to Logic Programming. IV, 125 pages. 1987.

Vol. 272: P. Treleaven, M. Vanneschi (Eds.), Future Parallel Computers. Proceedings, 1986. V, 492 pages. 1987.

Vol. 273: J.S. Royer, A Connotational Theory of Program Structure. V, 186 pages. 1987.

Vol. 274: G. Kahn (Ed.), Functional Programming Languages and Computer Architecture. Proceedings. VI, 470 pages. 1987.

Vol. 275: A.N. Habermann, U. Montanari (Eds.), System Development and Ada. Proceedings, 1986. V, 305 pages. 1987.

Vol. 276: J. Bézivin, J.-M. Hullot, P. Cointe, H. Lieberman (Eds.), ECOOP '87. European Conference on Object-Oriented Programming. Proceedings. VI, 273 pages. 1987.

Vol. 277: B. Benninghofen, S. Kemmerich, M.M. Richter, Systems of Reductions. X, 265 pages. 1987.

Vol. 278: L. Budach, R.G. Bukharajev, O.B. Lupanov (Eds.), Fundamentals of Computation Theory. Proceedings, 1987. XIV, 505 pages. 1987.

Vol. 279: J.H. Fasel, R.M. Keller (Eds.), Graph Reduction. Proceedings, 1986. XVI, 450 pages. 1987.

Vol. 280: M. Venturini Zilli (Ed.), Mathematical Models for the Semantics of Parallelism. Proceedings, 1986. V, 231 pages. 1987.

Vol. 281: A. Kelemenová, J. Kelemen (Eds.), Trends, Techniques, and Problems in Theoretical Computer Science. Proceedings, 1986. VI, 213 pages. 1987.

Vol. 282: P. Gorny, M.J. Tauber (Eds.), Visualization in Programming. Proceedings, 1986. VII, 210 pages. 1987.

Vol. 283: D.H. Pitt, A. Poigné, D.E. Rydeheard (Eds.), Category Theory and Computer Science. Proceedings, 1987. V, 300 pages. 1987.

Vol. 284: A. Kündig, R.E. Bührer, J. Dähler (Eds.), Embedded Systems. Proceedings, 1986. V, 207 pages. 1987.

Vol. 285: C. Delgado Kloos, Semantics of Digital Circuits. IX, 124 pages. 1987.

Vol. 286: B. Bouchon, R.R. Yager (Eds.), Uncertainty in Knowledge-Based Systems. Proceedings, 1986. VII, 405 pages. 1987.

Vol. 287: K.V. Nori (Ed.), Foundations of Software Technology and Theoretical Computer Science. Proceedings, 1987. IX, 540 pages. 1987.

Vol. 288: A. Blikle, MetaSoft Primer. XIII, 140 pages. 1987.

Vol. 289: H.K. Nichols, D. Simpson (Eds.), ESEC '87. 1st European Software Engineering Conference. Proceedings, 1987. XII, 404 pages. 1987.

Vol. 290: T.X. Bui, Co-oP A Group Decision Support System for Cooperative Multiple Criteria Group Decision Making. XIII, 250 pages. 1987.

Vol. 291: H. Ehrig, M. Nagl, G. Rozenberg, A. Rosenfeld (Eds.), Graph-Grammars and Their Application to Computer Science. VIII, 609 pages. 1987.

Vol. 292: The Munich Project CIP. Volume II: The Program Transformation System CIP-S. By the CIP System Group. VIII, 522 pages. 1987.

Vol. 293: C. Pomerance (Ed.), Advances in Cryptology – CRYPTO '87. Proceedings. X, 463 pages. 1988.

Vol. 294: R. Cori, M. Wirsing (Eds.), STACS 88. Proceedings, 1988. IX, 404 pages. 1988.

Vol. 295: R. Dierstein, D. Müller-Wichards, H.-M. Wacker (Eds.), Parallel Computing in Science and Engineering. Proceedings, 1987. V, 185 pages. 1988.

Lecture Notes in Computer Science

Edited by G. Goos and J. Hartmanis

351

J. Díaz F. Orejas (Eds.)

TAPSOFT'89

TAPSOFT '89

Proceedings of the International Joint Conference on Theory and Practice of Software Development
Barcelona, Spain, March 13–17, 1989

Volume 1:
Advanced Seminar on Foundations of Innovative Software Development I and Colloquium on Trees in Algebra and Programming (CAAP '89)

Springer-Verlag
Berlin Heidelberg New York London Paris Tokyo

Editors

Josep Díaz
Fernando Orejas
Facultat d'Informàtica
Universitat Politècnica de Catalunya
Pau Gargallo, 5, E-08028 Barcelona

CR Subject Classification (1987): D.2.1, D.2.10, D.3, F.3.1–3, F.4

ISBN 3-540-50939-9 Springer-Verlag Berlin Heidelberg New York
ISBN 0-387-50939-9 Springer-Verlag New York Berlin Heidelberg

Printing and binding: Druckhaus Beltz, Hemsbach/Bergstr.
2145/3140-543210 – Printed on acid-free paper

Preface

TAPSOFT'89 is the Third International Joint Conference on Theory and Practice of Software Development.

TAPSOFT'89 is being held from March 13 to March 17, 1989 in Barcelona. TAPSOFT'89 has been organized by Departament de Llenguatges i Sistemes Informàtics (Universitat Politècnica Catalunya) and Associació de Tècnics d'Informàtica, and has been supported by EATCS and BCS.

TAPSOFT'89 consists of three parts:

Advanced Seminar on Foundations of Innovative Software Development

The aim of the Advanced Seminar is to bring together leading experts in the various fields which form the foundations of the Software development and to provide a forum to discuss the possible integration of available theories and methods in view of their applications. The Advanced Seminar will consist of a number of invited talks and panel discussions.

Invited Speakers:

D. Bjørner (Lyngby)	G. Gonnet (Waterloo)
C.A.R. Hoare (Oxford)	B. Krieg-Brückner (Bremen)
J.L. Lassez (Yorktown Heights)	P. Lescanne (Nancy)
C.A. Vissers (Twente)	J. Wing (Pittsbugh)
M.Wirsing (Passau)	

Panels:

Software Environments and Factories. Chairman: H.Weber (Dortmund)
Effectiveness of EC Programs for Software Development. Chairman: J.Díaz (Barcelona)
Social Responsibility in Computer Science. Chairman: B. Mahr (Berlin)
Theory and Practice of Software Development. Chairman: H.Ehrig (Berlin)

Colloquium on Trees in Algebra and Programming

This is the sixteenth edition of these Colloquia. The preceding colloquia were held in France and Italy, except in 1985 in Berlin, where for the first time CAAP was integrated into TAPSOFT. This was repeated in 1987 in Pisa.

In keeping with the tradition of CAAP as well as with the overhall theme of TAPSOFT, the selected papers focus on the following topics:

- Algorithms
- Proving Techniques
- Algebraic Specifications
- Concurrency
- Foundations

The Program Committee for CAAP'89 is the following:

S. Abramsky (London)
A. Arnold (Bordeaux)*
A. Bertoni (Milano)*
M. Dauchet (Lille)*
P. Deagano (Pisa)*
J. Díaz (Barcelona)* (Chairman)
H. Ehrig (Berlin)*
N. Francez (Haifa)*
G. Gonnet (Waterloo)
U. Montanari (Pisa)
M.Nivat (Paris)
A. Pettorosi (Roma)*
M. Rodriguez-Artalejo (Madrid)*
G. Rozenberg (Leiden)*
U. Schöning (Koblenz)*
J.M. Steyaert (Palaiseau)*
M. Wirsing (Passau)*

Thirteen of them (the ones with *) attended the final Program Committee meeting.

Colloquium on Current Issues in Programming Languages

In keeping with the overall theme of TAPSOFT conferences, CCIPL focuses on those aspects of programming which are most important for innovative software development. Integratior of formal methods and practical aspects of software production are also stressed.

The selected papers cover the following topics:

- Programming Language Concepts
- Language Implementation
- Programming Paradigms
- Software Development

The Program Committee for CCIPL is the following:

E. Astesiano (Genova)*
D. Bjørner (Lyngby)*
H. Ganzinger (Dortmund)*
N. Jones (Copenhagen)*
G. Kahn (Sophia-Antipolis)*
J.L. Lassez (Yorktown Heights)
P. Lescanne (Nancy)*
G. Levi (Pisa)
B. Mahr (Berlin)*
T. Maibaum (London)*
J. Meseguer (Palo Alto)
F. Orejas (Barcelona)* (Chairman)
H. Weber (Dortmund)*

Ten of them (the ones with *) attended the final Program Committee meeting.

The TAPSOFT'89 Conference Proceedings are published in two volumes. The first volume includes the papers from CAAP, plus the most theoretical invited papers. The second volume includes the papers from CCIPL and the invited papers more relevant to CCIPL.

We would like to thank all the Program Committee members as well as the referees listed on the next page for their care in reviewing and selecting the submitted papers.

We also wish to express our gratitude to the members of the Organizing Committee: C.Alvarez, F.Bach, P.Botella (Chairman), X.Burgués, X. Franch, T.Lozano, R.Nieuwenhuis, R.Peña and C.Roselló. Without their help, the Conference would not have been possible.

Barcelona, March 1989

J.Díaz
F.Orejas

Referees for TAPSOFT'89

M. Abadi	H. Aida	F. Baiardi	F. Bellegarde
G. Berry	P. Bonizzoni	F.J. Brandenburg	M. Broy
H. Bruun	P.M. Bruun	J. Buntrock	R. Casas
R. Casley	I. Classen	G. Costa	B. Courcelle
J. Cramer	K. Culik II	R. J. Cunningham	O. Danvy
D. de Frutos	R. De Nicola	R. de Simone	S. Debray
Delahaye	M. Diaz	W. Ditt	S. Donatelli
J. Engelfriet	M.I Fernandez-Camacho	G.L. Ferrari	A.C.W. Finkelstein
D.A. Fuller	J. Gabarró	A. Geser	R. Giegerich
A. Giovini	R. Glas	M. Goedicke	M. Goldwurm
R. Gorrieri	V. Gruhn	O. Grumberg	A. Gündel
H. Gust	M. Hallmann	C. Hankin	H. Hansen
M. Hanus	K. Haveland	N. Heintze	R. Hennicker
D. Hoffmann	C.J. Hogger	C.K. Holst	H. Hormann
T. Hortalá	G. Huet	U. Hummert	H. Hussmann
H. Huwig	P. Inverardi	A. Itai	J.P. Jacquot
J. Jaffar	M. Jantzen	J. Jaray	M. Jeusfeld
A. Kaul	U. Kelter	H. Kirchner	J. Kleijn
J.W. Klop	L.B. Kova'cs	J. Kramer	H.J. Kreowski
A. Labella	A. Laville	M. Loewe	G. Longo
H.H. Løvengreen	S. Lynenskjold	A. Maggiolo-Schettini	K. Malmkjær
M. Martelli	N. Marti-Oliet	S. Martini	G. Mauri
P. Messazza	R. Milner	T. Mogensen	E. Moggi
M. Moriconi	L.S. Nielsen	M. Nielsen	R. Nieuwenhuis
J.F. Nilsson	C. Niskier	P. Nivela	F. Nürnberg
E.-R. Olderog	R. Orsini	P. Padawitz	C. Palamidessi
J.S. Pedersen	D. Pedresch	R. Peña	P. Pepper
L. Pomello	M. Proietti	A. Poigné	A. Quere
A.P. Ravn	G. Reggio	G.A. Ringwood	M. Ringwood
E. Ruohonen	N. Sabadini	M.R. Sadler	A. Salibra
K. Salomaa	F. Saltor	L. Schöpe	A. Skowron
M. Smyth	J. Staunstrup	A. Tarlecki	P. Taylor
W. Thomas	J. Torán	G. Vaglini	M. Venturini-Zilli
G. Vidal-Naquet	H. Vollmer	S.J. Vickers	H.Wagener
R. Weber	S. Wiebrock	T. Winkler	S. Wolf
S. Yoccoz	Zhou Chaochen	E. Zucca	

CONTENTS OF VOLUME I

Invited Lectures

Communications

CONTENTS OF VOLUME II

Invited Lectures

Communications

The Varieties of Programming Language

C.A.R. Hoare
Oxford University Computing Laboratory

In 1973, Christopher Strachey wrote a monograph with the above title [8]. It began

> There are so many programming languages in existence that it is a hopeless task to attempt to learn them all. Moreover many programming languages are very badly described; in some cases the syntax, or rather *most* of the syntax, is clearly and concisely defined, but the vitally important question of the semantics is almost always dealt with inadequately.

Since that time, there has been a further spectacular increase in the variety of languages, and several of them have even entered common use. Completely new programming paradigms have emerged, based on logical inference, on functions and relations, or on the interaction of objects. New language features include communication, non-determinism, type inheritance, type inference, and lazy evaluation. Many programmers have become fluent in several languages; and many large applications have been constructed from parts implemented in different programming notations. Programming language designers are attempting to alleviate the resulting problems by combining the merits of several paradigms into a single more comprehensive language; but the signs of success in such a combination are difficult to recognise.

A solution to these problems may emerge from a wider understanding and agreement about the nature of programming and the choices available to the programming language designer and user. The current state of our understanding seems little better now than when Strachey wrote

> Not only is there no generally accepted notation, there is very little agreement even about the use of words. The trouble seems to be that programming language designers often have a rather parochial

> outlook and appear not to be aware of the range of semantic possibilities for programming languages. As a consequence they never explain explicitly some of the most important features of a programming language and the decisions among these, though rarely mentioned (and frequently I suspect made unconsciously), have a very important effect on the general flavour of the language...
> The unsatisfactory nature of our understanding of programming languages is shown up by the fact that although the subject is clearly a branch of mathematics, we still have virtually no theorems of general application and remarkably few specialised results.

This paper is dedicated to pursuit of the goals that Strachey set himself in his own work.

> The main purpose of this paper is to discuss some features of the range of semantic possibilities in programming language... A second purpose is to advocate a more conventionally mathematical approach to the problem of describing a programming language and defining its semantics, and, indeed, to the problems of computation generally.

To lighten the task, this paper concentrates on the following varieties of programming language.

1. Deterministic, like LISP or PASCAL: the possible result of executing each program is fully controllable by its environment or user.
2. Non-deterministic, like occam or Dijkstra's language [2]: the result of execution of a program may be affected by circumstances beyond the knowledge or control of its environment or user.
3. Strict, like LISP or Dijkstra's language: each stage of a calculation must be completed successfully before the start of the next stage which will use its results.
4. Lazy, like KRC or Miranda[1] [9]: a calculation is started only when its result is found to be needed.
5. Interactive, like CSP or occam: the program communicates with its environment during execution.
6. Non-interactive, like the original FORTRAN: all the data required by the program may in principle be accumulated before the program starts; and all the results may be accumulated during the calculation for eventual output if and when the program terminates successfully.

---

[1] Miranda is a trade mark of Research Software Ltd.

Summary

The next section compares the merits of two methods of studying the mathematics of programming languages. The denotational method, due to Strachey, requires the construction of a mathematical meaning for each major component of each program expressed in each language. The various operators which combine the components are defined as functions, on the meanings of their operands. The algebraic approach avoids giving any meaning to the operators and operands; instead it formulates general equations describing the algebraic properties of the operators individually and in relation to each other. In some cases, this is sufficient to characterise the meaning of each operator, at least up to some form of equivalence or isomorphism. The algebraic approach seems to offer considerable advantages in the classification of the varieties of programming language, because it permits individual features and properties common to a range of languages to be characterised independently of each other.

The next important idea, due to Scott [7], is an ordering relation (approximation) between programs. This permits an elegant mathematical construction for the most essential of programming language features, namely recursion or iteration. Other benefits of the ordering include an elegant treatment of the phenomena of non-termination or even non-determinism. In software engineering, the same ordering can be used for stepwise development of specifications, designs, and programs.

For classifying the major varieties of programming language, the combination of the approximation ordering with an algebraic approach seems to offer the greatest benefits. Each variety is characterised by the different laws which they satisfy; and in many cases, the laws differ from each other only in the direction of the ordering between the two sides of an inequation. The final sections of this paper give several simple examples, relating to zero morphisms (**abort**), products (records, declarations), and coproducts (variants, cases).

The paper ends with a brief summary of related work, both completed and remaining to be started.

1 Denotational and Algebraic Approaches

A denotational semantics for a language defines the mathematical meaning of every syntactically valid program expressed in the language. A meaning is also given to every significant component of the program, in such a way that the meaning of the whole program is defined in terms of the meaning of its parts.

The study of denotational semantics has made a significant contribution to the development both of mathematics and of computing science. In the early days, Church's untyped lambda-calculus was selected as the branch of mathematics within which to formalise the meanings of programs. After the discovery of reflexive domains by Scott, a much more elegant and abstract basis for denotational semantics is found in domain theory.

In computing science, denotational semantics gives the programming language designer and theorist an excellent method of exploring the consequences of design decisions before committing the resources required to implement it, or the even greater resources required to test it in practical use. The semantics also provides a basis for the correct design and development of efficient implementations of the language, compatible across a variety of machines. It provides a conceptual framework for the user to specify programs before they are written, and to write them in a manner which reduces the danger of failing to meet that specification. And finally, it reveals most clearly the major structural variations between programming languages, even those that result from apparently quite minor decisions about individual features. When Dedekind, Frege and Russell gave interpretations within set theory of the long-familiar mathematical concept of number, they found similar major differences between the natural numbers, the integers, the fractions, the reals, and the complex numbers.

Because denotational semantics is so effective in revealing the differences between languages, it is correspondingly less effective in revealing their similarities. The essential similarities between the varieties of number are most clearly illuminated by listing the many algebraic properties which they share; the differences can then be isolated by listing the properties which they do not share. The dependence or independence of various selections of laws can be established by standard algebraic techniques. New kinds of mathematical structure can be discovered by combining algebraic laws in new ways. In some cases it is possible to prove that the meaning of each primitive concept and operation is defined uniquely by the laws which it obeys (perhaps up to some acceptable degree of isomorphism or equivalence). And finally, one may hope to classify all possible mathematical structures obeying a given basic set of reasonable laws.

All these properties of an algebraic approach are potentially beneficial to theorists of computing science. In view of the important role of algebra in the development, teaching and use of mathematics, it would be reasonable to expect similar benefits to the study of programming languages, their design, implementation, teaching and use. The algebraic approach is the one adopted in this paper for exploring the varieties of programming language.

The branch of algebra which seems to be most relevant for programming language theory is category theory. Its high level of abstraction avoids the clutter of syntactic detail associated with concrete programs, including declarations, variables, scopes, types, assignment and parameter passing. Its object structure neatly captures and takes advantage of the type structure of strictly typed languages, and the scope structure of languages with a concept of locality. And finally, it turns out that many of the features of a programming language are definable as categorical concepts in such a way that their effective uniqueness is guaranteed.

Exploration of the variety of programming languages requires consideration of categories which are enriched by an ordering relation [10]. These constitute a particularly simple kind of two-category, in which a preorder is given on each of its homsets. However, no knowledge of two-categories or even of category theory is required of the reader of this paper.

2 The Approximation Ordering

The approximation ordering between programs was originally introduced into programming language theory to provide an explanation of recursion. We shall use it as a general method of comparing the quality of any two programs written in the same language. If p and q are programs, $p \sqsubseteq q$ means that, for any purpose whatsoever and in any context of use, the program q will serve as well as program p, or even better. This is an ordering relation because it is reflexive and transitive

$$p \sqsubseteq q$$

$$\text{if } p \sqsubseteq q \text{ and } q \sqsubseteq r \text{ then } p \sqsubseteq r.$$

We define two programs to be equivalent if they approximate each other

$$p \equiv q \;\hat{=}\; p \sqsubseteq q \text{ and } q \sqsubseteq p.$$

Since we are willing to tolerate equivalent but non-identical programs, we do not require the approximation ordering to be antisymmetric.

This ordering relation may hold not only between programs but also between a specification and a program, or even between two specifications. In this case, $p \sqsubseteq q$ means that p and q are essentially the same specification, or that p is the more abstract or general specification and q is a more concrete or detailed specification, closer to the design of a program to meet the specification p. Every implementation of q is therefore an implementation of p, but not necessarily the other way round. In the extreme, q may itself be a program expressed in some efficiently executable programming language; this then is the final deliverable outcome of a development process that may have gone through several successive stages of more and more concrete design. By transitivity of $\sqsubseteq$, this final program meets the original specification with which the design process started. Thus the approximation ordering is as important for software engineering as it is for the theory of programming languages.

In a deterministic programming language the relation $p \sqsubseteq q$ holds if p always gives the same result as q, and q terminates successfully in all initial states from which p terminates, and maybe more; this is because non-termination is, for any purpose whatsoever, the least useful behaviour a program can display. In a non-deterministic language, an additional condition for $p \sqsubseteq q$ is that every possible outcome of executing q in a given environment should also be a possible outcome of executing p in the same environment; but maybe p is less deterministic, and might give some outcome that q will not. The program p is therefore less predictable, less controllable, or in short, a mere approximation to q. In a programming language like the typed lambda-calculus, all programs are terminating deterministic functions. One such program approximates another only if they compute the same function. As a result, the approximation ordering is an equivalence relation. The algebraic properties of such languages have been formulated directly in standard category theory; in this paper we will concentrate on languages with a non-trivial approximation ordering.

The $\sqsubseteq$ relation between programs is seen to have several different meanings, according to whether it is applied to a deterministic or a non-deterministic language. This is a characteristic feature of the algebraic method; for example, addition of complex numbers is very different from addition of integers; nevertheless the same symbol is used for similar purposes in both cases, for the very reason that the two meanings share their most important algebraic properties. The purpose of the $\sqsubseteq$ relation is to compare the quality of programs and specifications; and for this reason it must be transitive, and might as well be reflexive. Given that it has these properties, we are free to give the symbol the most appropriate interpretation for each individual language that we wish to study.

The *bottom* of an ordering relation (if it exists) is often denoted by $\perp$. It is defined (up to equivalence) by the single law

$$\perp \sqsubseteq q, \qquad \text{for all } q.$$

Proof. Let $\perp'$ be another bottom. Then $\perp \sqsubseteq \perp'$ because $\perp$ is a bottom; and $\perp' \sqsubseteq \perp$ because $\perp'$ is a bottom. Therefore $\perp \equiv \perp'$ □

The bottom program is for any purpose the most useless of all programs; for example, the program which never terminates under any circumstances whatsoever. In FORTRAN or machine code, the bottom is the tight loop

10 GO TO 10

In LISP it is the function *BOTTOM*, defined by a non-terminating recursion

(DEFUN BOTTOM X (BOTTOM X)).

In Dijkstra's language the bottom is called **abort**. This may fail to terminate; or being non-deterministic it may do even worse: it may terminate with the wrong result, or even the right one (sometimes, just to mislead you).

The *meet* of two elements p and q (if it exists) is denoted $p \sqcap q$. It is the best common approximation of both p and q. It is defined (up to equivalence) by the single law

$$r \sqsubseteq p \sqcap q \text{ iff } r \sqsubseteq p \text{ and } r \sqsubseteq q, \qquad \text{all } p, q, r.$$

Proof. Let $\sqcap'$ be another operator satisfying this law. By substituting $p \sqcap q$ for r,

$$\begin{aligned} p \sqcap q \sqsubseteq p \sqcap' q \quad &\text{iff} \quad p \sqcap q \sqsubseteq p \text{ and } p \sqcap q \sqsubseteq q \\ &\text{iff} \quad p \sqcap q \sqsubseteq p \sqcap q \end{aligned}$$

which is true by reflexivity of $\sqsubseteq$. The proof that $p \sqcap' q \sqsubseteq p \sqcap q$ is similar. □

Other algebraic properties of $\sqcap$ follow from the defining law. For example, it is associative, commutative, and idempotent, and

$$p \sqcap q \sqsubseteq p.$$

Proof. $p \sqcap q \sqsubseteq p \sqcap q$ implies $p \sqcap q \sqsubseteq p$ and $p \sqcap q \sqsubseteq q$ □

In a non-deterministic language, $p \sqcap q$ is the program which may behave like p or like q, the choice not being determined either by the programmer or by the environment of the program. In Dijkstra's language it would be written

if true → p

[] true → q

fi.

In this language, non-determinism is "demonic": if a program contains the non-deterministic possibility of going wrong, this is as bad as if it always goes wrong:

$$\bot \sqcap p \equiv \bot, \qquad \text{all } p.$$

If this law were not confirmed by ample experience of bugs in computer programs, it should still be adopted as a moral law by the engineer who undertakes to deliver reliable products.

Even a deterministic language has a meet. $(p \sqcap q)$ is the program that terminates and gives the same result as p and q when started from any initial state in which p and q both terminate *and both* give the *same* result. $(p \sqcap q)$ otherwise fails to terminate. More simply put, the graph of $p \sqcap q$ is the intersection of the graphs of p and q.

Although this description uniquely specifies the meaning of $(p \sqcap q)$, the operator $\sqcap$ is unlikely to be included explicitly in the notations of a deterministic programming language. Nevertheless it does exist, and can actually be programmed by

$$\mathit{meet}\,(p, q) \;\widehat{=}\; (\textbf{if } p = q \textbf{ then } p \textbf{ else } \bot).$$

A more efficient implementation of $(p \sqcap q)$ is simply to select an arbitrary member of the pair. By definition of $\sqcap$ this will be better than $(p \sqcap q)$, and an implementation should always be allowed to implement a program better (for all purposes) than the one written by the programmer. That is one of the motives for introducing the approximation ordering.

A program p is called *total* if it is a maximal element of the $\sqsubseteq$ ordering:

$$\text{if } p \sqsubseteq q \text{ then } q \sqsubseteq p, \qquad \text{for all } q.$$

A total program is one which is deterministic and always terminates. That is why it cannot be improved, either by extending the range of environments in which it terminates, or by reducing the range of its non-determinacy. Although we can reasonably point to a single worst program, unique up to equivalence, no reasonable programming language can have a single best program. If there were, it would be the best program for all purposes whatsoever, and there would be no point in using any other program of the language! Such a program would be a miracle (according to Dijkstra). But so far from solving all the problems of the world, its existence in programming language can only make that language futile.

The laws postulated so far (and those that can be deduced from them) are true in every programming language that admits the possibility of non-termination. It is not yet possible to distinguish varieties of language, not even to distinguish deterministic from non-deterministic languages. These and other important distinctions will be drawn in the next section.

3 Composition

Without much loss of generality, we can confine attention to languages in which there exists some method of composing programs p and q into some larger program called $(p; q)$. Execution of such a composite program usually (but not always) involves execution of both its components. In a procedural programming language like PASCAL or occam, we interpret this notation as sequential execution: q does not start until p has successfully terminated. In a functional language it denotes functional composition, such as might be defined in LISP

$$LAMBDA\ X(p(q\ X))).$$

Here is an example of an advantage of the algebraic approach: it ignores the spectacular difference in the syntax of composition in procedural and functional languages and concentrates attention on their essential mathematical similarities. Chief among the general properties of composition is associativity

$$p; (q; r) = (p; q); r.$$

Another property is the existence of a unit denoted I. It is uniquely defined by the equations

$$p; I = p = I; p, \qquad \text{for all } p.$$

Proof. Let I' be another unit. Because I' is a unit, $I = I'; I$. Because I is a unit, $I'; I = I'$. The result follows by transitivity of equality. □

In a procedural programming language, the unit is the trivial assignment $x := x$. In FORTRAN it is CONTINUE, in Dijkstra's language **skip**, and in LISP the identity function

$$(LAMBDA\ X\ X).$$

Another general property of composition is that it is monotonic, i.e., it respects the ordering of its operands:

$$\text{if } p \sqsubseteq q \text{ then } p; r \sqsubseteq q; r \text{ and } r; p \sqsubseteq r; q, \qquad \text{all } p, q, r.$$

If the preorder is regarded as giving a two-categorical structure to the programming language, monotonicity of composition is just a restatement of the interchange law. More generally, every operator of a programming language must be monotonic with respect to $\sqsubseteq$. Otherwise, there would be some context which would fail to be improved (or at least left the same) by replacing a worser component by a better. So any non-monotonic operator in a programming language would make the $\sqsubseteq$ relation useless for the purpose for which it is intended, namely to state that one program is better than another in all contexts of use. In algebra, that is one of the most convincing reasons for adopting a particular law as an axiom.

In a typed language, the composition of programs is undefined when the type of the result of the first component differs from that expected by the second component. This complication is elegantly treated in category theory by associating source and target objects with each arrow. Nevertheless, in this paper we will just ignore the complication, and assume that the source and target types of all the operands are the same. Where this is impossible, restrictions will be placed informally on the range of the variables.

It is well-known that the composition of two total functions is a total function. The same is true of total programs, which form a deterministic subset of a non-deterministic language. In fact all languages of interest will satisfy the law:

$$\text{if } p \text{ and } q \text{ are total, so is } (p; q), \qquad \text{all } p, q.$$

We have seen how the unit I of composition is uniquely defined. A zero of composition (if it exists) is denoted by Z. It is uniquely defined by the laws

$$p; Z = Z = Z; q, \qquad \text{for all } p \text{ and } q.$$

Proof. Let Z' also satisfy these laws. Then, because Z is a zero, $Z'; Z = Z$, and because Z' is a zero $Z' = Z'; Z$. The result follows by transitivity of equality. □

In Dijkstra's programming language, the zero is the bottom program **abort**. To specify the execution of q after termination of **abort** cannot redeem the situation, because **abort** cannot be relied on to terminate. To specify execution of p before abortion is equally ineffective, because the non-termination will make any result of executing p inaccessible and unusable. In other words, composition in Dijkstra's language is *strict* in the sense that it gives bottom if either of its arguments is bottom.

However, in a language which interacts with its environment, the account given in the last paragraph does not apply. The program p may engage in some useful communications (perhaps even forever) thus postponing or avoiding the fate of abortion. So $(p; \perp)$ can be strictly better than $\perp$. The same is true in a language with jumps, since p can avoid the abortion by jumping over it. A similar situation obtains in a logic programming language like PROLOG, in which the ordering of the clauses of a program is significant. If a query can be answered by the rules in the first part of the program p, it does not matter if a later part aborts; but abortion at the beginning is an irrecoverable error.

In these languages, the bottom is a *quasi-zero* z, in a sense defined up to equivalence by the laws

$$z; p \sqsubseteq z \sqsubseteq q; z, \quad \text{all } p \text{ and } q.$$

Proof. let z' also be a quasi-zero. Then because z is a quasi-zero, $z \sqsubseteq z';z$. Because z' is a quasi-zero $z'; z \sqsubseteq z'$. By transitivity of $\sqsubseteq$, $z \sqsubseteq z'$. A similar proof shows that $z' \sqsubseteq z$. The two quasi-zeroes are therefore equivalent. □

In a lazy functional programming language, the call of a function will not evaluate an argument unless the value of the argument is actually needed during evaluation of the body of the function. Such a mechanism of function call is said to be non-strict or lazy. Thus the function

$$DEFFUN\ K3 = LAMBDA\ X\ 3$$

can be successfully called by

$$(K3\ (\bot Y))$$

and will deliver the value 3, because no attempt is made to evaluate its argument $(\bot Y)$. However the call

$$\bot(K3\,Y)$$

will never succeed, no matter what the value of Y. This shows that $K3; \bot$ is actually worse than $\bot; K3$.

As a result, the bottom program in a lazy language is neither a zero nor a quasi-zero. In fact it is a quasi-cozero, in a sense defined up to equivalence by the laws

$$p; \bot \sqsubseteq \bot \sqsubseteq \bot; q, \qquad \text{for all } p \text{ and } q.$$

We have now explored the way in which composition interacts with the bottom program $\bot$. The question now arises how composition interacts with the $\sqcap$ operator. From the defining property of $\sqcap$ we can derive the following weak distribution law

$$r; (p \sqcap q); s \ \sqsubseteq\ (r; p; s) \sqcap (r; q; s).$$

Proof. $(p \sqcap q) \sqsubseteq p$ and $(p \sqcap q) \sqsubseteq q$. Because composition is monotonic, $r;(p \sqcap q);s \sqsubseteq r;p;s$ and $r;(p \sqsubseteq q);s \sqsubseteq r;q;s$. The law follows from the defining property of $\sqcap$. □

These laws hold in any programming language. However, in a truly non-deterministic language, the laws may be strengthened to an equation

$$r;(p \sqcap q);s = (r;p;s) \sqcap (r;q;s).$$

This strengthening is not valid in a functional or deterministic language.

The treatment of zeroes, quasi-zeroes and quasi-cozeroes is a simple example of the power of algebraic laws in classifying the varieties of programming language. Three clear varieties have emerged

(1) The strict non-interactive languages in which $\bot$ is a zero.

(2) The strict interactive languages in which $\bot$ is a quasi-zero.

(3) The non-strict languages in which $\bot$ is a quasi-cozero.

An orthogonal classification is that between deterministic languages and non-deterministic languages, in which composition distributes through $\sqcap$.

4 Products

If p and q are programs, we define their product $\langle p, q\rangle$ to be a program which makes a second copy of the current argument or machine state, and executes p on one of the two copies and q on the other one. The two results of p and q are delivered as a pair. This allows execution of p and q to proceed in parallel without interfering with each other. In a functional programming language with lists as a data structure, this can be defined:

$$\langle p, q\rangle \triangleq \lambda x.cons(px, qx).$$

In such a language, the duplication of the argument x is efficiently implemented by copying a machine address rather than the whole value of the argument. A procedural language has side-effects which update its initial machine state; so this implementation causes interference, and would not be valid. Instead, a completely fresh copy of the machine state is required; and this is what is provided by the *fork* feature of UNIX. The algebraic laws abstract from the radical differences of implementation, syntax, and general cultural environment between functional and parallel procedural programming paradigms. Some startling mathematical similarities are thereby revealed.

A stack-based language like PASCAL provides a restricted version of $\langle p, q\rangle$, where p is an expression delivering a reasonably small data value, and q represents the side-effect (if any) of evaluating the expression (usually, $q = I$). In such a language, the current machine state is a stack, and the effect of $\langle p, q\rangle$ is to pop the value of p onto the stack, making it available as the initial value of a new local variable to be accessed and updated by the body of the following block. By convention, we have put the top of the stack on the left.

If (x, y) is a pair of values, then we define the operation π to be one that selects the first of the pair (x), and μ selects the second of the pair (y). In LISP, these are the built-in functions *car* and *cdr* respectively. In a stack-based procedural language, μ has the effect of popping the stack, and is used for block exit. The operation π gives access to the most recently declared (innermost, local) variable. A combination of π and μ give access to variables declared in outer blocks; for example (1) $\mu;\pi$ gives the value of the next-to-local variable; (2) $\langle \mu;\pi, I\rangle$ makes a fresh copy of it as a new local variable; whereas (3) $\langle(\mu;\pi), \mu\rangle$ assigns it as a new value to the current most local variable. A high-level language uses identifiers to refer to these variables, and for practical programming this is a far better notation. However, in exploring the varieties of programming language, a notation closer to machine code seems paradoxically to be more abstract [1].

Each of the selectors π and μ would normally be expected to cancel the effect of a preceding operation of pairing, as described in the laws:

(1) $\langle p, q\rangle; \pi = p$

(2) $\langle p, q\rangle; \mu = q$

Furthermore, if $\langle \pi, \mu\rangle$ is applied to a pair, it will laboriously construct a pair from the first and second components of its argument, and thereby leave its argument unchanged. In general, if r is a program that produces a pair as result

(3) $\langle (r;\pi), (r;\mu)\rangle = r.$

These three laws are equivalent to the single biconditional law

$$(x;\pi = p \text{ and } x;\mu = q) \text{ iff } x = \langle p, q\rangle.$$

It is easy to calculate

$$\begin{aligned}
\langle \pi, \mu\rangle &= \langle I;\pi, I;\mu\rangle = I && \text{(from (3))}\\
r;\langle p, q\rangle &= \langle (r;\langle p,q\rangle;\pi), (r;\langle p,q\rangle;\mu)\rangle && \text{(from (3))}\\
&= \langle (r;p), (r;q)\rangle && \text{(from (1),(2))}
\end{aligned}$$

The laws given above define the concept of a product only up to isomorphism. Two programs p and q are said to be *isomorphic* if there exist an x and y such that

$$p; x = y; p,$$

where x and y are isomorphisms. An element x is said to be an *isomorphism* if there exists an $\breve{x}$ (known as the inverse of x) such that

$$x; \breve{x} = I = \breve{x}; x.$$

For example, I itself is an isomorphism ($I = \breve{I}$). Furthermore, if x and y are isomorphisms, so is $x; y (x; y; \breve{y}; \breve{x} = x; \breve{x} = I = \breve{y}; y = \breve{y}; \breve{x}; x; y)$. It follows that isomorphism between arrows is a reflexive relation ($p; I = I; p$), which is also transitive (if $p; x = y; q$ and $q; z = w; r$ then $p; x; z = y; q; z = y; w; r$), and symmetric (if $p; x = y; q$ then $\breve{y}; p; x; \breve{x} = \breve{y}; y; q; \breve{x}$, and hence $q; \breve{x} = \breve{y}; p$). In summary, isomorphism of programs is an equivalence relation.

If p is isomorphic to q ($p; x = y; q$), then the program p could equally well be implemented by the program q, merely preceding and following it by programs which are isomorphisms ($y; q; \breve{x}$). Similarly q could be implemented by means of p ($\breve{y}; p; x$). In either case, it would be impossible to distinguish the implementation from the original. That is why a collection of algebraic laws which defines a concept up to isomorphism should be accepted as an adequately strong definition in the theory of programming languages. It is certainly accepted as such in algebra or category theory.

To prove that products are defined up to isomorphism by laws (1) (2) and (3), we suppose that π', μ' and $\langle , \rangle'$ obey similar laws (1′),(2′) and (3′). We prove first that $\langle \pi, \mu \rangle'$ is the inverse of $\langle \pi', \mu' \rangle$:

$$\begin{array}{rcll} \langle \pi, \mu \rangle'; \langle \pi', \mu' \rangle & = & \langle ((\langle \pi, \mu \rangle'; \pi'), (\langle \pi, \mu \rangle'; \pi')) & \text{by } (3') \\ & = & \langle \pi, \mu \rangle & \text{by } (1'), (2') \\ & = & I. & \end{array}$$

The proof that $\langle \pi', \mu' \rangle; \langle \pi, \mu \rangle' = I$ is similar. Now the required isomorphism is established by

$$\begin{array}{rcll} \langle \pi', \mu' \rangle; \pi & = & \pi' & \text{by } (1) \\ \langle \pi', \mu' \rangle; \mu & = & \mu & \text{by } (2) \\ \langle p, q \rangle'; \langle \pi', \mu' \rangle & = & \langle ((\langle p, q \rangle'; \pi'), (\langle p, q \rangle'; \mu')) & \text{by } (3) \\ & = & \langle p, q \rangle. & \text{by } (1'), (2'). \end{array}$$

The above definition of a product is standard in category theory. Unfortunately, it is not valid in a strict programming language, for which an implementation may insist on evaluating both components of a pair before proceeding with execution of the next instruction of the program. If evaluation of either component fails to terminate, the product also fails. Thus, for example

$$\langle p, \bot \rangle = \bot = \langle \bot, q \rangle.$$

A non-strict language avoids this problem, because its implementation does not evaluate either component of a pair until it is known to be needed [3]. If the very next operation is a selection (π or μ), this necessarily discards one of the components without evaluating it. So it does not matter if such evaluation would have failed.

The solution to this problem is to weaken the definition of a product to that of a *quasi-product*, in which some of the equalities are replaced by inequalities

(1) $\langle p, q \rangle ; \pi \sqsubseteq p$

(2) $\langle p, q \rangle ; \mu \sqsubseteq q$

(3) $r \sqsubseteq \langle (r;\pi), (r;\mu) \rangle$.

On the usual assumption of monotonicity, these three laws are equivalent to the single law

$$(x;\pi \sqsubseteq p \text{ and } x;\mu \sqsubseteq q) \text{ iff } x \sqsubseteq \langle p, q \rangle.$$

It is easy to calculate

$$I \sqsubseteq \langle \pi, \mu \rangle$$
$$r; \langle p, q \rangle \sqsubseteq \langle (r;p), (r;q) \rangle.$$

We also postulate that I, π and μ are total, and that the pairing function ($\langle , \rangle$) maps total programs onto total programs. Under this hypothesis, we can show that the product is defined up to *quasi-isomorphism*, i.e., an isomorphism defined in terms of equivalence instead of equality.

Proof: very similar to the proof for products. Since $I, \pi, \mu, \pi', \mu', \langle \pi, \mu \rangle'$ and $\langle \pi', \mu' \rangle$ are all total, all inequations involving them are equivalences.

$$\begin{aligned} \langle \pi, \mu \rangle ; \langle \pi', \mu' \rangle &\equiv \langle ((\langle \pi, \mu \rangle'; \pi'), (\langle \pi, \mu \rangle'; \pi') \rangle \\ &\equiv \langle \pi, \mu \rangle \equiv I. \end{aligned}$$
$$\langle \pi', \mu' \rangle ; \pi \equiv \pi' \text{ and } \langle \pi', \mu' \rangle ; \mu \equiv \mu'.$$

For the last clause we derive two inequations with similar proofs

$$\langle p,q\rangle';\langle \pi',\mu'\rangle \sqsubseteq \langle p,q\rangle.$$
$$\langle p,q\rangle;\langle \pi,\mu\rangle' \sqsubseteq \langle p,q\rangle.$$

Apply "; $\langle \pi,\mu\rangle'$" to both sides of the first inequation to get

$$\langle p,q\rangle' \sqsubseteq \langle p,q\rangle;\langle \pi,\mu\rangle'$$

Together with the second inequation, this gives

$$\langle p,q\rangle;\langle \pi,\mu\rangle' \equiv \langle p,q\rangle'. \qquad \square$$

Unfortunately, in a non-deterministic programming language, the third law defining a quasi-product is not valid. Suppose r is a non-deterministic program, giving as answer either (3,3) or (4,4). Then $r;\pi$ can give answer 3 or 4, and so can $r;\mu$. Consequently, $\langle (r;\pi),(r;\mu)\rangle$ can give answers (3,3), (3,4), (4,3), or (4,4). This is certainly different from r, in fact more non-deterministic and therefore worse. In such a language, the inequality in the third law must be reversed

$$\langle (r;\pi),(r;\mu)\rangle \sqsubseteq r.$$

The consequences of such a reversal will not be explored here.

A similar analysis can be given for a coproduct. If p and q are programs, $[p,q]$ is a program whose execution involves execution of exactly one of p and q. The choice is made by testing some recognised "tag" component of its argument. In PASCAL this can be implemented by a **case** statement

case tag **of** $0: p;\ 1: q$ **end.**

Two operations are needed to put tags onto an argument: β puts on a 0 tag and γ puts on a 1 tag. As a result, we have two laws

(1) $\beta;[p,q] = p$

(2) $\gamma;[p,q] = q$.

These two laws are very similar to the first two laws defining the product, except that the order of the composition is reversed. The analogue of the third law is

(3) $[(\beta;r),(\gamma;r)] = r$.

In a strict language, this law is valid provided that r is an operation of a type that expects a tagged operand. $(\beta;r)$ selects the first alternative action of r, and $(\gamma;r)$ selects the second alternative. Putting these together again as a coproduct pair yields back the original r.

Unfortunately, this law is not valid in a non-strict programming language. In the case when evaluation of the tag field itself fails to terminate, the coproduct pair also fails to terminate. So if r is a non-strict function, which terminates even on a undefined argument, it is actually better than $[(\beta;r),(\gamma;r)]$. The third law must therefore be weakened to

$$[(\beta;r),(\gamma;r)] \sqsubseteq r.$$

Putting the three weaker laws together gives the biconditional law

$$(p \sqsubseteq \beta;x \text{ and } q \sqsubseteq \gamma;x) \text{ iff } [p,q] \sqsubseteq x.$$

A solution to these weakened equations is called a *quasi-coproduct.*

Discussion

This paper has suggested that many of the structural features of a programming language may be defined adequately by means of algebraic laws. The remaining features (basic types and primitive operations) can be given a conventional denotational semantics. The denotational and algebraic semantics can then be combined to give a complete semantics for the whole language. The combination is achieved by a familiar categorical construction: the language is defined as the free object over the denotational semantics in the variety of algebras which obey the algebraic equations. The construction needs slight adaptation to deal with inequations instead of equations.

In a language with strong typing, the operations of the language (both primitive and non-primitive) are defined only in contexts which satisfy the constraints of the type system. A type system can be conveniently represented in the object structure of a category. The free construction which defines the semantics of the language now yields a heterogeneous algebra, with a sort corresponding to each homset. If the language allows the programmer to define new types, the set of objects themselves have to be defined as a free (word) algebra over the primitive types of the language.

The advantage of the two-level denotational and algebraic semantics of a programming language is the potential simplification that results from separate treatments of the different features of the language. Another potential advantage is that the algebraic laws can be used directly in the optimisation of programs by means of correctness-preserving transformations. But perhaps the main advantage is that it permits the denotational semantics of the primitive types and operations of the language to be varied, while the algebraic semantics of the more structured features of the language are held constant. Thus each implementation of the language defines a different operational semantics, which can be proved correct with respect to the same denotational semantics. This proof is achieved by means of an abstraction function (or "retrieve" function of VDM), which plays the role of a natural transformation between functors in category theory.

It is hoped that further exploration of these topics will appear in the literature [4, 5, 6].

Acknowledgements.

The author is grateful to the Admiral R. Inman Chair of Computing Science at the University of Texas at Austin for providing an opportunity for an initial study of Category Theory. Also to He Jifeng for technical assistance with the mathematical basis of this article.

References

[1] G. Cousineau, P.L. Curien, M. Mauny. The Categorial Abstract Machine in Functional Languages and Computer Architecture. LNM 201 Springer-Verlag.

[2] E.W. Dijkstra. A Discipline of Programming. Prentice Hall.(1976).

[3] D.P. Friedman and D.S. Wise. Aspects of Applicative Programming for Parallel Processing. IEEE Trans. Comp. C-27 (1978) 289-296.

[4] C.A.R. Hoare and He, Jifeng. Natural Transformations and Data Refinement. Information Processing Letters (to appear).

[5] C.A.R. Hoare and He, Jifeng. Two-categorical Semantics for Programming Languages. (In preparation).

[6] C.A.R. Hoare and He, Jifeng. Data Refinement in a Categorical Setting. (In preparation).

[7] D.S. Scott. Data Types as Lattices. SIAM Journal of Computing 5 (1976) 552-587.

[8] C. Strachey. The varieties of Programming Language. PRG-10. (1973).

[9] D.A. Turner. Miranda, a non-strict Functional Language with Polymorphic Types. Lecture Notes in Computer Science 201 (1885) 1-16.

[10] M Wand. Fixed-Point Constructions in Order-Enriched Categories. THEOR. COM. 8(1) 13-30 (1979).

Independence of Negative Constraints

J.L. Lassez
IBM T.J. Watson Research Center
K. McAloon*
Brooklyn College and CUNY Graduate Center

Abstract

The independence of negative constraints is a recurring phenomenon in logic programming. This property has in fact a natural interpretation in the context of linear programming that we exploit here to address problems of canonical representations of positive and negative linear arithmetic constraints. Independence allows us to design polynomial time algorithms to decide feasibility and to generate a canonical form, thus avoiding the combinatorial explosion that the presence of negative constraints would otherwise induce. This canonical form allows us to decide by means of a simple syntactic check the equivalence of two sets of constraints and provides the starting point for a symbolic computation system. It has, moreover, other applications and we show in particular that it yields a completeness theorem for constraint propagation and is an appropriate tool to be used in connection with constraint based programming languages.

***Research partially supported by NSF Grant CCR-8703086**

1 Introduction

Programming with constraints is an increasingly important element in research on programming languages and implementations. In languages such as CONSTRAINTS [Steele,Sussman] and THINGLAB [Borning] constraint solving has been used as a tool for declarative programming and for knowledge representation. Constraint solving is also used to drive rule-based systems such as ISIS [Fox] and MOLGEN [Stefik].

The constraint paradigm has also emerged in a significant way in logic programming. Thus in [Jaffar, Lassez] the Constraint Logic Programming scheme of languages is introduced which provides a formal framework for constraint programming. Moreover, the constraint point of view represents a significant extension of the logic programming paradigm beyond the original resolution based framework and is an important extension of the range of declarative programming [Clark]. Several such languages have been successfully implemented. The language CLP(**R**) for constraint programming over the real numbers has been implemented by [Jaffar, Michaylov] and employed in applications to decision support systems [Huynh,C. Lassez]. Colmerauer and his group have developed Prolog III which supports boolean and linear arithmetic constraints [Colmerauer 2]. The constraint logic programming system CHIP [Dincbas et al] has had very successful industrial applications.

At the same time, there have been dramatic mathematical developments in constraint solving ranging from the work of Karmakar [Karmakar] and others on interior point methods in linear programming to new work on the existential theory of **R** [Canny],[Renegar],[Grigor'ev,Vorobjov]. New avenues of research have also been opened by the use of 'analog' techniques such as simulated annealing and neural nets, *e.g.* [McClelland,Rumelhart]. Recent papers such as [Davis], [Pearl], [Dechter,Pearl] contain new developments in AI motivated constraint based techniques and on connections between AI and OR work.

Clearly the design and implementation of languages of the CLP scheme will make heavy use of work on constraint solving from various areas. However, the requirements on an arithmetic constraint solver in the OR or AI context are quite different from those that are imposed on a constraint solver which is embedded in a general purpose programming language. Designing languages to solve constraint satisfaction problems or optimization problems is a fundamentally different task from using the constraint paradigm to design programming languages. Thus problems of considerable importance for a CLP system such as the equivalence of sets of constraints do not appear to have attracted much attention in Operations Research or Artificial Intelligence, the work of Bradley [Bradley] and of Adler [Adler] being notable exceptions. In this paper we build on experience from the fundamental concepts and algorithms of logic programming itself and show how the 'independence of negative constraints' can be used in the context of an extended class of linear arithmetic constraints to develop a polynomial time canonical form algorithm for use in CLP languages. Moreover, this analysis leads to a completeness theorem for constraint propagation that should prove applicable in other contexts as well.

2 Independence and Logic Programming

The *independence of negative constraints* is a recurrent phenomenon in logic programming which we can describe schematically as follows: we are given 'positive' constraints $P_1, \ldots, P_n$ and other 'positive' constraints $Q_1, \ldots, Q_m$ which have 'negative' counterparts $\overline{Q_1}, \ldots, \overline{Q_m}$ and are asked to determine the feasibility of the combined system of positive and negative constraints $P_1, \ldots, P_n, \overline{Q_1}, \ldots, \overline{Q_m}$. This combined system is not feasible iff any solution that simultaneously satisfies the $P_1, ..., P_n$ must also satisfy at least one of the Q_j; in logical notation, this system is not feasible iff $P_1, \ldots, P_n \models Q_1 \vee \ldots \vee Q_m$. We say that negative constraints are independent if whenever $P_1, \ldots, P_n \models Q_1 \vee ... \vee Q_m$ then for some j_0 we have $P_1, \ldots, P_n \models Q_{j_0}$. When negative constraints are independent in this sense, it follows that verifying the consistency or feasibility of $P_1, \ldots, P_n, \overline{Q_1}, \ldots, \overline{Q_m}$ reduces to the simultaneous verification of the feasibility of the constraints $P_1, ..., P_n, \overline{Q_j}$.

The phenomenon of independence of negative constraints is central to logic programming. In fact, if $P_1, ..., P_n$ are definite Horn clauses and if $Q_1, ..., Q_m$ are elements of the Herbrand base, then indeed $P_1, ..., P_n \models Q_1 \vee ... \vee Q_m$ if and only if for some j, we have $P_1, ..., P_n \models Q_j$. This follows since Horn formulas are closed under products of models; Historically, it was this very property of Horn formulas - preservation under products of models - that motivated the research of Horn, Keisler and others [Chang and Keisler].

Another example of this phenomenon was discovered by Colmerauer in his work on Prolog II. Colmerauer considered equations and inequations of the form $s_1 = t_1, ..., s_n = t_n, u_1 \neq v_1, ..., u_m \neq v_m$ where the t_i, s_i, u_j, v_j are terms. In [Colmerauer], the independence of inequations in this context is established and used to develop an algorithm for deciding feasibility of equations and inequations for Prolog II. In [Lassez et al.] the concept of *dimension* was introduced in the term algebra context which brought together the results of Colmerauer on the independence of negative constraints, the work of Robinson on idempotent mgu's, and algebraic equation solving. It is the notion of dimension that also applies in the context of linear arithmetic constraints and which gives us the requisite independence property.

3 Independence and Linear Arithmetic Constraints

The language of *generalized linear constraints* is comprised first of *positive* constraints which are equations $ax = \beta$ and weak inequalities $ax \leq \beta$. Here a denotes an n-dimensional vector of real numbers, x denotes an n-dimensional vector of variables, β denotes a real number and juxtaposition denotes inner product. A basic *negative* constraint is a disjunction of inequations $a_i x \neq \beta_i, i = 1, \ldots, n$. Using DeMorgan's Law and matrix notation, a negative constraint can be written $\overline{\{Ax = b\}}$ which denotes the set of points x which lie in the complement of the affine space defined by the equations $Ax = b$. Conjunctions of negative constraints can be written $\overline{\{A_j x = b_j\}}, j = 1, \ldots, n$; conjunctions of equality constraints will be written $Ax = b$ and similarly conjunctions of weak inequality constraints will be written $Ax \leq b$.

We also admit strict inequality constraints $ax < b$. In matrix form conjunctions of strict inequality constraints are written $Ax < b$. This is a hybrid form of constraint in that it can be reduced to the combined positive and negative constraints $Ax \leq b$, $\overline{\{a_i x = \beta_i\}}$, $i = 1, \ldots, n$.

Thus a set of generalized linear constraints consists of positive constraints $Ex = f$ and $Ax \leq b$, strict inequality constraints $Gx < h$ and negative constraints $\overline{\{C_j x = d_j\}}, j = 1 \ldots n$.

By way of example, the constraints $z \geq 0, x - y + z < 1, x \geq 0, y \leq 0, \overline{\{y = 0, z = 0\}}$ define a wedge shaped polytope with a facet and an edge removed.

We want to develop efficient algorithms for testing the feasibility of a set of generalized linear constraints and to generate a canonical representation of such constraint sets. The presence of negative constraints introduces new problems. The point sets defined by the constraints are no longer convex sets and so the methods of convex analysis and of linear programming do not apply directly. Negative constraints themselves are *disjunctions* of inequations which enhances the expressive power of the constraint sets but which complicates the combinatorics of the situation. The key to dealing with these problems is the independence of negative constraints.

As mathematical preliminaries, we require basic results on polyhedral sets [Schrijver]. We establish the independence of negative constraints as well as a theorem which states that the solution set defined by a system of generalized linear constraints has a unique factorization into positive and negative components. For both results, dimensionality arguments play an essential role.

Theorem 1 (Independence of Negative Constraints) *A system $Ex = f, Ax \leq b$, $\overline{C_j x = d_j}, j = 1 \ldots n$ of constraints is feasible if and only if for each j_0, the subsystem $Ex = f$, $Ax \leq b$, $\overline{\{C_{j_0} x = d_{j_0}\}}$ is feasible.*

The next result requires two definitions. Suppose that $Ex = f, Ax \leq b, \overline{\{C_j x = d_j\}}, j = 1 \ldots n$ is a feasible set of generalized linear constraints and that the positive constraints define the polyhedral set P; then a negative constraint $\overline{\{C_i x = d_i\}}$ is said to be *relevant* if $\{x : C_i x = d_i\} \cap P$ is not empty. Suppose that P is a polyhedral set defined by a system of positive constraints; a negative constraint $\overline{\{Cx = d\}}$ is said to be *P-precise* if $\overline{\{Cx = d\}}$ is relevant and $\{x : Cx = d\} = Aff(P \cap \{x : Cx = d\})$, where Aff denotes affine closure.

Theorem 2 (Unique Factorization) *Suppose that two systems of generalized linear constraints define the same non-empty solution set. Then the positive constraints in the two systems define the same polyhedral set P. Moreover if the negative constraints in both systems are P-precise, then the complements of the negative constraints in the two systems define the same union of affine sets.*

4 Canonical Representation

Next we define the canonical representation of a set of generalized linear constraints and describe the algorithms to compute it, including an algorithm to decide feasibility.

A set of linear equations

$$y_1 = a_{1,1}x_1 + \ldots + a_{n,1}x_n + c_1$$
$$\vdots$$
$$y_m = a_{m,1}x_1 + \ldots + a_{m,n}x_n + c_m$$

is said to be in *solved form* if the variables $y_1, \ldots, y_m$ and $x_1, \ldots, x_n$ are all distinct. The variables $y_1, \ldots, y_m$ are called the *eliminable variables* and the variables $x_1, \ldots, x_n$ are called the *parameters* of the solved form. Further, we will say that a negative constraint $\{Cx = d\}$ is in solved form if the complementary equality constraint $Cx = d$ is given in solved form.

A set of generalized linear constraints is in *canonical* form if it is non-redundant and consists of (1) a set of equations in solved form with parameters x which define the affine hull of the solution set (2) a set of inequality constraints $Ax \leq b$ which define a full dimensional polyhedral set P in the parameter space and (3) a set of P-precise negative constraints in solved form.

A system of generalized linear constraints in canonical form is thus partitioned into three modules (E,I,N) where E is a set of equality constraints, I is a set of weak inequality constraints and N is a set of negative constraints. What we develop is an algorithm CanForm that maps generalized linear constraints to triples of this form in such a way that if two constraint sets define the same solution set they are mapped to the same triple (E,I,N).

By way of example, consider the constraints in four variables $x_1 + x_3 \leq x_4$, $x_1 + x_3 \leq 10$, $x_4 \leq x_3$, $x_2 \leq x_3 + x_4$, $x_3 \leq x_1 + x_4$, $0 \leq x_2$, $\{x_4 = 0\}$. The CanForm procedure will return

$$E = \{x_1 = 0, x_3 = x_4\}$$
$$I = \{x_4 \leq 10, x_2 - 2x_4 \leq 0, -x_2 \leq 0\}$$
$$N = \{\{x_2 = 0, x_4 = 0\}\}$$

The constraints thus define a two dimensional point set, a triangle with a vertex removed.

In the definition of canonical form no mention is made of strict inequality constraints. As noted above, each strict inequality constraint $e_k x < \phi_k$ can be replaced by the pair $e_k x \leq \phi_k$, $\{e_k x = \phi_k\}$. From the algorithmic point of view, we can suppose that this transformation has been made throughout; we return to this point later and show how to restore the strict inequality constraints at the end of the simplification process.

The independence property allows us to avoid the combinatorial explosion that the presence of negative constraints would normally introduce and also allows for a large degree of parallelism in the treatment of the negative constraints. Moreover, we show that the positive and negative constraints can be separated out in the treatment of redundancy. This will serve to reduce eliminating redundancy among negative constraints to a parallel sieving process. An important step in the algorithm is the computation of the affine hull of the solution set defined by the constraints. This allows for the replacement of Linear Programming routines by Gaussian Elimination for part of the feasibility check in the algorithm. Also, going to the affine hull brings us to a full dimensional situation. In order to overcome the sequentiality that is typically found when eliminating redundancy [Karwan et al], we introduce a classification of redundant inequality constraints, which combined with affine hull arguments leads to a procedure with highly decomposable parallelism for eliminating redundancy. For further details see [Lassez, McAloon].

In the theorem that follows, the uniqueness of the equality constraints and negative constraints depends on a fixed ordering of the variables in the system.

Theorem 3 (Canonical Form Theorem) *If two sets of constraints define the same solution set the canonical form procedure returns the same equations to define the affine hull, the same inequality constraints (up to multiplication by positive scalars) and the same set of negative constraints.*

At this point, if strict inequality constraints are to be returned in the canonical form, let us note that a strict inequality corresponds to a pair $(ax \leq \beta, \overline{\{cx = \delta\}})$ where $ax \leq \beta$ is a positive weak inequality constraint in the canonical form and $\overline{\{cx = \delta\}}$ is a negative constraint such that the vector c, δ is a scalar multiple of the vector a, β. This pair can then be replaced by the strict inequality constraint $ax < \beta$. As in the use of linear programming and Gaussian elimination in the decision procedures of the CanForm algorithm, here too sufficient precision arithmetic is required. This variant on the canonical form algorithm is a natural one in the context of symbolic processing of generalized linear constraints and in the context of output constraints where strict inequality information can be significant.

If, in the example above, the constraint $0 \leq x_2$ is sharpened to $0 < x_2$, then after transforming this constraint into the pair $0 \leq x_2, \overline{\{x_2 = 0\}}$, the canonical form procedure will return

$$
\begin{aligned}
E &= \{x_1 = 0, x_3 = x_4\} \\
I &= \{x_4 \leq 10, x_2 - 2x_4 \leq 0, -x_2 \leq 0\} \\
N &= \{\overline{\{x_2 = 0\}}\}
\end{aligned}
$$

The negative constraint $\overline{\{x_2 = 0, x_4 = 0\}}$ has been eliminated because it is now redundant. Since the vector $(1,0,0)$ is a scalar multiple of $(-1,0,0)$ the constraints $-x_2 \leq 0$ and $\overline{\{x_2 = 0\}}$ can be replaced by the strict inequality constraint $-x_2 < 0$.

5 Computational Complexity

In this section we address considerations of computational complexity for the canonical form algorithm. We have

Theorem 4 *The canonical form procedure CanForm is a polynomial time algorithm.*

Moreover, in terms of the PRAM model of parallel computation, the processor complexity of the algorithm is bounded by the number of constraints and its time complexity by the sequential complexity of linear programming. Further, the number of arithmetic operations in the algorithm is determined by Gaussian elimination and linear programming; thus if a strongly polynomial algorithm were to be found for linear programming, then the canonical form problem would also enjoy a strongly polynomial solution. On the other hand, it is not to be expected that an NC algorithm can be found: the canonical form problem subsumes the Phase I or Feasibility Problem of linear programming and this is known to be P-Complete.

The algorithm can be applied to give polynomial time procedures for computing two normal forms for systems consisting of equality and weak inequality constraints. In [Schrijver] a normal form is defined for a set of positive linear constraints which involves computing the affine hull of the polyhedral set defined by the constraints and representing this set in terms of its facets; in the non-full-dimensional case, the facets are defined by means of supporting hyperplanes in the full space chosen orthogonal to the affine hull. We can also show that the normal form of [Adler] for linear programs with objective function is polynomial time computable using our algorithm. Adler's normal form, the *core* of a linear program involves both the program and its dual and both the primal and dual constraint sets are analyzed in the process.

6 Constraint Propagation

Finally, we consider the canonical form in the context of constraint propagation and constraint based programming.

Linear arithmetic constraints arise naturally in constraint programming and in AI situations. When determining the solvability and/or the solutions to a set of constraints, a standard strategy is to work forward in an incremental way by starting with the 'most constraining' conditions and propagating these constraints throughout the rest of the computation. Intuitively a 'more constraining' condition corresponds to a set of smaller dimension in the solution space, information which is in general not explicit in a given system S of constraints. The canonical form procedure returns in *explicit* form a representation of equality, inequality and negative information in packets of smallest possible dimension. In fact, the canonical form leads to a soundness and completeness theorem for constraint propagation.

Theorem 5 (Constraint Propagation Completeness Theorem) *Let S be a consistent system of generalized linear constraints with canonical form (E, I, N). Let y be the eliminable variables and x the parameters of (E, I, N). Then we have*

(1) $S \models ay = bx$ *iff* $E \models ay = bx$

(2) $S \models ax \leq \beta$ *iff* $I \models ax \leq \beta$

(3) $S \models \{Cx = d\}$ *iff* $N \models \{Cx = d\}$

where $\{Cx = d\}$ *is a precise negative constraint.*

This theorem also has application to constraint logic programming languages in the CLP class [Jaffar and Lassez]. In particular, for CLP(R) [Jaffar and Michaylov] and Prolog III, it shows how the canonical form is the analog in the constraint context of the most general unifier.

Bibliography

[Adler]
I. Adler, The Core of a Linear Program, Technical Report, Department of Operations Research, Berkeley

[Borning]
A. Borning, The Programming Language Aspects of THINGLAB, a Constraint Oriented Simulation Laboratory, *ACM Transactions on Programming Languages and Systems* 3 (1981) 252-387

[Bradley]
G. Bradley, Equivalent Integer Programs, *Proceedings of the Fifth International Conference on Operations Research*, ed. J. Lawrence, Venice 1969

[Canny]
J. Canny, Some Algebraic and Geometric Computations in PSPACE, *STOC*, 1988

[Chang and Keisler]
C.C. Chang and H.J. Keisler, *Model Theory*, North-Holland 1973

[Clark]
K. Clark, Logic Programming Schemes, *Proceedings of 1988 FGCS Conference*, Tokyo

[Colmerauer]
A. Colmerauer, Equations and Inequations on Finite and Infinite Trees, *Proceedings of 1984 FGCS Conference*, Tokyo

[Colmerauer 2]
A. Colmerauer, An Introduction to Prolog III, Technical Report, Groupe d'Intelligence Artificielle (1987)

[Davis]
E. Davis, Constraint Propagation, *Artificial Intelligence* 1988

[Dechter,Pearl]
R. Dechter and J. Pearl, Network-based Heuristics for Constraint Satisfaction Problems, *Artificial Intelligence*, 34 (1987) 1-38

[Dincbas et al]
M. Dincbas, P. Van Hentenryck, H. Simonis, A. Aggoun, T. Graf and F. Berthier, The Constraint Logic Programming Language CHIP, *Proceedings of the 1988 FGCS Conference*, Tokyo

[Fox]
M. Fox, *Constraint Directed Search: A Case Study of Job-Shop Scheduling*, Morgan Kaufmann, 1988

[Grigor'ev,Vorobjov]
D. Grigor'ev and N. Vorobjov, Solving Systems of Polynomial Inequalities in Subexpo-

nential Time, ***Journal of Symbolic Computation*** 5 (1988) 37-64
[Huynh,C. Lassez]
T. Huynh and C. Lassez, A CLP(R) Options Analysis System, ***Proceedings of the 1988 Logic Programming Symposium***
[Jaffar, Lassez]
J. Jaffar and J.L. Lassez, Constraint Logic Programming, ***Proceedings of POPL 1987***, Munich
[Jaffar, Michaylov]
J. Jaffar and S. Michaylov, Methodology and Implementation of a CLP System, ***Proceedings of the 1987 Logic Programming Conference***, Melbourne, MIT Press
[Karwan et al.]
M.H. Karwan, V. Lofti, J. Telgen and S. Zionts, ***Redundancy in Mathematical Programming***, Lecture Notes in Economics and Mathematical Systems 206, Springer-Verlag 1983
[Lassez et al.]
J.L. Lassez, M. Maher and K. Marriott, Unification revisited, ***Foundations of Deductive Databases and Logic Programming***, J. Minker editor, Morgan Kaufmann 1988
[Lassez, McAloon]
J-L. Lassez and K. McAloon, Applications of a Canonical Form for Generalized Linear Constraints, ***Proceeding of FGCS 1988***, Tokyo.
[McClelland,Rumelhart]
J. McClelland and D. Rumelhart, ***Explorations in Parallel Distributed Processing***, MIT Press, 1988
[Pearl]
J. Pearl, Constraints and Heuristics, ***Artificial Intelligence*** 1988
[Renegar]
J. Renegar, A Faster PSPACE ALgorithm for Deciding the Existential Theory of the Reals, ***FOCS 1988***, pp 291-285
[Schrijver]
A. Schrijver, ***Theory of Linear and Integer Programming***, Wiley 1986
[Steele,Sussman]
G. Steele and G. Sussman, CONSTRAINTS - A Language for Expressing Almost Hierarchical Descriptions, ***Artificial Intelligence*** 1980
[Stefik]
M. Stefik, Planning with Constraints (MOLGEN: Part 1), ***Artificial Intelligence*** 16 (1984) 111-140

J-L. Lassez
IBM T.J. Watson Research Center
P.O. Box 704
Yorktown Heights NY 10598

K. McAloon
Logic Based Systems Lab
Brooklyn College CUNY
Brooklyn NY 11210

Completion Procedures as Transition Rules + Control

Pierre LESCANNE*
Centre de Recherche en Informatique de Nancy
LORIA
Campus Scientifique, BP 239,
54506 Vandœuvre-lès-Nancy, France
email: lescanne@poincare.crin.fr

Abstract

A description of the completion of a set of identities by a set of inference rules has allowed recent progresses in proving its completeness. But there existed no attempt to use this description in an actual implementation. This paper shows that this is feasible using a functional programming language namely *CAML*. The implementation uses a toolkit, a set of transition rules and a short procedure for describing the control. A major role is played by the data structure on which both the transition rules and the control operate. Three versions of the classical Knuth-Bendix completion and two versions of the unfailing completion are proposed.

1 Completion procedures as sets of transitions rules

The interest of rewriting techniques in programming, algebraic and computer algebra specifications is well-known as is its ability to provide proof environments essentially based on completion procedures [FG84,GG88,KS83,Fag84,Les83]. In this introduction, I suppose the reader is familiar with this concept. Indeed my goal is not to present it, but to study how methods developed essentially with a theoretical purpose, namely proving completeness can be used to present simple short and understandable programs. This paper can also be seen as a set of exercises on the use of a functional language to program high level procedures and as a bridge between theory and practice. Readers who want to get more introductory informations are invited to look at Appendix A or to Dershowitz survey *Completion and its Applications* [Der87]. The completion procedure is a method used in equational logic to built from a set of identities an equivalent canonical set of rewrite rules i.e., a confluent, noetherian and interreduced set of rules used to compute normal forms. If one tracks the history of the presentation of this procedure, one can notice different methods of description. In their seminal paper [KB70] Knuth and Bendix describe essentially the procedure in natural language (see Appendix B), in [Hue80] Huet uses a style similar to Knuth's book, *The Art of Computer Programming*, in [Hue81] he uses a program structured by while loops, in [Kir84] H. Kirchner uses a recursive procedure and in [For84] Forgaard proposes an organization of the procedure around tasks to be performed. In the following, a

*The research was sponsored by PRC "programmation avancée et outils de l'intelligence artificielle", CNRS and INRIA

completion will be seen as a set of *inference rules* or more precisely a set of *transition rules* acting on a data structure. The idea of using *inference rules* when dealing with completion is not new and leads to the beautiful proofs of completeness proposed by Bachmair and Dershowitz [Bac87,BDH86,BD87] and their followers [GKK88,Gan87]. The *completeness* is the ability of the procedure to eventually generate a proof by normalization or a rewrite proof for every equational theorem. In this paper, I want to show how this description leads to actual, nice and elegant programs when used as a programming method and I illustrate that by an actual *CAML* implementation [FOR87b]. Appendix C gives the basic notions that are useful to understand the programs. Actually the inference rules one considers in completion are specific in the sense that they transform a *t*-uple of objects into a *t*-uple of objects with the same structure. This is why I refer to them as *transition rules.* Thus the basic components of such a procedure are four,

- **a data structure** on which the transition rules operate, sometimes called the universe,
- **a set of transition rules,** that are the basic operations on the data structure,
- **a control,** that is a description of the way the transition rules are invoked[1],
- **a toolkit** that is shared by several completion procedures.

When one wants to describe a specific completion procedure, usually one uses the following method. First one chooses the data structure, then one chooses transition rules and often at the same time the control. The toolkit is something that remains from one procedure to the other in many cases, it was partly borrowed from the "*CAML* Anthology" [FOR87c] as a natural attempt to reuse pieces of codes already debugged and tested. As we will see the control is typically data driven and can be easily expressed by rewrite rules. In the following, the influence of these choices on the efficiency of the procedure will be illustrated through three refinements of the Knuth-Bendix completion procedures and a two unfailing completions. Indeed, we will see how, starting from a naive implementation of the completion, improvements can be obtained by changing the data structure and consequently the transition rules and the control. These ideas are implemented in my program *ORME.*

2 The N-completion

In this section, I give a naive implementation of the completion, called the *N-completion*, where *N* stands for naive. It is already an improvement of the set of rules of Appendix A in order to take the computation of critical pairs into account. Its control part is fully given in Figure 1 and its data structure has three components, namely

- **E** is a set of identities, either critical pairs or given identities,
- **T** is a set of rules, the non marked rules in Huet's terminology [Hue81],
- **R** is a set of rules whose critical pairs have been computed, the marked rules.

In the procedure, *ordering* is a parameter which is a relation used to orient the identities into rules, by the way it is also a parameter of *Orientation*. There are three kinds of transition

[1] To give a gastronomic comparison [Ore83], the control is the recipe.

```
let rec N_Completion ordering (R,T,E) = let COMP = N_Completion ordering in
match (T,E) with
    [],[] -> (R,[],[]) (* success *)

  | (_::_),[] -> COMP (repeat_list [Simpl_left_T_by_T;Simpl_left_T_by_R;
                                     Simpl_left_R_by_T;Simpl_left_R_by_R;
                                     Simpl_right_T;Simpl_right_R]
                        (Deduction(R,T,E)))

  | _,(_::_) -> let (R',T',E') = repeat_list[Remove_trivial_E;Simpl_E](R,T,E) in
        (match E' with
                [] -> COMP(R',T',E')
              | (_::_) -> COMP(Orientation ordering (R',T',E')
                              ? failwith "non orientable equation"));;
```

Figure 1: *The N-completion*

rules, their names are taken according to Dershowitz [Der87] (see also Appendix A). *Deduction* computes critical pairs, in this case it computes critical pairs between one rule in T, usually the smallest one to be more efficient, and all the rules in R. *Orientation* chooses an identity that can be oriented by an *ordering* and produces a rule, if no identity is orientable, it fails. This requires an reduction ordering, currently *ORME* contains an ordering based on polynomial interpretations [BL87b], implementing other orderings would not be too difficult since the *CAML* Anthology [FOR87c] contains the recursive path ordering and a *CAML* implementation of the transformation ordering also exists [BL87a,Gal88]. *Remove_trivial_E* removes from E a trivial identity. The rules *Simpl_left_T_by_T*, *Simpl_left_T_by_R* etc. simplify terms in the rules or the identities. *repeat_list* repeats the application of of a list of inference rules until they all fail. The control given in Figure 1 has essentially three steps, namely *success,* when T and E are empty, *computing critical pairs* after simplification of the rules, when E is empty, and *orienting* an identity into a rule after simplification of the identities, when E is not empty. In the orientation part it could happen that by simplification all the identities disappear, in this case one does nothing, that is just translated by a recursive call to COMP with the same parameters. The recursive calls mean that one restarts the process. The completion terminates with success when E and T are empty. The system works as a machine where the identities enter E and proceeds through T and R. Its description is therefore really similar to this of an automaton.

3 The S-completion

Another name for rewrite systems is sometime *simplifying systems* and the theory of rewrite systems is a *theory of simplification*, that could be applied to many fields other than computer algebra or software specification. Therefore the main aim of orienting identities is to use them to simplify whenever it is possible. But as noted by Hsiang and Mzali [HM88], the *N-completion* makes a bad use of simplification. Indeed a rule is not used for simplification as soon as it has been generated. Thus in a better implementation, when an identity is oriented into a rule it enters a set S where it is used to simplify all the other identities and

```
let rec S_Completion ordering (R,T,S,E) = let COMP = S_Completion ordering in
match (T,S,E) with
    [],[],[] -> (R,[],[],[]) (* success *)

 | _,(_::_),_ -> (COMP (R',T'@S',[],E')
                    where R',T',S',E' =
                        repeat_list [Simpl_left_T_by_S;Simpl_right_T_by_S;
                                    Simpl_left_R_by_S;Simpl_right_R_by_S] (R,T,S,E))
 | (_::_),[],[] -> COMP (Deduction (R,T,S,E))

 | _,[],(_::_) -> let (R',T',S',E') = repeat_list[Remove_trivial_E;Simpl_E](R,T,S,E)
                  in (match E' with
                        [] -> COMP(R',T',S',E')
                      | (_::_) -> COMP (Orientation ordering (R',T',S',E')
                            ? failwith "non orientable equation"));;
```

Figure 2: *The S-completion*

rules. In the *S-completion*, the data structure is made of four components,

- **E** like in the *N-completion*,
- **S** a set of oriented identities or rules that are used to simplify others identities or rules and that I call the ***simplifiers***, during the completion S contains zero or one rule,
- **T** a set of rules already used for simplifying, but whose critical pairs are not yet computed,
- **R** like in the *N-completion*.

The only difference with the *N-completion* is the set S through which a rule has to go, before entering T. The step of simplification is clearly distinguished from the three others. It is performed when S is not empty. The completion process ends when there is no more identity or rule in E, S and T.

4 The ANS-completion

The *S-completion* can still be improved since it computes at the same time the critical pairs between all the rules in R and one rule in T. It should be better to compute the critical pairs between one rule in R and one rule in T at a time. As previously, S contain the simplifiers. In addition, a set C is created to contain one rule extracted from T with which critical pairs with rules of R are computed. To keep track of the rules whose critical pairs are computed with the rule in C, R is split into two sets A (for already computed) and N (for not yet computed). Thus the data structure contains,

- **E** like in the *S-completion*,
- **S** like in the *S-completion*,

- **T** is a set of rules coming from S and waiting to enter C,
- **C** is a set that contains one or zero rule and whose critical pairs are computed with one in N,
- **N** is the part of R whose critical pairs have not been computed with C but whose critical pairs with $A \cup N$ have been computed,
- **A** is a set whose critical pairs with $A \cup N \cup C$ have been computed.

The transition rules are adapted to work with this new data structure and three new rules are introduced. *Deduction* computes the critical pairs between the smallest rule in N and the rule in C. *Internal_Deduction* computes the critical pairs obtained by superposing the rule(s) in C on itself (themselves). *A_C2N* moves the rules in A and C into N to start a new "loop" of computation of critical pairs, according to the emptyness of the components of the data structure. The procedure has now clearly six parts, namely *success, simplification, orientation, deduction, internal deduction* and *beginning of a new loop* of computation of critical pairs. Typically this cannot be easily structured by a while loop because at each time the iteration on the computations of the critical pairs can be interrupted by a simplification. A data driven control is then much better (Figure 3).

5 The unfailing completion

The previous method may fail because at certain time no rule can be oriented, this is for instance the case if $(x * y = y * x) \in E$. A method called either *unfailing completion* or *unfailing Knuth-Bendix* or *UKB* has been proposed by Hsiang and Rusinowitch [HR87] and Bachmair, Dershowitz and Hsiang [BDH86] and is complete for proving equational theorems in equational theories. The idea is to refute the equality to be proved, thus variables become Skolem constants, terms become ground terms and the equality becomes a disequality i.e., a negation of an equality. One does not orient the identities. But because one works with an ordering total on ground terms, one knows that for any pair of ground terms one can tell which one is simpler and therefore one can tell whether a term that matches a side of an identity can be transformed in the other side in a decreasing way. In some sense, these new pairs of terms are not rules but "abstract" sets of rules on ground terms. To precise the difference with identities and rules, let me propose the word *likeness* for them. The aim of unfailing completion is then to make confluent the rewrite relation on ground terms defined by likenesses. Such a relation which is confluent on ground terms is called *ground confluent*. Although the likenesses are not oriented, one tries however to save generation of too many critical pairs by not keeping those of the form $\langle s, t \rangle$ obtained from a superposition u by $s \longmapsfrom u \longmapsto t$ if either $u < s$ or $u < t$, because this kind of equality will never be used for rewriting ground terms. This is what makes this procedure different from classical paramodulation. As a refutation procedure, at each step, an attempt to refute the negation of the disequality is performed. The data structure of this naive unfailing completion is a follow

- **E** is again the set of identities,
- **C** is a set that contains one or zero likenesse and whose critical pairs are computed with one in N,

```
let rec ANS_Completion ordering (A,N,C,T,S,E) =
    let COMP = ANS_Completion ordering in

match (N,C,T,S,E) with
   _,[],[],[],[] -> (A,N,C,T,S,E) (* success *)

 | _,_,_,(_::_),_ -> (COMP (A',N',C',T'@S',[],E')
                      where A',N',C',T',S',E' =
                        repeat_list [Simpl_left_A_by_S;Simpl_right_A_by_S;
                                     Simpl_left_N_by_S;Simpl_right_N_by_S;
                                     Simpl_left_C_by_S;Simpl_right_C_by_S;
                                     Simpl_left_T_by_S;Simpl_right_T_by_S]
                                    (A,N,C,T,S,E))

 | _,_,_,[],(_::_) -> let  A',N',C',T',S',E' = repeat_list[Remove_trivial_E;
                                                  Simpl_E](A,N,C,T,S,E) in
          (match E' with
              [] ->  COMP(A',N',C',T',S',E')
            | (_::_) ->  COMP(Orientation ordering (A',N',C',T',S',E')
                         ? failwith "non orientable equation"))

 | (_::_),[_],_,[],[] -> COMP (Deduction(A,N,C,T,S,E))

 | [],[_],_,[],[] -> COMP (A_C2N crit (Internal_Deduction (A,N,C,T,S,E)))

 | _,[],(_::_),[],[] -> (COMP([],A@N,[r],T',[],[]) where r,T' = least Size T);;
```

Figure 3: *The ANS-completion*

```
let rec Unfailing_Completion ordering (e,A,N,C,E) =
        let COMP = Unfailing_Completion ordering in
let e' = Gnormalize  ordering  (A @ N @ C @ E) e in
if matches <<x ~ x>> e'
   then (e',A,N,C,E)  (* refutation *)
   else match (N,C,E) with
      _,[],[] ->    (e',A,N,C,E) (* end of the completion *)

    | _,[],(_::_) -> let (A',N',C',E') = Clean_E (Simpl_E ordering
                                                  (Simpl_N ordering (A,N,C,E))))
        in (match E' with
           [] -> COMP (e',A',N',C',E')
         | eq :: E'' -> COMP(e', F_Subsumption ordering ([], A' @ N', [eq], E''))

    | (_::_),(_::_),_ -> COMP(e', (Deduction ordering (A,N,C,E)))

    | [],[_],_ ->  COMP(e', (A_C2N (Internal_Deduction ordering (A,N,C,E))));;
```

Figure 4: *The unfailing completion*

- **N** is a set f likeness whose critical pairs have not been computed with C but whose critical pairs with $A \cup N$ have been computed,
- **A** is a set whose critical pairs with $A \cup N \cup C$ have been computed.

It should be noticed that the idea of computing the critical pairs between only two pairs at a time is used, but not the idea of putting a high priority to simplification and since there is no simplification the identities enter directly C from E. With the disequality to be refuted, the procedure has five parameters. The last four ones remind the data structure of the classical completion. Obviously, there is no *Orientation* transition rule, but there are *Deduction* and *Internal_Deduction* as previously. << x ~ x >> is an external notation for the disequality whose both sides are x. There are four steps in this procedure, either success or, simplification, or deduction or, the beginning of a new "loop" of computation of critical pairs. If one runs this algorithm on examples, one quickly realizes that many generated identities are instances of existing identities or obtained by inserting in a same context sides of instances of identities and therefore do not carry new information. Rules *Subsumption* or *F_Subsumption* remove these useless identities. *Subsumption* filters the identities that matches another one and *F_Subsumption* tries to remove identities of the form $C[s] = C[t]$, where $C[\,]$ is a context, when the identity $s = t$ already exists. *Gnormalize* takes a ground term and returns its normal form using identities. However, when rewriting with identities as in refutation care must be taken with variables that can be introduced. The usual solution is to instantiate them by a new *least constant*.

6 An improved unfailing completion: the ER-completion

The previous unfailing completion has the advantage of being short and easy to understand. However its main drawback is that it makes no difference between non orientable identities and rules. This can be fixed by refining considerably the data structure, using ideas from the *ANS-completion*. The new data structure contains eleven components. They are obtained by splitting the corresponding components of the *ANS-completion* into two parts, a *E*-part and a *R*-part.

- **E** is not changed and is again the set of identities,
- **RS** the simplifiers obtained from rules *E* by orientation,
- **ES** the identities from *E* that cannot be oriented, they are used in ***F_Subsumption***,
- **RT** the *R*-part of *T*,
- **ET** the *E*-part of *T*,
- **RC** the *R*-part of *C*,
- **EC** the *E*-part of *C*,
- **RN** the *R*-part of *N*,
- **EN** the *E*-part of *N*,
- **RA** the *R*-part of *A*,
- **EA** the *E*-part of *A*,

The transition rules are changed accordingly and one gets a procedure I call the ***ER-unfailing completion*** (see Figure 5) which performs as a classical completion if all the identities can be oriented. The fact that everything which is orientable is actually oriented is a major improvement for the efficiency of the procedure. The structure of the completion procedure gets now more complex and requires studies on how to make it more modular.

7 Conclusion

The main idea of the approach presented here is to decompose the algorithm into basic actions and to describe some kind of abstract machine where these actions as the instructions. This may remind either Forgaard's description of ***REVE*** based on tasks [For84], or ***ERIL*** [Dic85] where users have access to the basic operations or Huet's first description [Hue80]. The rigorous and formal approach of this paper gives precision and concision and leads to a better understanding of the program and therefore to a better confidence. Since one is closer to the proof of completeness there are more chance that the implementation is both correct and complete. Another important aspect of this approach is that modifications and improvements are easily done. Basically this level of programming allows to study very high level optimizations [Ben82] and when an efficient procedure is discovered, a low level implementation can be foreseen. Here I made many implementation choices that still can be discussed, but since they are rather explicit this discussion is easy and changes can be quickly made. However, as well illustrated by the ***ER_completion*** compared with the ***unfailing completion***, it should also be noticed that the complexity of the completion procedures described by transition rules increases exponentially with the size of the number of components of the data structure, which implies that some kind of modularity has to be found.

```
let rec Unfailing_Completion ordering (e,((EA,RA,EN,RN,EC,RC,ET,RT,ES,RS,E) as STATE)) =
    let COMP = Unfailing_Completion ordering and ord = ordering in
let e' = Gnormalize ordering (EA @ EN @ EC @ ET @ ES @ E) e''
  where e'' = (normalize (RA @ RN @ RC @ RS @ RT) e) in
if matches <<x ~ x> e'
   then  (print_state "REFUTATION" (e',STATE);(e',STATE))  (* refutation *)
   else match (EN,RN,EC,RC,ET,RT,ES,RS,E) with
      _,_,[],[],[],[],[],[],[] ->  (e',STATE) (* end of the completion *)

    | _,_,_,_,_,_,_,(_::_),_ -> (COMP(e',EA',RA',EN',RN',EC',RC',ET',RT'@RS',ES',[],E')
      where (EA',RA',EN',RN',EC',RC',ET',RT',ES',RS',E') = Simp_by_RS (STATE))

    | _,_,_,_,_,_,(_::_),[],_ -> (COMP(e,EA',RA',EN',RN',EC',RC',ET'@ES',RT',[],RS',E')
      where (EA',RA',EN',RN',EC',RC',ET',RT',ES',RS',E') =
                                          F_Subsume_by_ES ord (STATE))

    | _,_,_,_,_,_,[],[],(_::_) -> let ((EA',RA',EN',RN',EC',RC',ET',RT',ES',RS', E')
                                                       as STATE') =
        Clean_E ord (STATE) in (match E' with
                     [] -> COMP (e',STATE')
                   | (_::_) -> COMP(e',Orientation ord (STATE')))

    | _,(_::_),[],[_],_,_,[],[],[] -> COMP(e', RN_RC_Deduction (STATE))

    | (_::_),_,[],[_],_,_,[],[],[] -> COMP(e', EN_RC_Deduction ord (STATE))

    | _,(_::_),[_],[],_,_,[],[],[] -> COMP(e', RN_EC_Deduction ord (STATE))

    | (_::_),_,[_],[],_,_,[],[],[] -> COMP(e', EN_EC_Deduction ord (STATE))

    | [],[],[],[_],_,_,[],[],[] -> COMP(e', A_C2N (RC_Internal_Deduction(STATE)))

    | [],[],[_],[],_,_,[],[],[] -> COMP(e', A_C2N (EC_Internal_Deduction ord (STATE)))

    | _,_,[],[],(_::_),[],[],[],[] -> (COMP(e',[],[],EA@EN,RA@RN,[e],[],ET',[],[],[],[])
                                       where e,ET' = (least Size ET))
    | _,_,[],[],[],(_::_),[],[],[] -> (COMP(e',[],[],EA@EN,RA@RN,[],[r],[],RT',[],[],[])
                                       where r,RT' = (least Size RT))
    | _,_,[],[],(_::_),(_::_),[],[],[] ->
      let r,RT' = least Size RT and e,ET' = least Size ET in
      if Size r <= Size e
      then COMP(e',F_Subsume_by_ES ord ([],[],EA@EN,RA@RN,[],[r],ET,RT',[],[],[]))
      else COMP(e',F_Subsume_by_ES ord ([],[],EA@EN,RA@RN,[e],[],ET',RT.[],[],[]))
```

Figure 5: *The ER unfailing completion*

Another interesting aspect of the programming by transition rules is that simple snapshots exist, therefore the process can easily be stopped after each rule and restarted on this state. Thus backtracking on the choice of the orderings as implemented in *REVE* or any kind of backtracking to insure fairness [DMT88], backups, breakpoints or integration of an already completed rewrite system in another equational theory can be easily handled.

But this approach does not address low level controls, for instance refinements that computes one critical pair at a time. This indeed requires a level of granularity in the actions that cannot be handled by the current form of the data structure. Attempts to fully formalize all the tasks, including substitutions and unifications should answer this question [GS88,HJ88].

All the procedures presented in this paper are a part of *ORME*, a set of *CAML* procedures that were run for completing a set of examples. Both the programs and the examples can be obtained from the author upon request.

I would like to thank Leo Bachmair, Françoise Bellegarde, Jieh Hsiang, Jean-Pierre Jouannaud, Jean-Luc Remy, Michael Rusinowitch and the EURECA group at CRIN who provided me with stimulating discussions, Gérad Huet who gave me access to the *CAML Anthology* and Alain Laville for wise advices on how to use *CAML*.

References

[Bac87] L. Bachmair. *Proof methods for equational theories.* PhD thesis, University of Illinois, Urbana-Champaign, 1987.

[BD87] L. Bachmair and N. Dershowitz. Completion for rewriting modulo a congruence. In *Proceedings Second Conference on Rewriting Techniques and Applications*, Springer Verlag, Bordeaux (France), May 1987.

[BDH86] L. Bachmair, N. Dershowitz, and J. Hsiang. Orderings for equational proofs. In *Proc. Symp. Logic in Computer Science*, pages 346–357, Boston (Massachusetts USA), 1986.

[Ben82] J. L. Bentley. *Writing Efficient Programs.* Prentice Hall, 1982.

[BL87a] F. Bellegarde and P. Lescanne. Transformation orderings. In *12th Coll. on Trees in Algebra and Programming, TAPSOFT*, pages 69–80, Springer Verlag, 1987.

[BL87b] A. BenCherifa and P. Lescanne. Termination of rewriting systems by polynomial interpretations and its implementation. *Science of Computer Programming*, 9(2):137–160, October 1987.

[Der87] N. Dershowitz. Completion and its applications. In *Proc. Colloquium on Resolution of Equations in Algebraic Structures*, MCC, 3500 West Balconies Center Drive, Austin, Texas 78759-6509, May 4-6 1987.

[Dic85] A.J.J. Dick. ERIL equational reasoning: an interactive laboratory. In B. Buchberger, editor, *Proceedings of the EUROCAL Conference*, Springer-Verlag, Linz (Austria), 1985.

[DMT88] N. Dershowitz, L. Marcus, and A. Tarlecki. Existence, uniqueness and construction of rewrite systems. *SIAM J. Comput.*, 17(4):629–639, August 1988.

[Fag84] F. Fages. *Le système KB. Manuel de référence, présentation et bibliographie, mise en œuvre.* Technical Report, Greco de Programmation, Bordeaux, 1984.

[FG84] R. Forgaard and J. Guttag. *REVE: A term rewriting system generator with failure-resistant Knuth-Bendix.* Technical Report, MIT-LCS, 1984.

[For84] R. Forgaard. *A program for generating and analyzing term rewriting systems.* Technical Report 343, Laboratory for Computer Science, Massachusetts Institute of Technology, 1984. Master's Thesis.

[FOR87a] Projet FORMEL. *The CAML Primer.* Technical Report, INRIA LIENS, 1987.

[FOR87b] Projet FORMEL. *CAML: the reference Manuel.* Technical Report, INRIA-ENS, March 1987.

[FOR87c] Projet FORMEL. *The CAML Anthology.* July 1987. Internal Document.

[Gal88] B. Galabertier. *Implémentation de l'ordre de terminaison par transformation.* Technical Report, CRIN, Septembre 1988.

[Gan87] H. Ganzinger. A completion procedure for conditional equations. In *Proc. 1st International Workshop on Conditional Term Rewriting Systems*, pages 62–83, Springer-Verlag, 1987. Extended version to appear in Journal of Symbolic Computation.

[GG88] S. Garland and J. Guttag. *An Overview of LP, The Larch Prover.* Technical Report, MIT, 1988.

[GKK88] I. Gnaedig, C. Kirchner, and H. Kirchner. Equational completion in order-sorted algebras. In M. Dauchet and M. Nivat, editors, *Proceedings of the 13th Colloquium on Trees in Algebra and Programming*, pages 165–184, Springer-Verlag, Nancy (France), 1988.

[GS88] J. Gallier and W. Snyder. Complete sets of transformations for general E-unification. *Journal of Theorical Computer Science*, 1988.

[HJ88] J. Hsiang and J-P. Jouannaud. General e-unification revisited. In *Proceedings of 2nd Workshop on Unification*, 1988.

[HM88] J. Hsiang and J. Mzali. *Algorithme de Complétion SKB.* Technical Report, LRI, Orsay, France, 1988. Submitted.

[HR87] J. Hsiang and M. Rusinowitch. On word problem in equational theories. In *Proceedings of 14th International Colloquium on Automata, Languages and Programming*, Springer-Verlag, Karlsruhe (West Germany), 1987.

[Hue80] G. Huet. *A complete proof of correctness of the Knuth-Bendix completion algorithm.* Technical Report 25, INRIA, August 1980.

[Hue81] G. Huet. A complete proof of correctness of the Knuth and Bendix completion algorithm. *Journal of Computer Systems and Sciences*, 23:11–21, 1981.

[KB70] D. Knuth and P. Bendix. *Simple Word Problems in Universal Algebra*, pages 263–297. Pergamon Press, 1970.

[Kir84] H. Kirchner. A general inductive completion algorithm and application to abstract data types. In R. Shostak, editor, *Proceedings 7th international Conference on Automated Deduction*, pages 282–302, Springer-Verlag, Napa Valley (California, USA), 1984.

[KS83] K. Kapur and G. Sivakumar. Experiments with an architecture of RRL, a rewrite rule laboratory. In *Proc. of an NSF Workshop on the Rewrite Rule Laboratory*, pages 33–56, 1983.

[Les83] P. Lescanne. Computer Experiments with the REVE Term Rewriting Systems Generator. In *Proceedings, 10th ACM Symposium on Principles of Programming Languages*, ACM, 1983.

[Ore83] F. Orejas. Good food considered helpful. *Bulletin of EATCS*, 20:14–22, June 1983.

A Introduction to completion procedures

Let us take a simple example namely the type *Lists* where the constructors are $[\,]$, $[\_]$, a, b and @ and satisfy the relations

$$[\,]@x \rightarrow x$$

$$x@[\,] \rightarrow x$$

$$(x@y)@z \rightarrow x@(y@z)$$

and a function *flatten* is given by:

$$flatten([\,]) \rightarrow [\,]$$

$$flatten(a) \rightarrow a$$

$$flatten(b) \rightarrow b$$

$$flatten(a@x) \rightarrow a@flatten(x)$$

$$flatten(b@x) \rightarrow b@flatten(x)$$

$$flatten([x]@y) \rightarrow flatten(x)@flatten(y)$$

The term $flatten([x]@[\,])$ can be rewritten into $flatten([x])$ by the second rule and into $flatten(x)$ by three rewrites, namely to $flatten(x)@flatten([\,])$ by the last rule, to $flatten(x)@[$ by the fourth rule and to $flatten(x)$ by the second rule. $flatten([x]@[\,])$ is called a ***superposition*** and $\langle flatten(x)@flatten([\,]), flatten([x])\rangle$ a ***critical pair***. If both parts of the critical pair rewrite to the same terms, the critical pair is said *convergent*, otherwise it is said *divergent*.

$$\langle (x_1@(x_2@y))@z, (x_1@x_2)@(y@z)\rangle$$

is a convergent critical pair and

$$\langle flatten(x)@flatten([\,]), flatten([x])\rangle$$

is a divergent critical pair. A completion procedure is a way to generate a rewrite system without such divergent critical pairs with the same proving power. It is based on inference rules like the following ones where one works on a data structure with two sets, namely E that contains the identities and R that contains the rules or oriented identities.

Delete: $E \cup \{s = s\}; R \vdash E; R$
Compose:$E; R \cup \{s \to t\} \vdash E; R \cup \{s \to u\}$ if $t \to_R u$
Simplify: $E \cup \{s = t\}; R \vdash E \cup \{s = u\}; R$ if $t \to_R u$
Orient: $E \cup \{s = t\}; R \vdash E; R \cup \{s \to t\}$ if $s > t$
Collapse: $E; R \cup \{s \to t\} \vdash E \cup \{u = t\}; R$ if $s \to_R u$ by a rule $l \to r \in R$ with $s \rhd l$
Deduce: $E; R \vdash E \cup \{s = t\}; R$ if $s \leftarrow_R u \to_R t$ for some u

Delete removes trivial identities from E. *Compose* reduces the right-hand side of a rule if it can be rewritten by a rule in R. *Simplify* simplifies an identity. *Orient* transforms an identity into a rule provided the left-hand side is greater than the right-hand side for a given ordering. *Collapse* transforms an identity into a rule when the left-hand side is rewritten. *Deduce* creates new identities from superpositions.

The inference rule are used as long as they apply and the procedure can stop because E is empty and no rule applies or can stop with failure when no identity can be oriented or can run forever. It is *complete* if given an identity $a = b$ to be proved there exists a step i such that the R_i-normal form of a is equal to the R_i-normal form of b, where R_i is the value of R at i^{th} step. Under some assumptions of fairness not given here the procedure is complete.

B Original description of the Knuth-Bendix procedure

The next paragraph is a strict quotation of the Knuth-Bendix paper [KB70]. I found intersting to give the actual description of the algorithm we work on for now close to two decades. The corollary which is mentioned describe the concept of critical pair and (6.1) shows the stability of the congruence generated by a set of identities after adjunction of an equational consequence.

> The following procedure may now be used to attempt to complete a given set of reductions.
>
> Apply the tests of the corollary to Theorem 5, for all λ_1, λ_2 and μ. If in every case $\sigma'_0 = \sigma''_0$, R is complete and the procedure terminates. If some choice of λ_1, λ_2, μ leads to $\sigma'_0 \neq \sigma''_0$, then either $\sigma'_0 > \sigma''_0$, $\sigma''_0 > \sigma'_0$ or $\sigma'_0 \# \sigma''_0$. In the latter case, the process terminates unsuccessfully, having derived an equivalent $\sigma'_0 \equiv \sigma''_0(R)$ for which no reduction [...] can be used. In the former cases, we add a new reduction (σ'_0, σ''_0) or (σ''_0, σ'_0), respectively, to R and begin the procedure again.
>
> Whenever a new reduction (λ', ρ') is added to R, the entire new set R is checked to make sure it contains only irreducible words. This means, for each reduction (λ, ρ) in R we find irreducible words λ_0 and ρ_0 such that $\lambda \xrightarrow{*} \lambda_0$ and $\rho \xrightarrow{*} \rho_0$, with respect to $R - \{(\lambda, \rho)\}$. Here it is possible that $\lambda_0 = \rho_0$ in which case by (6.1) we may remove (λ, ρ) from R. Otherwise we might have $\lambda_0 > \rho_0$ or $\rho_0 > \lambda_0$, and (λ, ρ) may be replaced by (λ_0, ρ_0) or (ρ_0, λ_0) respectively [...]. We might also find that $\lambda_0 \# \rho_0$, in which case the procedure terminates unsuccessfully as above.

C Some basic notions of *CAML*

CAML is a polymorphic functional language of the *ML* family. Its basic constructions used here are the following.

let introduces an identifiers and its definition by a subexpression that will replace each occurrence of the identifier in the body that follows and which is introduced by **in**.

where is similar to **let**, but is placed after the body.

match *pattern* **with** identifies a structure that will be checked for a use as a rewrite system in the part that follows the *with*. Each rule is introduced by a pattern and the corresponding computation follows the sign −>. The rules are separated by signs | and are evaluated with a priority according to their position. In a pattern, the sign _ means any value. For instance, (_ :: _) matches any non empty list and [_] matches any list with one element. The empty list is [].

failwith signals an exception to the a caller, such an exception is caught by a ?.

A full description appears in [FOR87b,FOR87a].

A Modular Framework for Specification and Implementation*

Martin Wirsing Manfred Broy

Fakultät für Mathematik und Informatik
Universität Passau
Postfach 2540
D-8390 Passau

Abstract

A modular framework for the formal specification and implementation of large families of sorts and functions is introduced. It is intended to express generation principles, to rename, combine and construct implementations of specifications in flexible styles. Parameterization is also included.
The main characteristics of this approach are the inclusion of predicates in the signature of specifications and the use of an ultra-loose semantics. Signatures are triples of sets of sorts, sets of function symbols and sets of predicate symbols; the latter contain among others also standard predicate symbols, in particular the equality symbols as well as predicate symbols expressing generation principles which hold for an object if and only if it can be denoted by a term of a specific signature. These standard predicate symbols lead to an ultra-loose semantics for specifications: models are not required to be term-generated; instead, the term-generated subalgebra of a model is required to satisfy the axioms. Main advantages of this approach are the simplicity of the notion of implementation and the simplicity of the corresponding language for writing structured specifications.

1. Introduction

Algebraic specifications provide a flexible and abstract way for the description of data structures. The basic idea of the algebraic approach consists in introducing names for the different sets of data and for the basic functions operating on the data and by giving logical requirements for the data that may be associated with those names. The requirements are described by equations or by first-order formulas using equations.
Writing large specifications in this basic way can be very time consuming. For their use in requirements engineering and software design applications it seems decisive that specifications can be modularized and manipulated by appropriate language constructs and be written in a parameterized way. In order to support formal software development, specifications have to be equipped with a notion of implementation which allows to verify the transitions from abstract descriptions to more concrete ones. Consequently, any formal framework for specifications should comprise language constructs for building structured specifications and a notion of implementation. The first of such frameworks was given by the ADJ-group [Goguen et al. 78] who introduced the

*This paper has been partially sponsored by the ESPRIT project METEOR.

initial algebra approach together with the notion "Forget-Restrict-Identify" (FRI) for implementation. This suggestion has been completed on the one hand by Goguen and Burstall with the languages CLEAR and OBJ ([Burstall, Goguen 80], [Futatsugi et al. 85]) and on the other hand by Ehrig and his group who added parameterization and structuring operators with the language ACT ONE [Ehrig, Mahr 85] and studied thoroughly the mathematical properties of the FRI notion [Ehrig, Kreowski 82], [Ehrig et al. 82].

The hierarchical approach to specifications with loose semantics [Wirsing et al. 83] has been incorporated in the language CIP-L [Bauer, Wössner 82]. There also the "FRI"-relation is the main notion of implementation. The language ASL ([Sannella, Wirsing 83], [Wirsing 86]) is also based on loose semantics but a much simpler notion of implementation is chosen, namely the model class inclusion. This is possible due to the power of the specification operators of ASL which comprise in particular an operator for behavioural abstraction. The latter notion represents a third actual research stream on the semantics of specifications showing a number of encouraging results [Hennicker 89]. Actual language developments such as ACT TWO [Ehrig, Weber 86], combine loose and initial semantics or elaborate better the notion of module (see ACT TWO, Larch [Guttag et al. 85], Extended ML [Sannella, Tarlecki 85]. However, all these powerful languages (including ASL) have the drawback that they do not well support the FRI-notion of implementation: the horizontal composition theorem for FRI-implementation does only hold under severe restrictions. A comparison of all these different approaches seems to indicate a principal difficulty: either a conceptually simple language (such as ACT ONE) is combined with a complex notion of implementation (such as FRI) or a simple notion of implementation (such as model class inclusion) is combined with powerful, but mathematically difficult operators (such as behavioural abstraction).

In the following we introduce a formal framework for writing, manipulating and combining specifications in which we try to avoid the above difficulty: the notion of implementation is model class inclusion, the language (similar to ACT ONE) comprises only three basic and mathematically simple operators which support syntactic manipulations of specifications such as export and renaming; horizontal and vertical composition of implementations holds without any restrictions. Technically, the novelty of the approach consists in the inclusion of standard predicate symbols in the signature of specifications and in the use of an ultra-loose semantics: the models of an algebraic specification with signature Σ and axioms E are all first-order Σ-structures which satisfy the axioms E. Therefore, the models are not restricted to term-generated structures - junk is allowed. However, properties of the term-generated elements are easily expressible due to the use of standard generating predicate symbols. Moreover, the interpretation of the equality symbol in a model A of a specification SP does not necessarily coincide with the equality between the elements of A; it has only to be an appropriate congruence relation. Hence, the models of SP coincide exactly with the "implementations" of SP.

The paper is organized as follows: In section 2, the notions of signatures and structures with predicates are introduced. In section 2.1 the basic definitions are given; in section 2.2 these notions

are extended to comprise the standard predicate symbols and the generating predicate symbols.
In section 3 many-sorted first-order formulas with standard predicates are studied. In particular, proof rules are given and it is shown how reachability is expressible by such formulas.
In section 4 flat specifications with constructors are introduced and the class of models of well-known examples such as total orderings, truth values and natural numbers is studied. It is shown that natural numbers are admitted as model (i.e. implementation) of truth values and similarly that integers are a model of natural numbers.
In section 5 the three basic structuring operators for specifications are introduced. The operators "translate" and "derive" are deduced from both directions of a signature morphism, the operator "+" combines two specifications without taking care of name clashes. Using "+" the operator "enrich" can be defined which is used for the structured specification of two examples, sets of natural numbers and sequences of integers.
In section 6 properties of the specification operators are studied. It is shown that exporting, hiding and copying can be defined using the basic operators; a number of algebraic identities are given which will be useful for proving properties of implementations such as horizontal and vertical composition of ERE-implementations (cf. section 8).
In section 7 model class inclusion is introduced as notion of implementation. It is shown that after extensions and renamings integers implement natural numbers, ordered sequences of integers implement sets of natural numbers. The most difficult part of these extensions is the axiomatization of the congruence to be implemented. On the other hand, for the verification of the implementation only the validity of the axioms has to be checked, semantic manipulations such as restrictions and identification as for FRI-implementations are not necessary.
In section 8 parameterization of specifications is defined by lambda-abstraction; partial and total implementations of parameterized specifications are introduced. It is shown that all specification operators are monotonic w.r.t the refinement and implementation relation which implies the horizontal and vertical composition theorem. Finally, the implementation relation is generalized to include export, rename and extension (as in the examples of section 6). For this so-called ERE-implementations vertical and horizontal composition theorems are proven as well.
In the concluding remarks some additional possible standard predicate symbols are shortly discussed.

2. Signatures and structures

The semantics of algebraic specifications is determined by the notions of signature and structure or algebra. In contrast to the classical approaches (cf. e.g. [Goguen et al. 82], [Guttag 75]), signatures do not consist only of sorts and function symbols but contain predicate symbols as well. Moreover, all signatures are supposed to contain a number of standard predicate symbols. For algebraic specifications the most important standard symbols are the equality symbols $. =_s .$ (indexed by sorts s) and the generating predicate symbols $. \in \Sigma$ (for expressing that an element is denotable by a ground Σ-term).

2.1 Signatures and structures with predicates

The syntactic structure of a data type D is determined by its signature. Usually, a signature Σ is given as a set S of names of different kinds of data (the sorts) and a family F of notations for distinguished data and operations (the function symbols). In our approach, also predicate symbols are included in the signature.

Def. 2.1.1 A **signature** Σ consists of a triple $< S,F,P >$ where S is a set (of **sorts**), F is a set (of **function symbols**), and P is a set (of **predicate symbols**) such that F is equipped with a mapping type: $F \to S^* \times S$ and P is equipped with a mapping type: $P \to S^+$. For any $f \in F$ (or $p \in P$, resp.), the value type(f) (or type(p), resp.) is the **type** of f (or p, resp.). We write sorts(Σ) to denote S, opns(Σ) to denote F, preds(Σ) to denote P, $f: w \to s$ to denote $f \in F$ with type(f)=w,s and p:w to denote $p \in P$ with type(p)=w.

Let X be an S-sorted set. For every sort $s \in S$ the set, $T(<S,F>, X)_s$, of **terms of sort** s (containing elements in X) is defined as usual (as the least set built using X and the function symbols $f \in F$). Terms without elements of X are called **ground terms** and the set $T(<S,F>, \emptyset)$ of **all ground terms** is denoted by $T(<S,F>)$.We also write $T(\Sigma,X)$ for $T(<S,F>,X)$ and similarly, $T(\Sigma)$ for $T(<S,F>)$.

A signature morphism is defined as usual as a mapping from one signature into another such that the types of function and predicate symbols are compatible with the mapping of the sorts, i.e. for signatures $\Sigma=< S,F,P >$ and $\Sigma'=< S',F',P' >$ a **signature morphism** $\sigma:\Sigma\to\Sigma$ is a triple $<\sigma_{sorts}, \sigma_{opns}, \sigma_{preds}>$ where $\sigma_{sorts}:S\to S'$. $\sigma_{opns}: F\to F'$ and $\sigma_{preds}: P\to P'$ are mappings such that for any $f: w\to s \in F$, $type(\sigma_{opns}(f))=\sigma^*(w),\sigma(s)$ and

for any $p: w \in P$, $type(\sigma_{preds}(p))=\sigma^*(w)$

where denotes $\sigma^*(s_1...s_n)$ the extension of σ to words, i.e. $\sigma^*(s_1...s_n) =_{def} \sigma_{sorts}(s_1)...\sigma_{sorts}(s_n)$ for $s_1, ..., s_n \in S$.

Any mapping between sorts, function and predicate symbols induces a number of signature morphisms: let $\varphi=<\varphi_{srt}, \varphi_{opn}, \varphi_{prd}>$ be a triple of mappings $\varphi_{srt}: S1\to S2$, $\varphi_{opn}: F1\to F2$, $\varphi_{prd}: P1\to P2$ where S1, S2 are sets of sorts, F1, F2 are sets of function symbols and P1, P2 sets of predicate symbols; moreover let $\Sigma=<S,F,P>$ be a signature. Then ${}_\Sigma\varphi: \Sigma\to\Sigma'$ denotes the signature morphism defined by

$$ {}_\Sigma\varphi(x) =_{def} \begin{cases} \varphi(x) & \text{if } x\in S1\cup F1\cup P1, \\ x & \text{otherwise} \end{cases} $$

where $\Sigma'=<S',F',P'>$ is a signature defined by

$S' =_{def} {}_{\Sigma}\varphi(S)$,

$F' =_{def} \{{}_{\Sigma}\varphi(f): ({}_{\Sigma}\varphi)^*(w) \rightarrow {}_{\Sigma}\varphi(s) \mid f: w \rightarrow s \in F\}$,

$P' =_{def} \{{}_{\Sigma}\varphi(p): ({}_{\Sigma}\varphi)^*(w) \mid p: w \in P\}$.

If the signature Σ'' contains Σ' (i.e. $\Sigma' \subseteq \Sigma''$) then ${}_{\Sigma}\varphi_{\Sigma''}: \Sigma \rightarrow \Sigma''$ is the signature morphism ${}_{\Sigma}\varphi$ with range Σ''. If instead of Σ the signature Σ'' is given, we define the signature morphism $\varphi_{\Sigma''}: \Sigma \rightarrow \Sigma''$ analogously. We often omit indices from ${}_{\Sigma}\varphi$, $\varphi_{\Sigma''}$ or ${}_{\Sigma}\varphi_{\Sigma''}$ if the indexed signatures are obvious from the context. Moreover, if $x_1, ..., x_n$ are pairwise different sorts, function and predicate symbols and φ maps x_i to y_i for i=1,...,n, we write $[x_1/y_1, ..., x_n/y_n]$ for ${}_{\Sigma}\varphi$ (and thus for all induced signature morphisms). If φ is injective, then we call it a **renaming**.

A general Σ-structure has a carrier set (the elements of the data type) for each sort, a function on these sets for each function symbol and a relation for each predicate symbol.

Def. 2.1.2 Let $\Sigma = \langle S, F, P \rangle$ be a signature. A **general Σ-structure** A consists of an S-sorted family of non-empty carrier sets $\{A_s\}_{s \in S^*}$ and, for each f: s1, ..., sn $\rightarrow$ s $\in$ F, of a total function $f^A: A_{s1} \times ... \times A_{sn} \rightarrow A_s$, and for each p: s1, ..., sn $\in$ P, of a relation $p^A \subseteq A_{s1} \times ... \times A_{sn}$.

The relationship between terms and structures is as usual given by the notion of interpretation. For any valuation v: X $\rightarrow$ A of an S-sorted set X into A we denote the **interpretation** (w.r.t. v) of a term t $\in$ T(Σ, X) by v*(t); if t is a ground term, then its interpretation does not depend on v and we write t^A instead of v*(t).

To any reachable Σ-structure A one may associate a term structure T(A) with carrier sets $T(\Sigma)_s$ as follows.

Def. 2.1.3 Let A be a reachable Σ-structure. The **Σ-term structure** T(A) associated with A is defined as follows:

(1) for each s $\in$ sorts(Σ), $T(A)_s =_{def} T(\Sigma)_s$;

(2) for each f: s1, ..., sn $\rightarrow$ s $\in$ F and each $t_1 \in T(A)_{s1}, ..., t_n \in T(A)_{sn}$, $f^{T(A)}(t_1, ..., t_n) =_{def} f(t_1, ..., t_n)$

(3) for each p: s1, ..., sn $\in$ P and each $t_1 \in T(A)_{s1}, ..., t_n \in T(A)_{sn}$, $\langle t_1,...,t_n \rangle \in p^{T(A)}$ iff $\langle t_1^A, ..., t_n^A \rangle \in p^A$.

2.2 Standard predicates

In the remainder of this paper it is assumed that any signature contains the following set St of standard predicate symbols which depends on the available sorts and function symbols.

Def 2.2.1 Let S be a set of sorts and F be an S-sorted set of function symbols. The set $St_{S,F}$ of **standard predicate symbols** for S, F consists of the following elements:

$. =_s . : s,s$ the **equality symbols** for all $s \in S$,

$. \in \langle S',F' \rangle_s : s$ the **generating predicate symbols** for any $S' \subseteq S$,

any subsignature $\langle S',F' \rangle$ of $\langle S,F \rangle$ and any $s \in S'$ such that $T(\langle S',F' \rangle)_s \neq \emptyset$.

In order to distinguish the equality symbols of different specifications, these symbols may get a name as additional index. Hence the set $St_{S,F}$ of standard predicate symbols contains an equality symbol for any sort of S and a family of generating predicate symbols for any subsignature of $\langle S,F \rangle$. Moreover, if the sort s can be derived from the context then we write often = instead of $=_s$.

A simple example for a signature with standard predicate symbols is the signature $\Sigma B0$ of the data structure of truth values, which consists of one sort, two constants, one unary function symbol and the set $St_{\langle SB0,FB0 \rangle}$ of standard predicate symbols.

$\Sigma B0 = \langle SB0, FB0, St_{\langle SB0,FB0 \rangle} \rangle$ is defined by
$SB0 =_{def} \{Bool\}$,
$FB0 =_{def} \{true, false: \to Bool, not: Bool \to Bool\}$,
$St_{\langle SB0,FB0 \rangle} =_{def} \{ =_{Bool} : Bool, Bool;$
$. \in \langle \{Bool\}, \{true, false, not\} \rangle_{Bool},$
$. \in \langle \{Bool\}, \{true, not\} \rangle_{Bool},$
$. \in \langle \{Bool\}, \{false, not\} \rangle_{Bool},$
$. \in \langle \{Bool\}, \{true, false\} \rangle_{Bool},$
$. \in \langle \{Bool\}, \{true\} \rangle_{Bool},$
$. \in \langle \{Bool\}, \{false\} \rangle_{Bool} : Bool \}$

Thus the generating predicate symbols contain all those subsignatures of < SB0, FB0 > which admit at least one ground term; for < {Bool}, {not} > no generating predicate symbol is defined. Notice the choice of "true, false, not" as boolean function symbols; in logic programming one would have considered them as predicate symbols.

The standard predicate symbols do not have arbitrary relations as interpretations: equality symbols must always be associated with congruence relations and the generating predicate symbols are associated with the corresponding sets of ground terms.

Def 2.2.2 Let $\Sigma = \langle S, F, P \cup St_{S,F} \rangle$ be a signature with standard symbols. A **Σ-structure** A is a general Σ-structure such that

(1) for any generating predicate symbol $\in \Sigma'_s$, for all $a \in A_s$:
$a(\in \Sigma'_s)^A$ iff there exists $t \in T(\Sigma')_s$ with $a = t^A$;

(2) $=^A$ is a Σ-congruence relation,
i.e. the following holds for all $a \in A_s$, $a_1,b_1 \in A_{s1}, \ldots, a_n,b_n \in A_{sn}$, where $s,s1,\ldots,sn \in S$,

(refl) $a =_s^A a$

(subst$_f$) $a_1 =_{s1}^A b_1 \wedge \ldots \wedge a_n =_{s1}^A b_n \Rightarrow f^A(a_1, \ldots, a_n) =_s^A f^A(b_1, \ldots, b_n)$
for each $f: s1, \ldots, sn \to s \in F$;

$$(\text{subst}_p) \quad a_1 =_s^A b_1 \wedge \ldots \wedge a_n =_{sn}^A b_n \wedge \langle a_1, \ldots, a_n\rangle \in p^A \Rightarrow \langle b_1, \ldots, b_n\rangle \in p^A$$

for each $p: s1, \ldots, sn \to s \in P$.

Notice that symmetry and transitivity of $=_s^A$ follow from reflexivity (refl) and the substitution property (subst_) for equality symbols: let p: s, s be $=_s$ and choose for proving symmetry $a_1 =_{def} a_2 =_{def} b_2 =_{def} a$, $b_1 =_{def} b$ and for proving transitivity $a_1 =_{def} b_1 =_{def} a$, $a_2 =_{def} b$, $b_2 =_{def} c$.

Given interpretations for sorts and function symbols the interpretation of the generating predicate symbols is uniquely determined whereas there may exist several congruence relations. Therefore, in the examples for structures usually the interpretation of the equality symbols is omitted if it coincides with the equality in the structure.

Examples for ΣB0-structures are the standard structure B of truth-values and the following two structures N1 and N2 with the natural numbers as carrier sets; we write "$\in \Sigma_{\{true,false\}}$" instead of "$\in <\{Bool\}, \{true,false\}>_{Bool}$" and analogously for the other generating predicate symbols.

$B_{Bool} =_{def} \{O,L\}$,
$true^B =_{def} L$, $false^B =_{def} O$, $not^B(L) =_{def} O$, $not^B(O) =_{def} L$
$x(\in\Sigma_{\{true,false\}})^B \Leftrightarrow x(\in\Sigma_{\{true,not\}})^B \Leftrightarrow x(\in\Sigma_{\{false,not\}})^B \Leftrightarrow x(\in\Sigma_{\{true,false,not\}})^B \Leftrightarrow x \in \{O,L\}$,
$x =^B y \Leftrightarrow x = y$;

$N1_{Bool} =_{def} \mathbb{N}$,
$true^{N1} =_{def} 1$, $false^{N1} =_{def} 0$, $not^{N1}(x) =_{def} x+1$,
$x(\in\Sigma_{\{true,false\}})^{N1} \Leftrightarrow x \in \{0,1\}$,
$x(\in\Sigma_{\{true,not\}})^{N1} \Leftrightarrow x \geq 1$, $x(\in\Sigma_{\{false,not\}})^{N1} \Leftrightarrow x(\in\Sigma_{\{true,false,not\}})^{N1} \Leftrightarrow x \in \mathbb{N}$,
$n =^{N1} m \Leftrightarrow n \bmod 2 = m \bmod 2$;

$N2_{Bool} =_{def} \mathbb{N}$,
$true^{N2} =_{def} 1$, $false^{N2} =_{def} 0$, $not^{N2}(0) =_{def} 1$, $not^{N2}(1) =_{def} 0$, $not^{N2}(n+2) =_{def} n+3$
$x(\in\Sigma_{\{true,false\}})^{N2} \Leftrightarrow x(\in\Sigma_{\{true,not\}})^{N2} \Leftrightarrow x(\in\Sigma_{\{false,not\}})^{N2} \Leftrightarrow x(\in\Sigma_{\{true,false,not\}})^{N2} \Leftrightarrow x \in \{0,1\}$,
$x =^{N2} y \Leftrightarrow x = y$.

For B and N2 the interpretation of the equality symbol coincides with the equality between the elements of the structure.

Def. 2.2.3 For any signature Σ with standard symbols a Σ-structure A is called

(1) **Σ-algebra** if for each $s \in S$ the equality symbol $=_s$ has the standard equational interpretation, i.e. for all $a, b \in A_s$, $a = b \Leftrightarrow a =_s^A b$;

(2) **Σ-reachable** if for each $s \in S$ and $a \in A_s$ there exists a ground Σ-term $t \in T(\Sigma)_s$ such that $a =_s^A t^A$

(3) **Σ-standard**, if A is a Σ-reachable Σ-algebra.

We will denote the class of all Σ-structures by **Struct(Σ)**, the class of all Σ-algebras by **Alg(Σ)** and the class of all Σ-standard structures by **Gen(Σ)**.

For example, the structure B is ΣB0-standard, the structure N1 is ΣB0-reachable but not a ΣB0-algebra, and the structure N2 is a ΣB0-algebra but it is not ΣB0-reachable.

A Σ-homomorphism is a mapping from a Σ-structure into another which is compatible with all function and predicate symbols.

Def. 2.2.4 Let Σ = <S,F,P> be a signature with standard symbols and let A, B be Σ-structures. A **Σ-homomorphism** h: A → B is a family of maps $\{h_s: A_s \to B_s\}_{s \in S}$ which is compatible with function and predicate symbols, i.e. for each f: s1, ..., sn -> s ∈ F, each p: s1, ..., sn ∈ P and each $a_1 \in A_{s1}, \ldots, a_n \in A_{sn}$,

$$h_s(f^A(a_1, \ldots, a_n)) =_s^B f^B(h_{s1}(a_1), \ldots, h_{sn}(a_n)) \text{ and}$$

$$<a_1, \ldots, a_n> \in p^A \Rightarrow <h_{s1}(a_1), \ldots, h_{sn}(a_n)> \in p^B.$$

Notice that for checking the homomorphism property it is sufficient to consider the non-standard predicate symbols and the equality symbols; then the conditions for the generating predicate symbols are automatically satisfied.

Fact 2.2.5

(1) Every Σ-structure A has an **associated Σ-algebra A/=** which is defined as the quotient of A by the Σ-congruence $=^A$. Moreover, it has an associated Σ-reachable substructure, the **least Σ-substructure** $<A>_\Sigma$ of A.

(2) There exists a unique (surjective) Σ–homomorphism from A onto A/= and there exist unique Σ–homomorphisms from $<A>_\Sigma$ into A and into A/=.

(3) For any reachable Σ-structure A there exists a unique Σ-homomorphism from the Σ-structure T(A) onto A.

Every signature morphism σ: Σ ->Σ´ (where Σ=<S,F,P> and Σ´=<S´,F´,P´>) induces an adjoint morphism on structures. Let A be a Σ´-structure; then A|σ, the **σ-reduct** of A, denotes the following Σ-structure:

$$(A|\sigma)_s =_{def} A_{\sigma(s)} \text{ for } s \in S, \quad f^{A|\sigma} =_{def} \sigma(f)^A \text{ for } f \in F, \quad p^{A|\sigma} =_{def} \sigma(p)^A \text{ for } p \in P.$$

If Σ ⊆ Σ´ and σ denotes the embedding from Σ into Σ´ (i.e. σ(x) = x for all x∈ Σ), then we write A|Σ for A|σ. For a class C of Σ ´- structures we denote {A|σ | A∈ C} by C|σ.

3. Formulas

Properties of many-sorted structures are expressed by many-sorted first-order formulas which are defined as usual.

Def. 3.1 The set **WFF(Σ)** of **Σ-formulas** is the least set satisfying the following properties:

(i) if $p \in P_{s1,\ldots,sn}$ and t_i, i=1,…,n, are Σ-terms of sort si, then $p(t_1,\ldots,t_n) \in WFF(\Sigma)$;

(ii) if $G, H \in WFF(\Sigma)$ then $(\neg G)$, $(G \wedge H) \in WFF(\Sigma)$;

(iii) if $x \in X_s$ and $G \in WFF(\Sigma)$ then $(\forall x{:}s.G) \in WFF(\Sigma)$.

Formulas of the form $p(t_1,\ldots,t_n)$ are called **atomic Σ-formulas**; in particular, an atomic formula $=_s(t_1,t_2)$ (often written $t_1=t_2$) is called **Σ-equation**. The generating predicate symbols are often used as restrictions for quantifications. We write

$\forall\, x{:}\, \Sigma_s.A$ for $\forall x{:}\ s.\ x \in \Sigma_s \Rightarrow A$ and

$\exists\, x{:}\, \Sigma_s.A$ for $\exists x{:}\ s.\ x \in \Sigma_s \wedge A$.

Further logical operators such as ∨, ⇒ and ∃ are defined as usual abbreviations. Superfluous brackets will be omitted. By V(G) we denote the set of all variables occurring in the Σ-formula G. A Σ-formula without free variables is called **Σ-sentence.** It is called **ground** if it is without variables at all.

Def. 3.2 For any Σ-structure A, valuation v:X→A and Σ-formula G, the relation **A satisfies G w.r.t. v**, written $A,v \models G$, is inductively defined as usual.

(i) $A,v \models p(t_1,\ldots,t_n)$ iff $<v^*(t_1),\ldots,v^*(t_n)> \in p^A$ (for $p \in P$),

(ii) $A,v \models (\neg G)$ iff $(A,v \models G)$ does not hold,

(iii) $A,v \models (G \wedge H)$ iff $(A,v \models G)$ and $(A,v \models H)$,

(iv) $A,v \models (\forall x{:}s.G)$ iff $(A,v_x \models G)$ for all valuations v_x: X→A with $v_x(z)=v(z)$ for z≠x.

The interpretation of the relativized quantification is a consequence of this definition:

Fact 3.3 Let Σ = < S, F, P > be a signature with standard symbols, Σ′ = < S′, F′ > with S′⊆S and F′⊆F, s∈S′. Then for any Σ-structure A, valuation v: X→A and Σ-formula G

$$A,v \models \forall x{:}\Sigma'_s.G \text{ iff } (A,v_t \models G) \text{ for all } t \in T(\Sigma')_s \text{ and all valuations } v_t{:}\ X \to A \text{ with}$$

$$v_t(z) =_{def} \begin{cases} t^A & \text{for } z=x, \\ v(z) & \text{otherwise.} \end{cases}$$

The structure **A satisfies G**, written $A \models G$, if $A,v \models G$ holds for all valuations v.

A Σ-formula G is **valid** in a class K of Σ-structures if each A∈K satisfies G.
The satisfaction relation is closed under isomorphism and w.r.t. the associated Σ-algebras.

Fact 3.4 Let A,B be two Σ- structures such that their associated Σ-algebras A /= and B /=

are isomorphic. Then the following four propositions are equivalent:

$A \models G$, $A/\!= \models G$, $B/\!= \models G$, $B \models G$.

Reachability can be expressed using "$\forall\exists$-formulas".

Fact 3.5 A Σ-structure A is Σ-reachable iff for each $s \in sorts(\Sigma)$, $A \models \forall x{:}s\ \exists y{:}\Sigma_s.x=y$.

With every class K of Σ-structures one may associate the **theory of** K, Th(K), i.e. the set of all Σ-formulas which are valid in K. For example as a consequence of fact 3.4., theories are invariant under isomorphism and transition of K to the class of Σ-algebras associated with K.
On the other hand, with every signature Σ and every set E of Σ-formulas one may associate the class Mod (Σ,E) of all **models** of E, i.e. of all Σ-structures which satisfy all formulas from E.

According to fact 3.5. for $E_0 =_{def} \{\forall x{:}s\ \exists y{:}\Sigma_s.x=y \mid s \in S\}$, $Mod(\Sigma, E_0)$ is the class of all Σ-reachable Σ-structures. Similarly, standard arithmetic can be axiomatized by Σ-formulas. Hence, Σ-formulas are able to express reachability and therefore there exists Σ_1, E_1 such that the theory of $Mod(\Sigma_1, E_1)$ is not recursively enumerable. Thus there cannot exist any formal system computing this theory, but there exists a **semi-formal** system for which a completeness result can be proven.

Let *F* be any formal system given by a set of **logical and nonlogical axioms** and a set of **inference rules** which is sound and complete for many-sorted first-order predicate logic with equality (cf. e.g. [Barwise 77]). Let "$\vdash$" denote the binary relation $E \vdash G$ which holds, if and only, if the Σ-formula G is **derivable** from E using *F*. Define a derivation relation $\vdash_I$ by adding for every Σ-formula G and every subsignature $\Sigma' = \langle S', F' \rangle$ (with $S' \subseteq S$ and $F' \subseteq F$) and every sort $s \in S'$ the following induction rule II_s to the rules of *F*:

(II_s) Infinite Induction

If $G[t/x]$ is derivable for all $t \in T(\Sigma')_s$, then so is $\forall x{:}\Sigma_s'.G$, i.e. for all sets E of Σ-formulas, $E \vdash_I G[t/x]$ for all $t \in T(\Sigma')_s$ implies $E \vdash \forall x{:}\Sigma'.G$.

The logical and non-logical axioms remain the same.
A Σ-formula is called **ω-derivable** from E if $E \vdash_I G$.

Theorem 3.6 (Completeness Theorem) Let E be a set of Σ-sentences. A Σ-formula G is ω-derivable from E (i.e. $E \vdash_I G$), if and only if, $Mod(\Sigma,E) \models G$.

4. Algebraic specifications

Following Hoare [Hoare 69], a specification is a formal documentation for a data type D which guarantees properties of implementations of D for use in proving correctness of programs. The syntax of a data type can be expressed by a signature Σ and the properties by Σ-formulas. This motivates the first part of the following definition.

Def. 4.1 Let $\Sigma = < S, F, P >$ be a signature with standard symbols.

(1) A (flat) **specification** SP consists of a pair <Σ, E> where E is a set of Σ-sentences.

(2) A specification SP = <Σ, E> is called **term-oriented** if all quantifiers occurring in E range over sets of ground terms; i.e. each quantified subformula in E has the form $\forall x{:}\Sigma_s'.G$ or $\exists x{:}\Sigma_s'.G$ with $\Sigma' = < S', F' >$ with $S' \subseteq S$ and $F' \subseteq F$, $s \in S'$.

(3) A specification SP = <Σ, E> is called **Horn-specification**, if SP is term-oriented and all axioms of E are Horn-sentences (of the form $\forall x1{:}\ W1 \ldots \forall xn{:}\ Wn.\ G_1 \wedge \ldots \wedge G_m \Rightarrow G$ where $G_1, \ldots, G_m$ are atomic Σ-formulas).

The notation "term-oriented" is motivated by the "term generation principle" (cf. e.g. [Bauer, Wössner 82]) which requires that every element of a data type can be represented by a term. Thus properties of data types can always be reduced to properties of sets of ground terms.
Our notion of specification differs in two respects from the usual one. First syntactically, also predicates are allowed so that the notion of "logic program" would be appropriate as well. We will call a specification **equational Horn-specification** if all predicates in its axioms are equality symbols. Second semantically, the models of specifications may contain arbitrary "junk" (i.e. non-reachable elements) for which nothing is required by the axioms, and, moreover, the (non-standard) interpretation of the equality symbols allows for multiple representants of congruence classes; in particular, the set T(Σ) of all terms is always an appropriate carrier set.
The signature of a specification SP = <Σ,E> will be denoted by **sig(SP)** and the class of all models of SP will be denoted by **Mod(SP)**, i.e. $\text{sig(SP)} =_{def} \Sigma$ and $\text{Mod(SP)} =_{def} \text{Mod}(\Sigma,E)$.
The operators "sig" and "Mod" induce an equivalence relation on specifications.

Def. 4.2 Two specifications SP1 and SP2 are called **equivalent**, written SP1 = SP2, if sig(SP1) = sig(SP2) and Mod(SP1) = Mod(SP2).

For example, a flat specification <Σ,E> is equivalent to the specification derived from the theory of their models.

Fact 4.3 Let <Σ,E> be a flat specification. Then <Σ,E> = <Σ,Th(Mod(<Σ,E>))>.

In the following we use the notation

spec SP ≡ **sorts** S **functions** F **predicates** P **axioms** E **endspec**

to denote a specification <Σ,E>, where $\Sigma =_{def} <S,F,P \cup St_{S,F}>$. Hence, the standard symbols are always implicitly contained in any signature Σ. We often omit the brackets "{" and "}" around sets of sorts, function symbols, predicate symbols and axioms.

Example 4.4

(1) **Total ordering**

The following specification consists of one sort s for which a total ordering relation $\leq$ is defined.

spec T0 ≡
  sort s
  predicates $\leq$: s,s
  axioms $\forall$ x,y,z : s.
    $x \leq x \ \wedge$
    $(x \leq y \wedge y \leq z \Rightarrow x \leq z) \wedge$
    $(x \leq y \wedge y \leq x \Rightarrow x = y) \wedge$
    $(x \leq y \vee y \leq x)$
endspec

All models of T0 are partial orderings with respect to the equality relation = which is implicitly declared. But note that in some of these models the interpretation of = does not coincide with the equality within the structure. For instance, the structure N3 defined by

$N3_s =_{def} \mathbb{N}$,

$n =^{N3} m \Leftrightarrow_{def} n \bmod 3 = m \bmod 3$, $n \leq^{N3} m \Leftrightarrow_{def} n \bmod 3 \leq m \bmod 3$,

is a model of T0 where the equality relation $=^{N3}$ does not coincide with the equality in $N3_s$.

(2) **Truth values**

The following specification BOOL0 consists of two (different) constants "true", "false" and the unary boolean function symbol "not".

spec BOOL0 ≡
  sort Bool
  functions true, false :$\rightarrow$Bool,
    not: Bool $\rightarrow$ Bool
  axioms true $\neq$ false,
    not(true) = false,
    $\forall x: \Sigma B0_{Bool}$. not(not(x)) = x
endspec

All models M of BOOL0 have at least two elements "$true^M$" and "$false^M$" representing "true" and "false". Moreover, the interpretation of "not" has the usual meaning on these two elements, but otherwise, it may have completely arbitrary values. E.g. the structures B, N1 and N2 (see section 2.2) are models of BOOL0. For reachable models of BOOL0, the two constants true and false are **constructors**: all ground terms $t \in T(\Sigma B0)$ are equivalent to one of these constants; i.e. the formula

$\forall x : \Sigma B0_{Bool} \exists y: < \{Bool\}, \{true,false\} >_{Bool}.\ x=y$

holds in BOOL0.

The third axiom is equivalent to the following infinite number of ground equations.

$$\text{not(not(t))} = t \quad \text{for } t \in \{\text{not}^k(\text{true}) \mid k \geq 0\} \cup \{\text{not}^k(\text{false}) \mid k \geq 0\}\ .$$

All these equations can be deduced from the second axiom together with the equation not(false)=true. Hence BOOL0 is equivalent to the following specification BOOL1.

```
spec BOOL1 ≡
        sort  Bool
        functions true, false:→Bool,
                  not: Bool→Bool
        axioms    true ≠ false,
                  not(true) = false,
                  not(false) = true
endspec
```

The following specification BOOL2 differs from BOOL0 just by quantifying over all elements, not only over terms.

```
spec BOOL2 ≡
        sort  Bool
        functions true, false:→Bool,
                  not: Bool→Bool
        axioms    true ≠ false,
                  not(true) = false,
                  ∀x: Bool. not(not(x)) = x
endspec
```

All models of BOOL2 are also models of BOOL1, but not vice versa: N2 is not a model of BOOL2. □

Quantification ranging over the set of all (interpretations of) constructor terms as in BOOL0 is common for algebraic specifications and deserves a special notation: the set C of constructor function symbols of a specification SP will be indicated by the keyword "**constructors**" and axioms of the form Q x: $<S,C>_s$.G with Q∈ {∀,∃} will be written as Q **cons** x:s.G. We write "$\mathbf{cons_{SP1}}$" if **cons** refers to the set of constructors of another specification SP1.

Example 4.5

(1) Truth values (continued)

The following is an equivalent notation for BOOL0:

```
spec BOOL3 ≡
        sort Bool
        constructors true, false :→ Bool
```

functions not: Bool → Bool
axioms true≠false,
not(true) = false,
∀ **cons** x: Bool. not(not(x)) = x
endspec

(2) Natural numbers with ordering

The following specification describes natural numbers together with the usual "less-or-equal"-relation.

spec NAT0 ≡
sort Nat
constructors 0 :→ Nat,
succ: Nat → Nat
functions .+. : Nat,Nat → Nat
predicate ≤ : Nat, Nat
axioms ∀ **cons** m,n:Nat.
0≠succ(n) ∧ (succ(n) = succ(m) ⇒ n=m) ∧
n + 0 = n ∧ n + succ(m) = succ(n+m) ∧
(n≤m ⇔ ∃ **cons** b: Nat . n+b=m)
endspec

The function symbols 0 and succ are constructors, i.e. the quantification

"∀ **cons** m,n:Nat" is equivalent to "∀ m,n: $<\{Nat\},\{0,succ\}>_{Nat}$".

The (up to isomorphism) only standard model N0 of NAT0 consists of the standard natural numbers together with the usual less-or-equal-relation:

$$N0_{Nat} =_{def} \mathbb{N},\ 0^{N0} =_{def} 0,\ succ^{N0}(n) =_{def} n+1,\ n+^{N0}m =_{def} n+m,\ n\leq^{N0}m \text{ iff } n\leq m .$$

Another (non-standard) model is the algebra Z0 of negative integers, i.e.

$$Z0_{Nat} =_{def} \mathbb{Z},\ 0^{Z0} =_{def} 0,\ succ^{Z0}(n) =_{def} n-1,\ n+^{Z0}m =_{def} n+m,\ n\leq^{Z0}m \text{ iff } |n|\leq|m|.$$

Notice that for Z0 the last axiom holds (for constructor terms) but it does not hold for all elements: $0 \leq^{Z0} 1$ but there does not exist a term t of the form $succ^k(0)$, $k\geq 0$, such that $0+t^{Z0} = 1$ since each t^{Z0} is a non-positive integer.

As a third example for a model of NAT0 consider natural numbers with a bottom element. The algebra N⊥ is defined as follows:

$$N\bot_{Nat} =_{def} \mathbb{N} \cup \{\bot\},\ 0^{N\bot} =_{def} 0,\ n\leq^{N\bot} m \text{ iff } n\neq\bot \wedge m\neq\bot \wedge n\leq m.$$

$succ^{N\bot}(n) =_{def} n+1$, if $n\neq\bot$; and ⊥, otherwise,

$n+^{N\bot} m =_{def} n+m$, if $n\neq\bot \wedge m\neq\bot$; and ⊥, otherwise.

(3) Integers

Integers can be specified similar to natural numbers. In two cases, quantification ranges over constructors for natural numbers, not for integers.

spec INT ≡
sort Int
constructors 0 :→ Int
pred, succ: Int→ Int
functions .+. : Int, Int→ Int

predicate $\leq$: Int, Int

axioms $\forall$n: <{Int}, {0,succ}>$_{Int}$. $0 \neq succ(n)$

$\forall$ **cons** y,z: Int.

pred(succ(z)) = z $\wedge$ succ(pred(z)) = z $\wedge$

z+0 = z $\wedge$

z + succ(y) = succ(z+y) $\wedge$

z+pred(y) = pred(z+y) $\wedge$

(y$\leq$z $\Leftrightarrow$ $\exists$n : T(<{Int}, {0,succ}>) . y+n = z)

endspec

As for natural numbers, the standard model Z1 of integers is up to isomorphism the only such model of INT.

$Z1_{Int} =_{def} \mathbb{Z}$,

$0^{Z1} =_{def} 0$, $succ^{Z1}(z) =_{def} z+1$, $pred^{Z1}(z) =_{def} z-1$, $y+^{Z1}z =_{def} y+z$,

$y \leq^{Z1} z$ iff $y \leq z$.

The algebra Z0 (which is a model of NAT0, cf. 4.5(2)) can be extended to a sig(INT)-algebra by choosing the obvious definition for $pred^{Z0}$. However, it is not a model of INT since it does not satisfy the last axiom.

Similar to $\mathbb{N}\perp$, one can define the algebra $\mathbb{Z}\perp$ of integers with bottom element:

$\mathbb{Z}\perp_{Int} =_{def} \mathbb{Z} \cup \{\perp\}$, $0^{\mathbb{Z}\perp} =_{def} 0$, $y \leq^{\mathbb{Z}\perp} z$ iff $y \neq \perp \wedge z \neq \perp \wedge y \leq z$,

$succ^{\mathbb{Z}\perp}(z) =_{def} z+1$, if $z \neq \perp$; and $\perp$, otherwise,

$pred^{\mathbb{Z}\perp}(z) =_{def} z-1$, if $z \neq \perp$; and $\perp$, otherwise,

$y +^{\mathbb{Z}\perp} z =_{def} y+z$, if $y \neq \perp \wedge z \neq \perp$; and $\perp$, otherwise

5. Structured specifications

For writing large specifications it is convenient to design specifications in a structured fashion by combining and modifying smaller specifications. This supports a modular decomposition into specifications of manageable size and helps to master the complexity originating from a large number of (function and predicate) symbols and axioms.

In the following we introduce three specification building operators which are derived from the operators of the specification language ASL [Sannella, Wirsing 83]. For this language an institution-independent semantics has been given by [Sannella, Tarlecki 85a] which makes it easy to derive (more exactly to instantiate) a semantics appropriate for our approach.

The first operator "derive from • by •" will be used for renaming, hiding, exporting and copying.

Def. 5.1 Let SP1 be a specification with signature $\Sigma 1$ and let $\sigma:\Sigma \rightarrow \Sigma 1$ be a signature morphism, then

derive from SP1 **by** σ

denotes the specification SP defined by

$sig(SP) =_{def} \Sigma$,

$Mod(SP) =_{def} \{A|\sigma \in Struct(\Sigma) \mid A \in Mod(SP1)\}$.

Using "derive", renaming of specifications can be defined as follows: if SP1 is a specification and $[x_1/y_1,\dots, x_n/y_n] : \Sigma \rightarrow sig(SP1)$ denotes a renaming (cf. section 2.1), then

rename SP1 **by** $[x_1/y_1,\dots, x_n/y_n] =_{def}$ **derive from** SP1 **by** $[x_1/y_1,\dots, x_n/y_n]$

denotes a specification SP which is the same as SP1 but with $y_1,\dots, y_n$ renamed into $x_1,\dots, x_n$. The second operator "translate . with ." is the converse of "derive" [Sannella, Tarlecki 87].

Def. 5.2. Let SP1 be a specification with signature $\Sigma 1$ and $\sigma: \Sigma 1 \rightarrow \Sigma$ be a signature morphism, then

translate SP1 **with** σ

denotes the specification SP defined by

$sig(SP) =_{def} \Sigma$,

$Mod(SP) =_{def} \{A \in Struct(\Sigma) \mid A|\sigma \in Mod(SP1)\}$.

The third operator, called "+", combines two specifications SP1 and SP2 (without taking care of name clashes).

Def. 5.3 Let SP1 and SP2 be two specifications with signatures $\Sigma 1$ and $\Sigma 2$ resp..
Then SP1 + SP2 denotes a specification SP defined by

$sig(SP) =_{def} \Sigma 1 \cup \Sigma 2$,

$Mod(SP) =_{def} \{A \in Struct(sig(SP)) \mid A|\Sigma 1 \in Mod(SP1) \text{ and } A|\Sigma 2 \in Mod(SP2)\}$.

The models of SP1+SP2 can be defined in terms of "translate":

Mod(SP1+SP2) = Mod(**translate** SP1 **with** ${}_{\Sigma 1}in$) $\cap$ Mod(**translate** SP2 **with** ${}_{\Sigma 2}in$)

where ${}_{\Sigma i}in$: $\Sigma i \rightarrow$ sig(SP1 + SP2), i=1,2, is the canonical embedding, defined by ${}_{\Sigma i}in(x)=x$ for $x \in \Sigma i$.

As an example for an operator which can be explicitly defined using "+" we introduce simple enrichments:
Let SP1 be a specification with signature $\Sigma 1 = < S1, F1, P1 >$ and let S,F,P,W be sets of sorts, function symbols, predicate symbols and formulas respectively.
Then

enrich SP1 **by** **sorts** S **functions** F **predicates** P **axioms** E

denotes the specification SP $=_{def}$ SP1+ $<\Sigma,E>$ where Σ is the union of the signature $\Sigma 1$ with the new symbols and with new implicitly defined equality symbols and generating predicate symbols for the new sorts, i.e. $\Sigma =_{def} < S1 \cup S,\ F1 \cup F,\ P1 \cup P \cup St_{S1 \cup S, F1 \cup F} >$.
Notice that the equality symbols of Σ are interpreted as Σ-congruence relations, whereas in SP1 the same equality symbols (on the sorts of S1) are interpreted as $\Sigma 1$-congruence relations.
Thus, for $=_s$, $s \in S1$, implicitly new substitution axioms have been added.
If, as before, the constructor notation is used, then in E "**cons**" denotes the set

$T(<S1 \cup S, C1 \cup C>)$ of ground terms where C1 is the set of constructor functions of SP1 and C is the set of constructor functions introduced by the enrichment. In the set E1 of axioms of SP1, cons denotes the set $T(< S1, C1 >)$ of ground SP1-constructor terms.

Example 5.4

(1) Finite sets of natural numbers

The following specification describes finite sets of natural numbers.

```
spec SETNAT ≡
        enrich NAT0 by
                sort Set
                constructors empty:→ Set,
                             add: Nat, Set → Set
                predicates   .ε. : Nat, Set
                axioms       ∀ cons m,n: Nat,  s : Set .
                             ¬(n ε empty) ∧
                             (nεadd(m,s) ⇔ (n=m ∨ nεs)) ∧
                             add(n,add(n,s)) = add(n,s) ∧
                             add(n,add(m,s)) = add(m,add(n,s))
endspec
```

The (up to isomorphism) only standard model S0 of SETNAT consists of finite sets of natural numbers; i.e.

$S0 \mid NAT0 =_{def} N0$, $S0_{Set} =_{def} \{s \subseteq \mathbb{N} \mid s \text{ is finite}\}$

$empty^{S0} =_{def} \emptyset$, $add^{S0}(m,s) =_{def} \{m\} \cup s$, $n\, \varepsilon^{S0}\, s$ iff $n \in s$.

The power set of integers and ordered sequences of natural numbers form two other models S1 and S2:

$S1 \mid NAT0 =_{def} Z0$, $S1_{Set} =_{def} \{s \mid s \subseteq \mathbb{Z}\}$

$empty^{S1} =_{def} \emptyset$, $add^{S1}(m,s) =_{def} \{m\} \cup s$, $n\, \varepsilon^{S1}\, s$ iff $n \in s$;

$S2 \mid NAT0 =_{def} N0$, $S2_{Set} =_{def} \{<n_1,\ldots,n_k> \mid k \geq 0 \wedge n_1 \leq \ldots \leq n_k\}$,

$empty^{S2} =_{def} <>$ "the empty sequence",

$add^{S2}(n,<n_1,\ldots,n_k>) =_{def} <n_1,\ldots,n_{i-1},n,n_i,\ldots,n_k>$ if $n_{i-1} \leq n \leq n_i$ for some $i \in \{1,\ldots,k\}$,

$n\varepsilon^{S2}<n_1,\ldots,n_k>$ iff there exists $i \in \{1,\ldots,k\}$: $n=n_i$.

For S2, the interpretation of the equality symbol $=_{Set}$ does not coincide with the equality of sequences:

for all $s,s' \in S2_{Set}$, $s(=_{Set})^{S2}s'$ iff for all $n \in \mathbb{N}$: $n\varepsilon^{S2}s \Leftrightarrow n\varepsilon^{S2}s'$.

(2) Finite sequences of integers

Finite sequences of integers can be specified as follows

```
spec SEQINT ≡
        enrich INT by
                sort Seq
                constructors emptyseq:→ Seq,
```

<.>: Int→ Seq,
.°.: Seq, Seq → Seq
functions first: Seq→Int,
rest: Seq→Seq
axioms ∀ **cons** x: Int, s,s1,s2 : Seq .
emptyseq ° s = s ∧
s ° emptyseq = s ∧
s ° (s1 ° s2) = (s ° s1) ° s2 ∧
first(<x> ° s) = x ∧
rest (<x> ° s) = s

endspec

SEQINT has infinitely many standard models since the values of "first(emptyseq)" and "rest(emptyset)" are not fixed by the axioms. The algebra $S\bot$ of sequences with bottom element is a standard model, whereas the structure PA of arrays with pointers is a non-standard model of SEQINT.

$S\bot \mid INT =_{def} \mathbb{Z}\bot$, $S\bot_{Seq} =_{def} \mathbb{Z}^* \cup \{\bot_{Seq}\}$,

$emptyseq^{S\bot} =_{def} \langle\rangle$,

$\langle z\rangle^{S\bot} =_{def} \langle z\rangle$, if $z\neq\bot$; and $\bot_{Seq}$, otherwise;

$s \circ^{S\bot} s' =_{def} \langle y_1,\ldots,y_k,z_1,\ldots,z_m\rangle$, if $s=\langle y_1,\ldots,y_k\rangle\in\mathbb{Z}^* \wedge s'=\langle z_1,\ldots,z_m\rangle\in\mathbb{Z}^*$; and $\bot_{Seq}$, otherwise;

$first^{S\bot}(s) =_{def} x$, if $s=\langle x,x_1,\ldots,x_k\rangle\in\mathbb{Z}^*$; and $\bot$, otherwise;

$rest^{S\bot}(s) =_{def} \langle x_1,\ldots,x_k\rangle$, if $s=\langle x,x_1,\ldots,x_k\rangle\in\mathbb{Z}^*$; and $\bot_{Seq}$, otherwise.

Each non-bottom element of the carrier set PA_{Seq} of arrays with pointers is represented by a pair, consisting of a natural number (called pointer) and a mapping $\alpha\in[\mathbb{N}_+ \to \mathbb{Z}]$ from the positive natural numbers $\mathbb{N}_+ =_{def} \mathbb{N}\backslash\{0\}$ into the integers.

$PA \mid INT =_{def} \mathbb{Z}\bot$, $PA_{Seq} =_{def} (\mathbb{N}\times[\mathbb{N}_+ \to \mathbb{Z}]) \cup \{\bot_{Seq}\}$

$emptyseq^{PA} =_{def} \langle 0,\lambda x.0\rangle$

$\langle z\rangle^{PA} =_{def} \langle 1,\alpha_2\rangle$, if $z\neq\bot$; and $\bot_{Seq}$, otherwise;

where $\alpha_2(x) =_{def} z$, if $x=1$; and 0, otherwise;

$s \circ^{PA} s' =_{def} \langle n+m,\gamma\rangle$, if $s=\langle n,\alpha\rangle$ and $s'=\langle m,\beta\rangle$; and $\bot_{Seq}$, otherwise;
where $\gamma(x) =_{def} \alpha(x)$, if $1\leq x\leq n$; and $\beta(x-n)$, if $n<x\leq n+m$; and 0, otherwise;

$first^{PA}(s) =_{def} z$, if $s=\langle n,\alpha\rangle$ and $\alpha(n)=z$; and $\bot$, otherwise;
$rest^{PA}(s) =_{def} \langle n-1,\alpha\rangle$, if $s=\langle n,\alpha\rangle$; and $\bot_{Seq}$, otherwise.

The equality symbol is interpreted in PA in a non-standard way:

$s\ (=_{Seq})^{PA}\ s'$ iff $s=s'=\bot_{Seq}$ or there exist n, α, β such that

$$s=\langle n,\alpha\rangle,\ s'=\langle n,\beta\rangle \text{ and } \forall x\in\{1,\ldots,n\}.\ \alpha(x)=\beta(x).$$

6. Properties of specification operators

A number of other specification building operators can be defined using <.,.>, + and derive.

Example 6.1

(1) Exporting and hiding can be explicitly defined as follows. Let SP1 be a specification with

signature $\Sigma 1 = \langle S1,F1,P1 \rangle$ and let $\Sigma = \langle S,F,P \rangle \subseteq sig(SP1)$ be a signature. Then

export Σ **from** SP1 $=_{def}$ **derive from** SP1 **by** $_{\Sigma}in$

where $_{\Sigma}in: \Sigma \to sig(SP1)$ is the canonical embedding from Σ into sig(SP1), i.e. $_{\Sigma}in(x)=x$ for all $x \in \Sigma$. Hiding can be explicitly defined in terms of "export".

Let $s_0 \in S1$ be a sort of SP1 and $f_0 \in F1 \cup P1$ be a function or predicate symbol of SP1. Then hiding s_0 means to forget s_0 and all function and predicate symbols with s_0 in their domain or range, whereas hiding f_0 means just to forget f_0.

hide s_0 **from** SP1 $=_{def}$ **export** $\Sigma 1\text{-}s_0$ **from** SP1,

hide f_0 **from** SP1 $=_{def}$ **export** $\Sigma 1 \backslash \{f_0\}$ **from** SP1,

where $\Sigma 1\text{-}s_0 =_{def} \langle S1^-, \{F1_{w,s}\}_{w \in (S1^-)^*, s \in S1^-}, \{P1_w\}_{w \in (S1^-)^*} \rangle$, and $S1^- =_{def} S1 \backslash \{s_0\}$.

(2) Exporting and renaming can be combined by the following specification operator which exports Σ after renaming $y_1,\ldots,y_n$ into $x_1,\ldots,x_n$.

export Σ **with** $[x_1/y_1,\ldots,x_n/y_n]$ **from** SP $=_{def}$ **derive from** SP **by** $_{\Sigma}[x_1/y_1,\ldots,x_n/y_n]_{sig(SP)}$

where $_{\Sigma}[x_1/y_1,\ldots,x_n/y_n]_{sig(SP)}: \Sigma \to sig(SP)$ denotes the signature morphism induced by $[x_1/y_1,\ldots,x_n/y_n]$ with domain Σ and range sig(SP) (cf. section 2.1).

(3) Non-injective signature morphisms allow for the copying of structures.

As example, we consider the specification INT of 4.4(3) with signature sig(INT).

We extend sig(INT) by a new sort Nat, function symbols 0: $\to$Nat, succ: Nat$\to$Nat and +: Nat,Nat$\to$Nat, predicate symbols $\leq$: Nat,Nat and standard symbols. Call this signature ΣINT-NAT and consider the signature morphism

copy: ΣINT-NAT$\to$sig(INT) defined by

copy(Nat) $=_{def}$ Int, copy(0) $=_{def}$ 0, copy(succ) $=_{def}$ succ, copy(+) $=_{def}$ +,

copy($\leq$) $=_{def}$ $\leq$, copy($=_{Nat}$) $=_{def}$ $=_{Int}$, copy(x) $=_{def}$ x for $x \in sig(INT)$.

Then

derive from INT **by** copy

denotes a specification with signature ΣINT-NAT. Each model M of this specification consis of a model of INT together with a copy of itself the carrier set of which is named "Nat" but where the operation pred is missing. Hence, the carrier set M_{Nat} is always isomorphic to a superset of the natural numbers, it is never the set $\mathbb{N}$ of all standard natural numbers. A predecessor function for natural numbers can be introduced as follows.

```
spec INTNAT ≡
    enrich
        derive from INT by copy
    by  functions pred: Nat→Nat
        axioms pred(0)=0, ∀cons_NAT0 x: Nat. pred(succ(x))=x
endspec
```

Here "$\forall cons_{NAT0}$ x" is an abbreviation for quantification over the constructors of Nat0, i.e. for $\forall x: \langle\{Nat\},\{0,succ\}\rangle_{Nat}$.

The specification operators satisfy a number of algebraic identities which are useful for transforming specification expressions. Such identities are given e.g. in [Sannella, Wirsing 83]. A comprehensive study of an algebra of specification expressions can be found in [Bergstra et al. 86].

For example, let the signature morphism $\sigma:\Sigma\rightarrow\Sigma'$ be given as well as the specifications SP and SP′ with signatures Σ and Σ' resp. Then we have

Mod(translate (derive from SP′ by σ) with σ) =

$\{A'\in \text{Struct}(\Sigma') \mid A'|\sigma \in$ Mod(**derive from SP′ by σ**)$\}$ =

$\{A'\in \text{Struct}(\Sigma') \mid A'|\sigma \in \{A|\sigma\in \text{Struct}(\Sigma) \mid A \in \text{Mod}(SP')\}\}$

This shows that Mod(**translate (derive from SP′ by σ) with σ**) $\supseteq$ Mod(SP′). We even have

translate (derive from SP′ by σ) with σ = SP′

provided σ is surjective. If σ is not surjective, then the carrier sets for sorts and the interpretations for function symbols that are not in the range of σ can be choosen arbitrarily.

Vice versa we have:

Mod(derive from (translate SP with σ) by σ) =

$\{A'|\sigma\in \text{Struct}(\Sigma) \mid A' \in$ Mod(**translate SP with σ**)$\}$ =

$\{A'|\sigma\in \text{Struct}(\Sigma) \mid A' \in \{A'\in \text{Struct}(\Sigma) \mid A'|\sigma \in \text{Mod}(SP)\}\}$ =

$\{A'|\sigma\in \text{Struct}(\Sigma) \mid A'|\sigma \in \text{Mod}(SP)\}$

This shows that Mod(**derive from (translate SP with σ) by σ**)$\subseteq$Mod(SP). We even have

(derive from (translate SP with σ) by σ) = SP

provided σ is injective. If σ is not injective then those models of SP are excluded in **(derive from (translate SP with σ) by σ)** in which sorts (or function symbols or predicates) that occur twice in the range are associated with different carrier sets (functions or predicates resp.).

Fact 6.2 Let SP, SP1 and SP2 be specifications.

(1) SP+SP = SP,

SP1+SP2 = SP2+SP1,

SP+(SP1+SP2) = (SP+SP1)+SP2,

(2) if $\sigma_1: \Sigma_1\rightarrow \text{sig}(SP)$, $\sigma_2: \Sigma_2\rightarrow\Sigma_1$ are signature morphisms, then

derive from (derive from SP by σ_1) by σ_2 = derive from SP by $\sigma_1 \circ \sigma_2$

(3) if σ is an injective mapping with range(σ)$\subseteq$sig(SP2) inducing the signature morphisms $_{\Sigma}\sigma: \Sigma\rightarrow \text{sig}(SP2)$, $_{\text{sig}(SP1)}\sigma: \text{sig}(SP1)\rightarrow\Sigma'$ (where $\Sigma' = {}_{\text{sig}(SP1)}\sigma(\text{sig}(SP1))$), $_{\text{sig}(SP1)\cup\Sigma}\sigma: \text{sig}(SP1)\cup\Sigma\rightarrow\Sigma'\cup\text{sig}(SP2)$ (for the notions see section 2.1), then

SP1+derive from SP2 by $_{\Sigma}\sigma$ =

derive from (translate SP1 with $_{\text{sig}(SP1)}\sigma$)+SP2 by $_{\text{sig}(SP1)\cup\Sigma}\sigma$

provided that the hidden symbols $(\text{sig}(SP2)\setminus{}_{\Sigma}\sigma(\Sigma))$ do not interfere with SP1 (i.e.

$(sig(SP1)\backslash\Sigma) \cap (sig(SP2)\backslash_{\Sigma}\sigma(\Sigma))=\emptyset)$.

Corollary 6.3 For all specifications SP, enrichments Δ, $\Delta1$, signatures Σ,$\Sigma1$, substitution lists α,$\alpha1$ such that the following expressions are well-defined, the following holds.

(1) **enrich (enrich SP by Δ) by $\Delta1$ = enrich SP by $\Delta \cup \Delta1$,**

(2) if $_{\Sigma}\alpha: \Sigma \rightarrow sig(SP)$ and $_{\Sigma1}\alpha1: \Sigma1 \rightarrow \Sigma$ are signature morphisms induced by α and $\alpha1$ (cf. section 2.1), then
export $\Sigma1$ with $\alpha1$ from (export Σ with α from SP) =
export $\Sigma1$ with $\alpha \circ \alpha1$ from SP,

(3) if $sig(SP) \cap sig(\alpha(\Delta))=\emptyset$, then
enrich (export Σ with α from SP) by Δ =
export $\Sigma \cup sig(\Delta)$ with α from (enrich SP by $\alpha(\Delta)$)
where $sig(\Delta)$ denotes the sorts, function and predicate symbols from Δ and where $\alpha(\Delta)$ denotes the substitution of $x_1,...,x_n$ by $y_1,...,y_n$ in Δ, i.e. $\alpha=[x_1/y_1, \dots, x_n/y_n]$.

Flat specifications are equivalent to their theory (cf. fact 4.3). The result of the combination of two flat specifications by the operator "+" is also a flat specification.

Fact 6.4 Let SP1 = < S1, F1, E1 > and SP2 = < S2, F2, E2 > be two flat specifications. Then

$SP1+SP2 = < \Sigma, E1 \cup E2 >$

where $\Sigma =_{def} < S1 \cup S2, F1 \cup F2, P1 \cup P2 \cup St_{S1 \cup S2, F1 \cup F2} >$.

On the other hand as a consequence of the "hiding facility" of the operator "derive", not all specifications (using derive) are equivalent to a flat specification. Examples for data structures that can be specified using "hidden functions" (i.e. with "derive") but not as flat specifications (with finitely many axioms) are given e.g. in [Majster 77], [Bergstra, Tucker 86] and [Wirsing 86]. In this case, a specification may not be equivalent to the theory of their models.

Fact 6.5 For all specifications SP, $Mod(SP) \subseteq Mod(Th(Mod(SP)))$.

There exist examples for which this inclusion is strict. Therefore, the theory operator is a closure operator which abstracts from the particular models of a specification and considers only the properties of the class of all models.

7. Implementation

The programming discipline of stepwise refinement suggests that a program be evolved from a high-level specification by working gradually via a series of successively more detailed lower-level intermediate specifications. A formalization of this approach requires a precise definition of the concept of refinement, i.e. of the implementation of a specification by another. A specification SP´ implements another specification SP if it has the same signature and all its models are models of SP.

Def 7.1 Let SP and SP´ be two specifications. SP´ is an **implementation** of SP, written SP ~~~> SP´, if sig(SP)=sig(SP´) and Mod(SP´)$\subseteq$Mod(SP).

This notion of implementation (used e.g. by [Sannella, Wirsing 83], [Sannella, Tarlecki 87]) is conceptually simpler than other notions known from the literature such as "forget-restrict-identify"-implementations (cf. e.g. [Ehrig et al. 82]). Our notion represents the idea that a specification describes those properties which have to be respected by the implementations. For flat specifications, the implementation relation can be characterized as follows.

Fact 7.2 Let SP=<Σ,E> and SP´=<Σ,E´> be two flat specifications with the same signature. Then the following properties are equivalent:

(i) SP ~~~> SP´,

(ii) $E' \vdash_I e$ for all $e \in E$

(iii) Mod(SP´)$\subseteq$Mod(SP).

In the following we give a number of simple examples for implementations. Here it is often necessary to define congruence relations w.r.t the signature of the specification to implement. In order to avoid writing standard axioms we introduce the following notation. Let SP be a specification with signature Σ. We write

equalities $=_{SP,s}$: s,s for s$\in$ sorts(Σ)

as abbreviation for the following predicate symbols and axioms:

predicates $=_{SP,s}$: s,s for s$\in$ sorts(Σ)
together with the additional axioms for a Σ-congruence:
$\forall$**cons**$_{SP}$ x:s. x$=_{SP,s}$x, for all s$\in$ S,
$\forall$**cons**$_{SP}$ x_1,y_1:s1, ... ,x_n,y_n:sn. $x_1 =_{SP,s1} y_1 \wedge ... \wedge x_n =_{SP,sn} y_n \Rightarrow f(x_1,...,x_n) =_{SP,s} f(y_1,...,y_n)$
for all f: s1,...,sn$\to$s $\in$ opns(Σ),
$\forall$**cons**$_{SP}$ x_1,y_1:s1, ... ,x_n,y_n:sn. $x_1 =_{SP,s1} y_1 \wedge ... \wedge x_n =_{SP,sn} y_n \wedge p(x_1,...,x_n) \Rightarrow p(y_1,...,y_n)$
for all p: s1,...,sn $\in$ preds(Σ).

Example 7.3 Implementations

(1) Truth-values by natural numbers

The specification BOOL3 of truth-values can be implemented by extending the specification NAT0, renaming Nat into Bool and exporting the signature of BOOL3.

```
spec BOOL_by_NAT ≡
    export ΣB0 with [Bool/Nat, true/0, not/succ, =Bool/=BOOL3,Nat] from
        enrich NAT0
        by functions  false: →Nat
           equalities =BOOL3,Nat: Nat, Nat
```

axioms $false = succ(0)$,
$\forall$**cons** x: Nat. $succ(succ((x)) =_{BOOL3,Nat} x$

endspec

The equality symbol for truth values has to be added to NAT0. It is <u>not</u> a congruence w.r.t. the signature of NAT0; in particular, the predicate "≤" does not satisfy the substitution property w.r.t. $=_{BOOL3,Nat}$. Nevertheless, it is easy to see that all axioms of BOOL3 hold in BOOL_by_NAT. Hence, BOOL3 ~~~> BOOL_by_NAT holds.

(2) <u>Natural numbers by integers</u>

Natural numbers are implemented by integers in the obvious way.

spec NAT_by_INT ≡
export sig(NAT0) **with** [Nat/Int]
from INT
endspec

Using the "export-after-rename"-operator the sort "Int" is renamed into "Nat" and the operation "pred" is forgotten. The equality relation for integers is the same as for natural numbers. The axioms of NAT0 hold in NAT_by_INT: NAT0 ~~~> NAT_by_INT.

(3) <u>Finite sets by ordered sequences</u>

Finite sets of natural numbers can be implemented by ordered sequences of integers as follows.

spec SET_by_ORDSEQINT ≡
export sig(SETNAT) **with** [Set/Seq, Nat/Int, $=_{SET}/=_{SETNAT}$] **from**
enrich SEQINT
by functions empty: →Seq,
add: Int, Seq→Seq
predicates .ε.: Int, Seq
equalities $=_{SETNAT,Seq}$: Seq, Seq,
$=_{SETNAT,Int}$: Int, Int
axioms empty=emptyseq,
$\forall$**cons** x,y: Nat, s: Set.
$add(x, emptyseq) = \langle x \rangle \wedge$
$(x \leq y \Rightarrow add(x, \langle y \rangle \circ s) = \langle x \rangle \circ \langle y \rangle \circ s) \wedge$
$(y \leq x \wedge y \neq x \Rightarrow add(x, \langle y \rangle \circ s) = \langle y \rangle \circ add(x,s)) \wedge$
$\neg(x \; \varepsilon \; emptyseq) \wedge$
$[x \; \varepsilon \; (\langle y \rangle \circ s) \Leftrightarrow (x=y \vee (y \leq x \wedge x \; \varepsilon \; s))]$,
$\forall x,y{:}\langle\{Int\},\{0,succ\}\rangle_{Int}.\; x =_{SETNAT,Int} y \Leftrightarrow x=y$,
$\forall x{:}\langle\{Int\},\{0,succ\}\rangle_{Int},\; s{:}\langle\{Seq,Int\},\{0,succ,empty,add\}\rangle_{Seq}.$
$add(x,add(x,s)) =_{SETNAT,Seq} add(x,s)$

endspec

Natural numbers are represented by positive integers, sets are represented by ordered sequences with repetition. As a consequence, the specification SET_by_ORDSEQINT satisfies the axiom

(*) add(x,add(y,s)) = add(y,add(x,s))

for all ordered sequences of integers representing the left-hand side or (equivalently) the right-hand side of this equation. The other characteristic equation for sets

add(x,add(x,s)) = add(x,s)

does not hold in the enrichment part of SET_by_ORDSEQINT for the equality between sequences; but according to the last axiom of SET_by_ORDSEQINT the equation

$\text{add}(x,\text{add}(x,s)) =_{\text{SETNAT,Seq}} \text{add}(x,s)$

holds for all those sequences which represent sets of natural numbers, i.e. for all (interpretations of) ordered sequences of positive integers. Therefore, the equation (*) holds in SET_by_ORDSEQINT (after the renaming for all constructor terms of SETNAT). All other axioms of SETNAT can easily be derived from the other axioms of SET_by_ORDSEQINT. Hence SET_by_ORDSEQINT is an implementation of SETNAT, i.e.

SETNAT ~~~> SET_by_ORDSEQINT.

The implementation relation is a partial ordering on the class of all specifications and the specification operators are monotonic.

Theorem 7.4

(1) The implementation relation is a partial ordering w.r.t. the equivalence = of specifications: for all specifications SP, SP1 and SP2,

(i) SP ~~~> SP

(ii) SP ~~~> SP1 and SP1 ~~~> SP implies SP = SP1

(iii) SP ~~~> SP1 and SP1 ~~~> SP2 implies SP ~~~> SP2

(2) The specification operators are monotonic; i.e. for all signatures Σ, sets of Σ-formulas E1, E2, specifications SP1, SP2, SP1´, SP2´ and signature morphisms $\sigma:\Sigma\rightarrow \text{sig(SP1)}$, $\sigma':\text{sig(SP1)}\rightarrow\Sigma'$ the following holds:

(i) $E1\subseteq E2$ implies $<\Sigma,E1>$ ~~~> $<\Sigma,E2>$,

(ii) SP1 ~~~> SP1´ implies (**derive from** SP1 **by** σ) ~~~> (**derive from** SP1´ **by** σ),

(iii) SP1 ~~~> SP1´ implies (**translate** SP1 **with** σ') ~~~> (**translate** SP1´ **with** σ'),

(iv) SP1 ~~~> SP1´ and SP2 ~~~> SP2´ implies SP1+SP2 ~~~> SP1´+SP2´.

Hence in particular, the implementation relation is "horizontally" and "vertically" composable [Burstall, Goguen 80], since vertical composition corresponds to the transitivity and horizontal composition corresponds to the monotonicity of the implementation relation.

Common techniques for defining the notion of implementation of algebraic specifications can be seen as particular case of our notion. We call a specification SP' an <u>FRI-implementation</u> of SP (FRI stands for "<u>f</u>orget-<u>r</u>estrict-<u>i</u>dentify") if sig(SP)$\subseteq$sig(SP') and if there exists a sig(SP)-congruence relation $\sim_i$ on T(sig(SP)) such that for all standard models A´$\in$Gen(SP´), the Σ-standard structure

$<A'>_\Sigma/\sim_i$ is a model of SP; SP′ is called an <u>FIR-implementation</u> of SP, if in the second condition $\sim_i$ is a sig(SP′)-congruence on T(sig(SP′)) and $<A'/\sim_i>_\Sigma$ is a model of SP. Obviously, every FIR-implementation is also an FRI-implementation.

Fact 7.5 Let SP = <Σ,E> and SP′=<Σ′,E′> be two flat <u>algebraic</u> specifications (cf. def. 4.1). If SP′ is a FRI-implementation of SP, then the specification SP′′ is an implementation of SP where

spec SP′′ ≡
  export Σ **from**
    enrich SP′ **by**
      equalities $=_{SP,s}$: s,s for s∈sorts(Σ)}
      axioms { $t =_{SP,s} t'$ | t,t′∈T(Σ)$_s$, s∈sorts(Σ), $t \sim_i t'$}

and $\sim_i$ denotes the Σ-congruence associated with FRI.

8. Parameterization

Parameterization is the process of encapsulating a piece of software and abstracting from some names (or more generally, from some subexpressions) occurring in it in order to replace them in other contexts by different actual parameters.

Parameterization of structured specifications is done by means of λ-abstraction: a parameterized specification is considered as a mapping taking specifications as arguments and giving a specification as a result. This is formalized by a version of typed λ-calculus, the so-called λπ-calculus which has been introduced by [Feijs 89].

Def. 8.1 A **(<u>structured</u>) parameterized specification** PSP is a λ-expression of the form $\lambda X{:}SP_{par}.SP_{body}[X]$ where SP_{par}, $SP_{body}[X]$ are specification expressions built using the specification operators <.,.>, **derive from . by .**, **translate . with .**, .+., and where X is an identifier not occurring in SP_{par}. Moreover, it is required that $SP_{body}[SP_{par}/X]$ forms a well-defined specification. SP_{par} is called **formal requirement specification** and $SP_{body}[X]$ is called **target specification**.

For the semantics, the "refinement" relation for specifications is considered.

Def. 8.2 For any two specifications SP1 and SP2, SP2 is a **refinement** of SP1, written SP1⊑SP2, if sig(SP1)⊆sig(SP2) and Mod(SP2)|sig(SP1) ⊆ Mod(SP1).

Fact 8.3 The refinement relation is a partial ordering on the set of specification expressions (w.r.t "=") which is compatible with the specification operators, i.e. for all specifications SP, SP1, SP2,

(1) SP⊑SP (reflexivity),

(2) $SP \sqsubseteq SP1$ and $SP1 \sqsubseteq SP2$ implies $SP \sqsubseteq SP2$ (transitivity),

(3) $SP \sqsubseteq SP1$ and $SP1 \sqsubseteq SP$ implies $SP = SP1$ (antisymmetry),

(4) for all signature morphisms $\sigma: \Sigma \to sig(SP1)$,
$SP \sqsubseteq SP1$ implies **derive from** SP **by** $\sigma|(\sigma^{-1}(sig(SP)) \sqsubseteq$ **derive from** $SP1$ **by** σ,

(5) for all signature morphisms $\sigma: sig(SP1) \to \Sigma$,
$SP \sqsubseteq SP1$ implies **translate** SP **with** $\sigma|sig(SP) \sqsubseteq$ **translate** $SP1$ **with** σ,

(6) for all specifications $SP1'$, $SP2'$,
$SP1 \sqsubseteq SP1'$ and $SP2 \sqsubseteq SP2'$ implies $SP1+SP2 \sqsubseteq SP1'+SP2'$.

The semantics of any λ-expression is defined by function application.

Def. 8.3 For any specification SP and parameterized specification $\lambda X{:}SP_{par}.SP_{body}[X]$, the expression

$(\lambda X{:}SP_{par}.SP_{body}[X])(SP)$

is well-defined, if $SP_{par} \sqsubseteq SP$. In this case, it denotes the specification $SP_{body}[SP/X]$.

Example 8.4

(1) The specification SETNAT of 5.4.(1) can be parameterized by replacing the natural numbers by arbitrary data.

pspec SET $\equiv \lambda X{:}ELEM.SET_{ELEM}[X]$,

where

spec ELEM $\equiv$ **sort** Elem **endspec**

and where $SET_{ELEM}[X] =_{def}$ SETNAT[X/NAT0, Elem/Nat].

(2) Similarly, the specification SEQINT can be parameterized by replacing INT by arbitrary data.

pspec SEQ $\equiv \lambda X{:}ELEM.SEQ_{ELEM}[X]$,

where $SEQ_{ELEM}[X] =_{def}$ SEQINT[X/INT, Elem/Int].

(3) Derived operators can be considered as parameterized specifications. E.g. for all extensions Δ, substitution lists α, and signatures Σ, let

pspec $enr_\Delta \equiv \lambda X{:}R_1.$**enrich** X **by** Δ,

pspec $expren_{\Sigma,\alpha} \equiv \lambda X{:}R_2.$**export** Σ **with** α **from** X

where $R_1 =_{def} \langle free(\Delta),\emptyset\rangle$ denotes the specification formed by the symbols occurring in Δ but <u>not</u> declared in Δ and where $R_2 =_{def}$ **translate** $\langle\Sigma,\emptyset\rangle$ **with** $_\Sigma\alpha: \Sigma \to sig(X)$. Note that Δ, α, Σ are expressions which may depend on X.

The notion of implementation generalizes straightforwardly to parameterized specifications (cf. [Sannella, Wirsing 82], [Wirsing 86]).

Def 8.5 Let $SP1 = \lambda X{:}SP1_{par}.SP1_{body}$ and $SP2 = \lambda X{:}SP2_{par}.SP2_{body}$ be two parameterized specifications.

(1) SP2 is called **(partial parameterized) implementation** of SP1, written SP1~~~> SP2, if $SP1_{par} \sqsubseteq SP2_{par}$ and for all actual parameters SP of SP2 (with $SP2_{par} \sqsubseteq SP$) the application SP2(SP) of SP2 to SP is an implementation of the application SP1(SP) of SP1 to SP, i.e. SP1(SP) ~~~> SP2(SP).

(2) A partial parameterized implementation SP2 of SP1 is called **total**, if every actual parameter of SP1 is also an actual parameter of SP2, i.e. if $SP2_{par} \sqsubseteq SP1_{par}$.

Example 8.6 The parameterized specification

pspec SET_by_ORDSEQ ≡
    λX: **rename** T0 **by** [Elem/s].
    export sig(SET(X)) **with** [Set/Seq, $=/=_{SET(X)}$] **from**
    enrich X
    by Δ[Elem/Int, SET(X)/SETNAT, $cons_X$/{0, succ}, $cons_{SET(X)}$/{0, succ, empty, add}]
endspec

(where T0 is the specification of total orderings of example 4.4(1) and where Δ denotes the enrichment of SEQINT in example 7.3(3)) is a partial parameterized implementation of SET. It is not total because the formal parameter of SET_by_ORDSEQ requires a total ordering relation whereas the formal parameter of SET is trivial.

In order to prove a horizontal composition theorem for implementations we generalize the notion of refinement to parameterized specifications.

Def 8.7 Let $SP1 = \lambda X{:}SP1_{par}.SP1_{body}$ and $SP2 = \lambda X{:}SP2_{par}.SP2_{body}$. Then SP2 is called **refinement** of SP1, written $SP1 \sqsubseteq SP2$, if

(1) $SP2_{par}$ is a refinement of $SP1_{par}$, i.e. $SP1_{par} \sqsubseteq SP2_{par}$, and

(2) for all actual parameters SP of SP2 (i.e. $SP2_{par} \sqsubseteq SP$) SP2(SP) is a refinement of SP1(SP), i.e. $SP1(SP) \sqsubseteq SP2(SP)$.

For the horizontal composition theorem for refinements we need the following lemma.

Lemma 8.8 Let $SP1 = \lambda X{:}SP1_{par}.SP1_{body}$ be a parameterized specification and let SPA and SPB be actual parameters of SP1 (i.e. $SP1_{par} \sqsubseteq SPA$ and $SP1_{par} \sqsubseteq SPB$). If SPB is a refinement of SPA, then SP1(SPB) is a refinement of SP1(SPA).

Proof by structural induction on the form of $SP1_{body}$ by using the monotonicity of the specification operators w.r.t. ⊑ (see 8.3). □

Theorem 8.9 (Horizontal composition of refinements)

Let $SP1 = \lambda X{:}SP1_{par}.SP1_{body}$ and $SP2 = \lambda X{:}SP2_{par}.SP2_{body}$ be parameterized specifications and

let SPA be an actual parameter of SP1 and SPB be an actual parameter of SP2 (i.e. $SP1_{par} \sqsubseteq SPA$ and $SP2_{par} \sqsubseteq SPB$).
If SP2 is a refinement of SP1 and SPB is a refinement of SPA, then SP2(SPB) is a refinement of SP1(SPA).

Proof Since SPB is a refinement of SPA it is also an actual parameter of SP1. By lemma 8.8 we have $SP1(SPA) \sqsubseteq SP1(SPB)$. Since SP2 is a refinement of SP1, $SP1(SPB) \sqsubseteq SP2(SPB)$ holds. Thus by transitivity of $\sqsubseteq$ (see 8.3(2)) we have $SP1(SPA) \sqsubseteq SP2(SPB)$. □

As a corollary we get the horizontal composition theorem for implementations.

Corollary 8.10 Let SP1, SP2 and SPA, SPB be as in theorem 8.9. If SP2 is an implementation of SP1 and SPB is an implementation of SPA, then SP2(SPB) is an implementation of SP1(SPA).

Application of a parameterized specification $\lambda X{:}SP_{par}.SP_{body}[X]$ to an actual parameter SP as defined above requires in particular that the signature of SP contains the signature of SP_{par}. But this is almost <u>never</u> satisfied as e.g. the examples 8.4, (1) and (2), show: nor $ELEM \sqsubseteq NAT$ neither $ELEM \sqsubseteq INT$ do hold. In order to be able to perform parameter passing for arbitrary actual parameters, it is necessary to rename the requirement specification appropriately.

Def 8.11 Let SP be a specification and $\lambda X{:}SP_{par}.SP_{body}[X]$ be a parameterized specification.

(1) A signature morphism $\rho{:}sig(SP_{par}) \rightarrow sig(SP)$ is called **parameter passing morphism**, if (**translate** SP_{par} **with** $\rho) \sqsubseteq SP$.

(2) Function application takes two arguments, the actual parameter SP and parameter passing morphism ρ, and is defined as follows.

$$(\lambda X{:}SP_{par}.SP_{body}[X])(SP,\rho) =_{def} (\lambda X{:}\rho\bullet SP_{par}.\rho\bullet SP_{body}[X])(SP)$$

where $\rho\bullet SP_{par} =_{def}$ **translate** SP_{par} **with** ρ_X: $sig(SP_{par}) \rightarrow sig(X)$,

$\rho\bullet SP_{body}[X] =_{def}$ **translate** $SP_{body}[X]$ **with** $_{\Sigma[X]}\rho$,

$\Sigma[X] =_{def} sig(SP_{body}[X])$.

Notice that it may be necessary to rename also parts of $SP_{body}[X]$ in order to avoid name clashes with SP. The examples for implementations (in section 7) suggest the following definition [Sannella, Tarlecki 87].

Def. 8.12 For any two specifications SP and SP1 and parameterized specification γ, SP1 is called an **implementation of SP via** γ, written $SP \rightsquigarrow_{\gamma} SP1$, if SP ~~~> γ(SP1,ρ) for some appropriate parameter passing morphism ρ.

In many examples of implementations, γ has the form "$expren_{\Sigma,a} \circ enr_{\Delta}$", e.g. for

BOOL_by_NAT (cf. 7.3.(1)), γ=$\text{expren}_{\Sigma B0,\alpha 0}$ o $\text{enr}_{\Delta 0}$ where $\alpha 0 =_{def}$ [Bool/Nat, true/0, not/succ, $=_{Bool}/=_{BOOL3,Nat}$] and $\Delta 0 =_{def}$ <false: $\rightarrow$Nat, $=_{BOOL3,Nat}$: Nat,Nat, {false=succ(0), $\forall$**cons** x: Nat. succ(succ(x)) $=_{BOOL3,Nat}$ x}>. All examples of implementations in this paper are of this form. In the following we will call such implementations ERE-implementations (where ERE stands for enrich-rename-export).

Def. 8.13

(1) A specification SPB is called **ERE-implementation** of the specification SPA, written SPA $\sim\sim\sim>_{ERE}$ SPB, if there exists an enrichment Δ, a substitution list α and signature Σ such that SPB is an implementation of SPA via $\text{expren}_{\Sigma,\alpha}$ o enr_{Δ}.

(2) A parameterized specification SP2 = λX:$SP2_{par}$.$SP2_{body}$ is called **(partial parameterized) ERE-implementation** of the parameterized specification SP1 = λX:$SP1_{par}$.$SP1_{body}$, written SP1 $\sim\sim\sim>_{ERE}$ SP2, if $SP1_{par} \sqsubseteq SP2_{par}$ and there exists an enrichment Δ, a substitution list α and a signature Σ such that for each actual parameter SP of SP2, SP2(SP) is an implementation of SP1(SP) via $\text{expren}_{\Sigma[SP/X],\alpha[SP/X]}$ o $\text{enr}_{\Delta[SP/X]}$.

Theorem 8.14 (Vertical composition of ERE-implementations)

For any specifications SP1, SP2 and SP3, if SP1 $\sim\sim\sim>_{ERE}$ SP2 and SP2 $\sim\sim\sim>_{ERE}$ SP3, then SP1 $\sim\sim\sim>_{ERE}$ SP3.

Proof Let SP2 be an implementation of SP1 via $\text{expren}_{\Sigma 2,\alpha 2}$ o $\text{enr}_{\Delta 2}$ and let SP3 be an implementation of SP2 via $\text{expren}_{\Sigma 3,\alpha 3}$ o $\text{enr}_{\Delta 3}$ such that sig($\text{enr}_{\Delta 3}$(SP3))$\cap$sig(α3(Δ2))=$\emptyset$. Notice, that α3 can always be chosen in such a way that this syntactic condition is satisfied. Then by transitivity of the implementation relation, SP1 $\sim\sim\sim>$ γ (SP3) holds where γ = $\text{expren}_{\Sigma 2,\alpha 2}$ o $\text{enr}_{\Delta 2}$ o $\text{expren}_{\Sigma 3,\alpha 3}$ o $\text{enr}_{\Delta 3}$. According to corollary 6.3 we have

γ = [by 6.3(3)]

$\text{expren}_{\Sigma 2,\alpha 2}$ o $\text{expren}_{\Sigma 3\cup sig(\Delta 2),\alpha 3}$ o $\text{enr}_{\alpha 3(\Delta 2)}$ o $\text{enr}_{\Delta 3}$ = [by 6.3(1)-(2)]

$\text{expren}_{\Sigma 2,\alpha 3 \circ \alpha 2}$ o $\text{enr}_{\alpha 3(\Delta 2)\cup\Delta 3}$.

Hence SP3 is a ERE implementation of SP1. □

The full specification language supports the following version of a horizontal composition theorem.

Theorem 8.15 (Horizontal composition for ERE-implementations, I)

Let SP1 = λX:$SP1_{par}$.$SP1_{body}$ and SP2 = λX:$SP2_{par}$.$SP2_{body}$ be parameterized specifications, let SPA be an actual parameter of SP1 and SPB be an actual parameter of SP2.

If SP2 is an ERE-implementation of SP1 and SPB is an ERE-implementation of SPA using the substitution list α and the enrichment Δ, then SP2(**rename** enr_{Δ}(SPB) **by** α) is an ERE-implementation of SP1(SPA).

Proof Let SP1 ~~~> $\text{expren}_{\Sigma 2,\alpha 2}$ o $\text{enr}_{\Delta 2}$(SP2) and SPA ~~~> $\text{expren}_{\Sigma,\alpha}$ o enr_{Δ}(SPB) where α and Δ are given above. Then by monotonicity of SP1, SP1(SPA) ~~~> SP1(SP´) where SP´ $=_{def}$ $\text{expren}_{\Sigma,\alpha}$ o enr_{Δ}(SPB). Because of SP´⊑SP´´ for SP´´ $=_{def}$ **rename** (enr_{Δ}(SPB)) **by** α, we get by monotonity w.r.t.⊑

SP1(SP´)⊑SP1(SP´´).

Then

SP1(SP´´) ~~~> $\text{expren}_{\Sigma 2[SP´´],\alpha 2[SP´´]}$ o $\text{enr}_{\Delta 2[SP´´]}$ o SP2(SP´´)

where Γ[SP´´] $=_{def}$ Γ[SP´´/X, sig(SP´´)/sig($SP2_{par}$)] for $\Gamma \in \{\Sigma 2, \alpha 2, \Delta 2\}$. Hence by transitivity of ⊑, SP1(SPA) ⊑ $\text{expren}_{\Sigma 2[SP´´],\alpha 2[SP´´]}$ o $\text{enr}_{\Delta 2[SP´´]}$ o SP2(SP´´). Cutting down the export signature of the right-hand side to $\Sigma´ =_{def}$ sig(SP1(SPA)) we get the expected result:

SP1(SPA) ~~~> $\text{expren}_{\Sigma´,\alpha 2[SP´´]}$ o $\text{enr}_{\Delta 2[SP´´]}$ o SP2(SP´´).

Therefore SP2(SP´´) is an ERE-implementation of SP1(SPA). □

It is possible to strengthen this result by merging the renaming α of SPB with the renaming of SP2, if SP2 commutes with α: the parameterized specification SP2 is said to **commute with the substitution list** α if for each actual parameter SP of SP2,

SP2(**rename** SP **by** α) = **rename** (SP2(SP)) **by** α.

For example, if SP2 is built only by using the operators <..>, enrich, +, then SP2 commutes with any α. Renamings and exports may also occur in SP2 if they are invariant w.r.t. the signature of the formal parameter and if there are no name clashes of α with symbols declared in the body of SP2, i.e. in sig(SP2)\sig($SP2_{par}$).

Corollary 8.16 (Horizontal composition for ERE-implementations, II)

Let SP1, SP2, SPB and SPA as in theorem 8.14.

If SP2 is an ERE-implementation of SP1 (via $\text{expren}_{\Sigma 2,\alpha 2} \circ \text{enr}_{\Delta 2}$), SPB is an ERE-implementation of SPA (via $\text{expren}_{\Sigma,\alpha} \circ \text{enr}_{\Delta}$) and α commutes with SP2, then SP2(enr_{Δ}(SPB)) is an ERE-implementation of SP1(SPA)

(via $\text{expren}_{\Sigma´,\alpha 2[SP´]o\alpha} \circ \text{enr}_{\Delta 2[SP´]}$ where $\Sigma´ =_{def}$ sig(SP1(SPA)), SP´ $=_{def}$ enr_{Δ}(SPB) and Γ[SP´] $=_{def}$ Γ[SP´/X, sig(SP´)/sig($SP2_{par}$)] for $\Gamma \in \{\alpha 2, \Delta 2\}$).

9. Concluding remarks

In the previous sections a general framework for algebraic specification has been developed including a semantic basis, language constructs for structuring specifications and an appropriate notion of implementation. The key idea was the use of standard predicate symbols and of an ultra-loose semantics based on total algebras.

For other semantic approaches different standard symbols are considered: e.g. for the partial algebra approach (cf. e.g. [Wirsing et al. 83]) a so-called "definedness" predicate symbol D is used which holds in a structure A for a term t, if and only if the interpretation of t in A is defined. For monotonic and continuous specifications partial ordering predicate symbols are used instead of equality symbols (cf. e.g. [Möller 87]). For observation-oriented specifications Hennicker uses a so-called observability predicate "Obs" [Hennicker 89]. The generating predicate symbols of section 2.2 have a simple form. It may also be interesting to consider predicate symbols "$\in$ <S,F,X>" where $x \in$ <S,F,X> holds in a structure A if x is the interpretation of a term $t \in T(\Sigma,X)$ with variables in X. More generally, it may be interesting to consider a binary predicate symbol ":". For a structure A the expression x:W holds, if and only if (the interpretation of) x is an element of (the interpretation of) W where W is an arbitrary set of terms, this leads to an approach to algebraic specifications where sets are "first-class citizens": there exist not only function and predicate symbols for elements but also for sets of elements. For example, there may exist polomorphic sorts (as e.g. in ML), sort-building function symbols (as e.g. in ASL [Wirsing 86]), predicate symbols on sorts (such as subsorting in OBJ [Futatsugi et al. 85]), or predicate symbols relating elements with sets (such as ":" above). Due to the modularity of our approach which allows to choose (relatively) freely the standard symbols there is some hope that it will not be difficult to integrate many of these additional features.

Acknowledgement

Thanks go to Rolf Hennicker for reading carefully drafts of this paper and pointing out many inconsistencies and also to Stefan Gastinger for carefully typing this paper.

References

[Barwise 77] J.K. Barwise (ed.): Handbook of Mathematical Logic. Amsterdam: North Holland, 1977, 1091-1132.

[Bauer, Wössner 82] F.L. Bauer, H. Wössner: Algorithmic Language and Program Development. Berlin: Springer, 1982.

[Bergstra, Tucker 86] J.A. Bergstra, J.V.Tucker: Algebraic specificatios of computable and semicomputable data types. Mathematical Centre, Amsterdam, Dept. of Computer Science Research Report CS-R8619, 1986.

[Bergstra et al. 86] J.A. Bergstra, J. Heering, P. Klint: Module algebra. Math. Centrum Amsterdam, Report CS-R8615.

[Broy et al. 86] M. Broy, B. Möller, P. Pepper, M. Wirsing: Algebraic implementations preserve program correctness. Sci. Comput. Programming 7 ,1986, 35-53

[Burstall, Goguen 80] R.M. Burstall, J.A. Goguen: The semantics of CLEAR, a specification language. Proc. Advanced Course on Abstract Software Specifications, Copenhagen, Lecture Notes in Computer Science 86. Berlin: Springer, 1980, 292-232.

[Ehrig, Kreowski 82] H. Ehrig, H.-J.Kreowski: Parameter passing commutes with implementation of parameterized data types. Proc. 9th Int. Colloquium on Automata, Languages and Programming, Aarhus, Lecture Notes in Computer Science 140. Berlin: Springer, 1982, 197-211.

[Ehrig, Mahr 85] H. Ehrig, B. Mahr: Foundations of Algebraic Specifications I: Equations and Initial Semantics. Berlin: Springer, 1985.

[Ehrig, Weber 86] H. Ehrig, H. Weber: Programming in the large with algebraic module specifications. In: H. Kugler (ed.): Proc. 10th IFIP World Congress, Dublin, 1986.

[Ehrig et al. 82] H. Ehrig, H.-J. Kreowski, B. Mahr and P. Padawitz: Algebraic implementation of abstract data types. Theoret. Comput. Sci. 20 ,1982, 209-263.

[Feijs et al. 89] L.M.G. Feijs: The calculus $\lambda\pi$. To appear in: Algebraic methods: Theory, Tools and Applications. Lecture Notes in Computer Science. Berlin: Springer 1989.

[Futatsugi et al. 85] K. Futatsugi, J.A. Goguen, J.-P. Jouannaud, J. Meseguer: Principles of OBJ2. Proc 12th ACM Symp. on Principles of Programming Languages, New Orleans, 1985, 52-66.

[Goguen et al. 78] J.A. Goguen, J.W. Thatcher and E.G. Wagner: An initial algebra approach to the specification correctness, and implemantation of abstract data types, IBM Research Rept. RC-6487, also in: R.T. Yeh, ed., Current Trends in Programming Methodology, Vol. 4: Data Structuring. Englewood Cliffs: Prentice Hall, 1978, 80-149.

[Guttag 75] J.V. Guttag: The specification and application to programming of abstract data types. Ph.D. Thesis, Univ. of Toronto, 1975.

[Guttag et al. 85] J.V. Guttag, J.J. Horning, J.M. Wing: Larch in fife easy pieces. Digital Systems Research Center, Palo Alto, Technical Report 5, 1985.

[Hennicker 89] R. Hennicker: Observational implementation. To appear in Proc. STACS 89, Paderborn, Lecture Notes in Computer Science. Berlin: Springer, 1989.

[Hoare 69] C.A.R. Hoare: An axiomatic basis for computer programming. Comm. ACM 12, 1969, 576-583.

[Majster 77] M.E.Majster: Limits of the "algebraic" specification of abstract data types, ACM-Sigplan Notices 12, October 1977, 37-42.

[Möller 87] B. Möller: Higher-order algebraic specifications. Fakultät für Mathematik und Informatik der TU München, Habilitationsschrift, 1987.

[Mosses 89] P. Mosses: Unified algebras and modules. Computer Science Dept. Aarhus University, Technical Report DAIMIPB-266, also in: Proc. Symp. on Principles of Programming Languages 89. To appear.

[Sannella, Tarlecki 85a] D.T. Sannella and A. Tarlecki: Building specifications in an arbitrary institution, Internat. Symp. on Semantics of Data Types, Sophia-Antipolis, Lecture Notes in Computer Science 173. Berlin: Springer 1985, 337-356

[Sannella, Tarlecki 85b] D.T. Sannella, A. Tarlecki: Program specification and development in Standard ML. Proc. 12th ACM Symposium on Principles of Programming Languages, New Orleans, 1985, 67-77.

[Sannella, Tarlecki 87] D.T. Sannella, A. Tarlecki: Toward formal development of programs from algebraic specifications: implementations revisited. In: H. Ehrig et al. (eds.): TAPSOFT ´87, Lecture Notes in Computer Science 249. Berlin: Springer, 1987, 96-100.

[Sannella, Wirsing 82] D.T. Sannella, M.Wirsing: Implementation of parameterised specifications, Rept. CSR-103-82, Dept. of Computer Science, Univ. of Edinburgh; extended abstract in: Proc. 9th Internat. Coll. on Automata, Languages and Programming, Aarhus, Denmark, Lecture Notes in Computer Science 140. Berlin: Springer, 1982, 473-488.

[Sannella, Wirsing 83] D.T. Sannella, M.Wirsing: A kernel language for algebraic specification and implementation, Coll. on Foundations of Computation Theory, Linköping, Sweden; 1983, Lecture Notes in Computer Science 158. Berlin: Springer, 1983, 413-427.

[Wirsing 86] M. Wirsing: Structured algebraic specifications: a kernel language, Theoretical Computer Science 42, 1986, 123-249.

[Wirsing et al. 83] M. Wirsing, P. Pepper, H. Partsch, W. Dosch, M. Broy: On hierarchies of abstract data types. Acta Inform. 20, 1983, 1-33.

ON THE EXISTENCE OF INITIAL MODELS FOR PARTIAL (HIGHER-ORDER) CONDITIONAL SPECIFICATIONS

Egidio Astesiano and Maura Cerioli

Dipartimento di Matematica - Università di Genova
Via L.B. Alberti 4 - 16132 Genova Italy

Abstract. *Partial higher-order conditional specifications may not admit initial models, because of the requirement of extensionality, even when the axioms are positive conditional. The main aim of the paper is to investigate in full this phenomenon.*
If we are interested in term-generated initial models, then partial higher-order specifications can be seen as special cases of partial conditional specifications, i.e. specifications with axioms of the form $\wedge\Delta \supset \varepsilon$, *where* Δ *is a denumerable set of equalities,* ε *is an equality and equalities can be either strong or existential. Thus we first study the existence of initial models for partial conditional specifications.*
The first result establishes that a necessary and sufficient condition for the existence of an initial model is the emptiness of a certain set of closed conditional formulae, which we call "naughty". These naughty formulae can be characterized w.r.t. a generic inference system complete w.r.t. closed existential equalities and the above condition amounts to the impossibility of deducing those formulae within such a system. Then we exhibit an inference system which we show to be complete w.r.t closed equalities; the initial model exists if and only if no naughty formula is derivable within this system and, when it exists, can be characterized, as usual, by the congruence associated with the system.
Finally, applying our general results to the case of higher-order specifications with positive conditional axioms, we obtain necessary and sufficient conditions for the existence of term-generated initial models in that case.

0 An overview

Partial algebraic specifications ([BW1, R, WB]) are one of the most interesting specification paradigms. Originally proposed as a support to the stepwise refinement procedure, they have found more recently interesting applications to the specification of concurrency and of formal semantics of languages (see eg [BW2, AR1, AR2]).
Higher-order specifications creep in naturally in the specification activity because of the application requirements (see [AR1]) and of methodological considerations (see eg [M, MTW]).
Unfortunately higher-order specifications present a rather different situation from the first-order case, due to the extensionality requirement. The most striking feature is that the existence of initial model is not guaranteed, even if the axioms are in a form which looks a natural extension of the one that guarantees the existence in the first-order case. Let us consider this problem in some more detail.
A classical result [BW1] states that the initial partial model exists (and is term-generated) in the class of all partial models, if the specification is positive conditional, i.e. the axioms have the form

$$\wedge_{i=1,\dots,n}\, t_i =_e t'_i \supset t = t'$$

where $=_e$ and $=$ denote respectively existential (the sides are both defined and equal) and strong (the sides are either existentially equal or are both undefined) equality.
This result can be generalized in two directions: in the axioms the conjunction may be infinitary and the consequence may be an inequality; this situation is also the most general guaranteeing the existence of the initial term-generated model (see [T] for references and the most general results in this direction).

This work has been partly supported by CNR-Italy (Progetto Strategico "Software: ricerche di base e applicazione") and MPI-40%

Consider now the case of higher-order specifications, where starting from some basic sorts we have inductively functional sorts of the form $s_1 \times \ldots \times s_n \rightarrow s$, the corresponding carriers in a model are required to be sets of partial functions and the axioms may of course contain equalities between terms of functional sort. It is quite instructive to analyse a simple example, the specification T_0:

basic sorts:	s	operation symbols:	e: $\rightarrow$ s f, g: s $\rightarrow$ s
functional sorts:	s $\rightarrow$ s	axioms:	$D(f), D(g), f =_e g \supset D(f(e))$

where we use definedness predicate D, and D(t) is equivalent to $t =_e t$.

Recall now that the term generated initial model I of T_0 (and of any partial specification), if it exists, is such that an existential equality holds in I iff it holds in any model; in particular I is minimally defined: a term is defined in I iff it is defined in every model.

Since clearly there exist models of T_0 where f(e) is undefined and the same happens for g(e), both f(e) and g(e) should be undefined and hence equal in I. By partiality and extensionality, I being term-generated, we should conclude that f = g holds in I and this clearly contradicts I being a model, since from the axioms we should conclude that f(e) is defined in I. It is not difficult to understand that all the trouble comes from the fact that extensionality amounts to requiring that the term-generated models satisfy the supplementary axiom

$$\wedge_{t \in W_\Sigma} f(t) = g(t) \supset f = g$$

where in the premise the equality is strong and this makes the axiom essentially non-positive conditional (essentially, since neither the definedness of f(t) nor of g(t) can be logically derived).

Notice also that T_0 does not admit an initial model tout-court, even if we drop the condition about being term-generated (see Remark in sec. 4).

The T_0 example suggests that the anomaly shown is not peculiar of higher-order specifications but is typical instead of conditional specifications, i.e. specifications where the axioms have the form $\wedge \Delta \supset \varepsilon$, where Δ is a set of possibly strong equalities and ε is an equality.

Then if we confine ourselves to consider term-extensional models, i.e. models where two functions are equal if (and only if) they are equal on all term-generated arguments, we can express the (term-)extensionality axioms in the form

$$\wedge_{\bar{t} \in W_\Sigma} f(\bar{t}) = g(\bar{t}) \supset f = g$$

where $\bar{t} = (t_1,\ldots,t_n)$

which is conditional. Moreover higher-order specifications can be reduced to first-order specifications (see, e.g. [MTW]) by using the apply functions, so that a term f(t) is an abbreviation of apply(f,t). Thus we are led to consider the mentioned initiality problem in the (more general) framework of partial first-order conditional specifications, with axioms of the form $\wedge \Delta \supset \varepsilon$, where Δ is a denumerable set of equalities (not necessarily existential) and ε is an equality, and of course sorts are now only sorts of individual elements.

We present in this paper a complete answer to the initiality problem for partial conditional specifications, giving necessary and sufficient conditions for the existence of the initial model (and as usual characterizing that model, if it exists, as the quotient algebra w.r.t. the congruence associated with an inference system which is complete w.r.t. closed equalities). Since the detailed mathematical treatment is technically rather difficult, we explain briefly the main underlying ideas, with the help of some examples.

Anomalies similar to the one shown by the example T_0 can be seen also in the following two examples (discussed in a more technical context at the end of sec. 1):

spec T_1	sorts: s	opns: a, b: $\rightarrow$ s	axioms:	$a = b \supset D(b)$
spec T_2	sorts: s	opns: a, b, c, d: $\rightarrow$ s	axioms:	$D(c), D(d), a = b \supset c = d$.

The lack of the initial model for T_1 is due to the particular form of the axiom $a = b \supset D(b)$: since there exist models where a is undefined and the same happens for b, if the initial model I exists, then a^I and b^I are both undefined and equal, so that the axiom, which would imply D(b), cannot be met. A partly similar situation is shown by T_2; the combination of the three axioms D(c), D(d) and $a = b \supset c = d$ produces a situation analogous to the one just discussed: if the initial model I exists, then not only a^I and b^I would be undefined and equal, but also c^I and d^I should be different, otherwise $c^I = b^I$, being an existential equality, should hold in every model what is patently false; hence the axiom $a = b \supset c = d$, which would imply c = d, cannot be met. Note that the axiom $a = b \supset c = d$ in itself does not prevent the existence of an initial model (see remarks at the end of Sec. 1). Analogously the formula $a = b \supset D(b)$ of T_1 would not create any problem if we could deduce from the axioms, say, that D(a) holds in every model, which would imply that $a = b \supset D(b)$ is logically equivalent to a positive conditional axiom.

Hence formulae like the two ones above are causing trouble in connection with what we can logically deduce from the axioms (equivalently, what is true in every model); consequently we call them "naughty" w.r.t. the specification.

The ***first main result*** of the paper consists indeed in capturing precisely and formalizing the notion of "naughty formula" for a specification and establishing (theorem 1.6) that the initial model exists if and only if the set of naughty formulae is empty. Then naughty formulae can be characterized as a particular subset of the conditional formulae deducible within any inference system complete w.r.t. closed existential equalities, and so the initial model exists iff we cannot derive any of these naughty formulae in such a system. (***second main result*** , theorem 2.8).

All the above motivates our ***third main result*** showing an inference system complete w.r.t. closed (strong and existential) equalities. First of all note that since the axioms may be infinitary (and this is essential, since we want to handle also the term-extensionality axioms), the inference system is within infinitary logic (where no general completeness result is available as for first-order theories). Moreover our aim has been to obtain a system which reduces, for positive conditional specifications, to the one naturally associated with positive conditional axioms; and indeed we add to that system only one inference rule which is rather intuitive in the finitary case. Let us consider a finitary significant example.

Consider the specification T_4 defined by

sort: s opns: a, b, c: $\to$ s axioms: $D(a) \supset a = b$, $D(b) \supset a = b$, $a = b \supset D(c)$.

It is not difficult to show that T_4 admits an initial model I, consisting of a one point carrier containing just the value of c, with a and b undefined. Moreover $a = b$ and D(c) hold in every model.

However *the usual inference system which is complete for positive conditional axioms (see def. 2.2 for an infinitary version) is not powerful enough* to deduce $a = b$ nor D(c) and hence the quotient algebra w.r.t. the congruence associated with the system is not a model.

This limitation disappears if we can apply the following clearly sound rule (see Sec. 3).

$$* \quad \frac{\wedge(\Theta_1 \cup \{D(t)\}) \supset \alpha,\ \wedge(\Theta_2 \cup \{D(t')\}) \supset \alpha,\ \wedge(\Theta_3 \cup \{t = t'\}) \supset \alpha}{\wedge(\Theta_1 \cup \Theta_2 \cup \Theta_3) \supset \alpha}$$

where Θ_1, Θ_2, Θ_3 are finite sets of closed equations and definedness assertions, α is either a closed equation or a closed definedness assertion and t, t' are closed terms of the same sort.

Indeed recalling that $a = b \supset a = b$ and instantiating * we can deduce $a = b$ and hence D(c). Our complete system adds to the usual conditional system just an infinitary version of rule * (see Def. 3.1); please note that, in order to handle the extensionality axiom which is infinitary, then we need to work within infinitary logic.

The first two main results completely settle the initiality question for conditional specifications. We can now apply these results to positive conditional higher-order specifications, accordingly to our previous discussion. Restricting ourselves to consider term-extensional models, the extensionality axiom is just an infinitary conditional axiom, the only one non-positive (for a positive conditional higher-order specification). Thus we get our ***fourth main result*** (theorem 4.5) which gives interesting necessary and sufficient conditions for the existence of the (term-generated) initial model.

The paper is organized as follows. The basic elementary definition about partial algebras are collected in appendix. In the first section we present the basic properties of conditional specifications, introduce the notion of naughty formula and give the first main result. In the second, after defining conditional inference systems and presenting one of them, we characterize the naughty formulae w.r.t. a conditional system and state the second main result. The third section introduces the extra rule to be added to the conditional system of sec. 2 and proves in outline the completeness w.r.t. closed equalities of the resulting system. Finally the fourth and last section applies the results to higher-order specifications and contains our fourth main result.

Due to lack of room, with the exception of an outline of the proof of the completeness result of sec. 3, the proofs are omitted and will be found in a full version of this paper ([AC1]) more generally dealing with the existence of free objects and full equational deduction, generalizing and obtaining as a special case the corresponding Meseguer-Goguen [MG] completeness result for total algebras.

1 Conditional partial specifications and initial models

In the following we assume some familiarity with partial algebras (see [BW1, R, B]); however all the relevant definitions, concepts and results needed here are reported in appendix.

Def. 1.1. Let $\Sigma = (S,F)$ be a signature and X be a family of S-sorted variables.

- The set $EForm(\Sigma,X)$ of *elementary formulae* over Σ and X is the set
  $\{ D(t) \mid t \in W_\Sigma(X)|_s \} \cup \{ t = t' \mid t,t' \in W_\Sigma(X)|_s \}$,
  where D denotes the definedness predicate (one for each sort; but sorts are omitted).
  Note that $D(t)$ can be equivalently expressed by $t =_e t$, where $=_e$ denotes existential equality: $=_e$ holds iff both sides are defined and equal; hence elementary formulae are just equalities either strong or existential.
- The set $WForm(\Sigma,X)$ of *conditional formulae* is the set
  $\{ \wedge \Delta \supset \varepsilon \mid \Delta \subseteq EForm(\Sigma,X), \Delta \text{ is countable}, \varepsilon \in EForm(\Sigma,X) \}$.
  If Δ is the empty set, then $\wedge \Delta \supset \varepsilon$ is an equivalent notation for the elementary formula ε. ($\wedge \Delta$ is a notation for the couple $(\wedge,\Delta)$; see [K])
- For every formula φ let $Var(\varphi)$ denote the set of all variables which appear in φ. A formula φ is called *closed* iff $Var(\varphi)$ is empty.
- A *positive conditional formula* is a conditional formula $\wedge \Delta \supset \varepsilon$ s.t. for every $t = t'$ belonging to Δ either $D(t)$ or $D(t')$ belongs to Δ.
- If A is a partial algebra, φ is a formula and V is a valuation for $Var(\varphi)$ in A, then we say that φ holds for V in A (equivalently: is satisfied for V by A) and write $A \models_V \varphi$ accordingly to the following definitions.
  - $A \models_V D(t)$ iff $t^{A,V}$ is defined;
  - $A \models_V t = t'$ iff $t^{A,V}$ and $t'^{A,V}$ are either both defined and equal or both undefined;
  - $A \models_V \wedge \Delta \supset \varepsilon$ iff $A \models_V \delta$ for all $\delta \in \Delta$ implies that $A \models_V \varepsilon$.

  We write $A \models \varphi$ for a formula φ and say that φ holds in (equivalently: is satisfied by, is valid in) A iff $A \models_V \varphi$ for all valuation V for $Var(\varphi)$ in A. □

In the following a generic elementary formula will be denoted by ε or η or γ or δ, while a generic conditional formula will be denoted by φ or ϑ or ψ; moreover for all conditional formulae $\varphi = (\wedge \Delta \supset \varepsilon)$ we denote Δ by $prem(\varphi)$ and ε by $cons(\varphi)$; finally we will use some equivalent notations:

$\wedge \Delta_1 \wedge \ldots \wedge\wedge \Delta_n \supset \varepsilon$ is the same as $\wedge (\cup_{i=1\ldots n} \Delta_i) \supset \varepsilon$;

$\varepsilon_1 \wedge \ldots \wedge \varepsilon_n \supset \varepsilon$ is the same as $\wedge \{ \varepsilon_1,\ldots,\varepsilon_n \} \supset \varepsilon$;

where $\Delta_1,\ldots, \Delta_n$ are sets of elementary formulae and $\varepsilon_1,\ldots, \varepsilon_n, \varepsilon$ are elementary formulae.

Def. 1.2.

- A *conditional type* (also called *conditional specification*) consists of a signature Σ and of a set Ax of conditional formulae over Σ. A generic conditional type will be denoted by T; the formulae belonging to Ax are called usually the axioms of T and are denoted by α.
- A *positive conditional type* is a conditional type s.t. all its axioms are positive conditional formulae; a generic positive conditional type will be usually denoted by PT.
- For every conditional type $T = (\Sigma,Ax)$, $PMod(T)$ denotes the class of all *models* of T, ie the class of Σ-algebras satisfying every formula of Ax, ie $PMod(T) = \{ A \mid A \in PA(\Sigma), A \models \alpha, \forall\, \alpha \in Ax \}$.
- For every conditional type $T = (\Sigma,Ax)$, $PGen(T)$ denotes the class of all term-generated models of T, ie $PGen(T) = Gen(PMod(T))$. □

Prop. 1.3. If C is a non-empty subclass of $PMod(T)$ and either T is a positive conditional type, or $C = MDef(C)$, then W_Σ/K^C is a model of T. □

Theorem 1.4.

1) If PT is a positive conditional type, then $W_\Sigma/K^{PMod(PT)}$ is initial in $PMod(PT)$.
2) If T is a conditional type s.t. $MDef(PGen(T))$ is not empty, then $W_\Sigma/K^{MDef(PGen(T))}$ is initial in $MDef(PGen(T))$.

3) A model I is initial in PMod(T) iff it is initial in PGen(T).

4) If I is initial in PMod(T), then $I \in$ MDef(PGen(T)). □

Remarks.

Conditional types are rather pathological w.r.t. positive conditional types. Remarks 1 and 2 below show the key features of this pathological behaviour. In particular the remark 3 shows the intrinsic irreducibility of conditional types to positive conditional types, even in the case when an initial model exists. remark 4 from same pathological examples extracts the essential idea for a solution of the problem of initiality, which is presented in the next section.

1. *If* T *is a conditional type, then* MDef(PGen(T)) *may be empty and thus there may not be an initial model in* PGen(T) *and hence in* PMod(T), as the following example shows.
   Let $\Sigma = (S,F)$ be the signature defined by $S = \{ s \}$, $F = \{ a, b: \rightarrow s \}$, Ax consist only of the axiom $a = b \supset D(b)$ and T_1 be the conditional type (Σ, Ax).
   We can define two models A and B by:
   $s^A = \{ 1 \}$, $a^A = 1$, b^A undefined; $s^B = s^A$, $b^B = 1$, a^B undefined.
   Therefore, if C is an algebra s.t. $t^C \in s^C$ iff ($t^B \in s^B$ for all $B \in PMod(T_1)$), then a^C and b^C are undefined; thus $C \models a = b$ and $C \not\models D(b)$, and hence C is not a model.
2. *If* MDef(PGen(T)) *is not empty, then there exists a model* I *initial in* MDef(PGen(T))*; however in general this model is not initial in* PGen(T), as the following example shows.
   Let $\Sigma = (S,F)$ be the signature defined by $S = \{ s \}$ and $F = \{ a, b, c, d: \rightarrow s \}$, Ax be the set of conditional formulae $\{ D(c), D(d), a = b \supset c = d \}$ and T_2 be the conditional type (Σ, Ax).
   For any minimally defined model A, a^A and b^A are undefined and thus $A \models a = b$. Thus all term-generated and minimally defined models are isomorphic to I, defined by:
   $s^I = \{ 1 \}$, $c^I = 1$, $d^I = 1$, a^I, b^I undefined,
   Therefore I is initial in MDef(PGen(T_2)), but not in PGen(T_2), because there exist models $A \in PGen(T_2)$ s.t. $A \not\models c = d$.
3. *There exist classes* PMod(T) *admitting an initial model which are not definable by only positive conditional formulae*, as the following example shows.
   Let Σ be the signature $(\{ s \},\{ a, b, c, d: \rightarrow s \})$, Ax be the singleton set $\{ a = b \supset c = d \}$, T be the type (Σ, Ax). Then there exists a model I of T, which is initial in PMod(T) and is defined by:
   $s^I = \emptyset$; a^I, b^I, c^I, d^I undefined.
   In order to show that there does not exist a positive conditional type PT s.t. PMod(PT) = PMod(T), it is sufficient to show that there exists a subclass C of PMod(T) s.t. $W_\Sigma/K^C \notin PMod(T)$, since $W_\Sigma/K^{C'} \in PMod(PT)$ for all positive conditional types PT and all subclass C' of PMod(PT), because of prop. 1.3.
   Let C be the class $\{ A,B \}$, where A and B are defined by:
   $s^A = \{ \bullet, @ \}$; $a^A = \bullet$, b^A undefined, $c^A = @$, d^A undefined;
   $s^B = s^A$; a^B undefined, $b^B = \bullet$, $c^B = @$, d^B undefined.
   Both are models of T. Then W_Σ/K^C is isomorphic to the algebra C defined by:
   $s^C = \{ \bullet \}$, a^C, b^C, d^C undefined, $c^C = \bullet$.
   Thus W_Σ/K^C is not a model of T and hence there does not exist a positive conditional type PT s.t. PMod(PT) = PMod(T).
   This counterexample could be presented using a different terminology. Note that $s^C = s^A \cap s^B$ and hence the class of models of T is not closed w.r.t. the intersection of congruences and so it is not a quasi-variety, contrary to the fact that the classes of models of positive conditional types are quasi-varieties (see [T] for references and for a generalization of this results to arbitrary abstract algebraic institutions).
4. Let us consider again the types T_1 and T_2 of remarks 1 and 2 above. The lack of the initial model for T_1 is due to the particular form of the axiom $a = b \supset D(b)$: since there exist models where a is undefined and the same happens for b, if the initial model I exists, then a^I and b^I are both undefined and equal, so that the axiom, which would imply D(b), cannot be met. A partly similar situation is shown by T_2; the combination of the three axioms D(c), D(d) and $a = b \supset c = d$ produces a situation analogous to the one just discussed: if the

initial model I exists, then not only a^I and b^I would be undefined and equal, but also c^I and d^I should be different, otherwise $c^I = d^I$, being an existential equality, should hold in every model what is patently false; hence the axiom $a = b \supset c = d$, which would imply $c = d$, cannot be met. Note that the axiom $a = b \supset c = d$ in itself does not prevent the existence of an initial model as it is shown in remark 3. The formula $a = b \supset c = d$ is causing trouble, we call it "naughty", in connection with the other axioms. Now we will try to capture and formalize the notion of "naughty formula" for a specification and to show that the absence of naughty formulae is a necessary and sufficient condition for the existence of the initial model. □

Def. 1.5. Let be given a type $T = (\Sigma, Ax)$.

- The set $SEEq(T)$ is the set
  $\{ D(t) \mid t \in W_\Sigma \} \cup \{ t = t' \mid t,t' \in W_\Sigma$, either $A \models D(t)\ \forall A \in PMod(T)$ or $A \models D(t)\ \forall A \in PMod(T) \}$
- The set $SNF(T)$, where SNF stands for Semantic Naughty Formulae, consists of all closed conditional formulae $\wedge \Delta \supset \varepsilon$ s.t.
  1. $\wedge \Delta \supset \varepsilon$ is $\alpha[t_x/x \mid x \in Var(\alpha)]$ for some $\alpha \in Ax$ and some $t_x \in W_\Sigma$ s.t. $A \models D(t_x)$ $\forall A \in PMod(T)$;
  2. $A \models \delta$ for all $\delta \in \Delta \cap SEEq(T)$ and all $A \in PMod(T)$;
  3. $\varepsilon \in SEEq(T)$ and there exists $A \in PMod(T)$ s.t. $A \not\models \varepsilon$. □

The notation $SEEq$ stands for Semantic Existential Equations and is justified by the fact that we have recalled in sec. 1, that $D(t)$ also can be expressed as an existential equation $t =_e t$.

Theorem 1.6. *(Main Theorem 1)* For every type T, there exists a model initial in $PMod(T)$ iff $SNF(T) = \emptyset$. □

2 Initiality and logical deduction

In the following when referring to generic formulae and inference systems we consider formulae and inference systems within an infinitary logic which extends first-order logic by admitting denumerable conjunctions (, disjunctions) and quantification over denumerable sets of variables (see e.g.[K]). However we will show that we can restrict ourselves to consider only conditional formulae.

Def. 2.1. For a conditional type $T = (\Sigma, Ax)$, a *conditional system* $L(T)$, in the following abbreviated to c-system, is an inference system $L(T)$ s.t.:

cs_1 $\quad L(T) \vdash \alpha$ for all $\alpha \in Ax$;

cs_2 $\quad$ the family $\equiv^{L(T)} = \{ \equiv^{L(T)}_s \}_{s \in S}$, where
$\equiv^{L(T)}_s = \{ (t,t') \mid t, t' \in W_{\Sigma|s}, L(T) \vdash D(t), L(T) \vdash D(t'), L(T) \vdash t = t' \}$,
is a strict congruence over W_Σ s.t. $Dom(\equiv^{L(T)}) = \{ t \mid t \in W_\Sigma, L(T) \vdash D(t) \}$;

cs_3 $\quad$ for any countable set of elementary formulae Δ, any elementary formula η, any family X of variables, and any closed term t_x of appropriate sort
$L(T) \vdash \wedge \Delta \supset \eta$ and $L(T) \vdash D(t_x)$ for all $x \in X_s$ and $s \in S$ implies
$L(T) \vdash \wedge \{ \delta[\{t_x/x \mid x \in X_s, s \in S\}] \mid \delta \in \Delta \} \supset \eta[\{t_x/x \mid x \in X_s, s \in S\}]$;

cs_4 $\quad$ for any countable sets of elementary closed formulae Θ, Γ, Θ_γ and any elementary closed formula ε
$L(T) \vdash \wedge \Theta \wedge \wedge \Gamma \supset \varepsilon$ and $L(T) \vdash \wedge \Theta_\gamma \supset \gamma$ for all $\gamma \in \Gamma$ implies
$L(T) \vdash \wedge \Theta \wedge \wedge (\cup_{\gamma \in \Gamma} \Theta_\gamma) \supset \varepsilon$;

cs_5 $\quad$ is sound, ie for any formula φ, $L(T) \vdash \varphi$ implies $M \models \varphi$ for all $M \in PMod(T)$. □

In order to make the presentation more concrete and to prepare the way to a completeness result, we introduce, for the moment just as an example, a particular c-system, which is reminiscent of systems found in the literature (see, e.g. [WB]); however the peculiar form of the axioms and inference rules will play a very important technical role when dealing with completeness in the next section.

Def. 2.2. The *canonical* c-system for a conditional type $T = (\Sigma, Ax)$, denoted by $CS(T)$, consists of the axioms Ax and of the following axioms:

1	$t = t$	$t \in W_{\Sigma|s}$
2	$t = t' \supset t' = t$	$t, t' \in W_{\Sigma|s}$
3	$t = t' \wedge t' = t'' \supset t = t''$	$t, t', t'' \in W_{\Sigma|s}$
4	$t_1 = t'_1 \wedge \ldots \wedge t_n = t'_n \supset op(t_1, \ldots, t_n) = op(t'_1, \ldots, t'_n)$	$t_i \in W_{\Sigma|s_i}$, $i = 1 \ldots n$, $op: s_1 \times \ldots \times s_n \to s$
5	$D(op(t_1, \ldots, t_n)) \supset D(t_i)$	$t_i \in W_{\Sigma|s_i}$, $i = 1 \ldots n$, $op: s_1 \times \ldots \times s_n \to s$
6	$D(t) \wedge t = t' \supset D(t')$.	$t, t' \in W_{\Sigma|s}$
7	$\dfrac{\wedge \Theta \wedge \wedge \Gamma \supset \varepsilon,\ \{\wedge \Theta_\gamma \supset \gamma \mid \gamma \in \Gamma\}}{\wedge \Theta \wedge \wedge (\cup_{\gamma \in \Gamma} \Theta_\gamma) \supset \varepsilon}$	Θ, Θ_γ, Γ are arbitrary, countable subsets of $EForm(\Sigma, \varnothing)$, $\varepsilon \in EForm(\Sigma, \varnothing)$.
8	$\dfrac{\wedge \Delta \supset \eta}{\wedge\{D(t_x) \mid x \in X_s, s \in S\} \wedge \wedge \{\delta[\{t_x/x \mid x \in X_s, s \in S\}] \mid \delta \in \Delta\} \supset \eta[\{t_x/x \mid x \in X_s, s \in S\}]}$	Δ is an arbitrary, countable subset of $EForm(\Sigma, Var)$, $\eta \in EForm(\Sigma, Var)$, $t_x \in W_{\Sigma|s}$ for all $x \in X_s$.

Remarks.

1 Notice that $CS(T)$ is really a c-system. First of all it is trivial to verify properties cs_1, cs_2, cs_3, cs_4. So, in order to show that $CS(T)$ is a c-system for T, we only have to show that it is sound. Since the soundness of rules 1 to 7 is obvious, we only show the soundness of rule 8.
Let A be a model of T, ψ^* denote, for all formulae ψ, the formula $\psi[\{t_x/x \mid x \in X_s, s \in S\}]$, Y be the set $Var(\wedge \Delta \supset \eta)$, Y^* be $Var(\wedge \{ D(t_x) \mid x \in X \} \wedge \wedge \{ \delta^* \mid \delta \in \Delta \} \supset \eta^*)$, ie, since $t_x \in W_\Sigma$ for all $x \in X$, $Y^* = Var(\wedge \{ \delta^* \mid \delta \in \Delta \} \supset \eta^*) = Y - X$ and V be a valuation for Y^* in A. Let us assume that $A \models_V \delta^*$ for all $\delta \in \Delta$ and $A \models_V D(t_x)$ for all $x \in X$ and show that $A \models_V \eta^*$. Let V' be defined by $V'(y) = V(y)$ for all $y \in Y^*$ and $V'(x) = t_x{}^A$ for all $x \in X \cap Var(\wedge \Delta \supset \eta)$. First of all we show that V' is a valuation for Y in A. Since $X \cap Y^* = \varnothing$, V' is a (partial) function from X into A; so we only have to show that it is total. Let $y \in Y$; then either $y \in Y^*$, and in this case $V'(y) = V(y)$ and so $V'(y) \in A$, since V is a valuation, or $y \in X$, and in this case $V'(y) = t_y{}^A$ and so, since we have assumed $A \models_V D(t_y)$, $t_y{}^A \in A$.
By definition of V' and V, we also have that $A \models_V \psi^*$ iff $A \models_{V'} \psi$ for all formulae ψ and hence, since we have assumed that $A \models_V \delta^*$ for all $\delta \in \Delta$, we also have that $A \models_{V'} \delta$ for all $\delta \in \Delta$; moreover, by inductive hypothesis, $A \models_{V'} \wedge \Delta \supset \eta$ and hence $A \models_{V'} \eta$ ie $A \models_V \eta^*$.

2. Notice that the variables only may appear in the axioms of Ax and in rule 8. Thus in order to eliminate variables from a formula we must apply rule 8, which disposes of the problems of unsoundness for many-sorted deduction noted by Goguen and Meseguer (see [MG]). □

Def. 2.3. For a given conditional type T, a c-system $L(T)$

- is *complete* w.r.t. a set Θ of formulae iff for any $\vartheta \in \Theta$, if $M \models \vartheta\ \forall M \in PMod(T)$, then $L(T) \vdash \vartheta$.
- is EEq-*complete* iff it is complete w.r.t. the set SEEq(T); equivalently $L(T)$ is EEq-complete iff it is complete w.r.t the set

$$EEq(L(T)) = \{D(t) \mid t \in W_{\Sigma|s}\} \cup \{t = t' \mid t, t' \in W_{\Sigma|s}, \text{ either } L(T) \vdash D(t) \text{ or } L(T) \vdash D(t')\}. \ \square$$

Remark. Notice that for every c-system $L(T)$ completeness w.r.t. $EEq(L(T))$ and w.r.t. SEEq(T) are really equivalent. Indeed, since every c-system $L(T)$ is sound, if $L(T) \vdash D(t)$, then $A \models D(t)$ for all $A \in PMod(T)$

$A \in PMod(T)$ and hence $EEq(L(T)) \subseteq SEEq(T)$. Conversely assume that $L(T)$ is complete w.r.t. $EEq(L(T))$; thus in particular if $A \models D(t)$ for all $A \in PMod(T)$, then $L(T) \vdash D(t)$ and hence

$$\{ t = t' \mid t, t' \in W_\Sigma, \text{ either } A \models D(t)\ \forall A \in PMod(T) \text{ or } A \models D(t)\ \forall A \in PMod(T) \} \subseteq$$
$$\{ t = t' \mid t, t' \in W_{\Sigma|s}, \text{ either } L(T) \vdash D(t) \text{ or } L(T) \vdash D(t') \},$$

ie $SEEq(T) \subseteq EEq(L(T))$. □

Notice now that, since $\equiv^{L(T)}$ is a congruence because of condition cs_2 of def. 2.1, we can define the algebra $W_\Sigma/\equiv^{L(T)}$. Thus we state a proposition which is useful in the following and is a slight generalization of well-known results for total algebras.

Prop. 2.4. The algebra $W_\Sigma/\equiv^{L(T)}$ is a model of T iff it is initial in PMod(T). □

It is now convenient to give a notion of naughty formula related to a c-system, since it allows us to connect the initial model with logical inference systems.

Def. 2.5. For a given T and a c-system $L(T)$, the set $NF(L(T))$ (NF for Naughty Formulae) consists of all closed conditional formulae φ s.t.

- nf_1 φ is $\alpha[t_x/x \mid x \in Var(\alpha)]$ for some $\alpha \in Ax$ and $t_x \in W_\Sigma$ s.t. $L(T) \vdash D(t_x)$;
- nf_2 $L(T) \vdash \delta$ for all $\delta \in prem(\varphi) \cap EEq(L(T))$;
- nf_3 $L(T) \nvdash cons(\varphi)$ and $cons(\varphi) \in EEq(L(T))$. □

Prop. 2.6. For all EEq-complete c-systems $L(T)$ we have $NF(L(T)) = SNF(T)$. □

Theorem 2.7. Let $L(T)$ be a c-system. The set $NF(L(T))$ is empty iff $W_\Sigma/\equiv^{L(T)}$ is a model of T. □

Putting together prop. 2.4, prop. 2.6 and theorem 2.7 we get our second main result.

Theorem 2.8. *(Main theorem 2)* Let T be a conditional type. For every c-system $L(T)$ the following conditions are equivalent:

1) the set $NF(L(T))$ is empty;
2) the algebra $W_\Sigma/\equiv^{L(T)}$ is a model of T;
3) the algebra $W_\Sigma/\equiv^{L(T)}$ is initial in PMod(T).

Moreover, if $L(T)$ is EEq-complete then each one of the above conditions is equivalent to

4) there exists an initial model in PMod(T). □

It is easy to obtain the well-known initiality result for positive conditional types [BW1] as a corollary of the above results; for that we state an intermediate result.

Prop. 2.9. Let $T = (\Sigma, Ax)$ be a conditional type, $L(T)$ be a c-system and A be the algebra $W_\Sigma/\equiv^{L(T)}$; then $A \models \alpha$ for all positive conditional axioms α of Ax. □

Corollary 2.10 [BW1]. Let $PT = (\Sigma, Ax)$ be a positive conditional type, $L(PT)$ be a c-system; then the algebra $W_\Sigma/\equiv^{L(PT)}$ is initial in PMod(PT). □

Remark. If we consider a positive conditional type PT, then we have seen that the algebra $W_\Sigma/\equiv^{L(PT)}$ is the initial model of PT for every c-system $L(PT)$ for PT. However *if we consider a conditional type* T, *even if there exists an initial model in* PMod(T), *a generic c-system* $L(T)$ *is too poor for* $W_\Sigma/\equiv^{L(T)}$ *being the initial model of* T; for instance, even if there exists an initial model in PMod(T), $W_\Sigma/\equiv^{CS(T)}$ is not in general a model of T, as the following example shows.

Let T_4 be the conditional type (Σ, Ax), where Σ is the signature $(\{ s \}, \{ a, b, c: \to s \})$ and Ax is the set $\{ D(a) \supset a = b,\ D(b) \supset a = b,\ a = b \supset D(c) \}$.

Then the formulae $a = b$ and $D(c)$ hold in every model of T_4, while $CS(T_4) \nvdash a = b$ and hence $CS(T_4) \nvdash D(c)$. Moreover there exists an initial model of T_4, which is isomorphic to the algebra A defined by: $s^A = \{ \bullet \}$, a^A, b^A are undefined, $c^A = \bullet$. □

Another interesting consequence of the above results is the following proposition.

Prop. 2.11. Let T be the conditional type (Σ, Ax), $L(T)$ be an EEq-complete c-system and T^+ be the conditional type $(\Sigma, Ax \cup Ax^+)$, where Ax^+ is the set $EForm(\Sigma, \emptyset)$-$SEEq(T)$.

1. The following conditions are equivalent:
   a_1 MDef(PMod(T)) is not empty;
   a_2 $L(T) \vdash D(t)$ iff $L(T^+) \vdash D(t)$ for all $t \in W_\Sigma$;
   a_3 $W_\Sigma^{L(T^+)}$ is initial in MDef(PMod(T)).
2. There exists an initial model in PMod(T) iff ($L(T) \vdash \varepsilon$ iff $L(T^+) \vdash \varepsilon$ for all $\varepsilon \in$ SEEq(T)). □

Remark. We shortly show that the well-known theory of total types, ie of the types whose models are only total algebras, and its results of initiality can be seen as a particular case of partial types. First of all we note that the (total) models of a total type $TT = (\Sigma, Ax)$ are exactly the (partial) models of the type $Par(TT) = (\Sigma, Ax \cup Ax^{Tot})$, where $Ax^{Tot} = \{ D(op(x_1,...,x_n)) \mid op \in F_{(s_1...s_n,s)} \}$ and x_i are variables of suitable sort. Moreover if A and B are total algebras, the condition in order to $p = \{ p_s: s^A \to s^B \}$ be a homomorphism, since $op^A(a_1,...,a_n)$ is always defined, can be rewritten as follows: $p_{s_{n+1}}(op^A(a_1...a_n)) = op^B(p_{s_1}(a_1)...p_{s_n}(a_n))$; thus every homomorphism between total algebras is really a homomorphism as defined in the theory of total algebras (see [MG,pg.465]). Therefore the study of the initial model of TT in the framework of total algebras is completely equivalent to the study of initial model of Par(TT) in this framework. Finally note that for every total type TT and every c-system $L(Par(TT))$ for Par(TT), because of Ax^{Tot}, $L(T) \vdash D(t)$ for all closed term t, and hence NF(L(Par(TT)) is empty; therefore the algebra $W_\Sigma/\equiv^{L(Par(TT))}$ is the initial model in PMod(Par(TT)), because of theorem 2.7. □

3 A complete conditional system

This section is devoted to the third main result of the paper. We exhibit a complete conditional system and thus, instantiating the second main result, we can say that a necessary and sufficient condition for the existence of the initial model is the absence of formulae which are naughty w.r.t. the exhibited system. A most interesting feature of this system is that it is obtained by adding just one new rule to the canonical conditional system of the previous section. This new rule takes a very intuitive and simple form in the case of axioms with finitary premises, while the generalization to infinitary premises is rather subtle and tricky. So we introduce the basic ideas by discussing a finitary example.

Consider again the simple example T_4 already seen (remark after corollary 2.10). Let Σ be the signature ({ s },{ a, b, c: → s }), Ax be the set { $D(a) \supset a = b$, $D(b) \supset a = b$, $a = b \supset D(c)$ } and T_4 be the conditional type (Σ, Ax). Then $A \models a = b$ for every model A of T, since either a and b are both undefined, and hence equal, or at least one of them is defined and hence, because of the axioms, both are defined and equal. Thus we can think of adding to the canonical c-system the following rule

$$* \quad \frac{\wedge(\Theta_1 \cup \{D(t)\}) \supset \varepsilon,\ \wedge(\Theta_2 \cup \{D(t')\}) \supset \varepsilon,\ \wedge(\Theta_3 \cup \{t = t'\}) \supset \varepsilon}{\wedge(\Theta_1 \cup \Theta_2 \cup \Theta_3) \supset \varepsilon}$$

$t, t' \in W_{\Sigma|s}$, $\Theta_1, \Theta_2, \Theta_3$ are arbitrary finite subsets of EForm$(\Sigma,\varnothing)$, $\varepsilon \in$ EForm$(\Sigma,\varnothing)$.

Rule * holds in first order logic since from the premises we can infer

$$\wedge(\Delta_1 \cup \{ D(t) \}) \vee \wedge(\Delta_2 \cup \{ D(t') \}) \vee \wedge(\Delta_3 \cup \{ t = t' \}) \supset \varepsilon$$

and thus also $\wedge (\Delta_1 \cup \Delta_2 \cup \Delta_3) \wedge (\vee \{ D(t), D(t'), t = t' \}) \supset \varepsilon$ and finally, since $\vee \{ D(t), D(t'), t = t' \}$ is logically valid, $\wedge (\Delta_1 \cup \Delta_2 \cup \Delta_3) \supset \varepsilon$. If we add * to rules 1...8 of $CS(T_4)$ then clearly we get $*, CS(T_4) \vdash a = b$ which we could not get simply in $CS(T_4)$.

However, since we want also to handle the extensionality axiom which is intrinsically infinitary, we have to work within infinitary logic and hence we generalize rule * to the case of infinitary premises.

Def. 3.1. The inference system associated with $T = (\Sigma, Ax)$, denoted $CL(T)$ (or simply CL when there is no ambiguity) consists of the axioms and inference rules of $CS(T)$ and of the following inference rule:

$$9 \quad \frac{\{\wedge\Theta_j \wedge \wedge\Gamma_j \supset \varepsilon \mid j \in J\}}{\wedge (\cup_{j \in J}\, \Theta_j) \supset \varepsilon}$$

J is an arbitrary set (possibly more than countable), Θ_j, Γ_j are arbitrary countable subsets of EForm$(\Sigma,\varnothing)$, $\varepsilon \in$ EForm$(\Sigma,\varnothing)$.
$\forall\ \Psi \in$ FullInter(Γ) $\exists\ t, t' \in W_\Sigma$ s.t. $D(t), D(t'), t = t' \in \Psi$, where FullInter$(\Gamma) = \{ \Psi \mid \Psi \subseteq (\cup_{j \in J} \Gamma_j), \Psi \cap \Gamma_j \neq \varnothing, \forall j \in J \}$. □

Important Remark. Note that if all the axioms have finitary premises, then rule 9 can be replaced by rule *.

Prop. 3.2. The inference system $CL(T)$ is a c-system for T. □

Remark. Notice that rule 9 is clearly a generalization of rule * given above. Now we show an example of use of rule 9 in an infinitary case.

Let Σ be the signature $(\{ s_1, s_2, s_3 \}, \{ a_j: \to s_j; f_i, g_i: s_i \to s_i \mid i = 1,2, j = 1,2,3 \})$, x_i belong to Var_{s_i} for $i = 1,2$, Ax be the set $\{ \alpha_1, \alpha_2, \alpha_3 \}$, where $\alpha_1 = (D(x_1) \wedge D(x_2) \supset D(a_3))$, $\alpha_{i+1} = (\wedge \{ f_i^n(a_i) = g_i^n(a_i) \mid n \in \mathbf{N} \} \supset D(a_3))$ for i=1,2. From α_1, by rule 7, we deduce

$\vartheta^1_{i,j} = (D(f_1^i(a_1)) \wedge D(f_2^j(a_2)) \supset D(a_3))$;

$\vartheta^2_{i,j} = (D(f_1^i(a_1)) \wedge D(g_2^j(a_2)) \supset D(a_3))$;

$\vartheta^3_{i,j} = (D(g_1^i(a_1)) \wedge D(f_2^j(a_2)) \supset D(a_3))$;

$\vartheta^4_{i,j} = (D(g_1^i(a_1)) \wedge D(g_2^j(a_2)) \supset D(a_3))$;

for all $i, j \in \mathbf{N}$. Thus from α_2, α_3 and $\{ \vartheta^k_{i,j} \mid k = 1,\dots,4, i, j \in \mathbf{N} \}$, by rule 9, we deduce $D(a_3)$; indeed let J be the set $\{ prem(\alpha_2) \} \cup \{ prem(\alpha_3) \} \cup \{ prem(\vartheta^k_{i,j}) \mid k = 1,\dots,4; i, j \in \mathbf{N} \}$ and Ψ belong to $FullInter(\{\Gamma \mid \Gamma \in J\})$ then, since, by definition of $FullInter(\{\Gamma \mid \Gamma \in J\})$, $\{ prem(\varepsilon_2) \} \cap \Psi \neq \emptyset$ and $\{ prem(\varepsilon_3) \} \cap \Psi \neq \emptyset$, there exist $m, n \in \mathbf{N}$ s.t. $(f_1^m(a_1) = g_1^m(a_1)), (f_2^n(a_2) = g_2^n(a_2)) \in \Psi$; moreover, since, by definition of $FullInter(\{\Gamma \mid \Gamma \in J\})$, $prem(\vartheta^k_{m,n}) \cap \Psi \neq \emptyset$ for $k = 1,\dots,4$, we have that:

$\{ D(f_1^m(a_1)), D(f_2^n(a_2)) \} \cap \Psi \neq \emptyset$, $\{ D(f_1^m(a_1)), D(g_2^n(a_2)) \} \cap \Psi \neq \emptyset$,

$\{ D(g_1^m(a_1)), D(f_2^n(a_2)) \} \cap \Psi \neq \emptyset$, $\{ D(g_1^m(a_1)), D(g_2^n(a_2)) \} \cap \Psi \neq \emptyset$;

thus either $\{ D(f_1^m(a_1)), D(g_1^m(a_1)) \} \subseteq \Psi$ or $\{ D(f_2^n(a_2)), D(g_2^n(a_2)) \} \subseteq \Psi$ and hence either $\{ D(f_1^m(a_1)), D(g_1^m(a_1)), f_1^m(a_1) = g_1^m(a_1) \} \subseteq \Psi$ or $\{ D(f_2^n(a_2)), D(g_2^n(a_2)), f_2^n(a_2) = g_2^n(a_2) \} \subseteq \Psi$. □

We now proceed to show in outline the relative completeness of $CL(T)$ w.r.t. the closed elementary formulae. We need some definitions and preliminary results (whose proofs are omitted).

Def. 3.3.

- For a given conditional type T, the set PNF(T) (for Possibly Naughty Formulae) consists of all closed conditional formulae φ s.t.
  - $CL(T) \vdash \varphi$;
  - $CL(T) \nvdash cons(\varphi)$.
- An *r-choice* (for resolving choice) C is a set of closed elementary formulae s.t.
  for all $\varphi \in PNF(T)$, if $(prem(\varphi) \cap EEq(CL(T))) \subseteq C$, then either $cons(\varphi) \in C$, or there exists $(t = t') \in prem(\varphi) - EEq(CL(T))$ s.t. $(t = t'), (t' = t) \notin C$ and either $D(t)$ or $D(t')$ belongs to C.
- The set of all r-choices is denoted by *R-Choice*. □

Prop. 3.4. *(Deduction Theorem)* Let T be the conditional type (Σ, Ax) and Γ be a set of elementary closed formulae. Then $CL(\Sigma, Ax \cup \Gamma) \vdash \bigwedge \Theta \supset \varepsilon$ iff $CL(T) \vdash \bigwedge \Theta \wedge \bigwedge \Delta \supset \varepsilon$, for an opportune $\Delta \subseteq \Gamma$. □

Prop. 3.5. For all conditional types $T = (\Sigma, Ax)$ and all r-choices C, we have $NF(CL(\Sigma, Ax \cup C)) = \emptyset$. □

Prop. 3.6. If $T = (\Sigma, Ax)$ is a conditional type and ε is an elementary closed formula s.t. $CL(T) \nvdash \varepsilon$, then there exists an r-choice C s.t. $CL(\Sigma, Ax \cup C) \nvdash \varepsilon$. □

Theorem 3.7. *(Main theorem 3)* The system $CL(T)$ is complete w.r.t. the elementary closed formulae.

Proof. Let ε be an elementary closed formula and assume that $CL(T) \nvdash \varepsilon$. We divide the proof in two cases.

a Let ε belong to $EEq(CL(T))$; we show that there exists a model A of $T = (\Sigma, Ax)$ s.t. $A \nvDash \varepsilon$.
  - If $NF(CL(T))$ is empty, then $A = W_\Sigma/{\equiv^{CL(T)}}$ is a model, because of the theorem 2.7, and, by construction of $\equiv^{CL(T)}$, $A \nvDash \varepsilon$.
  - Otherwise there exists an r-choice C s.t. $CL(\Sigma, Ax \cup C) \nvdash \varepsilon$, because of prop. 3.6. Moreover $NF(CL(T'))$ is empty, where $T' = (\Sigma, Ax \cup C)$, because of prop. 3.5; thus $A = W_\Sigma/{\equiv^{CL(T')}}$ is a model of T', because of theorem 2.7. Finally A belongs to PMod(T), since $PMod(T') \subseteq PMod(T)$ by definition of T', and $A \nvDash \varepsilon$, by definition of A.

b Let ε have the form $t = t'$, $CL(T) \nvdash D(t)$, and $CL(T) \nvdash D(t')$; if there exists a conditional type T' s.t.
  1. $PMod(T') \subseteq PMod(T)$;
  2. $\varepsilon \in EEq(CL(T'))$;
  3. $CL(T') \nvdash \varepsilon$;

then, because of a and of conditions 2 and 3, there exists a model A of T' s.t $A \not\models \varepsilon$ and hence, because of 1, A is also a model of T which does not satisfy ε. Therefore we only have to show that there exists such a T'. Let T_1 be the type $(\Sigma, Ax \cup \{D(t)\})$ and T_2 be the type $(\Sigma, Ax \cup \{D(t')\})$; we show that either $CL(T_1) \not\vdash t = t'$ or $CL(T_2) \not\vdash t = t'$. By contradiction we assume that $CL(T_1) \vdash t = t'$ and $CL(T_2) \vdash t = t'$ and prove that $CL(T) \vdash t = t'$. By the absurd hypothesis, because of prop. 3.4, we have $CL(T) \vdash D(t) \supset t = t'$ and $CL(T) \vdash D(t') \supset t = t'$; moreover by rule 2 we have that $CL(T) \vdash t = t' \supset t' = t$ and $CL(T) \vdash t' = t \supset t = t'$ and hence by rule 7 we also have that $CL(T) \vdash t = t' \supset t = t'$. Thus, applying rule 9 to the set $\{ D(t) \supset t = t', D(t') \supset t = t', t = t' \supset t = t' \}$, we have $CL(T) \vdash t = t'$. Therefore either $CL(T_1) \not\vdash t = t'$, and in this case let T' be T_1, or $CL(T_2) \not\vdash t = t'$, and in this case let T' be T_2. In any case T' satisfies conditions 1, 2, 3 by definition. □

Putting together theorem 2.8 and theorem 3.7, we get the following conclusive result about initiality.

Theorem 3.8. The following conditions are equivalent:

1. the set NF(*CL*(T)) is empty;
2. the algebra $W_\Sigma/\equiv^{CL(T)}$ is a model of T;
3. the algebra $W_\Sigma/\equiv^{CL(T)}$ is initial in PMod(T);
4. there exists a model which is initial in PMod(T). □

4 Higher-order types

In this section we apply the results to higher-order specifications. After reducing higher-order specifications to particular classes of first-order specifications (see [MTW]), we consider positive conditional higher-order types, ie higher-order specifications where the only non-positive axioms are the axioms of term-extensionality: two functions are equal iff they coincide over all arguments which are values of terms (clearly term-extensionality coincides with ex-tensionality for term-generated models). As an application of our previous main results, we obtain necessary and sufficient conditions for the existence of initial models of higher-order specifications (theorem 4.5 and corollary 4.7).

Def. 4.1.

- If S is a set, then the set $S^{\to}$ of *functional sorts* over S is inductively defined by: $S \subseteq S^{\to}$ and if $s_1,\dots,s_n,s_{n+1} \in S^{\to}$, then $s = (s_1 \times \dots \times s_n \to s_{n+1}) \in S^{\to}$ for all $n \geq 1$.
- A *higher-order signature* consists of a set S of *basic sorts* and of a family F of sets of *operation symbols*, $F = \{ F_{(\Lambda,s)} \}_{s \in S^{\to}}$. A generic higher-order signature will be denoted by FΣ.
- Let FΣ = (S,F) be a higher-order signature. The *associated extended signature* is the (first-order) signature defined by $E(F\Sigma) = (S^{\to}, F \cup F^{apply})$, where F^{apply} is the family
  $\{ F^{apply}_{(s\, s_1 \dots s_n, s_{n+1})} \}_{n \geq 1, s = (s_1 \times \dots \times s_n \to s_{n+1}) \in S^{\to}}$, with $F^{apply}_{(s s_1 \dots s_n, s_{n+1})} = \{ apply_s \}$.
  We will often use the infix notation for the $apply_s$ operators, ie we will write $f(a_1,\dots,a_n)$ for $apply_s(f,a_1,\dots,a_n)$, dropping the sort indexes where there is no ambiguity.
- Let FΣ = (S,F) be a higher-order signature. A *higher-order partial algebra* A on FΣ is a partial algebra on E(FΣ) which satisfies the following *extensionality condition*:
  for all $s = (s_1 \times \dots \times s_n \to s_{n+1}) \in S^{\to}$, with $n \geq 1$ and for all $f, g \in s^A$,
  if for all $a_i \in s_i^A$, $i=1,\dots,n$, $f(a_1,\dots,a_n) = g(a_1,\dots,a_n)$, then $f = g$.
  A higher-order partial algebra on a higher-order signature FΣ is called an FΣ-algebra. We denote by PFA(FΣ) the class of all FΣ-algebras. □

Remarks. Let PT be a positive (higher-order) conditional type (E(FΣ),Ax) and Fun-Mod(PT) be the class PMod(PT) ∩ PFA(FΣ).

1 *In general there does not exist an* E(FΣ)*-algebra initial in* Fun-Mod(PT), as the following example shows.
Let PT be the positive type $(E(F\Sigma_1),Ax_1)$, where $F\Sigma_1$ is the signature $(\{s\},\{e{:}\to s;\ f,g{:}\to (s\to s)\})$ and Ax_1 is the set $\{D(f), D(g), D(f)\wedge f=g\supset D(f(e))\}$.
Assume by contradiction that I is initial in Fun-Mod(PT).
Let F and G be the $F\Sigma_1$-algebras defined by

$s^F=\{\bullet\};\ (s\to s)^F=\{\varphi,\xi\}$, where $\varphi(\bullet)$ is undefined and $\xi(\bullet)=\bullet$; $e^F=\bullet$; $f^F=\xi$, $g^F=\varphi$.
$s^G=s^F$; $(s\to s)^G=(s\to s)^F$; $e^G=\bullet$; $f^G=\varphi$; $g^G=\xi$.

Both F and G belong obviously to Fun-Mod(PT); thus there exist two homomorphisms $p^F{:}\ I\to F$ and $p^G{:}\ I\to G$.
Since homomorphisms are total functions, for all $a\in s^I$, $p^F(a)=\bullet$ and hence $g^I(a)$ must be undefined, since $g^F(\bullet)$ is undefined and $g^I(a)\in s^I$ implies, by definition of homomorphism, $p^F(g^I(a)) =_e g^F(p^F(a)) =_e g^F(\bullet)$. Analogously $p^G(a)=\bullet$ and hence, since $f^G(\bullet)$ also is undefined, $f^I(a)$ is undefined too. Thus for all $a\in s^I$ we have that both $f^I(a)$ and $g^I(a)$ are undefined and hence $f^I(a)=g^I(a)$ for all $a\in s^I$, ie $f^I=g^I$. Therefore $f^I=g^I$ and, since f^I is totally undefined, $(f(e))^I$ is undefined, contrary to the assumption that I, belonging to Fun-Mod(PT), satisfies the axiom $D(f)\wedge f=g\supset D(f(e))$.

2 *In general there does not exist a conditional type* T *s.t.* Fun-Mod(PT) = PMod(T); indeed, if A is a higher-order algebra, then in general $W_{E(F\Sigma)}/K^A$ is not higher-order (consider for instance the algebras F and G of the example above) while for every conditional type $T=(\Sigma,Ax)$ and every model B of T the algebra W_Σ/K^B is a model of T too.
Therefore we restrict the class of higher-order algebras to the only algebras satisfying a stronger condition of extensionality; this condition is the weakest compatible wih the requirement that, for any PT, the models of PT satisfying this condition are exactly all the partial models of an suitable conditional type T. □

Def. 4.2. An FΣ-algebra A is *term–extensional* iff
for any $f,g\in s^A$ with $s=(s_1\times\ldots\times s_n\to s_{n+1})$ and $n\geq 1$,
if for all $t_i\in W_{E(F\Sigma)|s_i}$, with $i=1,\ldots,n$, $f(t_1{}^A,\ldots,t_n{}^A)=g(t_1{}^A,\ldots,t_n{}^A)$, then $f=g$. □

Def. 4.3.

- A *(positive) conditional higher-order type* (also *higher-order specification*) (P)FT = (FΣ,Ax) consists of a higher-order signature FΣ and a set Ax of (positive) conditional axioms over E(FΣ).
A generic (positive) higher-order type will be denoted by (P)FT.
- Let FT be the conditional higher-order type (FΣ,Ax) and T be the conditional type $(E(F\Sigma),Ax\cup Ax^{ext})$, where Ax^{ext} is the set
$\{\ \wedge\ \{\ f(t_1,\ldots,t_n)=g(t_1,\ldots,t_n)\mid t_i\in W_{\Sigma|s_i},\ i=1,\ldots,n\ \}\supset f=g\mid (s_1\times\ldots\times s_n\to s_{n+1})\in S^{\to}\ \}$, and f, g are variables of sort $(s_1\times\ldots\times s_n\to s_{n+1})$. Then the class FMod(FT) of the ***higher-order models*** of FT is defined by FMod(FT) = PMod(T), ie it is the class of all term-extensional models of FT. □

Remarks. Although the positive conditional higher-order types are a very special case of conditional types, since they have only one kind of (non-positive) conditional axioms, they have all the limitations of conditional types, as the following examples show.
Let PFT be a positive conditional higher-order type.

1. *In general there does not exist an initial model in* Gen(FMod(PFT)). Indeed, with the notations of the previous remark, let PFT be the higher-order type $(F\Sigma_1,Ax_1)$. Since F and G are term-generated, then F and G are term-extensional and hence F, G ∈ Gen(FMod(PFT)). Thus, since we have seen that for any algebra I if there exist two homomorphisms $p^F{:}\ I\to F$ and $p^G{:}\ I\to G$, then I does not satisfy the axiom $D(f)\wedge f=g\supset D(f(e))$, there does not exist an initial model in Gen(FMod(PFT)).
2. *Even if there exists a model* I *initial in* MDef(Gen(FMod(PFT))), *in general this model is not initial in* Gen(FMod(PFT)).
Let PFT_{MD} be the higher-order type $(F\Sigma_{MD},Ax_{MD})$, where $F\Sigma_{MD}$ is the signature (S_{MD},F_{MD}), $S_{MD}=\{\ s\ \}$, $F_{MD}=\{\ e_1,e_2{:}\to s;\ f,g{:}\to (s\to s)\ \}$ and Ax_{MD} is the set $\{\ D(e_1), D(e_2), D(f), D(g), D(f)\wedge f=g\supset e_1=e_2\ \}$.

For any minimally defined higher-order model A and any closed term t, f(t) and g(t) are undefined; thus $A \models f = g$. Therefore every minimally defined model is isomorphic to I, defined by:

$s^I = \{ 1 \}$, $(s \to s)^I = \{ \varphi \}$, where $\varphi(1)$ is undefined, $e_1{}^I = 1$, $e_2{}^I = 1$, $f^I = \varphi$, $g^I = \varphi$,

which is initial in $MDef(Gen(Mod(PFT_{MD})))$, but not in $Gen(Mod(PFT_{MD}))$, because there exist models A s.t. $A \not\models f = g$ and so $A \not\models e_1 = e_2$. □

Thus we explore the existence of initial models for positive conditional higher-order types.

Def. 4.4. Let FT be the conditional higher-order type (FΣ,Ax) and T be the conditional type $(E(F\Sigma), Ax \cup Ax^{ext})$.

- A *c-system* for FT is a c-system for T. In the following a generic c-system will be denoted by *FL*(FT).
- The system *FCL*(FT) is the system *CL*(T). □

Theorem 4.5*(Main theorem 4)*. Let PFT be a positive conditional higher-order type and *FL*(PFT) be a c-system for PFT. The following conditions are equivalent.

1. The algebra $W_{E(F\Sigma)}/\equiv^{FL(PFT)}$ is initial in FMod(PFT).
2. For all $f, g \in W_{E(F\Sigma)}|_{(s_1 \times \ldots \times s_n \to s_{n+1})}$, $n \geq 1$, s.t. $FL(PFT) \vdash D(f)$ and $FL(PFT) \vdash D(g)$ either $FL(PFT) \vdash f = g$, or there exist $t_i \in W_{E(F\Sigma)}|_{s_i}$, $i=1,\ldots,n$, s.t. $FL(PFT) \nvdash f(t_1,\ldots,t_n) = g(t_1,\ldots,t_n)$ and (either $FL(PFT) \vdash D(f(t_1,\ldots,t_n))$ or $FL(PFT) \vdash D(g(t_1,\ldots,t_n))$).

Moreover if *FL*(PFT) is EEq-complete, then the conditions above are also equivalent to the following:

3) there exists a model of PFT which is initial in FMod(PFT). □

Since we know that *FCL*(PFT) is complete we can instantiate theorem 4.5

Prop. 4.6. Let PFT be a positive conditional higher-order type. The following conditions are equivalent.

1) for all $f, g \in W_{E(F\Sigma)}|_{(s_1 \times \ldots \times s_n \to s_{n+1})}$, $n \geq 1$, s.t. $FCL(PFT) \vdash D(f)$ and $FCL(PFT) \vdash D(g)$ either $FCL(PFT) \vdash f = g$, or there exist $t_i \in W_{E(F\Sigma)}|_{s_i}$, $i=1,\ldots,n$, s.t. $FCL(PFT) \nvdash f(t_1,\ldots,t_n) = g(t_1,\ldots,t_n)$ and (either $FCL(PFT) \vdash D(f(t_1,\ldots,t_n))$ or $FCL(PFT) \vdash D(g(t_1,\ldots,t_n))$);
2) the algebra $W_{E(F\Sigma)}/\equiv^{FCL(PFT)}$ is initial in FMod(PFT);
3) there exists a model of PFT which is initial in FMod(PFT). □

Remark. Note that for any functional type PFT = (FΣ,Ax) and any couple $\varphi, \xi \in W_{E(F\Sigma)}|_{(s_1 \times \ldots \times s_n \to s_{n+1})}$ of total functions, ie of functions s.t. $FCL(PFT) \vdash D(\varphi(x_1,\ldots,x_n))$ and $FCL(PFT) \vdash D(\xi(x_1,\ldots,x_n))$, the condition 1 of prop. 4.6 is satisfied, since for all $t_i \in W_{E(F\Sigma)}|_{s_i}$, $i=1,\ldots,n$, $FCL(PFT) \vdash D(\varphi(t_1,\ldots,t_n))$ and $FCL(PFT) \vdash D(\xi(t_1,\ldots,t_n))$. Thus, in particular, if we want to give a specification for a set of total functions, then we have that condition 1 of prop. 4.6 holds for all $f, g \in W_{E(F\Sigma)}|_{(s_1 \times \ldots \times s_n \to s_{n+1})}$, $n \geq 1$, s.t. $FCL(PFT) \vdash D(f)$ and $FCL(PFT) \vdash D(g)$ and hence there exists an initial model of the specification.

Acknowledgements. We wish to thank Y.Gurevich for a long and helpful discussion of the content of this paper and M.Borga for some useful information on related mathematical logic issues.

References

[AC1] Astesiano, E.; Cerioli, M. "Free objects and equational deduction for partial (higher-order) conditional specifications", (Technical report, February 1989).

[AR1] Astesiano, E.; Reggio, G. "SMoLCS-Driven Concurrent Calculi", (invited paper) *Proc. TAPSOFT'87*, vol.1, Berlin, Springer Verlag, 1987 (Lecture Notes in Computer Science n. 249), pp. 169–201.

[AR2] Astesiano, E.; Reggio, G. "An Outline of the SMoLCS Methodology", (invited paper) *Mathematical Models for the Semantics of Parallelism, Proc. Advanced School on Mathematical Models of Parallelism* (Venturini Zilli, M. ed.), Berlin, Springer Verlag, 1987 (Lecture Notes in Computer Science n. 280), pp. 81-113.

[B] Burmeister, P. *A Model Theoretic Oriented Approach to Partial Algebras*, Berlin, Akademie-Verlag, 1986, pp. 1-319.

[BW1] Broy, M.; Wirsing, M. "Partial abstract types", *Acta Informatica* 18 (1982), 47-64.

[BW2] Broy, M.; Wirsing, M. "On the algebraic specification of finitary infinite communicating sequential processes", *Proc. IFIP TC2 Working Conference on "Formal Description of Programming Concepts II"*, Garmisch 1982.

[K] Keisler, H.J. *Model Theory for Infinitary Logic*, Amsterdam - London, North-Holland Publishing Company, 1971, pp. 1-208.

[M] Möller, B. "Algebraic Specification with Higher-Order Operations", *Proc. IFIP TC 2 Working Conference on Program Specification and Transformation, Bad Tolz F.R.G. 1986* (Meertens, L.G.L.T. ed.), Amsterdam-New York-Oxford-Tokyo, North-Holland Publ. Company, 1987.

[MG] Meseguer, J.; Goguen, J.A. "Initiality, Induction and Computability", *Algebraic Methods in Semantics*, Cambridge, edited by M.Nivat and J.Reynolds, Cambridge University Press, 1985, pp.459-540.

[MTW] Möller B., Tarlecki A., Wirsing M. "Algebraic Specification with Built-in Domain Constructions", *Proceeding of CAAP '88 (Nancy France, March 1988)*, edited by Dauchet M. and Nivat M., Berlin, Springer-Verlag, 1988, pp. 132-148.

[R] Reichel H. *Initial Computability, Algebraic Specifications, and Partial Algebras*, Berlin (D.D.R.), Akademie-Verlag, 1986.

[T] Tarlecki A. "Quasi-varieties in Abstract Algebraic Institutions", Journal of Computer and System Science, n. 33 (1986), pp. 333 - 360.

[WB] Wirsing, M.; Broy, M. *An analysis of semantic models for algebraic specifications*, International Summer School Theoretical Foundation of Programming Methodology, Munich. Germany 28/7 - 9/8, 1981.

Appendix: Basic definitions and results on partial algebras

We start with a short collection of basic notions and results, which are well known. However we need to report them here, in order to fix the notation and also because sometimes there are subtle differences; for example, the notion of congruence differs from the one in [B]. The notation here coincides more or less with that used by Goguen and Meseguer [MG] and Broy and Wirsing [BW1].
Proofs are omitted since they are straightforward adaptation of well known proofs for total algebras (see anyway [B], [MG]).
In the following the symbol $=$ will always denote strong equality, ie if p and q are expressions in the metalanguage, then $p=q$ holds iff either both p and q are undefined, or both are defined and equal. Moreover the word "family" will stay in general for "indexed family", where the indexes are clear from the context.

A ***signature*** (S,F) consists of a countable set S of ***sorts*** and of a family $F=\{F_w\}_{w\in S^*\times S}$ of sets of ***operation symbols***. We also write $op: s_1\times...\times s_n \to s$ for $op\in F_{(s_1...s_n,s)}$. A generic signature will be denoted by Σ.

A ***partial algebra*** A on a signature $\Sigma=(S,F)$ consists of a family $\{s^A\}_{s\in S}$ of sets, the ***carriers***, and of a family $\{op^A\}_{op\in F_w, w\in S^*\times S}$ of partial functions, the ***interpretations of operation symbols***, s.t. if $w=(\Lambda,s)$, with $s\in S$, then either op^A is undefined or $op^A\in s^A$, and if $w=(s_1...s_n,s_{n+1})$, where $n\geq 1$, then $op^A: s_1{}^A\times...\times s_n{}^A \to s_{n+1}{}^A$. Often we denote the partial algebra A by the couple $(\{s^A\},\{op^A\})$, omitting the quantifications about s and op which are associated with the signature. A partial algebra over a signature Σ is called a Σ-algebra. We denote by $PA(\Sigma)$ the class of all Σ-algebras.

A particular example of algebra is the ***term-algebra***, defined in the usual way; in the following we will denote the term-algebra over a signature Σ and a family X of variables by $W_\Sigma(X)$, or shortly W_Σ if X is the empty set, and $W_\Sigma(X)|_s$ will be called the ***set of the terms*** of sort s.

Let A and B be two Σ-algebras and p be a family of total functions $p=\{p_s\}_{s\in S}$, s.t. $p_s: s^A \to s^B$. Then p is a ***homomorphism*** iff for any $op\in F_{(s_1...s_n,s_{n+1})}$, $n\geq 0$, and any $a_i\in s_i{}^A$, $i=1,...,n$,

$$op^A(a_1,...,a_n)\in s_{n+1}{}^A \text{ implies } p_{s_{n+1}}(op^A(a_1,...,a_n)) = op^B(p_{s_1}(a_1),...,p_{s_n}(a_n)).$$

Note that the homomorphisms are composable and that the identity is always a homomorphism; thus we can define a category C having Σ-algebras as its objects and homomorphisms as its morphisms; composition and identity are composition and identity as maps.

Two algebras A and B are ***isomorphic*** iff there exist two homomorphisms $\{p_s\}_{s\in S}$ from A into B and $\{q_s\}_{s\in S}$ from B into A, s.t. $p_s{\cdot}q_s = id_{s^B}$ and $q_s{\cdot}p_s = id_{s^A}$ for all $s\in S$.

Let A be a Σ-algebra, $X = \{ X_s \}_{s \in S}$ be a family of S-sorted variables and $V = \{ V_s: X_s \to s^A \}_{s \in S}$, be a family of total functions, called a ***valuation*** for X in A. Then the ***natural interpretation*** of terms w.r.t. A and V, denoted by $eval^{A,V}$, is defined by the following clauses, where we write $t^{A,V}$ for $eval^{A,V}(t)$:

- $x^{A,V} = V_s(x)$, for all $x \in X_s$; $op^{A,V} = op^A$, for all $op \in F_w$, with $w = (\Lambda, s)$;
- $(op(t_1,\ldots,t_n))^{A,V} = op^A(t_1^{A,V},\ldots,t_n^{A,V})$ for all $op \in F_{(s_1 \ldots s_n, s_{n+1})}$, $n \geq 1$, and all $t_i \in W_\Sigma(X)_{|s_i}$.

When restricted to W_Σ, $eval^{A,V}$ is denoted by $eval^A$ and, correspondingly, $t^{A,V}$ becomes t^A. If $eval^A$ is surjective, then A is called ***term-generated***. Note that it is easy to show, by structural induction, that if A and B are two Σ-algebras and $p: A \to B$ is a homomorphism, then $p(t^A) = t^B$ for all closed terms $t \in W_{\Sigma|s}$ s.t. $t^A \in s^A$ and hence if A is term-generated there exists at most one homomorphism from A into B.

Let A be a Σ-algebra. The ***kernel*** of the natural interpretation of closed terms, denoted by K^A, is the family $\{ K^A{}_s \}_{s \in S}$, where $K^A{}_s = \{ (t,t') \mid t, t' \in W_{\Sigma|s}, t^A, t'^A \in s^A \text{ and } t^A = t'^A \}$.

Given a signature $\Sigma = (F,S)$ and a Σ-algebra A, a ***congruence*** $\equiv$ over A is a family of binary relations $\{ \equiv_s \}_{s \in S}$ satisfying the following conditions (where we omit the obvious quantifications over sorts):

C1) $\equiv_s \subseteq s^A \times s^A$, and $\equiv_s$ is symmetric, transitive and relatively reflexive, ie if $(a,a') \in \equiv_s$, then $(a,a) \in \equiv_s$; in the following we denote by $Dom(\equiv_s)$ the set $\{ a \mid (a,a) \in \equiv_s \}$ and we define $a \equiv^D{}_s a'$ iff either $a \equiv_s a'$ or $a, a' \notin Dom(\equiv_s)$;

C2) for any $w = (s_1 \ldots s_n, s_{n+1})$, $op \in F_w$ and $a_i, a_i' \in s_i{}^A$, $i = 1 \ldots n$, if $a \equiv_{s_i} a_i'$ for $i = 1 \ldots n$, then $op^A(a_1,\ldots,a_n) \equiv^D{}_{s_{n+1}} op^A(a'_1,\ldots,a'_n)$.

A congruence is ***strict*** if it also satisfies

C3) for any $op \in F_w$, with $w = (s_1 \ldots s_n, s_{n+1})$ and $n \geq 1$, and any $a_i \in s_i{}^A$ for $i=1,\ldots,n$, if $op^A(a_1,\ldots,a_n) \in Dom(\equiv_{s_{n+1}})$, then $a_i \in Dom(\equiv_{s_i})$, for $i=1,\ldots,n$.

Let $\equiv$ be a congruence over a Σ-algebra A; let [a] denote the equivalence class of a in $\equiv_s$ for all $s \in S$ and all $a \in s^A$. The ***quotient algebra*** of A w.r.t. $\equiv$, denoted by $A/\!\equiv$, is defined by:

- $s^{A/\equiv} = \{ [a] \mid a \in Dom(\equiv_s) \}$, for all $s \in S$;
- $op^{A/\equiv}([a_1],\ldots,[a_n]) = [op^A(a_1,\ldots,a_n)]$ if $op^A(a_1,\ldots,a_n) \in Dom(\equiv_{s_{n+1}})$, otherwise $op^{A/\equiv}([a_1],\ldots,[a_n])$ is undefined, for all $op \in F_{(s_1 \ldots s_n, s_{n+1})}$, $a_i \in Dom(\equiv_{s_i})$, $i = 1 \ldots n$.

For every family $\equiv = \{ \equiv^i \}_{i \in I}$ of congruences over a Σ-algebra A we denote by $\cap(\equiv)$ the family $\{ \cap_{i \in I} \equiv^i{}_s \}_{s \in S}$. Notice that for all families $\equiv = \{ \equiv^i \}_{i \in I}$ of (strict) congruences over a Σ-algebra A the family $\cap(\equiv)$ is a (strict) congruence over A, as it is easy to check.

Notice also that if $\equiv$ is a strict congruence, then for every term $t \in W_\Sigma(X)_{|s}$ and every valuation V for X in $A/\!\equiv$ we have that $[t^{A,V'}] = t^{A/\equiv,V}$, where V' is a valuation for X in A s.t. $[V'(x)] = V(x)$, as it easy to show by structural induction. Thus if A is term-generated, then also $A/\!\equiv$ is term-generated.

Let C be a non-empty subclass of $PA(\Sigma)$.

- MDef(C) is the subclass of C defined by:
  $\{ A \mid A \in C \text{ s.t. for all } s \in S, \text{ for all } t \in W_{\Sigma|s}, \text{ if } t^A \in s^A, \text{ then } t^B \in s^B\ \forall B \in C \}$.
- Gen(C) is the subclass of C defined by: $\{ A \mid A \in C$ s.t. A is term-generated $\}$.
- K^C is the congruence defined by $\cap(\{ K^A \mid A \in C \})$.
- A Σ-algebra I is ***initial*** in C iff $I \in C$ and $\forall\ B \in C\ \exists$ a unique homomorphism from I into B.

Now we can state some results on the existence and the characterization of the initial model for a class C of algebras.

Prop. A. Let C be a non-empty subclass of $PA(\Sigma)$.

(1) If I is initial in C, then $K^I = K^C$, ie for all $s \in S$

  (i1) $\forall\, t \in W_{\Sigma|s}$, $t^I \in s^I$ iff ($t^B \in s^B$ for all $B \in C$), or, equivalently, $I \in MDef(C)$.

  (i2) $\forall\, t, t' \in W_{\Sigma|s}$ s.t. $t^I, t'^I \in s^I$, $t^I = t'^I$ iff ($t^B = t'^B$ for all $B \in C$).

(2) If there exists an algebra $I \in C$ isomorphic to W_Σ/K^C, then it is initial in C.

(3) If for all $A \in C$ there exists $B \in C$ isomorphic to W_Σ/K^A, then the following conditions are equivalent

  (a) I is initial in C;

  (b) I is initial in Gen(C);

  (c) I is isomorphic to W_Σ/K^C and $I \in C$. □

Terms and infinite trees as monads over a signature

Eric BADOUEL
Irisa – Campus de Beaulieu – 35042 Rennes cedex – France

Abstract

In this paper, we prove that the usual construction of terms and infinite trees over a signature is a particular case of a more general construction of monads over the category of sets. In this way, we obtain a family of semantical domains having a tree-like structure and appearing as the completions of the corresponding finite structures. Though it is quite different in its technical developments our construction should be compared with the one of De Bakker and Zucker which is very similar in spirit and motivation. We feel that one outcome of the present approach is that, due to its connection with Lawvere's algebraic theories, it should provide an interesting framework to deal with equational varieties of process algebras.

1 Introduction

A major concern in denotational semantics is to provide a meaning to recursively defined expressions ; then, we must be able to express and to solve systems of equations in various domains. Such a requirement leads us to supply those domains with three fundamental algebraic operations, namely builing up t-uple of elements, extracting an element from such a t-uple and making substitution of a t-uple of elements for a t-uple of variables. Lawvere's algebraic theories ([Law63]) are categories that embody those basic manipulations in a uniform manner. An alternative presentation of algebraic theories is that of monad (or following Manes [Man76] *algebraic theories on monoid form*). Monads found applications in automata theory ([AM75]), tree processing ([Ala75]) they also have been used to study recursion schemes ([BG87]) and more recently Petri Nets ([MM88]) .

In both presentation of algebraic theories term substitution plays a central role and allows for the statement of equations (we shall make this statement precise in the case of monads). In analogy to Elgot's iterative algebraic theories we may define the iterative monads as those monads for which all non degenerate equations (such as x=x) admits a unique solution.

In this paper, we prove that the usual construction of terms and infinite trees over a signature Σ is a particular case of a more general construction of monads over the category of sets. Moreover, the monad corresponding to the infinite trees is proved iterative ; and, as such, constitute a potential semantical domain. In this way, we obtain a family of iterative monads associated to an extended notion of signature called ω**-signature.** As typical example we present the monad of synchronization trees ([Mil80]) but other similar examples of algebraic structures may be considered. Roughly speaking such structures are tree-like and the infinite objects are obtained through a completion process from the finite ones. Though quite different in its technical developments our present construction should be compared with the one of De Bakker and Zucker in [dBZ82]. Actually, in that paper they gave a variety of process domains having such a branching structure and enabling one to deal with various concepts arising in the semantics of concurrency such as *parallel composition*, *synchronization* and *communication*. The novelty of the present approach, compared with De Bakker and Zucker's, is that, due to its connection with Lawvere's algebraic theories, it should moreover provide an interesting framework to deal with equational varieties of process algebras.

Concerning monads we refer to the books *Algebraic Theories* ([Man76]) by E. Manes and *Toposes, Triples and Theories* ([BW85]) by M. Barr and C. Wells ; nevertheless we shall recall along this paper all definitions and results we need about monads. The reader is just expected familiar with the basic notions of category theory such as limit and colimit of diagrams, functor categories and adjunctions.

This paper is organized as follows : section 2 gives a construction of terms and infinite trees for a signature, a signature being there some endofunctor of the category **Set**, providing a generalization of the usual construction where the signature corresponds to a ranked alphabet. In sections 3 and 4 respectively we supply the sets of terms and infinite trees with substitution operations by forming the corresponding monads ; moreover the monad of infinite trees is proved iterative. Section 5 is the conclusion.

2 Terms and infinite trees for a signature

We first recall some definitions from universal algebra. A (finitary) signature or ranked alphabet Σ is given by a set whose elements are operator symbols together with a mapping $a : \Sigma \rightarrow \mathbf{N}$ which assigns to each operator f a natural number a(f) called its arity. We denote Σ_n the set of operators of arity n. If Σ' is another signature, a morphism $\varphi : \Sigma \rightarrow \Sigma'$ of signatures is any mapping from Σ to Σ' that preserves the arities i.e. $f \in \Sigma_n \Rightarrow \varphi(f) \in \Sigma'_n$.

Definition 1 *a Σ-algebra is a pair (D,δ) where D is a non empty set (the carrier or domain of the algebra) and $\delta = \{\delta_f \,;\, f \in \Sigma\}$ is a set of functions $\delta_f : D^{a(f)} \rightarrow D$ associated to each operator f in Σ. And a morphism φ between two Σ-algebras*

(D,δ) and (D',δ') is any mapping between their respective domains such that :

$$\varphi(\delta_f(a_1,\ldots,a_n)) = \delta'_f(\varphi(a_1),\ldots,\varphi(a_n))$$

A ranked alphabet Σ can be interpreted as the functor $\Sigma : \mathbf{Set} \to \mathbf{Set}$ defined on objects and on arrows as

$$\Sigma A = \amalg_{f\in\Sigma} A^{a(f)} \quad and \quad \Sigma\varphi = \amalg_{f\in\Sigma} \varphi^{a(f)}$$

a Σ-algebra is then any mapping $\Sigma A \xrightarrow{a} A$ and a morphism between two Σ-algebras (A,a) and (B,b) is any mapping $\varphi : A \to B$ such that : $\varphi \circ a = b \circ \Sigma\varphi$. Σ-algebras and their morphisms constitute a category denoted Σ-Alg. Since the coproduct of sets are their disjoint union we get a more intuitive representation of ΣA as $\Sigma A = \{ f[< a_1 >,\ldots,< a_n >] \,/\, f \in \Sigma_n$ and $a_1,\ldots,a_n \in A\}$ (where <,>,[,] are special symbols) and, in this way, $\Sigma\varphi(f[< a_1 >,\ldots,< a_n >]) = f[< \varphi(a_1) >,\ldots,< \varphi(a_n) >]$.

More generally, if F is an endofunctor in a category C we let F-Alg, F-co-Alg and F-fp denote the categories whose objects are respectively the F-algebras (i.e. arrows $FA \xrightarrow{a} A$), F-co-algebras (i.e. arrows $A \xrightarrow{a} FA$) and F-fixed-points (i.e. isomorphisms $FA \xrightarrow{a} A$) and whose arrows from (A,a) to (B,b) are those arrows $\varphi : A \to B$ such that $\varphi \circ a = b \circ F\varphi$ for algebras and fixed-points and $F\varphi \circ a = b \circ \varphi$ for co-algebras.
Our purpose, in this section, is to provide a generalization of the construction of terms and infinite trees corresponding to a signature, a signature being now some endofunctor of the category Set. For this, to any endofunctor Σ of Set we associate an endofunctor $\mathcal{F}$ of the functor category $C = Func(\mathbf{Set},\mathbf{Set})$. $\mathcal{F}$ is the endofunctor of C defined on objects (functors F) and on arrows (natural transformations τ) by : [1]

$$\mathcal{F}F = 1_C + \Sigma F \quad and \quad (\mathcal{F}\tau)_X = 1_X + \Sigma\tau_X$$

We recall the category of sets admits initial and terminal elements being respectively the empty set ∅ and any singleton set, say {Ω}. Let 0 and 1 denote the constant endofunctors of Set whose respective values are ∅ and {Ω}. They are the initial and terminal objects of C, moreover C is, as the category Set, complete and co-complete. Let Δ and ▽ be the following chains on C :

$$\Delta = 0 \xrightarrow{\varphi_0} \mathcal{F}0 \xrightarrow{\mathcal{F}\varphi_0} \mathcal{F}^2 0 \ldots \mathcal{F}^n 0 \xrightarrow{\mathcal{F}^n\varphi_0} \mathcal{F}^{n+1} 0 \ldots$$

$$\nabla = 1 \xleftarrow{\psi_0} \mathcal{F}1 \xleftarrow{\mathcal{F}\psi_0} \mathcal{F}^2 1 \ldots \mathcal{F}^n 1 \xleftarrow{\mathcal{F}^n\psi_0} \mathcal{F}^{n+1} 1 \ldots$$

(where φ_0 and ψ_0 are uniquely defined by universality of 0 and 1)
and let $(T,j) = colim(\Delta)$ and $(T^\infty,\pi) = lim(\nabla)$ be their colimiting cone and

---

[1] **If x is an object in a category we let 1_x stands for the identity arrow in x ; in particular 1_C is the identity functor from C to itself.**

limiting cone respectively. Concerning limits and colimits in functor categories we recall from [Sch72] the following result : if $\mathcal{D}$ is complete (resp. co-complete) the functor category $Func(\mathcal{C},\mathcal{D})$ is also complete (co-complete) and the constructioon is *pointwise* which means that for every object c of $\mathcal{C}$ the evaluation functor $E_c : Func(\mathcal{C},\mathcal{D}) \to \mathcal{D}$ defined by $E_c(F) = F(c)$ and $E_c(\tau) = \tau_c$ preserves the limits (co-limits). It follows that $(TX, \jmath_X) = colim(\Delta X)$ and $(T^\infty X, \pi_X) = lim(\nabla X)$ where ΔX and ∇X stands for the chains gotten by evaluating Δ and ∇ at X. [2]
We observe that :

$$\Delta X = \emptyset \xrightarrow{\varphi_{0,X}} \mathcal{F}_X\emptyset \xrightarrow{\mathcal{F}_X(\varphi_{0,X})} \mathcal{F}_X^2\emptyset \ldots \mathcal{F}_X^n\emptyset \xrightarrow{\mathcal{F}_X^n(\varphi_{0,X})} \mathcal{F}_X^{n+1}\emptyset \ldots$$

$$\nabla X = \{\Omega\} \xleftarrow{\psi_{0,X}} \mathcal{F}_X\{\Omega\} \xleftarrow{\mathcal{F}_X(\psi_{0,X})} \mathcal{F}_X^2\{\Omega\} \ldots \mathcal{F}_X^n\{\Omega\} \xleftarrow{\mathcal{F}_X^n(\psi_{0,X})} \mathcal{F}_X^{n+1}\{\Omega\} \ldots$$

where $\mathcal{F}_X$ is the endofunctor of **Set** defined on sets and mappings as :

$$\mathcal{F}_X Y = X + \Sigma Y \quad \mathcal{F}_X \varphi = 1_X + \Sigma\varphi$$

Now let us have a look to TX and T^∞X in the particular case where Σ is the functor associated to a ranked alphabet. First, we notice that in the chain

$$\Delta X = \emptyset \xrightarrow{\varphi_{0,X}} X + \Sigma\emptyset \xrightarrow{\varphi_{1,X}} X + \Sigma(X + \Sigma\emptyset) \xrightarrow{\varphi_{2,X}} \ldots$$

each set is an actual subset of its follower and the mappings $\varphi_{i,X}$ are the inclusion maps. The colimit TX is then their set-theoretic union i.e. the *least* set (regarding inclusion) containing X and such that if $f \in \Sigma_n$ and $t_1,\ldots,t_n \in TX$ then $f[t_1,\ldots,t_n]$ is also an element of TX. We then meet the usual definition of the set of terms corresponding to a signature. Now, as $T^\infty X$ is the limit of the chain

$$\nabla X = 1 \xleftarrow{\psi_0} X + \Sigma 1 \xleftarrow{\psi_1} X + \Sigma(X + \Sigma 1) \xleftarrow{\psi_2} \ldots$$

an element of $T^\infty X$ is a sequence $(u_n)_{n\in\mathbf{N}}$ such that $u_n \in \mathcal{F}_X^n 1$ and for every integer n one has $u_n = \psi_{n,X}(u_{n+1})$. Note that the limiting cone $\pi_X : T^\infty X \to \nabla X$ satisfies the following :

$$\begin{aligned}
\pi_{X,0}(t) &= \Omega \\
\pi_{X,n+1}(< x >) &= < x > \\
\pi_{X,n+1}(f[t_1,\ldots,t_n]) &= f[\pi_{X,n}(t_1),\ldots,\pi_{X,n}(t_n)]
\end{aligned}$$

We then have the usual construction of infinite tree, where a tree is represented by the sequence of its nth-sections.

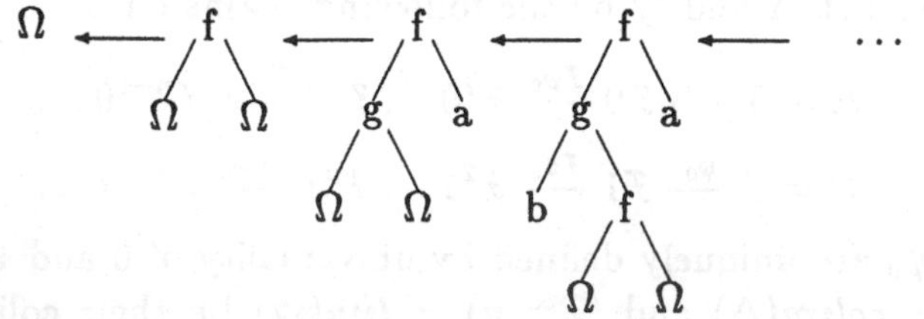

[2] There is a slight abuse of notation justified by the isomorphisms $Func(\omega, Func(\mathbf{Set},\mathbf{Set})) \cong Func(\mathbf{Set}, Func(\omega,\mathbf{Set}))$ and $Func(\omega^{op}, Func(\mathbf{Set},\mathbf{Set})) \cong Func(\mathbf{Set}, Func(\omega^{op},\mathbf{Set}))$

We usually assume that Σ owns at least one operator of arity 0 otherwise the set of closed terms (i.e. $T\emptyset$) should be empty ; under this hypothesis we may prove that the set of term TX is a dense (in a topological sense) subset of $T^\infty X$. This result will generalize if we make some additional assumptions ; those considerations lead to the following definition that summarize the hypothesis on Σ which are necessary so as to make our construction work.

Definition 2 *A signature Σ is any endofunctor of* **Set** *such that for any set X, ΣX is a non-empty set ; and, if X is a subset of Y with inclusion map $i : X \to Y$, ΣX is a subset of ΣY with inclusion map Σi.*

3 The term monad over an ω-signature

As we stressed in the introduction, in both presentations of algebraic theories term substitution plays a central role ; it is modelled by composition in Lawvere's algebraic theories and by a natural transformation (called structure map : μ) in a monad . Incidently, another natural transformation (called embedding of generators : η) allows substitutions to take place without explicit mention to variables.

Definition 3 *a monad on a category C is a triple (T, η, μ) where T is an endofunctor on C and $\eta : I \to T$ and $\mu : T^2 \to T$ are natural transformations (I is the identity endofunctor of C) satisfying the following commuting diagrams :*

$$
\begin{array}{ccc}
TTT & \xrightarrow{\mu T} & TT \\
{\scriptstyle T\mu}\downarrow & & \downarrow{\scriptstyle \mu} \\
TT & \xrightarrow{\mu} & T
\end{array}
\qquad
\begin{array}{ccccc}
T & \xrightarrow{\eta T} & TT & \xleftarrow{T\eta} & T \\
 & {\scriptstyle 1_T}\searrow & \downarrow{\scriptstyle \mu} & \swarrow{\scriptstyle 1_T} & \\
 & & T & &
\end{array}
$$

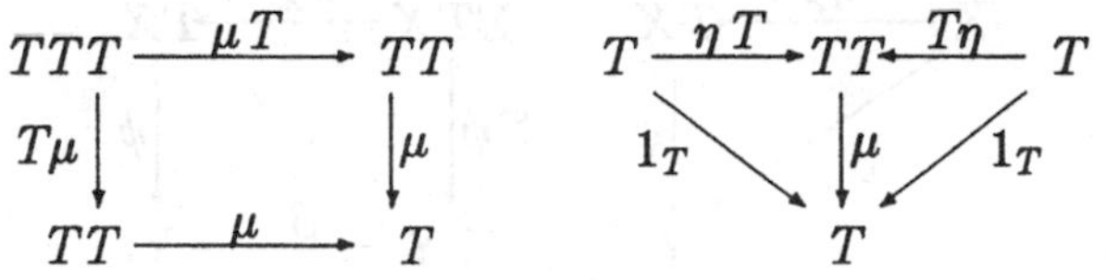

Those axioms, that should be compared with the axioms of a monoid (the associative law and the two unit laws), provide the minimal conditions expected for a substitution operation namely being associative and well-behaved regarding the embedding of elements.

In order to supply T and T^∞ with monad structures, we make the additional assumption that the functor $\mathcal{F}$, associated to Σ, is both ω-co-continuous and ω^{op}-continuous ; we shall say, in such a case, that Σ is an ω-**signature**. We know (see [SP82]) that, under those hypothesis, we obtain an initial $\mathcal{F}$-algebra and a terminal $\mathcal{F}$-co-algebra whose respective carriers are T and T^∞. More precisely, $\Phi : \Delta \to \mathcal{F}T$ defined by $\Phi_n = \mathcal{F}j_n \circ \mathcal{F}^n\varphi_0$ is a co-cone ; let φ be the mediating arrow $\Phi = \varphi \circ j$; thanks to ω-co-continuity of $\mathcal{F}$ φ is an isomorphism and (T, φ^{-1}) is the initial $\mathcal{F}$-algebra. In the same way, we obtain $\psi : \mathcal{F}(T^\infty) \to T^\infty$ making (T^∞, ψ^{-1}) the terminal $\mathcal{F}$-co-algebra ; moreover, (T, φ^{-1}) and (T^{∞}, ψ) are, as well, the initial and terminal elements of the category of $\mathcal{F}$-fixed points. Now, for our particular case, we observe that $\varphi^{-1} : \mathcal{F}T = I + \Sigma T \longrightarrow T$ splits into $\varphi^{-1} = [\eta, \sigma]$ where $\eta : I \dot{\rightarrow} T$ and $\sigma : \Sigma T \dot{\rightarrow} T$. In the same way $\psi = [\eta^{\infty}, \sigma^{\infty}]$ where $\eta^{\infty} : I \dot{\rightarrow} T^{\infty}$ and $\sigma^{\infty} : \Sigma T^{\infty} \dot{\rightarrow} T^{\infty}$. Since all previous constructions were made componentwise, $(TX, [\eta_X, \sigma_X])$ and $(T^{\infty}X, [\eta_X^{\infty}, \sigma_X^{\infty}])$ are actually the respective initial $\mathcal{F}_X$-algebra and terminal $\mathcal{F}_X$-co-algebra. So

$$\boxed{(TX, [\eta_X, \sigma_X]) \textit{ is the initial } \mathcal{F}_X \textit{ algebra}}$$

Spelled out, for any $\mathcal{F}_X$-algebra $(Y, [\alpha, \beta])$ there exists a unique mapping $\psi : T_X \rightarrow Y$ such that

$$\begin{array}{ccc} X + \Sigma TX & \xrightarrow{[\eta_X, \sigma_X]} & TX \\ {\scriptstyle 1_X + \Sigma\psi} \downarrow & & \downarrow {\scriptstyle \psi} \\ X + \Sigma Y & \xrightarrow{[\alpha, \beta]} & Y \end{array} \qquad \text{commutes.}$$

In other words, splitting this diagram into two parts, for a given set Y and a couple of mappings $\alpha : X \rightarrow Y$ and $\beta : \Sigma Y \rightarrow Y$ there exists a unique mapping $\psi : T_X \rightarrow Y$ such that the two following diagrams commute.

$$\begin{array}{ccc} X & \xrightarrow{\eta_X} & TX \\ & {\scriptstyle \alpha} \searrow & \downarrow {\scriptstyle \psi} \\ & & Y \end{array} \qquad \begin{array}{ccc} \Sigma TX & \xrightarrow{\sigma_X} & TX \\ {\scriptstyle \Sigma\psi} \downarrow & & \downarrow {\scriptstyle \psi} \\ \Sigma Y & \xrightarrow{\beta} & Y \end{array}$$

So we have an adjunction **Set** $\overset{<F,U>}{\longleftrightarrow}$ **Σ-Alg** where U is the forgetful functor ; F sends X to the Σ-algebra (TX, σ_X) and is defined on arrows as follows : given a mapping $\varphi : X \rightarrow Y$, $F\varphi$ is the unique morphism of Σ-algebras from (TX, σ_X) to (TY, σ_Y) such that $UF\varphi \circ \eta_X = \eta_Y \circ \varphi$. Since $T\varphi$ is the underlying mapping of a Σ-algebra morphism from (TX, σ_X) to (TY, σ_Y) (by naturality of σ) and that $T\varphi \circ \eta_X = \eta_Y \circ \varphi$ (by naturality of η) it follows that $T\varphi = UF\varphi$; and then $T = UF$. Now, we know (see, for example [BW85]) that to each adjunction (U, F, η, ϵ) (where F is left adjoint to U with unit η and counit ϵ) corresponds the monad $(UF, \eta, U\epsilon F)$. So if we let $\mu = U\epsilon F$, the triple $(T = UF, \eta, \mu)$ so obtained is a monad ; we shall call it **the term monad over the signature Σ**.
As particular case if (Y,y) is a Σ-algebra, there exists a unique T-algebra (Y, y^{*}) extending it in the following sense

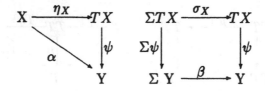

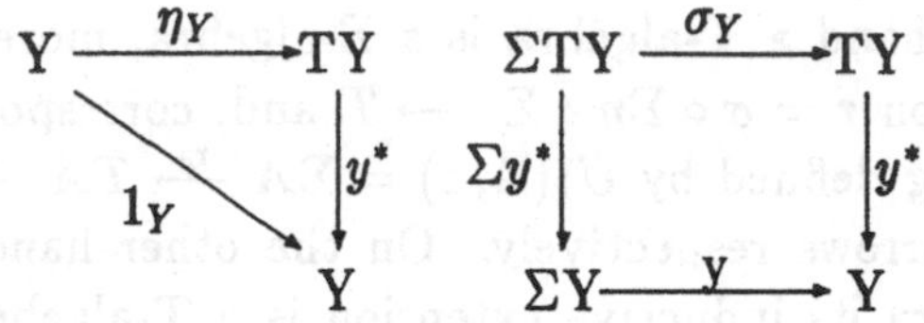

We shall call y^* the inductive extension of y.
We recall that a **T**-algebra is a T-algebra such that evaluation commutes with substitution ; more precisely

Definition 4 *If* $\mathbf{T}=(T,\eta,\mu)$ *is a monad on a category* $\mathcal{C}$ *a* **T**-*algebra is a pair (A,a) where A is an object of* $\mathcal{C}$ *and* $a : TA \to A$ *, an arrow (its structure map) such that both following diagrams commute.*

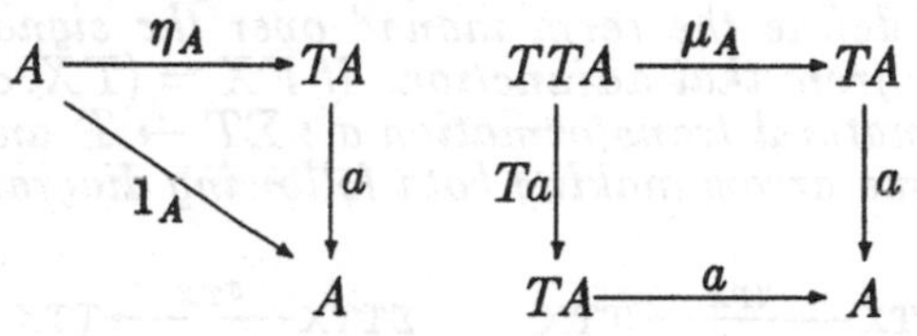

As for Σ-algebras, morphisms of **T**-algebras (A,a) and (B,b) are any arrow $\varphi : A \to B$ such that $\varphi \circ a = b \circ T\varphi$. **T**-algebras and their morphisms constitute a category denoted $\mathcal{C}^{\mathbf{T}}$; it is the Eilenberg-Moore category associated to **T**. Now we prove the

Proposition 1 *for any set Y,* $\mu_Y = U\epsilon FY : TTY \dot{\to} TY$ *is the inductive extension of* $\sigma_Y : \Sigma TY \dot{\to} TY$*. Moreover, if (Y,y) is a* Σ*-algebra then* (Y,y^*) *is a* **T**-*algebra.*

Proof
A set X and a Σ-algebra $\mathcal{Y}=(Y,y)$ being given, let us denote $g^\sharp : FX \to \mathcal{Y}$ the morphism of Σ-algebras corresponding, via the adjunction, to the mapping $g : X \to Y$.

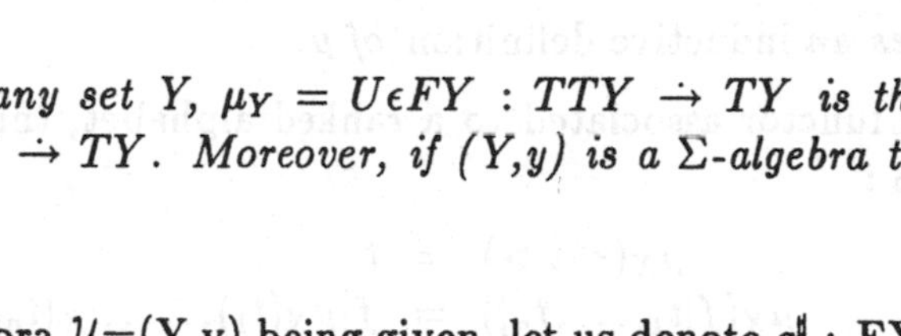

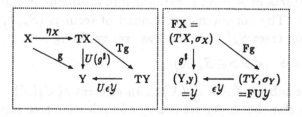

The component of the counit $\epsilon : FU \dot{\to} I$ of the adjunction in $\mathcal{Y}$ satisfies : $g^\sharp = \epsilon\mathcal{Y} \circ Fg$ and thus $Ug^\sharp = U\epsilon\mathcal{Y} \circ Tg$. Now if we take X=Y and $g = 1_Y$ we obtain the inductive extension of y as $y^* = Ug^\sharp = U\epsilon\mathcal{Y}$ and, in the particular case where $\mathcal{Y} = (TY,\sigma_Y) = FY$, we have $\sigma_Y^* = U\epsilon FY = \mu_Y$.
Now applying the naturality of the co-unit to the Σ-morphism $1_Y^\sharp : (TY,\sigma_Y) \to \mathcal{Y}$ leads to $1_Y^\sharp \circ \epsilon_{(TY,\sigma_Y)} = \epsilon_{\mathcal{Y}} \circ FU(1_Y^\sharp)$ and then, by applying U, $y^* \circ U\epsilon_{FTY} = U\epsilon_{\mathcal{Y}} \circ Ty^*$ i.e. $y^* \circ \mu_{TY} = y^* \circ Ty^*$. $\square$

Remark : On one hand a **T**-algebra is a Σ-algebra, more precisely we have a natural transformation $\tau = \sigma \circ \Sigma\eta : \Sigma \dot{\longrightarrow} T$ and, corresponding to it a functor U^τ : **Set**$^{\mathbf{T}}$ $\longrightarrow$ Σ-**Alg** defined by $U^\tau(A,a) = \Sigma A \xrightarrow{\tau_A} TA \xrightarrow{a} A$ and $U^\tau(f) = f$ on **T**-algebras and arrows respectively. On the other hand we just prove that if (A,a) is a Σ-algebra its inductive extension is a **T**-algebra ; actually we have a functor called the **Eilenberg Moore comparison functor** ϕ : Σ-**Alg** $\longrightarrow$ **Set**$^{\mathbf{T}}$ defined by $\phi(A,a) = (A,a^*)$ and $\phi(f) = f$. Those two functors are inverse isomorphisms (as readily verified). Axioms for **T**-algebras impose evaluation to respect compositionality and that is why a **T**-algebra is the same data as a Σ-algebra.

We can summarize our results in the following

Proposition 2 *Let Σ be a signature, the forgetful functor U :* Σ-**Alg** $\rightarrow$ **Set** *has a left adjoint F. We define the term monad over the signature Σ as the monad (T,η,μ) which results from that adjunction. If $FX = (TX, \sigma_X)$ the arrows σ_X are the components of a natural transformation $\sigma : \Sigma T \dot{\rightarrow} T$ and the component of μ at a set X is the unique arrow making both following diagrams commute.*

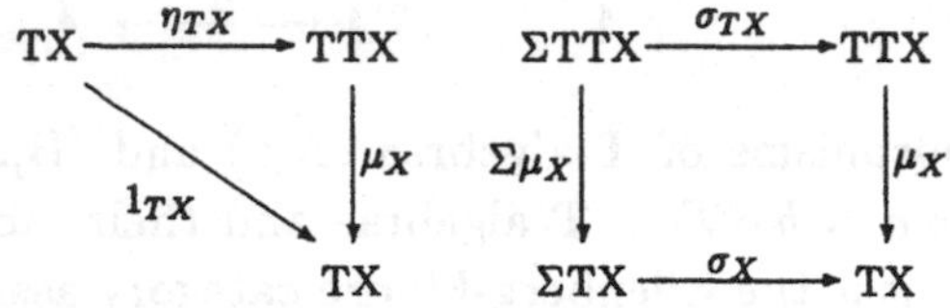

This provides an inductive definition *of μ.*

When Σ is the functor associated to a ranked alphabet, this inductive definition is equivalent to :

$$\begin{aligned} \mu_X(<t>) &= t \\ \mu_X(f[t_1,\ldots,t_n]) &= f[\mu_X(t_1),\ldots,\mu_X(t_n)] \end{aligned}$$

Another example is the following. If A is a set of actions we define ΣX as the set of all finite subsets of $A \times X$, it extends into a functor (we define Σ on mappings by extension). The corresponding monad of term is (S_A, η, μ) where the set of synchronization trees $S_A(X)$ is the least set verifying :

- $x \in X \quad \Longrightarrow \quad <x> \in S_A(X)$
- any *finite* subset of $A \times S_A(X)$ is an element of $S_A(X)$
  (in particular the empty set is a synchronization tree sometimes denoted *nil*)

$S_A(X)$ is the set of finitely branching, non-deterministic and unordered finitary trees whose branches are labelled by elements in A and leaves by elements in X. For example :

$$\{<a,\{<c,<x>>\}>;<b,\{<d,\emptyset>;<e,<z>>\}>\} \quad =$$

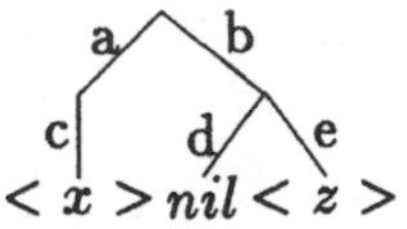

Its mapping function is defined for $\varphi : X \rightarrow Y$ by $S_A(\varphi)(<x>) = <\varphi(x)>$ and if
$T = \{<a_i, T_i> \ / \ i \in I\} \subset S_A(X)$ then $S_A(\varphi)(T) = \{<a_i, S_A(\varphi)(T_i)> \ / \ i \in I\}$
The embedding of generators $\eta_X : X \rightarrow S_A(X)$ is given by : $\eta_X(x) = <x>$
and the structure map $\mu_X : S_A(S_A(X)) \rightarrow S_A(X)$ by $\mu_X(<t>) = t$ and if
$T = \{<a_i, T_i> \ / \ i \in I\} \subset S_A(S_A(X))$ then $\mu_X(T) = \{<a_i, \mu_X(T_i)> \ / \ i \in I\}$

Since the forgetful functor $U : \Sigma$-**Alg** $\longrightarrow$ **Set** has a left adjoint an ω-signature is a particular case of **input process** as defined by Manes ([Man76]) ; the following proposition then follows from the similar and more general result on input processes. (see also [Ala75])

Proposition 3 *the term monad* (T, η, μ) *is the free monad generated by* Σ *; that is to say, the natural transformation* $\tau = \sigma \circ \Sigma\eta : \Sigma \dot{\rightarrow} T$ *is such that for every monad* (T', η', μ') *and natural transformation* $\lambda : \Sigma \dot{\longrightarrow} T'$ *there exists a unique monad morphism* φ *such that* $\lambda = \varphi \circ \tau$.

4 The monad of infinite trees

Concerning T^∞ we already have an embedding of generators (η^∞) it remains to define the structure map : $\mu^\infty : T^\infty T^\infty \dot{\longrightarrow} T^\infty$. For this, we define a morphism of ω^{op}-chains $\xi_X : \nabla T^\infty X \longrightarrow \nabla X$ as follows :

$$\begin{array}{ccccccc}
\{\Omega\} & \xleftarrow{\psi_{T^\infty X,0}} & T^\infty X + \Sigma\{\Omega\} & \longleftarrow \cdots & (\nabla T^\infty X)_n & \xleftarrow{\psi_{T^\infty X,n}} & T^\infty X + \Sigma(\nabla T^\infty X)_n \\
\downarrow \xi_{X,0} & & \downarrow \xi_{X,1} & & \downarrow \xi_{X,n} & & \downarrow [\pi_{X,n+1} \ ; \ \Sigma\xi_{X,n}] \\
\{\Omega\} & \xleftarrow{\psi_{X,0}} & X + \Sigma\{\Omega\} & \longleftarrow \cdots & (\nabla X)_n & \xleftarrow{\psi_{X,n}} & X + \Sigma(\nabla X)_n
\end{array}$$

- $\xi_{X,0} = 1_{\{\Omega\}}$ is the identity mapping on $\{\Omega\}$
- for every integer n : $\xi_{X,n+1} = [\pi_{X,n+1} \ ; \ \Sigma\xi_{X,n}]$.

proving that ξ_X is a morphism of ω^{op}-chains amounts to proving that each elementary square commutes which is an easy verification.
We then define $\mu^\infty_X : T^\infty T^\infty X \longrightarrow T^\infty X$ as the mediating morphism between the cone $\xi_X \circ \pi_{T^\infty X} : T^\infty T^\infty \longrightarrow \nabla X$ and the limiting cone $\pi_X : T^\infty X \longrightarrow \nabla X$. ξ_X is clearly natural in X, the naturality of μ^∞ then follows from the fact that the components of π are limiting cones and consequently left-cancelable arrows of the category $Func(\omega^{op}, \mathbf{Set})$.

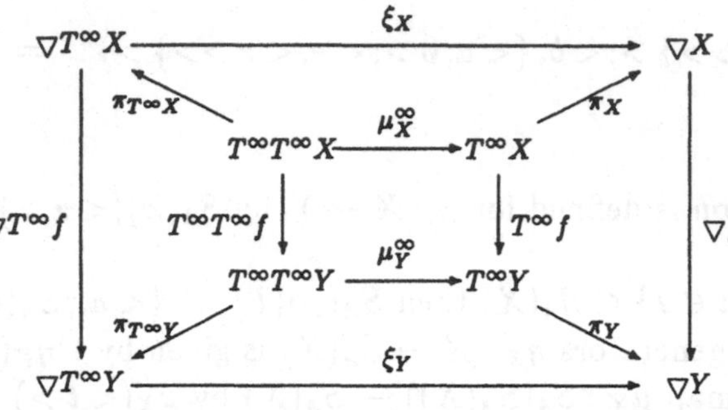

And we prove :

Proposition 4 *$(T^\infty, \eta^\infty, \mu^\infty)$ is a monad.*

In order to establish this result, a first stage consists in verifying by diagram chasing the following lemma.

Lemma 1 *For every set X the following three diagrams (that we shall name respectively $monad_1(X)$, $monad_2(X)$ and $monad_3(X)$) commute.*

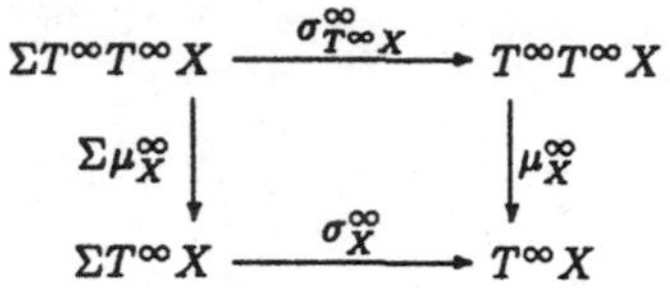

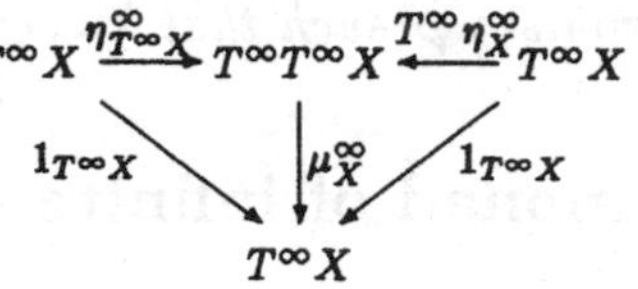

Proof :
Consider the diagram (in $Func(\omega^{op}, \mathbf{Set})$)

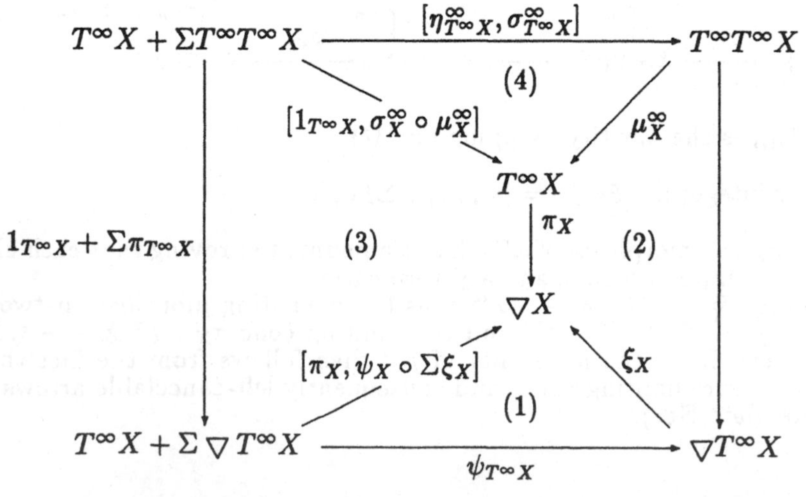

Since $\psi_{X,n} \circ \pi_{X,n+1} = \pi_{X,n}$ it follows $[\pi_{X,n}; \psi_{X,n} \circ \Sigma\xi_{X,n}] = \psi_{X,n} \circ [\pi_{X,n+1}; \Sigma\xi_{X,n}] = \xi_{X,n} \circ \psi_{T^\infty X,n}$ i.e. (1) commutes. (2) commute by definition of μ^∞ and since $\psi_X \circ \Sigma\pi_X = \pi_X \circ \sigma^\infty_X$

$$\begin{aligned}
[\pi_X; \psi_X \circ \Sigma\xi_X] \circ (1_{T^\infty X} + \Sigma\pi_{T^\infty X}) &= [\pi_X; \psi_X \circ \Sigma\xi_X \circ \pi_{T^\infty X})] \\
&= [\pi_X; \psi_X \circ \Sigma(\pi_X \circ \mu^\infty_X)] \\
&= [\pi_X; \pi_X \circ \sigma^\infty_X \circ \Sigma\mu^\infty_X] \\
&= \pi_X \circ [1_{T^\infty X}; \sigma^\infty_X \circ \Sigma\mu^\infty_X]
\end{aligned}$$

i.e. (3) commutes. Since diagrams (1), (2) and (3) commute, as well as the outer rectangle and because π_X is a left-cancellable arrow in $Func(\omega_{op}, \mathbf{Set})$ (as a limiting cone) it follows that diagram (4) commutes i.e. $monad_1(X)$ and $monad_2(X)$ commute. Now, consider the following diagram :

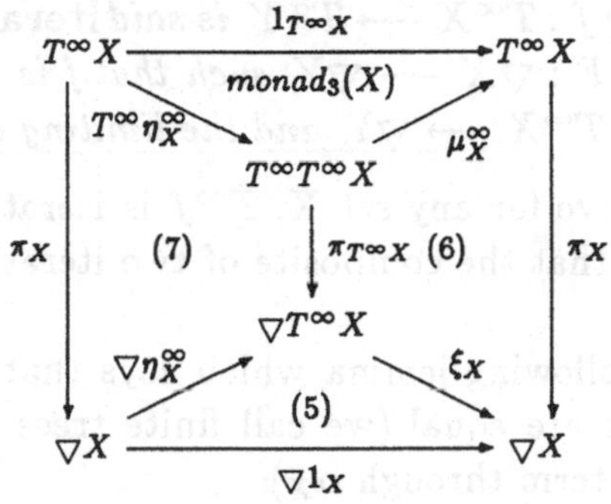

We prove $\xi_X \circ \nabla\eta^\infty_X = \nabla 1_X$ by induction, (6) commutes by definition of μ^∞ ; (7) and the outer rectangle commute by naturality of π. As previously, it follows that the upper triangle (i.e. $monad_3(X)$) commutes.
□
Now,to deduce the associativity law from $monad_1$ we shall need some topological arguments expressing that TX is (in a topological sense) a dense subset of $T^\infty X$. Firstly we need to define the embedding of TX into $T^\infty X$; since $(T, [\eta, \mu])$ and $(T^\infty, [\eta^\infty, \mu^\infty])$ are the respective initial and terminal $\mathcal{F}$-fixed points we know there exists a unique natural transformation $\alpha : T \dot{\longrightarrow} T^\infty$ such that

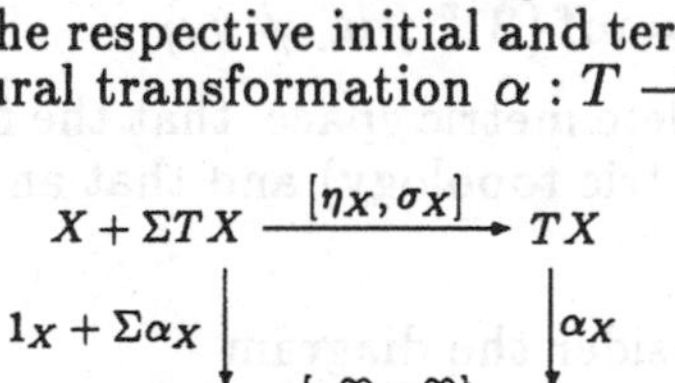

commutes.

In other words, splitting this diagram into two parts, both following diagrams (respectively named $morph_1(X)$ and $morph_2(X)$) commute.

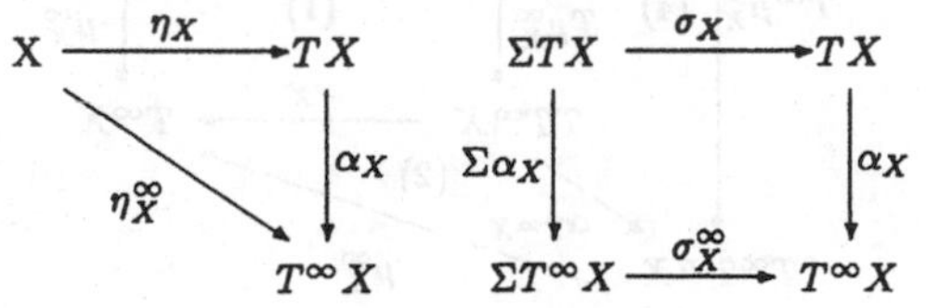

Lemma 2 *The inductive extension* $\nu_X : TT^\infty X \longrightarrow T^\infty X$ *of* $\sigma_X^\infty : \Sigma T^\infty X \longrightarrow$
$T^\infty X$ *is given by* $\nu_X = \mu_X^\infty \circ \alpha_{T^\infty X}$.

Actually both diagrams below commute since (1) is $morph_1(T^\infty X)$, (2) is $morph_2(T^\infty$
(3) is $monad_2(X)$ and (4) is $monad_1(X)$.

$$
\begin{array}{ccc}
T^\infty X & \xrightarrow{\eta_{T^\infty X}} & TT^\infty X \\
 & {}_{(1)} & \downarrow{\scriptstyle\alpha_{T^\infty X}} \\
\eta_{T^\infty X}^\infty \nearrow & & T^\infty T^\infty X \\
1_{T^\infty X} \searrow & {}_{(3)} & \downarrow{\scriptstyle\mu_X^\infty} \\
 & & T^\infty X
\end{array}
\qquad
\begin{array}{ccc}
\Sigma TT^\infty X & \xrightarrow{\sigma_{T^\infty X}} & TT^\infty X \\
{\scriptstyle\Sigma\alpha_{T^\infty X}}\downarrow & {}_{(2)} & \downarrow{\scriptstyle\alpha_{T^\infty X}} \\
\Sigma T^\infty T^\infty X & \xrightarrow{\sigma_{T^\infty X}^\infty} & T^\infty T^\infty X \\
{\scriptstyle\Sigma\mu_X^\infty}\downarrow & {}_{(4)} & \downarrow{\scriptstyle\mu_X^\infty} \\
\Sigma T^\infty X & \xrightarrow{\sigma_X^\infty} & T^\infty X
\end{array}
$$

Definition 5 *A mapping* $f : T^\infty X \longrightarrow T^\infty Y$ *is said* **iterative** *(w.r.t.* T^∞ *) if there
exists a chain morphism* $F : \nabla X \longrightarrow \nabla Y$ *such that* f *is the mediating morphism
between the cone* $F \circ \pi_X : T^\infty X \longrightarrow \nabla Y$ *and the limiting cone* $\pi_Y : T^\infty Y \longrightarrow \nabla Y$.

For example μ_X^∞ is iterative for any set X, $T^\infty f$ is iterative for any mapping $f :$
$X \to Y$; note, moreover that the composite of two iterative mappings is, as well,
iterative.

Then we can state the following lemma which says that two iterative mappings
that agree on finite trees are equal (we call finite trees those elements of $T^\infty X$
which are image of some term through α_X)

Lemma 3 (Density lemma) *if* $f, g : T^\infty X \longrightarrow T^\infty Y$ *are two iterative mappings
such that* $f \circ \alpha_X = g \circ \alpha_X$ *then* $f = g$.

Sketch of proof : we recall that the elements of $T^\infty X$ are the sequences $(u_n)_{n \in \mathbb{N}}$
such that $u_n \in \mathcal{F}_X^n 1 = (\nabla X)_n$ and for every integer n, $u_n = \psi_n(u_{n+1})$; we supply
$T^\infty X$ with an ultrametric distance as follows :

$$ d(u, v) = \inf\{2^{-n} \; ; \; u_n \neq v_n\} $$

We verify that $(T^\infty X, d)$ is a complete metric space, that the finite trees constitute
a dense subset of $T^\infty X$ (for the metric topology) and that an iterative mapping is
continuous for that topology. $\square$

Proof of the proposition : Consider the diagram

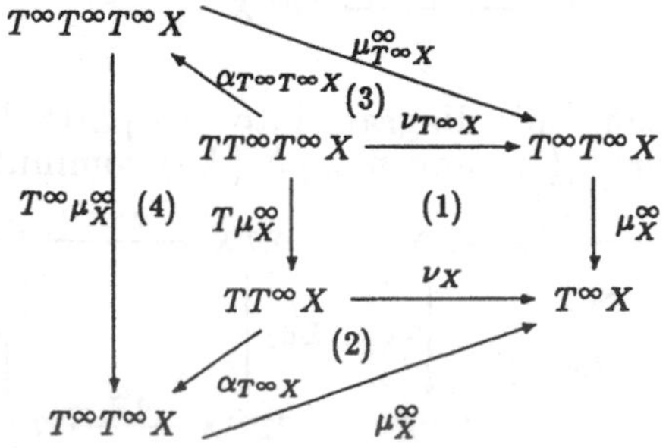

monad$_1(X)$ told us that $\mu_X^\infty : (T^\infty T^\infty X, \sigma_{T^\infty X}^\infty) \longrightarrow (T^\infty X, \sigma_X^\infty)$ is a morphism of Σ-algebras ; thanks to the Eilenberg-Moore comparison functor $\phi : \Sigma\text{-}\mathbf{Alg} \longrightarrow \mathbf{Set}^{\mathbf{T}}$ we know that μ_X^∞ is, as well, a morphism of **T**-algebras between the corresponding inductive extensions i.e. (1) commutes. (2) and (3) commute thanks to lemma 2 and (4) commutes because α is a natural transformation. Then it follows :

$$(\mu_X^\infty \circ \mu_{T^\infty X}^\infty) \circ \alpha_{T^\infty T^\infty X} \quad = \quad (\mu_X^\infty \circ T^\infty \mu_X^\infty) \circ \alpha_{T^\infty T^\infty X}$$

Since $\mu_X^\infty \circ \mu_{T^\infty X}^\infty$ and $\mu_X^\infty \circ T^\infty \mu_X^\infty$ are iterative mappings thanks to the density lemma they are equal. Which completes the proof □

We recall that a morphism of monads $\alpha : (T, \eta, \mu) \longrightarrow (T', \eta', \mu')$ is a natural transformation $\alpha : T \dot{\rightarrow} T'$ preserving the structure i.e. verifying

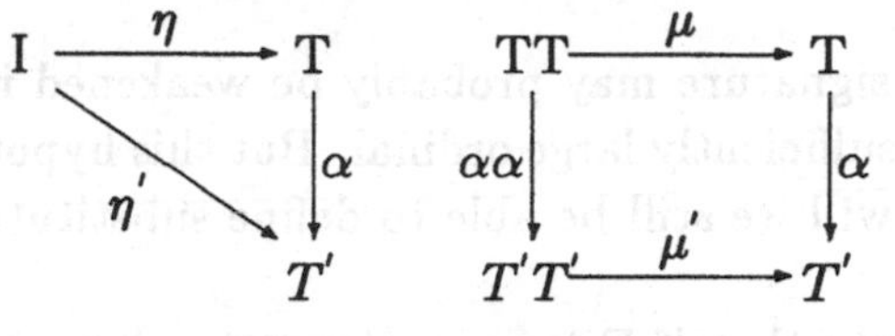

where $\alpha\alpha = \alpha T' \circ T\alpha = T'\alpha \circ \alpha T$ is the vertical composition of natural transformations . Now we can state the following

Proposition 5 *$\alpha : T \dot{\longrightarrow} T^\infty$ is a monad morphism.*

Proof :

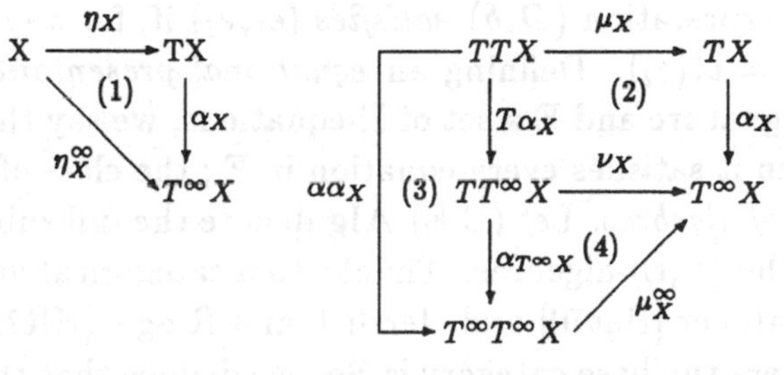

(1) is *morph*$_1(X)$, (2) follows from *morph*$_2(X)$ thanks to the Eilenberg Moore comparison functor, (3) is the definition of the vertical composition of natural transformations and (4) is lemma 2. □

To end this section, we prove that the monad of infinite trees is iterative which means, in analogy with Elgot's terminology, that all non degenerated equations have a unique solution. To make this precise, we first give some definitions. Let $\mathbf{T} = (T, \eta, \mu)$ be a monad over **Set**, for a given valuation $v : X \rightarrow TY$ we let $v^* = \mu_Y \circ Tv : TX \rightarrow TY$. $v^*(t)$ is the term t in which each occurrence of a variable $x \in X$ has been replaced by its value $v(x) \in TY$; so we shall sometimes denote $v^*(t)$ as t[v]. If $x \in X$ and $u \in TY$ let v=(u/x) denote the valuation $v : X \rightarrow TY$ such that v(x)=u and v(y)=$\eta(y)$ for $y \neq x$.

Definition 6 *A monad* $\mathbf{T} = (T, \eta, \mu)$ *is said to be* **algebraically closed** *whenever* x *is a variable in* X, t *a term in* TX *and* $t \neq < x >$ *there exists a term* $v \in T(X \backslash \{x\})$ *such that* $t[v/x] = v$. *If moreover this solution is unique* $\mathbf{T}$ *is said to be* **iterative.**

Proposition 6 *For every* ω*-signature* Σ, *the monad* $(T^\infty, \eta^\infty, \mu^\infty)$ *of infinite trees over that signature is iterative.*

Sketch of proof : with the hypothesis of the above definition, the mapping $\mathbf{subs}_{x,t} : T^\infty X \to T^\infty X$ defined by $\mathbf{subs}_{x,t}(v) = t[v/x]$ is contractive and then admits, thanks to Banach's fixed point theorem, a unique fixpoint in $T^\infty X$. □

5 Conclusion

The restriction to ω-signature may probably be weakened if, in our construction, we admit chains of a sufficiently large ordinal. But this hypothesis was essential for the density lemma ; will we still be able to define substitution for infinite objects in that case ?

To conclude we note that if Σ is an ω-signature, we can deal with equational varieties of Σ-algebras defined as follows. Let (D,δ) be a **T**-algebra (where **T** is the term monad corresponding to Σ) and $v : V \to D$ be a mapping (called a valuation). Let $v^*_\delta = \delta \circ Tv : TV \longrightarrow D$, it assigns each term in TV to its *value* in the *valued interpretation* (D,δ;v). Now let a Σ-*equation* be a pair (e_1, e_2) of elements in TV, we say that an interpretation (D, δ) *satisfies* (e_1, e_2) if, for any valuation v: V $\to$ D, one has $v^*_\delta(e_1) = v^*_\delta(e_2)$. Defining an *equational presentation* as a pair (Σ,E) where Σ is an ω-signature and E a set of Σ-equations, we say that a Σ-algebra is a (Σ,E)-algebra when it satisfies every equation in E ; the class of all (Σ,E)-algebras is called a *variety of algebras.* Let (Σ,E)-**Alg** denote the full subcategory of Σ-**Alg** corresponding to the (Σ,E)-algebras. Thanks to a categorical version of Birkhoff's theorem due to Hatcher [Hat70] and Herrlich and Ringel [HR72] used here in the particular case where the base category is **Set** we deduce that the forgetful functor from (Σ,E)-**Alg** to **Set** admits a left adjoint and the category of (Σ,E)-algebras is isomorphic as a category of sets with structure to the category of $\mathbf{T}_{\Sigma,E}$-algebras where $\mathbf{T}_{\Sigma,E}$ is the monad induced by this adjunction. And moreover we are able to characterize the full subcategory of $\mathbf{T}^\infty$-algebras verifying a set E of Σ-equations as an equational variety (of $\mathbf{Set}^{\mathbf{T}^\infty}$) by intersecting (a pull-back construction) the two subcategories (Σ, E)-**Alg** and $\mathbf{Set}^{\mathbf{T}^\infty}$ of Σ-algebras. For example we can describe a variety of process algebras by mean of equations over finite synchronization trees and consider the free algebras corresponding to that set of axioms. We feel that an interesting outcome of the construction described in the present paper is to set up a link between the models (e.g. process algebras) and the construction of domains those models are based upon.

Acknowledgements : I would like to thank Philippe Darondeau for his comments on earlier drafts of this paper.

References

[Ala75] Suad Alagić. Natural state transformations. *JCSS*, 10, 1975.

[AM75] M. Arbib and E. Manes. A categorist's view of automata and systems. *Lecture notes in computer science*, 25, 1975.

[BG87] D.B. Benson and I. Guessarian. Algebraic solutions to recursion schemes. *JCSS*, 35, 1987.

[BW85] M. Barr and C. Wells. *Toposes, triples and theories.* Springer Verlag, 1985.

[dBZ82] J.W. de Bakker and J.I. Zucker. Processes and the denotational semantics of concurrency. *Information and Control*, 54, 1982.

[Hat70] W.S. Hatcher. Quasiprimitive categories. *Math. Ann.*, 190, 1970.

[HR72] H. Herrlich and C.M. Ringel. Identities in categories. *Canadian Math. Bull.*, 15, 1972.

[Law63] F. W. Lawvere. Functorial semantics of algebraic theories. *Proc. Nat. Acad. Scien. USA*, 50(5), 1963.

[Man76] E. G. Manes. *Algebraic theories.* Springer Verlag, 1976.

[Mil80] Robin Milner. *A calculus of communicating systems.* Springer Verlag LNCS, n 92, 1980.

[MM88] J. Meseguer and U. Montanari. Petri nets are monoids. *Comp. Sci. Lab. SRI international*, 1988.

[Sch72] Horst Schubert. *Categories.* Springer Verlag, 1972.

[SP82] Smyth and Plotkin. The category-theoretic solution of recursive domain equations. *Siam J. Compt.*, 11(4), 1982.

The Subsequence Graph of a Text

(Preliminary version)

Ricardo A. Baeza-Yates
Data Structuring Group
Department of Computer Science
University of Waterloo
Waterloo, Ontario, Canada N2L 3G1 *

Abstract

We define the directed acyclic subsequence graph of a text as the smallest deterministic partial finite automaton that recognizes all possible subsequences of that text. We define the size of the automaton as the size of the transition function and not the number of states. We show that it is possible to build this automaton using $O(n \log n)$ time and $O(n)$ space for a text of size n. With this structure, we can search a subsequence in logarithmic time. We extend this construction to the case of multiple strings obtaining a $O(n^2 \log n)$ time and $O(n^2)$ space algorithm, where n is the size of the set of strings. For the later case, we discuss its application to the longest common subsequence problem improving previous solutions.

1 Introduction

Given a text, a subsequence of that text is any string such that its symbols appear somewhere in the text in the same order. Subsequences arise in data processing and genetic applications, being the longest common subsequence problem (LCS) the most important problem. They are used in data processing to measure the differences between two files of data, and in genetic research to study the structure of long molecules (DNA).

The first interesting question to answer, is the membership problem. That is, if a given string is a subsequence of another string. This can be expressed as a regular expression (see [1] for the standard notation). For example, if the

***This work was supported by the Institute of Computer Research of the University of Waterloo and by the University of Chile.**

subsequence is $x_1 x_2 \cdots x_r$, and t is the text, then the problem may be expressed as

$$t \in \theta^* x_1\, \theta^* x_2\, \theta^* \cdots\, \theta^* x_r\, \theta^* \ ?$$

where θ is the don't care symbol and $*$ the star operator or Kleene closure. Clearly, we can answer this question in linear time. However, we are interested in answer this question in optimal time, by allowing the text to be preprocessed.

A natural question is which is the size of the deterministic finite automaton that given a text, recognizes any possible subsequence of that text. We allow the automaton to be *partial*, that is, each state need not to have a transition on every symbol. As all the states of this automaton are accepting, it can be viewed as a directed acyclic graph, which we call the Directed Acyclic Subsequence Graph (DASG). This problem is analogous to the Directed Acyclic Word Graph (DAWG) in where we are interested in subsequences instead of subwords [3].

In section 2 we introduce the DASG, and in section 3 we show how to build it in $O(n \log n)$ time and space for arbitrary alphabets, and in $O(n \log |\Sigma|)$ time and space for finite alphabets, where Σ denotes the alphabet. With this structure, we can test membership in $O(|s| \log n)$ time for arbitrary alphabets and $O(|s|)$ time for finite alphabets, where s is the subsequence that we are testing. One interesting thing to point out is that the DAWG recognizes all possible $O(n^2)$ subwords using $O(n)$ space, while the DASG recognizes all possible 2^n subsequences using $O(n \log n)$ space. In section 4 we show that is possible to reduce the space required to $O(n)$, but having a $O(|s| \log n)$ searching time for any alphabet.

In section 5 we extend the DASG to the case of multiple strings and we use it to solve the longest common subsequence problem and variations of it [2]. Our algorithm improves upon previous solutions of this problem for more than two strings, running in time $O(n^2 \log n)$ using $O(n^2)$ space. Previous solutions to the general case used $O(n^L)$ time and space for L strings [7] by using dynamic programming, or $O(n^3)$ time and $O(n^2)$ space [6] using an approach similar to the one developed in this paper.

2 Building the DASG

We can define the DASG recursively in the size of the text. The DASG of a text of size n must recognize all possible subsequences of the last $n-1$ symbols of the text, and all possible subsequences that start with the first symbol. As a regular expression this is:

$$S_n = (\epsilon + t_1) S_{n-1} \ \text{ and } \ S_0 = \epsilon$$

where ϵ is the empty word and $t = t_1 t_2 \cdots t_n$ is the text. The size of the regular expression S_n is linear on n, and so is the non-deterministic finite automaton equivalent to S_n. Suppose that all the symbols of the text are different. The "deterministic" version of S_n for this case is

$$S_n = \epsilon + t_1 S_{n-1} + t_2 S_{n-2} + \cdots + t_n S_0$$

Clearly, the size of S_n is $O(n^2)$. Figure 1 shows the DASG for the text *abcd*. This automaton has $n+1$ states (all of them are final states) and $n(n+1)/2$ transitions. The number of states is minimal because we have to recognize the complete text (the longest subsequence). The number of edges (given the minimal set of states) is also minimal, because in the position i of the text we have to recognize any subsequence starting with t_j for $j = i+1, ..., n$. It is not difficult to generalize this for the case of repeated symbols.

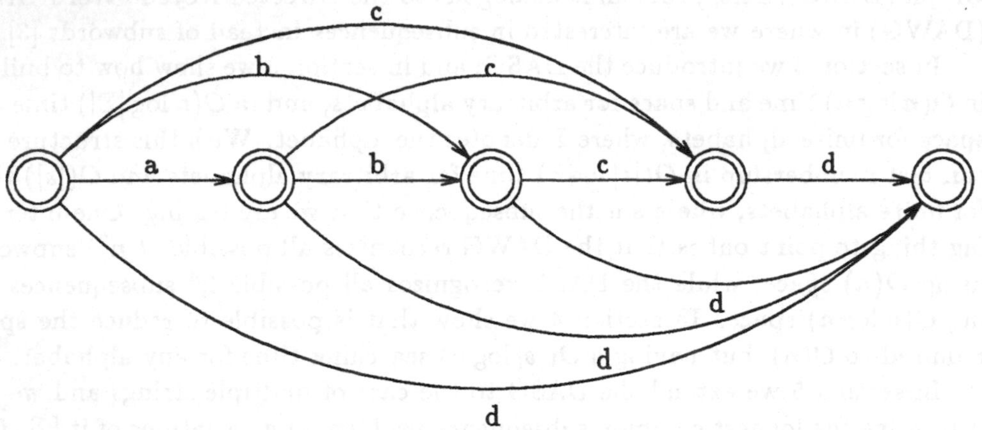

Figure 1: Minimal state DASG for the text *abcd*.

Definition: Let Σ be the alphabet. We define the effective size of Σ as $c = \min(|\Sigma|, n)$.

To build the DASG in $O(cn)$ time and space we use an incremental algorithm scanning the text from the *right to the left*. At each step we maintain a dictionary that contains all different symbols of the already scanned text, and the state in which the first skeleton transition labeled with that symbol appears. Hence, the algorithm is

1. Create state s_n and create an empty dictionary D.

2. For each symbol in the text t_i scanning from the right to the left do:

   (a) Create state s_{i-1}.

(b) Insert the pair $(t_i, i-1)$ in D. If t_i is already in D, its associated state is updated to $i-1$.

(c) For each symbol in D (d_j), append a transition labeled with d_j to state s_{k+1}, where k is the state associated to d_j in D.

Step (a) takes constant time. The insertion, step (b), can be performed in $\log c$ time, because the size of D is $O(c)$. For the same reason, step (c), the traversal of D takes $O(c)$ time. The cycle is performed n times. Then, the total time is $O(n(c + \log c))$. If we apply the same algorithm scanning the text from the left, we obtain the DASG of the reversed text. For this DASG, we can test the membership of a subsequence s using s reversed.

A membership query in this DASG takes $O(|s| \log c)$ time, where the $\log c$ term is the time to search for the appropriate transition in each state. Using a complete table for small alphabets, a $O(|s|)$ worst case time is achieved. For larger alphabets, we can obtain $O(|s|)$ average time by using hashing.

3 The Smallest Automaton

It is possible to reduce the time and space requirements? The answer is yes. The main problem is that the number of edges is $O(n^2)$ while the number of states is linear. Here we are not interested in the minimal set of states, we are interested in minimal space and that means a *minimal number of edges*. In other words, the *smallest transition function* for the automaton. To the best of our knowledge, this is first time that such concept is given.

Definition: The smallest deterministic partial finite automaton A that recognizes the regular language $L(r)$ defined by the regular expression r, is such that does not exist other automaton that recognizes $L(r)$ with less transitions than A.

We shall show that minimal number of states it is not, in general, equivalent to the smallest automaton. In [3] is claimed that the DAWG is the smallest automaton that recognizes all the subwords of a text. However, they show that is the smallest in the sense of minimal number of states. Intuitively, the DAWG may be the smallest automaton, because the number of states and the number of edges only differ in $n + O(1)$. In our problem, it is not the case, and we introduce a method that we call *encoding*, since it basically encodes the alphabet used.

To achieve the previous goal we will balance the number of states and the number of edges. For that we encode each symbol using $k < c$ digits. This means $\log_k c$ digits per symbol. Hence, our skeleton will have $O(n \log_k c)$ states, each one with at most k edges. Then, the total space is $O(nk \log_k c)$.

Intuitively, what happens is that the encoding permits to share transitions. We can see this by noting that the skeleton representing a symbol has k transitions times all the transitions of a skeleton one state short. That is

$$T_s = kT_{s-1}$$

and $T_1 = k$. But the length of the skeleton for each symbol is $\log_k c$. Thus, $T(\log_k c) = k^{\log_k c} = c$ different transitions per state. That is, the number of transitions per state in the $O(cn)$ DASG. Note that each transition in the previous version of the DASG, is simulated by the encoded DASG in $O(\log c)$ steps.

The optimal choice for k is 3. However, for practical obvious reasons we want an integer power of two. In that case, the best integer choices are 2 and 4. Thus, using $k = 2$ (typically most inputs are already encoded in binary) we have at least 2 edges per state and $n\lceil\log_2 c\rceil + 1$ states. Of these states, $n + 1$ are final. However, we do not have to distinguish them, because any input must be of length multiple of $\lceil\log_2 c\rceil$. This leads to the following theorem:

THEOREM **3.1** *The smallest deterministic partial finite automaton that recognizes all possible subsequences of a text of size n over an alphabet of effective size c, has at most* $n\lceil\log_2 c\rceil + 1$ *states and at most* $(2n - (\lceil\log_2 c\rceil + 1)/2)\lceil\log_2 c\rceil$ *transitions.*

Proof: It is only necessary to prove the result in the number of edges. Clearly, any state has at most 2 edges. However, the last state has no transitions and the previous $\lceil\log_2 c\rceil$ states only can have 1 transition because they represent the last symbol. For the same reason, the skeleton representing the symbol $n - i$ has at most i states with 2 transitions for any $i \leq \lceil\log_2 c\rceil$. ∎

These upper bounds can be slightly improved using $k = 3$. This result is optimal, because the length of the encoded text is $O(n \log c)$, and then we need at least $O(n \log_2 c)$ transitions to recognize the complete text (the longest subsequence).

Figure 2 shows the encoded version for the text *abcd*. This DASG does not have less transitions that the one presented in Figure 1. However, this only happens for small n or periodic strings (for example a^n).

Again, to construct this version of the DASG, we use an incremental algorithm scanning from the right to the left. Now, we need two auxiliary structures. One that given a symbol tell us its encoding (encoding dictionary/function) and another that given a prefix of a symbol code, returns the position of the first symbol (in the previously scanned text) with that prefix (analogous to the D dictionary of the previous algorithm). For the last data structure we use a binary

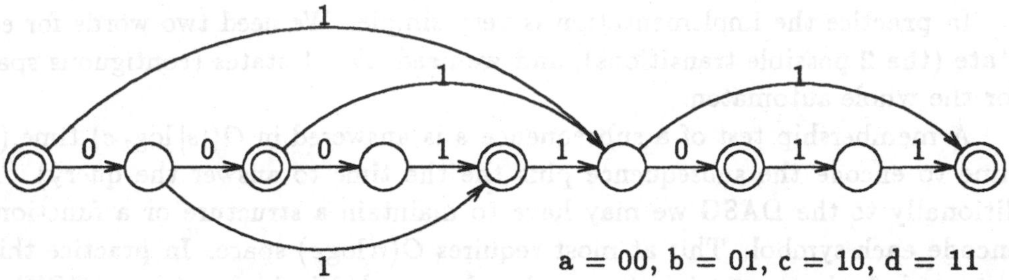

Figure 2: DASG for the text *abcd* (encoded).

trie (for example a Patricia tree [8]), where in each node we store the position (state) needed. Let b be $\lceil \log_2 c \rceil$. If c is not known in advance, we may use $c = n$ or we compute it using $O(n \log c)$ time. The detailed steps of the algorithm are:

1. Create state s_{nb+1} and create an empty binary trie D.

2. For each symbol in the text t_i scanning from the right to the left do:

   (a) Set the root as the actual position in D.

   (b) Create state $s_{(i-1)b}$.

   (c) Encode t_i.

   (d) For every bit x_j (0 or 1) in the encoding of t_i do:

      i. Create state $s_{(i-1)b+j}$ if $j < b$.

      ii. Append a transition labeled x_j between states $s_{(i-1)b+j-1}$ and $s_{(i-1)b+j}$.

      iii. If the $\overline{x}_j$ (complement of x_j) child of the current trie node exist, append a transition labeled $\overline{x}_j$ from state $s_{(i-1)b+j}$ to state k where k is the state stored in the child.

      iv. Set the x_j child of the current trie node as the new position in D and update its value (state) to $(i-1)b + j$. If the child does not exist, we create it.

All the steps in the internal loop takes constant time, and the internal loop is repeated nb times. Hence, the total time is $O(n \log c)$. The extra space is $O(c \log c)$ for the trie and $O(n \log c)$ for the encoding structure (if we do not have a function or table). This leads to the following theorem:

THEOREM 3.2 *It is possible to construct the DASG of a text of size n using $O(n \log n)$ worst case time and space for arbitrary alphabets, and using $O(n \log |\Sigma|)$ worst case time and space for finite alphabets.*

In practice the implementation is very simple. We need two words for each state (the 2 possible transitions), and we need $nb + 1$ states (contiguous space) for the whole automaton.

A membership test of a subsequence s is answered in $O(|s| \log_2 c)$ time (the time to encode the subsequence plus the the time to answer the query). Additionally to the DASG we may have to maintain a structure or a function to encode each symbol. This at most requires $O(n \log c)$ space. In practice this is not needed, since most inputs are already encoded in binary (e.g. ASCII). By keeping all the states visited during the search we can obtain where the subsequence started and where it finished.

The previous result proves the following (almost obvious) lemma:

Lemma 3.1 *The minimal state (partial) DFA and the minimal transition (partial) DFA are not equivalent.*

The lemma is also true for non-partial DFAs because the space complexities for our problem are the same for this case. The meaning behind this lemma is that to share part of a transition function in 2 "similar" states we need additional states. Encoding is one technique to share states. However, it is possible that the general problem of finding the smallest transition function is NP-complete based on related problems presented in [4,5]. Further research is being done in this problem and in local techniques to minimize space in finite automata.

The next lemma gives a necessary condition to have an encoding that may reduce the size of the automaton:

Lemma 3.2 *Given a minimal state partial DFA with s states, where s_0 of them do not have outgoing transitions, and t transitions, then encoding may reduce the size of the automaton only if $t > 2(s - s_0)$.*

Proof: If we apply encoding, each state is at least transformed in 2 states. That means that the number of transitions of the automaton of the encoded text is at least $2(s - s_0)$ transitions, because each new state must have at least one transition, $s - s_0$ of the original number of states also must have one transition and it is not necessary to encode symbols representing states without transitions. Hence, the new automaton may have less transitions if $t > 2(s - s_0)$. ∎

For example, any DAWG such that $t \leq 2s - 2$ ($s_0 = 1$ for this case) cannot be reduced using encoding. We have not found a single example where $t > 2s - 2$ for a DAWG. Based in the results presented in [3] we know that $t < 3s - 6$.

4 A Linear Space Representation

In section 3 we showed that we can transform the DASG of $O(n^2)$ transitions and $O(n)$ states, to a DASG with $O(n \log n)$ transitions and states. In this section we will describe how to simulate the $O(n^2)$ space DASG using only $O(n)$ space, but $\log n$ time per transition, independently of the alphabet size.

Instead of representing the transitions for each state, we will store all the states associated to the transitions of a given symbol. Let enumerate the states in the DASG defined in section 2 from 0 to n, or in other words, by using the position of each symbol in the text. For each symbol x we store, in order, all states s such that

$$\delta(i, x) = s$$

for any state i (in fact, $i < s$), where δ is the transition function. That is, we store all the positions in the text in where x appears. Let S_x be the ordered list of positions associated to x. To simulate $\delta(i, x)$, we look in S_x for the minimum state s in S_x such that $s > i$. Because the list is ordered, this takes $O(\log n)$ time (a sorted array suffices). To know where S_x is, we use an auxiliary index that tells us this information for each x.

Because there are n positions in the text, the space necessary for all the ordered lists is $O(n)$. The time necessary to construct this representation is $O(n \log n)$ to sort the lists, and $O(n \log c)$ to build the auxiliary index and to lookup all the symbols. This leads to the following theorem:

THEOREM **4.1** *It is possible to construct an implicit representation of the DASG of a text of size n using $O(n)$ space and $O(n \log n)$ worst case time, in where each transition is simulated in $O(\log n)$ steps.*

To test membership of a subsequence s, we need $O(|s| \log c)$ time to lookup each symbol, and $O(|s| \log n)$ time to simulate the transitions. That is, $O(|s| \log n)$ time, regardless of the alphabet size. Therefore, for finite alphabets we tradeoff space for search time. Table 1 shows a summary of the space and time complexities.

5 The DASG for a Set of Strings

Now we want to solve the following problem: Is a given string a subsequence of a string in a set of strings? Again, we can express the problem as a regular expression. To do this, we need first some additional notation.

Let S be a set of L strings, and s_i be the ith string of the set. We assume that no string is a subsequence of any other string (this implies that at least there are

DASG	Space	Searching time	Building time
Section 2	nc	$\|s\|$	nc
Section 3	$n \log c$	$\|s\| \log c$	$n \log c$
Section 4	n	$\|s\| \log n$	$n \log n$

Table 1: Summary of time and space complexities

two different symbols in S). Let $n = \sum_{i=1}^{L} |s_i|$ be the total number of symbols. Let $T(S)$ be the set of distinct symbols in S ($2 \leq |T(S)| \leq c = \min(|\Sigma|, n)$).

Definition: We define (as in [6]) a *matched point* of S as a j-tuple of pairs $([i_1, p_1], [i_2, p_2] ..., [i_j, p_j])$ ($1 \leq j \leq L$) which denotes a match of a symbol at positions p_1 in string s_{i_1}, p_2 in string s_{i_2}, ..., p_j in string s_{i_j}. A matched points is *maximal*, if the symbol matched does not appear in the $L - j$ remaining strings.

For example, all the maximal matched points for $S = \{aba, aab, bba\}$ are

$$([1,1],[2,1],[3,3]), ([1,1],[2,2],[3,3]), ([1,2],[2,3],[3,1]),$$

$$([1,2],[2,3],[3,2]), ([1,3],[2,1],[3,3]), \text{ and } ([1,3],[2,2],[3,3]).$$

Definition: We define the *initial maximal matched point* ($IM(S, x)$) in the set S for a given symbol x as the smallest maximal matched point (in a lexicographical sense) that matches x. That is, the maximal matched point with the smaller position p_i in each string that belongs to the matched point.

For the previous example, $IM(S, a)$ is $([1,1],[2,1],[3,3])$ and $IM(S, b)$ is $([1,2],[2,3],[3,1])$.

We denote by $R(S,$ *matched point*$)$ (*right set*) the set of non null substrings that are to the right (higher positions) of a matched point (we also eliminate any substring that is a subsequence of other substring). For the previous example, $R(S, IM(S, a)) = \{ba, ab\}$. Now, the regular expression that defines all possible common subsequences of S is recursively defined by

$$Subseq(S) = \sum_{t_i \in T(S)} t_i Subseq(R(S, IM(S, t_i)))$$

and $Subseq(\emptyset) = \epsilon$. This definition generates the subsequence automaton, and then allow us to count the number of states and edges needed by this automaton:

$$States(S) \leq 1 + \sum_{t_i \in T(S)} States(R(S, IM(S, t_i)))$$

and
$$Edges(S) \leq |T(S)| + \sum_{t_i \in T(S)} Edges(R(S, IM(S, t_i)))$$

Both results are not equalities, because identical right sets may appear (duplicated partial results). An example is given in Figure 3.

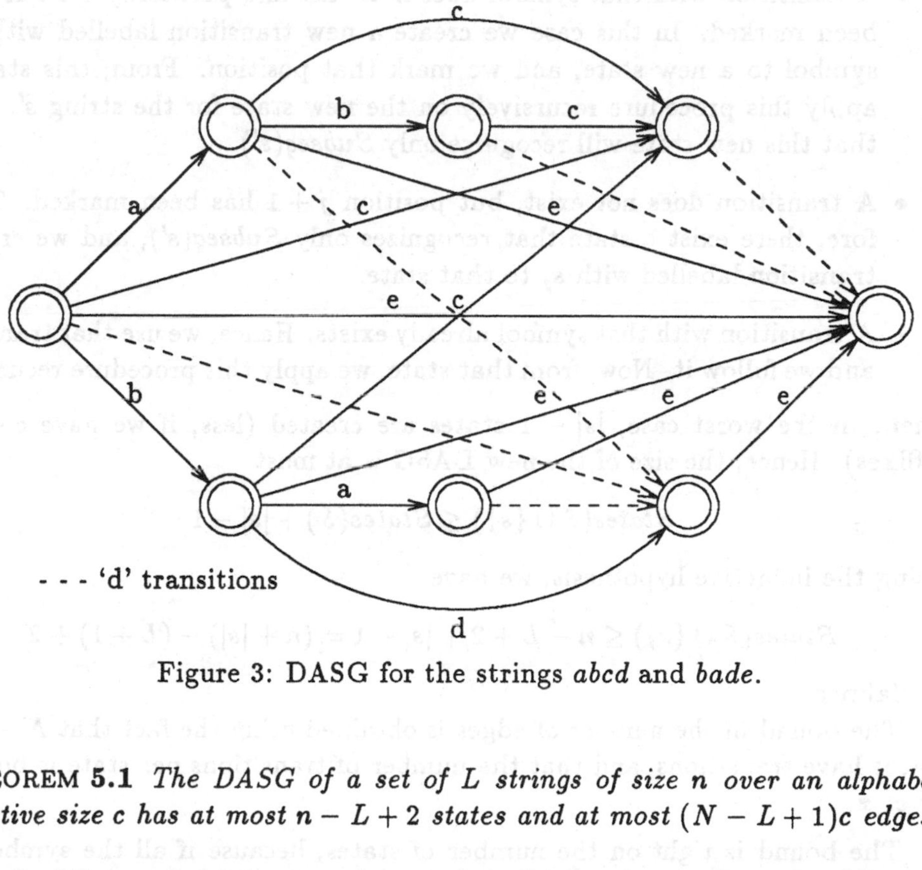

Figure 3: DASG for the strings *abcd* and *bade*.

THEOREM **5.1** *The DASG of a set of L strings of size n over an alphabet of effective size c has at most $n - L + 2$ states and at most $(N - L + 1)c$ edges.*

Proof: We use induction on the number L of strings. From section 2, the theorem is true for $L = 1$ ($n + 1$ states are necessary and sufficient).

Now, we will see what happens when we try to include a new string s in a DASG of a set S of L strings of size n. We will show that for each position in s (except one) we need to create at most one state. If we create a state for a transition labelled with s_j, we mark that position j in the string. We show that if position j has been marked, then there exists a state that recognizes $Subseq(s')$ and nothing else, where $s' = s_{j+1}...s_{|s|}$. Note that the last position will be never marked, because $Subseq(\epsilon)$ exists already in the DASG of S (last final state or *sink* state).

Then, for each position j in s (the order is not important) we need a transition from the initial state labelled with that symbol (s_j). For the last position, if there is no transition from the initial state labelled with that symbol, we create a transition from the initial state to the sink state. For the other positions, we have three cases:

- A transition with that symbol does not exist and position $j+1$ has never been marked. In this case we create a new transition labelled with that symbol to a new state, and we mark that position. From, this state we apply this procedure recursively on the new state for the string s'. Note, that this new state will recognize only $Subseq(s')$.

- A transition does not exist, but position $j+1$ has been marked. Therefore, there exist a state that recognizes only $Subseq(s')$, and we create a transition labelled with s_j to that state.

- A transition with that symbol already exists. Hence, we use that transition, and we follow it. Now, from that state, we apply this procedure recursively.

Then, in the worst case, $|s|-1$ states are created (less, if we have common suffixes). Hence, the size of the new DASG is at most

$$States(S \cup \{s\}) \leq States(S) + |s| - 1$$

Using the inductive hypothesis, we have

$$States(S \cup \{s\}) \leq n - L + 2 + |s| - 1 = (n + |s|) - (L+1) + 2$$

as claimed.

The bound in the number of edges is obtained using the fact that $N - L + 1$ states have transitions, and that the number of transitions per state is bounded by c. ∎

The bound is tight on the number of states, because if all the symbols are different, $n - L + 2$ states and $O(n^2/L)$ transitions are needed.

For this case, it is not possible to use the encoding technique of the previous section. If not, it would be possible to solve the LCS problem ($L = 2$) in $O(n \log n)$ comparisons for an arbitrary alphabet. This is a contradiction with the $O(n^2)$ lower bound in the comparison model presented in [2]. Therefore, the size of the DASG must be $O(n^2)$ for this case. In fact, the encoding technique fails for this case, because now we have more than one skeleton.

The only structure that resembles our automaton is the *ICS* tree of Hsu and Du [6], which is used to solve the LCS problem for a set of strings. In that case, only matched points between all the strings are considered.

Now, we will present the main ideas behind the algorithm that builds the DASG for this case. The algorithm must find, very efficiently, all possible different symbols at any state. For this, we first build the DASG (first version given) for each individual string. With this, for each string, we can find all different symbols after a given position. The algorithm recursively generates states until all possible symbols belongs to one string (or there are no more symbols left in that position). After that, the individual DASG is used. To keep track of how much of this DASG we have used we have a list of L positions D that indicates from where the DASG of each string is already available.

To find if a right set has been already generated we need two structures. First, a structure that maps common suffixes to one representative. For this we use a suffix tree of the strings (using $O(n)$ space and time). Second, to remember all the right sets (partial results) we use a dictionary that given a right set tell us where it is or if does not exist.

The algorithm is:

1. Create the last state F.
2. Create the right set remember dictionary, inserting the empty right set and its associated state F.
3. Initialize $D_j = |s_j|$ for $j = 1, ..., L$.
4. Set up the table of representatives.
5. Create the DASG for each string ($DASG_j$) using F as common last state.
6. Call *Merge* with pairs $(j, 1)$ for $j = 1, ..., L$. *Merge* will return the initial state.
7. Remove the first $D_j - 1$ states of each $DASG_j$.

The procedure *Merge* does almost all the work, merging the strings from position i_j for all j in the set of pairs P. Namely:

1. If $|P| = 0$ then return state F (no symbols left).
2. Else look up on the right set remember structure. If it is there, return the appropriate state.
3. Else if $|P| = 1$ then we can use the individual DASG (or a copy of it). Let j be the position different to 0. Return state i_j of $DASG_j$ and if $i_j < D_j$, set i_j as the new value of D_j.

4. **Otherwise**
   (a) Create a new state N
   (b) Insert P in the right set remember structure, and its associated state N.
   (c) Look for all different symbols in state i_j of $DASG_j$ for all values of j in P. For each new symbol x, create a transition labeled by x between state N and the state returned from $Merge$ called with pairs Q, where Q is defined as the pairs in P with all the positions updated to $\delta(i_j, x)$, where δ stands for the transition function. If $\delta(i_j, x)$ does not exist, or is the state F, we remove the pair corresponding to string j.
   (d) Return state N.

Step (b) takes time $O(\log States(S)) = O(L \log n)$ and we need a bit of care in step (c). To look for all different symbols defined by the set of positions P, we assume that there is a lexicographical order between the symbols, and that the edges of each $DASG_j$ are ordered. Then, we can look all the edges in time proportional to the number of edges to obtain all possible different symbols. If d is the number of different symbols, then at most Ld edges are inspected. Because d edges will be then generated, the time used is proportional to L for each new state created. At the same time that we inspect the edges, we build the new set of pairs Q. The size of the stack is at most $L \max_i(|s_i|) = O(Ln)$.

The time for each call to merge in the worst case is then $O(L \log n)$. There are as many calls to merge as matches between the strings (or edges created by *Merge*). The construction of the individual DASGs takes time $O(\sum_i |s_i|^2)$ and the construction of the table of representatives takes time and space $O(n)$. Hence, the total time is $O(L|Edges(S)| \log n + n^2/L)$. This leads to the following theorem:

THEOREM **5.2** *It is possible to construct the DASG for a set of L strings using* $O(Ln^2 \log n)$ *worst case time and* $O(n^2)$ *space for arbitrary alphabets, or using* $O(L|\Sigma| n \log n)$ *worst case time and* $O((L + |\Sigma|)n)$ *space for finite alphabets.*

The time to test membership of a subsequence s is in the worst case $O(|s| \log c)$, and $O(|s|)$ for small alphabets by using an array in each state. For arbitrary alphabets, we can achieve $O(|s|)$ average time by using a hashing table.

5.1 An Application: The Longest Common Subsequence

Additionally to fast searching of subsequences in a text or a set of strings, we can also use the DASG to solve the longest common subsequence problem, and some of its variants. For that purpose, we append to each edge (transition) the number of strings that are represented by that edge. Then, to know which is the longest common subsequence problem between $k \leq n$ strings, we search for the longest sequence of edges belonging to k or more strings. The LCS of all the strings is when $k = n$. This LCS can also be computed while the automaton is being built. In all these cases, we find all common subsequences in optimal time.

Hence, the DASG can be used to solve the LCS problem and many variants of it in time $O(n^2 \log n)$ using $O(Ln^2)$ space. This improves over the solution presented in [6] that uses $O(nc + Lc|P|)$ time and $O(nc + |P|)$ space, where $|P|$ is the total number of matched points between all the strings. Because $|P|$ may be as big as $O(n^2)$, this algorithm runs in $O(n^3)$ worst case time for arbitrary alphabets ($O(|\Sigma|n^2)$ for finite alphabets) using $O(n^2)$ space.

6 Concluding Remarks

We define the DASG of a text, giving different algorithms to build it. If the alphabet is known and finite, the DASG presented in section 3 uses only $O(n \log |\Sigma|)$ space and preprocessing time, and $O(|s| \log |\Sigma|)$ searching time for a subsequence s. To achieve this, we have introduced encoding as a technique to reduce the number of transitions in an automaton.

For arbitrary alphabets, the implicit representation of section 4 uses only $O(n)$ space, but $O(|s| \log n)$ searching time.

We used the number of transitions to measure the size of an automaton, and this problem shows that a minimal state automaton is in general not a minimal space automaton.

Remains as open problems the uniqueness of the minimal DASG and the complexity of transition function minimization in a DFA for a general case.

A related problem, is to search for a sequence of substrings. Using a Patricia tree [8], where each internal node have an ordered list of all the positions associated to the corresponding prefix, we can solve this problem in logarithmic time, using $O(n \log n)$ space on average, for a text of size n.

We extended the definition of the DASG to a set of strings, and we use it to solve the LCS problem between those strings, and several variations of it using $O(n^2 \log n)$ time and $O(n^2)$ space improving previous solutions for the case of more than two strings. Other application of this DASG is related to subset membership problems.

Acknowledgements

We wish to thanks the helpful comments of Gaston Gonnet.

References

[1] Aho, A., Hopcroft, J. and Ullman, J. *The Design and Analysis of Computer Algorithms*, Addison-Wesley, Reading, Mass., 1974.

[2] Aho, A., Hirschberg, D. and Ullman, J. "Bounds on the Complexity of the Longest Common Subsequence Problem", *JACM* 23 (1976), 1-12.

[3] Blumer, A., Blumer, J., Haussler, D., Ehrenfeucht, A., Chen, M.T., and Seiferas, J. "The Smallest Automaton Recognizing the Subwords of a Text", *Theoretical Computer Science*, 40 (1985), 31-55.

[4] Garey, M. and Johnson, D. *Computers and Intractability, A Guide to the Theory of NP-Completeness*, Freeman, New York, 1979.

[5] Hopcroft, J. and Ullman, J. *Introduction to Automata Theory, Languages, and Computation*, Addison-Wesley, 1969.

[6] Hsu, W. and Du, M. "Computing a longest common subsequence for a set of strings", *BIT* 24 (1984), 45-59.

[7] Itoga, S. "The string merging problem", *BIT* 21 (1981), 20-30.

[8] Morrison, D. "PATRICIA-Practical algorithm to retrieve information coded in alphanumeric", *JACM 15*, 4 (Oct 1968), 514-534.

SYNTACTICAL PROPERTIES OF UNBOUNDED NETS OF PROCESSORS

J. BEAUQUIER [(1)], A. CHOQUET [(1)], A. PETIT [(1,2)], G. VIDAL-NAQUET [(1,3)]

[(1)] L.R.I. Bât. 490, Université Paris-Sud, 91405 ORSAY FRANCE.
[(2)] L.I.F.O., Université d'Orléans, 45067 ORLEANS FRANCE.
[(3)] Ecole Supérieure d'Electricité, 91190 GIF SUR YVETTE FRANCE.

Abstract. We present a formal description of the logical links in an unbounded net of processors. This description is provided by a finite transducer. We prove some syntactical properties of the net : we give decision algorithms for its coherence and connectivity.

0. Introduction

In almost every paper on nets of processors, the number of considered processors is bounded. For instance, in systolic or neural architectures (cf. [Quinton...86] and [Kohonen...88] for an introduction), the number of processors, even if arbitrarily large, is nearly always considered to be given. Moreover, the topology of the nets is always very simple and regular : lines, grids,...
In this paper, we consider nets, in which the number of processors, although finite, has no known bound. In fact, that involves the assumption that the net is possibly infinite. And we deal with nets of arbitrary (regular) topology.
In such a net, a processor is physically connected to a finite number of other processors in a "crystalline" way. For example, on a plan : each processor is linked to four others ; in three dimensional space, it is linked to six, as shown in the figure 1.

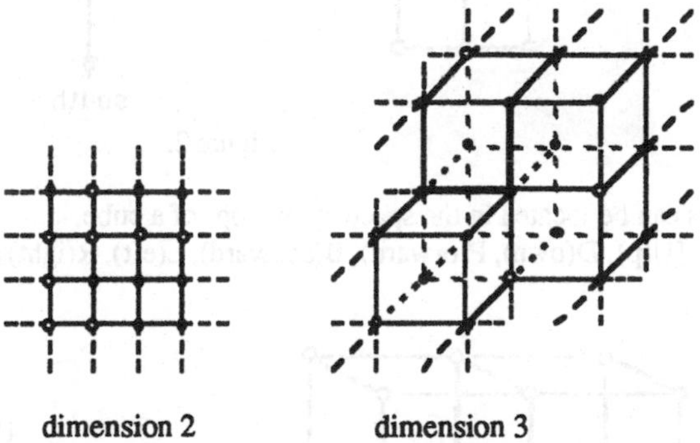

Figure 1.

The aim of this paper is to study some systems in which two processors that are physically connected can be logically connected in different ways. For instance, two processors can be not logically connected, or connected in a synchronous or an asynchronous way. We want to develop a specification allowing to formally prove properties about such systems. The model must take into account the unboundness of the number of processors and, at the opposite, provide a finite specification in order to be operational. The problem is thus to describe a possibly infinite object by a finite mechanism.
The method that we use associates a logical path, i.e. a sequence of logical connections, to each physical path. Such a mechanism is known in the literature as a transducer (cf. [Berstel...79]) and our requirements make that this transducer must be finite. Namely, we associate to a particular point of the net, called the starting point, a finite transducer that maps physical paths starting from him onto logical paths.
For instance, in a two dimensional space, a physical path from a particular point may be described as : north, north, west, north and could be related to a logical path as : (north, synchronous), (north, asynchronous), (west, asynchronous), (north, not connected).

Clearly enough, such a finite transducer does not always specify in an unique way the logical links of a net of processors: since several paths from the starting point can use a same physical link, the logical descriptions of this link can be different according to the considered path. This remark leads to impose some coherence constraints to the specification. We propose some of them and we prove that they can all be decided from the finite transducer.

1. Basic definitions

Throughout the paper, we will assume the classical definitons of formal language theory to be known by the reader. Also the notations are classical (for more details, see e.g. [Berstel...79] or [Hopcroft...69]).

A - Physical and logical links.

We introduce an alphabet of directions. A letter indicates which direction is taken, a word on this alphabet corresponds to a sequence of moves.
Such an alphabet is of size 2q, $\mathcal{D} = \{dir_1,...,dir_{2q}\}$ and for each i, dir_{2i-1} and dir_{2i} will be interpreted later as opposite directions. In other words : if we move in direction dir_{2i-1} and then in direction dir_{2i}, we come back to the start point, and conversely.

Examples.
We will give 2 examples for $\mathcal{D}$:
▷ In plane programming, each processor is located on a point on a plan.
In this case, $\mathcal{D}$ = {N(orth), S(outh), E(ast), W(est)}.

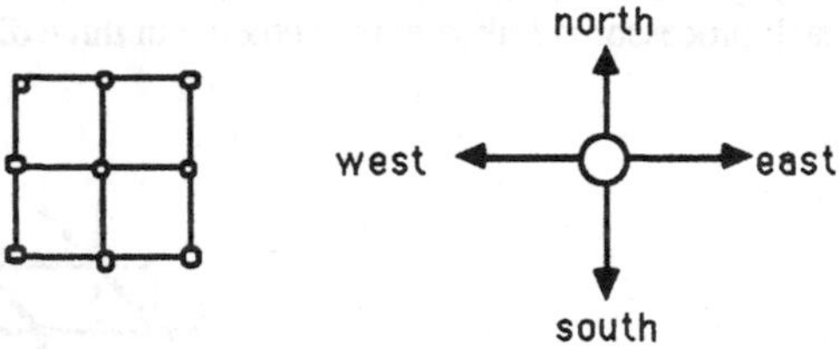

Figure 2.

▷ The processors can be located in the space at the tops of a cube.
In this case, $\mathcal{D}$ = {U(p), D(own), F(orward), B(ackward), L(eft), R(ight)}

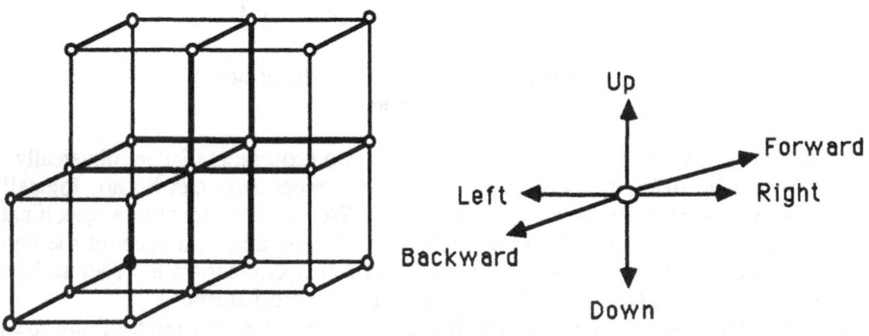

Figure 3.

On the other hand, we have a connexion alphabet $\mathcal{N}$ that describes the nature of the logical links between two processors that are physically connected.
Let us also give two examples for $\mathcal{N}$.
▷ The basic and simplest example is an alphabet of size 2, that just says if the connexion is open or closed : $\mathcal{N}$ = {C(onnected), D(isconnected)}.

▷ We can also make distinctions between the different kinds of connexions. For instance we can have synchronous or asynchronous connexions :
$\mathcal{N}$ = {D(isconnected), A(synchronous), S(ynchronous)}.

To a physical link we want to associate one (at least) logical nature. But there are an infinity of different physical links so this association cannot be done with a table. As stated in the introduction the mapping between physical and logical links is described by a (finite) strictly alphabetic transducer.

Example.
$\mathcal{D}$ = {n, s, e, w}, $\mathcal{N}$ = {c, d} and the transducer is given by the following scheme.

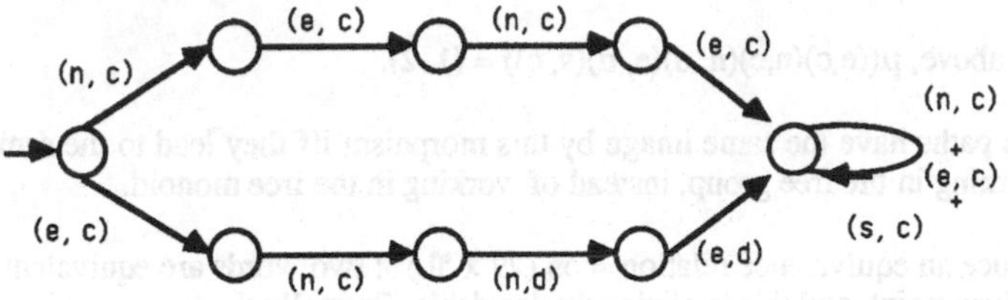

Figure 4.

I t is well known that to give a (finite) strictly alphabetic transducer, which transform a word of $\mathcal{D}^*$ in a word of $\mathcal{N}^*$, is equivalent to give a rational language of $(\mathcal{D} \times \mathcal{N})^*$.

In the example above, the associated rational language is :
$L = \{(n, c)(e, c)(n, c)(e, c) + (e, c)(n, c)(n, d)(e, d)\}\{(n, c) + (e, c) + (s, c)\}^*$.

Throughout this paper, we will quite always consider, instead of the finite strictly alphabetic tranducer, the associated rational language. Moreover, this language will be supposed to be prefix-closed.

B - Some morphisms on languages.

In order to study properties of physical and logical paths induced by this language, we have to introduce some morphisms over $(\mathcal{D} \times \mathcal{N})^*$.

a - The first projection on $\mathcal{D}^*$ gives the physical path that is followed.

$$p_{\mathcal{D}} : \quad (\mathcal{D} \times \mathcal{N})^* \longrightarrow \mathcal{D}^*,$$
$$(dir_1, nat_1)\ldots(dir_r, nat_r) \longrightarrow dir_1\ldots dir_r.$$

For instance, in the example above, $p_{\mathcal{D}}((e,c)(n,c)(n, d)(e, d)(s, c))$ = ennes.

b - The second projection on $\mathcal{N}^*$ give the nature of the connexions found along the followed path.

$$p_{\mathcal{N}} : \quad (\mathcal{D} \times \mathcal{N})^* \longrightarrow \mathcal{N}^*,$$
$$(dir_1, nat_1)\ldots(dir_r, nat_r) \longrightarrow nat_1\ldots nat_r.$$

In the example above, $p_{\mathcal{N}}((e,c)(n,c)(n, d)(e, d)(s, c))$ = ccdds.

c - The Parikh's (or commutative) image [Parikh...66] of this path gives the number of moves in each direction.

$$P : \quad (\mathcal{D} \times \mathcal{N})^* \longrightarrow \mathbb{N}^{2q},$$
$$w \longrightarrow (|p_{\mathcal{D}}(w)|_{dir_1}, \ldots, |p_{\mathcal{D}}(w)|_{dir_{2q}}).$$

In the example above, P((e,c)(n,c)(n, d)(e, d)(s, c)) = (2, 1, 2, 0). Two words have the same Parikh's ima when their projections on $\mathcal{D}$ are permutations of each other.

d - The morphism μ gives the coordinates of the point in $\mathbf{Z}^q$ reached after a path w. It reduces the Parikh image by eliminating each occurrence of any direction that is counterbalanced by an occurrence of opposite direction.

$$\mu: \quad (\mathcal{D} \times \mathcal{N})^* \longrightarrow \mathbf{Z}^q,$$

$$w \longrightarrow (|p_{\mathcal{D}}(w)|_{dir_1} - |p_{\mathcal{D}}(w)|_{dir_2}, \ldots, |p_{\mathcal{D}}(w)|_{dir_{2q-1}} - |p_{\mathcal{D}}(w)|_{dir_{2q}}).$$

In the example above, μ((e,c)(n,c)(n, d)(e, d)(s, c)) = (1, 2).

Intuitively, two paths have the same image by this morphism iff they lead to the same point. This means t we are now working in the free group, instead of working in the free monoid.

We then introduce an equivalence relation $\approx$ on $(\mathcal{D} \times \mathcal{N})^*$: two words are equivalent if they lead to the san point, from a given point, and this is obviously decidable. Formally :

$$u \approx v \Leftrightarrow \mu(u) = \mu(v).$$

In the example above, (e,c)(n,c)(n, d)(e, d)(s, c)(n, c)(s, c) $\approx$ (e,c)(n,c)(e, d).
This relation is clearly an equivalence relation ; moreover, it is a congruence.

We will also use the morphism ϕ, from $\mathbf{N}^{2q}$ on $\mathbf{Z}^q$, defined by $\mu = \phi \circ P$.

Let L be a language over $(\mathcal{D} \times \mathcal{N})^*$, and (dir, nat) an element of $(\mathcal{D} \times \mathcal{N})$.
$L(dir, nat)^{-1} = \{f \in (\mathcal{D} \times \mathcal{N})^* / f(dir, nat) \in L\}$ is the set of all words of L that can be completed, in L, by (dir, nat).

C - Semi-linear sets.

Let us recall the definition of a semi-linear set.

Definition 1.
$P \subseteq \mathbf{Z}^q$ is linear iff there are $a_o, a_1, \ldots, a_k$ in $\mathbf{Z}^q$ such that
$P = a_o + \mathbf{N}.a_1 + \ldots + \mathbf{N}.a_k$.
P is semi-linear iff P is a finite union of linear sets.

The link between context-free (and rational) languages and semi-linear sets is given by the following result.

Proposition 1.[Ginsburg…66]
Let L be a context-free language. Then P(L) and μ (L) are semi-linear sets that can be effectively constructe

In the sequel, we will use the following fundamental results on semi-linear sets several times.

Proposition 2.[Ginsburg…66]

Let S and S' be semi-linear sets.

1 - The emptiness of S is a decidable problem.
2 - $S \cap S'$ is a semi-linear set that can be effectively constructed.

2 - Notions of coherence

With the above morphims we can define precisely the way the logical links of the net are specified by the language of $(\mathcal{D} \times \mathcal{N})^*$. Let M be a point of the physical net defined by its components α in $\mu((\mathcal{D} \times \mathcal{N})^*)$ and dir a direction in $\mathcal{D}$. The physical link issued from M and of direction dir can have "nat" as logical value if there is some w(dir,nat) in L such that $\mu(w) = \alpha$.

A - Definitions.

Clearly enough, any language does not necessarily provide an unique logical nature for a given physical link. Since two distinct physical paths can use a same link, this link does not necessarily receive the same logical interpretation in the images of the two paths given by the language.
Now we want to determine some properties on the specification (i.e. the language) that yield properties of the logical links. The strongest property that we can require is the 0-strong coherence: an unique logical link is associate to a given physical link. It is a rather strong property and languages satisfying it can be hard to obtain. So that, in order to make the specification simpler to express, we introduce less constrained forms of coherence. Languages satisfying them are easier to build, but a finite number of physical links may have several associated logical natures.
Strong coherence specify a net everywhere, excepted in the neighbourhood of the starting point.

Definition 2.
Let k be an integer and L a language over $(\mathcal{D} \times \mathcal{N})^*$. We say that L is *strongly coherent with respect to k,* and denote it by $L \in SC(k)$ if and only if :

$$[\forall u, v \in L, u(dir, nat) \in L, v(dir, nat') \in L, u \approx v, |u| \geq k, |v| \geq k] \Rightarrow nat = nat'.$$

We define $SC = \bigcup_{k=0}^{\infty} SC(k)$.

A less constrained form of coherence can be defined in the following way:

Definition 3.
Let k be an integer and L a language over $(\mathcal{D} \times \mathcal{N})^*$. We say that L is *weakly coherent with respect to k,* and denote it by $L \in wc(k)$ iff :

$$\left.\begin{array}{l} u_o(dir, nat_1)u_1(dir, nat_2)\ldots(dir, nat_k)u_k(dir, nat) \in L \\ v_o(dir, nat'_1)v_1(dir, nat'_2)\ldots(dir, nat'_k)v_k(dir, nat') \in L \\ u_o \approx v_o,\ (dir, nat_i)u_i \approx (dir, nat'_i)v_i \approx \varepsilon,\ \forall i \in 1\ldots k. \end{array}\right| \Rightarrow nat = nat'$$

(where ε denotes the empty word).

We define $wc = \bigcup_{k=0}^{\infty} wc(k)$.

Remark. Obviously, $SC(0) = wc(0)$.

As direct consequences of the definitions, we have the following hierarchical relations on these sets :

Proposition 3.
$L \in SC(k) \Rightarrow L \in SC(k+1)$.
$L \in wc(k) \Rightarrow L \in wc(k+1)$.

Note that the reverse implications are false, as shown by the two following examples.

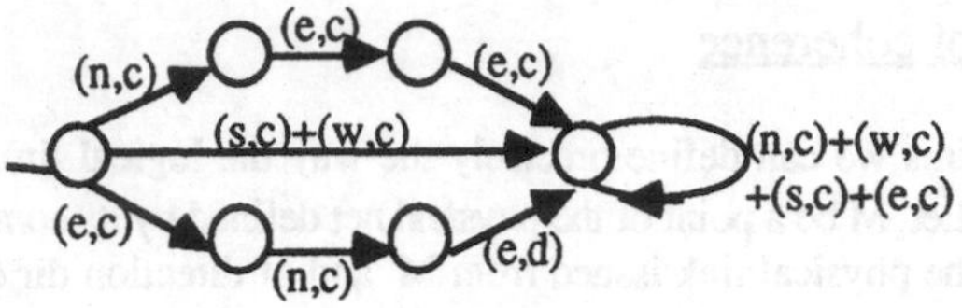

Figure 5.

The language accepted by the above transducer (every state is terminal) is 3-SC, but not 2-SC, since (n, c)(e, c)(e, c) ∈ L, (e, c)(n, c),(e, d) ∈ L, and (n, c)(e, c) ≈ (e, c)(n, c).

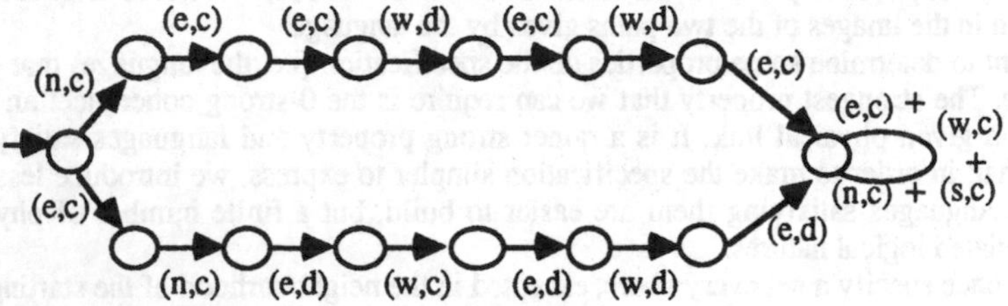

Figure 6.

The language accepted by this transducer (every state is terminal) is 3-wc but not 2-wc.

We also have a relation between the weak and the strong coherence.

Proposition 4.
L∈ SC(k) ⇒ L ∈ wc(k div 2).

Indeed, (dir, nat$_i$)u$_i$ ≈ε implies |(dir, nat$_i$)u$_i$ | ≥ 2, thus |u$_o$(dir, nat$_1$)u$_1$…(dir, nat$_{kdiv2}$)u$_{kdiv2}$| ≥ k, and the same holds for v.
But the reverse implication is false, as shown by the following counter-example :

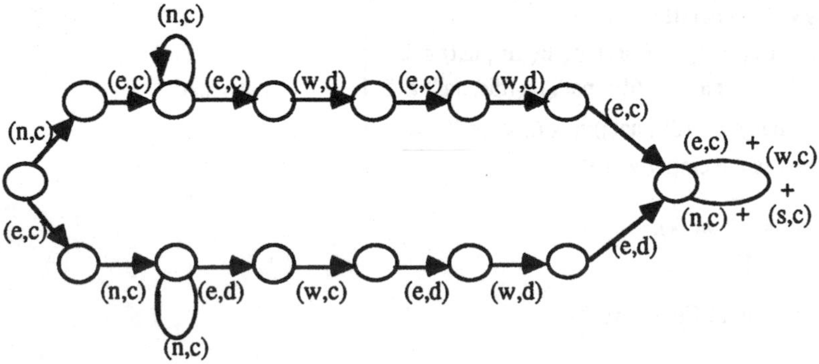

Figure 7.

This language is still 3-wc, but is not k-SC, for any k.

The natural question that arises then is : is it possible to decide, for a given k, whether a language is k weak strong coherent. Furthermore, can we decide if there is some k for which the language is k weak or stro coherent. Our purpose is now to show that we are able to decide most of these properties.

B. The strong coherence.

We always assume that L is a rational set, but the results of this section can be extended to context free languages.

We are able to decide whether the language L is strongly coherent or not.

Theorem 1.
Let L be a rational language of $(\mathcal{D} \times \mathcal{N})^*$. It is decidable whether L belongs to SC(0) or not.

Proof.

$$L \in SC(0) \Leftrightarrow \left| \begin{array}{l} u(dir, nat) \in L \\ v(dir, nat') \in L \\ u \approx v \end{array} \right| \Rightarrow nat = nat'$$

$$\Leftrightarrow [nat \neq nat' \Rightarrow \text{There is no couple } (u, v) \text{ such that } \left| \begin{array}{l} u(dir, nat) \in L. \\ v(dir, nat') \in L. \\ u \approx v. \end{array} \right.$$

$$\Leftrightarrow [nat \neq nat' \Rightarrow \mu\,(L(dir, nat)^{-1}) \cap \mu\,(L(dir, nat')^{-1}) = \emptyset\,].$$

$L(dir, nat)^{-1}$ is a rational set then $\mu\,(L(dir, nat)^{-1})$ is a semi-linear set (Proposition 1). So we have reduce the problem to the emptiness of the intersection of two semi-linear sets. As we recall it, this problem is decidable (Proposition 2). ■

It is well known that the emptiness of the intersection of two (semi) linear sets is an NP-complete problem [Karp...72]. One can ask if there exists a solution to our problem with a lower complexity. In fact, our problem is equivalent to the above problem of emptiness. We have the following proposition.

Proposition 5.
The problem of deciding whether a rational language is in SC(0) is NP-complete.

Proof.
Let $A = a_o + \mathbb{N}.a_1 + \ldots + \mathbb{N}.a_k$ and $B = b_o + \mathbb{N}.b_1 + \ldots + \mathbb{N}.b_{k'}$ be two linear sets. In order to simplify the notations, we will suppose that a_i and b_i are in $\mathbb{N}^2$. A proof in the general case can be straightforward deduced from the proof in the case n = 2.
Thus we define $a_i = (x_i, y_i)$ and $b_j = (z_j, t_j)$ with x_i, y_i, z_j and t_j in $\mathbb{N}$.
We introduce a direction alphabet $\mathcal{D}$ = {n, s, e, w, u, p} and a connexion alphabet $\mathcal{N}$ = {nat, nat', nat"}.
We will construct a transducer (and thus a language L on $\mathcal{D} \times \mathcal{N}$) such that the 0- strongly coherence of L is equivalent to the emptiness of $A \cap B$.
Let us consider the following transducer :

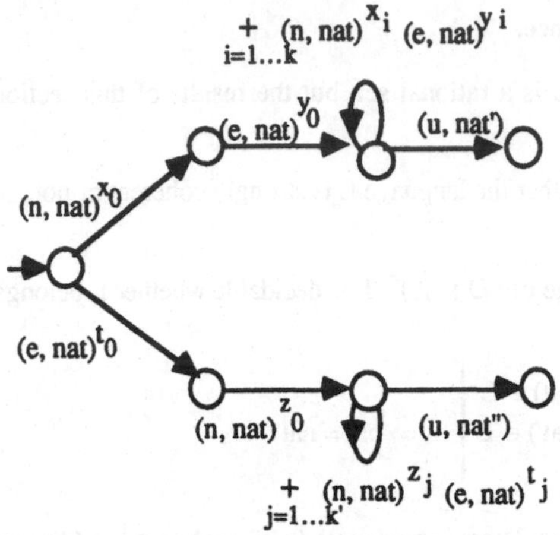

Figure 8.

It is then obvious that the language defined by this transducer is 0-strongly coherent if and only if $\mu\,[L(u, nat')^{-1}]$ and $\mu\,[L(u, nat")^{-1}]$ have an empty intersection. In other words, L is 0-strongly coherent if and only if $A \cap B = \emptyset$ and thus, the Proposition 5 is proved. ◆

Corollary.
It is decidable whether L belongs to SC(k) for a given integer k, and this is also an NP-complete problem.

Proof.
Indeed, we have $L \in SC(k) \Leftrightarrow [(\mathcal{D} \times \mathcal{N})^k]^{-1}L \in SC(0)$. ◆

Now suppose that only the language is given. Can we find an integer k such that L belongs to SC(k) ? That to say, can we prove that the language belongs to SC ? Again, we can decide the belonging to SC or not.

Theorem 2.
Let L be a rational language of $(\mathcal{D} \times \mathcal{N})^*$. It is decidable whether L belongs to SC or not.

Proof.

$L \notin SC \Leftrightarrow \forall k \geq 0,$ $\exists\,(nat, nat') \in \mathcal{N}^2, nat \neq nat'.$ $\exists\, dir \in \mathcal{D}.$ $\exists\,(u, v) \in [(\mathcal{D} \times \mathcal{N})^*]^2.$ such that

[I] $u(dir, nat) \in L.$ $v(dir, nat') \in L.$ $u \approx v.$ $|u| \geq k, |v| \geq k.$

$\Leftrightarrow$ [II] (since $\mathcal{N}$ and $\mathcal{D}$ are finite) $\exists\,(nat, nat') \in \mathcal{N}^2, nat \neq nat', \exists\, dir \in \mathcal{D}$ such that, $\forall k \geq 0, \exists\,(u, v) \in [(\mathcal{D} \times \mathcal{N})^*]^2$ such that [I] holds.

We need to introduce the following sets :
for $(dir, nat, nat') \in (\mathcal{D} \times \mathcal{N}^2)$ with $nat \neq nat'$, we define $E_{dir,nat,nat'} = \mu\,(L(dir, nat)^{-1}) \cap \mu\,(L(dir, nat')^{-1})$.

→ 1^{rst} case: ∃ (dir, nat, nat') such that $E_{dir,nat,nat'}$ is infinite.
There exists then an infinite sequence $(m_1,..,m_r,...)$ in $E_{dir,nat,nat'}$ such that $\|m_i\| \leq \|m_{i+1}\|$ for every i (with, if $m = (\mu_1,...,\mu_q)$, $\|m\| = \sum_{i=1...q} |\mu_i|$).
Thus, we can find two sequences (u_i) in L(dir, nat)$^{-1}$ and (v_i) in L(dir, nat')$^{-1}$ such that $\mu(u_i) = \mu(v_i) = m_i$. We have $|u_i| \geq \|m_i\|$, $|v_i| \geq \|m_i\|$, and $u_i \approx v_i$.
Thus, $\forall k \in \mathbb{N}$, $\exists u_k \in$ L(dir, nat)$^{-1}$, $\exists v_k \in$ L(dir, nat')$^{-1}$, $|u_k| \geq k$, $|v_k| \geq k$, $u_k \approx v_k$. It follows that L is not SC.

→ 2nd case: Every $E_{dir,nat,nat'}$ is finite.
For every $E_{dir,nat,nat'}$, we define, for each $m \in E_{dir,nat,nat'}$, $D_1 = P(L(dir, nat)^{-1}) \cap \phi^{-1}(m)$ and $D_2 = P(L(dir, nat')^{-1}) \cap \phi^{-1}(m)$. These two sets are semi-linear and actually computable. Moreover, the number of $E_{dir,nat,nat'}$ is finite and each of them is a finite set, thus, there is a finite number of sets D_1 and D_2.

We can then distinguish two sub-cases :

→ 1st subcase : There is some m in some $E_{dir,nat,nat'}$ such that D_1 and D_2 are infinite.
Let (t_k) and (t'_k) be two increasing (for ‖ ‖) sequences respectively of D_1 and D_2. Then, there exists two sequences (u_k) in L(dir, nat)$^{-1}$ and (v_k) in L(dir, nat')$^{-1}$ such that :

$t_k = P(u_k)$.
$t'_k = P(v_k)$.
$|u_k|$ and $|v_k|$ are increasing sequences.
$u_k \approx v_k \approx m$.

And thus, L ∉ SC.

→ 2nd sub-case : For each $E_{dir,nat,nat'}$, for each m in $E_{dir,nat,nat'}$, D_1 or D_2 are finite. We will prove by absurd that L is in SC.
In fact, suppose that L ∉ SC : Let (dir, nat, nat') be such that [II] holds, and let (u_k) and (v_k) be two sequences such that [I] holds. Since $E_{d,n,n'}$ is finite, there is some m and some subsequences (u'_k) and (v'_k) such that [I] holds, $u'_k \approx v'_k \approx m$, and $|u'_k]$ and $|v'_k|$ are strictly increasing sequences. Then the sequences $(P(u'_k))$ and $(P(v'_k))$ are strictly increasing to, and $P(u'_k) \in D_1$, $P(v'_k) \in D_2$, that are therefore both infinite, which is a contradiction with the hypothesis.

C. The weak coherence.

In this section, we will suppose that the language L is given by a finite deterministic automaton with N states, $\mathcal{A} = (Q, (\mathcal{D} \times \mathcal{N}), \delta, q_o, F)$.

Theorem 3.
Let L be a rational language of $(\mathcal{D} \times \mathcal{N})^*$. It is decidable whether L belongs to wc(k) or not.

Proof.
Let q_1 and q_2 be two states of the automaton $\mathcal{A}$, and d a direction of $\mathcal{D}$. We introduce the set of paths leading from q_1 to q_2, starting with the direction d.
$L(q_1, q_2, d) = \{(dir, nat)w \,/\, \delta(q_1, (dir,nat)w) = q_2\}$. This set is obviously rational.
Suppose that there exist two words u and v implying that L is not k-weakly coherent.

$u = u_o(dir, nat_1)u_1(dir, nat_2)...u_{k-1}(dir, nat_k)u_k(dir,nat)$.
$v = v_o(dir, nat'_1)v_1(dir, nat'_2)...v_{k-1}(dir, nat'_k)v_k(dir,nat')$.

with

$$(\mathrm{I})\quad \left|\begin{array}{l} u_0 \approx v_0 \\ (dir, nat_i)u_i \approx (dir, nat'_i)v_i \approx \varepsilon \\ nat \neq nat' \end{array}\right.$$

Then there exist two (k+1)-tuples of states $(q_1, q_2,\ldots,q_{k+1})$ and $(q'_1, q'_2,\ldots,q'_{k+1})$, such that we have:

$$\begin{array}{l} u = u_0(dir, nat_1)u_1(dir, nat_2)\ldots u_{k-1}(dir, nat_k)u_k(dir,nat). \\ \quad\uparrow\ \ \uparrow \qquad\quad \uparrow \qquad\qquad\qquad \uparrow \qquad\qquad\quad \uparrow \\ \quad q_0\ \ q_1 \qquad\quad q_2 \qquad\qquad\qquad q_k \qquad\qquad\quad q_{k+1} \end{array}$$

$$\begin{array}{l} v = v_0(dir, nat'_1)v_1(dir, nat'_2)\ldots v_{k-1}(dir, nat'_k)v_k(dir,nat'). \\ \quad\uparrow\ \ \uparrow \qquad\quad \uparrow \qquad\qquad\qquad \uparrow \qquad\qquad\quad \uparrow \\ \quad q'_0\ \ q'_1 \qquad\quad q'_2 \qquad\qquad\qquad q'_k \qquad\qquad\quad q'_{k+1} \end{array}$$

Moreover, the conditions (I) imply:

$$(\mathrm{II})\quad \left|\begin{array}{l} \mu(\{u \,/\, \delta(q_0, u) = q_1\}) \cap \mu(\{u \,/\, \delta(q_0, u) = q'_1\}) \neq \emptyset \\ 0 \in \mu(L(q_i, q_{i+1}, dir)) \\ 0 \in \mu(L(q'_i, q'_{i+1}, dir)) \end{array}\right.$$

Conversely, if two (k+1)-tuples of states $(q_1, q_2,\ldots,q_{k+1})$ and $(q'_1, q'_2,\ldots,q'_{k+1})$ verify that there exist direction dir and two natures nat and nat', such that:

$$(\mathrm{III})\quad \left|\begin{array}{l} (\mathrm{II}) \\ \delta(q_{k+1}, (dir, nat)) \text{ is defined} \\ \delta(q'_{k+1}, (dir, nat')) \text{ is defined} \end{array}\right.$$

it is obvious that the language L is not k-weakly coherent.

But if elements $(q_1, q_2,\ldots,q_{k+1})$, $(q'_1, q'_2,\ldots,q'_{k+1})$, dir, nat and nat' are given, the conditions (III) ar decidable (Proposition 2). Moreover, there is a finite number of such elements, thus we can decide whether t language L is k-weakly coherent or not. ■

In the case where L is a rational language (as it is assumed in this section), the increasing sequence (wc(k)) stationary:

Proposition 6.

$\forall\, k \geq N + 1, L \in wc(k) \Rightarrow L \in wc(N)$. (where N is the number of states of $\mathcal{A}$)

Proof.

Let us consider a word u such that :

$$\begin{array}{l} u = u_0(dir, nat_1)u_1(dir, nat_2)\ldots u_{N-1}(dir, nat_N)u_N(dir,nat). \\ \qquad\ \uparrow \qquad\qquad \uparrow \qquad\qquad\qquad \uparrow \qquad\qquad\quad \uparrow \\ \qquad\ q_1 \qquad\qquad q_2 \qquad\qquad\qquad q_N \qquad\qquad\quad q_{N+1} \end{array}$$

We run through the states $q_1, q_2,\ldots,q_{N+1}$. We find, since there are only N different states, two integers i < such that $q_i = q_j$.

Moreover, we have $\mu((dir, nat_i)u_i\ldots(dir, nat_{j-1})u_{j-1}) = \varepsilon$. Therefore, this loop can be iterated, so that we get a word :

$\hat{u} = u_0(dir, nat_1)\ldots u_{i-1}[(dir, nat_{i-1})u_i\ldots u_{j-1}]^s(dir, nat_j)\ldots u_{N-1}(dir, nat_N)u_N(dir,nat)$ with $N + (s - 1)(j - i) \geq k$

We proceed in the same way with v.

From the k-weak confluence, we then get nat = nat', and thus, L is N- wc. ■

Thus, to decide whether a rational language L is weakly coherent, it suffices to test whether L is in $\cup_{i=1...N}wc(i)$, which is decidable from Theorem 3. Thus, we have proved:

Theorem 4.
Let L be a rational language in $(\mathcal{D} \times \mathcal{N})^*$. Then, it is decidable whether L is weakly coherent or not.

3 - Coherent components

In this section we will refine our results. When a transducer is built, it can be possibly incorrect, i.e. that it does not meet any of the previous coherences. Nevertheless, some of these transducers can provide some useful information : they give in fact several (namely a finite number of) specifications. We can then partition the language, each part corresponding to a specification. Let us consider the following example :

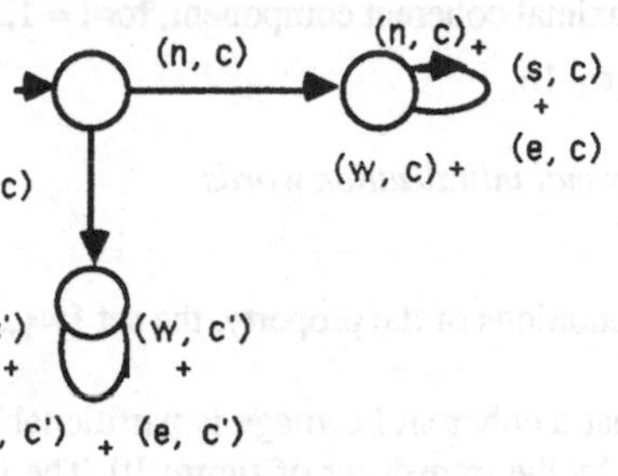

Figure 10.

We have $\mathcal{D}$ = {n, s, e, w} and $\mathcal{N}$ = {c, c'}. This transducer is neither strong nor weak coherent. But, the associated rational language is the union of two coherent languages : $(n, c)(\mathcal{D} \times \{c\})^* \cup (s, c')(\mathcal{D} \times \{c'\})^*$. For the description of such a transducer, we will introduce some notions. In the following, "coherent" will mean "strongly coherent with respect to 0" (Definition 2).

Definition 4.
Let L be a language over $(\mathcal{D} \cup \mathcal{N})^*$, and w be a word in L. The *w - component* induced by w, denoted by Init(w, L), is the set of the left factors of the words in L that start with w. It is defined by :
$Init(w, L) = LF(w) \cup \{wu / wu \in L\}$.

A language will then be said to be partitionable if it is the union of a finite number of w- components. This means that a finite number of choices (that may be connected to some test instructions in a program corresponding to the transducer) will determine which specification has to be considered among the finite number of specifications described by the transducer.

Definition 5.
A language L over $(\mathcal{D} \cup \mathcal{N})^*$ is *partitionable* iff there exist a finite number of words $w_1,\ldots,w_p$ such that :

1 - $Init(w_i, L)$ is coherent for $i = 1\ldots p$.
2 - $L = \cup_{i=1\ldots p} Init(w_i, L)$.

The words w_i are then called *initialization words*, and the sets $Init(w_i, L)$ *coherent components* of L.

Remark.
$w' <_l w \Rightarrow Init(w, L) \subseteq Init(w', L)$.
It follows that the component Init(w, L) can be replaced by Init(w', L) (of course if Init(w', L) is also coherent) in the decomposition of L.

We want to have the optimal decomposition, i.e with maximal components.

Definition 6.

A coherent component induced by w is said to be a *maximal coherent component* iff :

1 - Init(w, L) is coherent.
2 - $w' \triangleleft_l w, w' \neq w \Rightarrow$ Init(w', L) is not coherent.

We want to get a decomposition of a language with the smallest number of components, and such that t initialization words are the shortest ones. For that purpose, we will use the following characterization partitionable languages :

Property.
L is partitionable iff there exist a finite number of words $w_1,\ldots,w_q$ such that

1 - Init(w_i, L) is a maximal coherent component, for $i = 1,\ldots,q$.
2 - $L = \cup_{i=1\ldots q}$ Init(w_i, L).

The words w_i are called *maximal initialization words.*

Remark.
We can notice that, in the conditions of the property, the set $\{w_1,\ldots,w_q\}$ is a prefix set.

One can of course notice that a coherent language is partitionable (because it is equal to Init(ε, L)). But t converse is false, as shown by the transducer of figure 10. The language is partitionable, but is not cohere (and also not k - weak or strong coherent).
Furthermore, there exist languages that are not partitionable. Let us consider the next example for illustrati this :

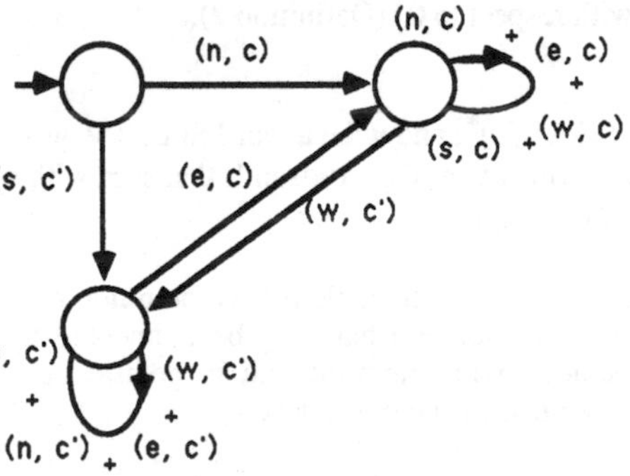

Figure 11.

We then want to know whether a language is partitionable or not. In order to prove that this property decidable we need the following lemma:

Lemma.
Let $w = w_1w_2w_3$ be a word such that :
1 - Init(w, L) is a maximal coherent component.
2 - w contains a loop, i.e $\delta(q_o, w_1) = \delta(q_1, w_2) = q_1$.
Then, for $p \in \mathbf{N}^*$, Init($w_1w_2^pw_3$, L) is a maximal coherent component.

Proof.
Let us first show that Init($w_1w_2^pw_3$, L) is coherent. Suppose that :

$$\left|\begin{array}{l} w1w2^p w3u(dir, nat) \in L \\ w1w2^p w3u(dir, nat) \in L \end{array}\right.$$

with $\mu(w_1w_2{}^pw_3u(dir, nat)) = \mu(w_1w_2{}^pw_3u(dir, nat)$

Since L is deterministic and w_2 corresponds to a loop,we have :

$$\left|\begin{array}{l} w1w2w3\ u\ (dir, nat) \in L \\ w1w2w3\ v\ (dir, nat') \in L \end{array}\right.$$

with $\mu(w_1w_2w_3\ u\ (dir, nat)) = \mu(w_1w_2w_3\ v\ (dir, nat'))$.

And this implies that nat = nat', according to the fact that Init(w, L) is a coherent component. Thus, $Init(w_1w_2{}^pw_3, L)$ is coherent.

Moreover, it is a maximal coherent component. Indeed, let us suppose that a left factor w' is also an initialization word. We will show that w admits then also a left factor that is an initialization word :

1 - $w' = w_1w_2{}^pw_3'$, with $w_3' <_l w_3$, $w_3 \neq w'_3$. Let us suppose that Init(w', L) is coherent.

▷ $w_1w_2{}^pw'_3u\ (dir, nat) \in L \Rightarrow w_1w_2w'_3u\ (dir, nat) \in L$

▷ $w_1w_2{}^pw'_3v\ (dir, nat') \in L \Rightarrow w_1w_2w'_3v\ (dir, nat') \in L$

▷ $[\mu(w_1w_2w_3\ u\ (dir, nat)) = \mu(w_1w_2w_3\ v\ (dir, nat'))]$

$$\Rightarrow [\mu(w_1w_2w'_3u\ (dir, nat)) = \mu(w_1w_2w'_3v\ (dir, nat'))].$$

As Init(w', L) is coherent, this implies then that nat = nat'. And thus, $Init(w_1w_2w'_3, L)$ is coherent too. But then, Init(w, L) is not a maximal coherent component, which is in contradiction with our hypothesis. Thus, $Init(w_1w_2{}^pw_3, L)$ is a maximal coherent component.

2 - $w' = w_1w_2{}^qw_2'$, with $q < p$ and $w_2' <_l w_2$, and $w_2 \neq w'_2$. Applying the same reasoning, we see that the coherence of $Init(w_1w_2{}^qw_2', L)$ implies that $Init(w_1w_2', L)$ is also coherent, which is impossible since Init(w, L) is maximal.Thus, $Init(w_1w_2{}^pw_3, L)$ is a maximal coherent component.

3 - $w' = w'_1$, with $w'_1 <_l w_1$. Then if $Init(w'_1, L)$ is coherent, Init(w, L) is not maximal, which is impossible according to our hypothesis.

Thus, no left factor of $w_1w_2{}^pw_3$ is an initialization word, and $Init(w_1w_2{}^pw_3, L)$ is a maximal coherent component. ∎

With this lemma we can prove the decidability of the partitionable property:

Theorem 6.
Let L be a rational language of $(\mathcal{D} \cup \mathcal{N})^*$. It is decidable whether L is partitionable or not.

Proof.

We recall that the language L is given by a finite deterministic transducer with N states, $\mathcal{A} = (Q, (\mathcal{D} \times \mathcal{N}), \delta, q_o, F)$.

The lemma proves that if a word containing a loop induces a maximal coherent component, there is an infinity of maximal components, and the language is not partitionable. Indeed, suppose that the language is partitionable, and that a word w containing a loop is a maximal initialization word. Let $\alpha_1,\ldots,\alpha_p$ be the initialization words of L. Let $k = Max_{i=1\ldots p}|\alpha_i| + 1$, and q be such that $|w_1w_2{}^qw_3| > k$. Since $w_1w_2{}^qw_3 \in L$, there is a word α_i such that $w_1w_2{}^qw_3 \in Init(\alpha_i, L)$.

As $|\alpha_i| < |w_1w_2{}^qw_3|$, we have $\alpha_i <_l w_1w_2{}^qw_3$, but then $Init(w_1w_2{}^qw_3, L)$ is not a maximal coherent component. According to the precedent lemma, this is impossible.

Thus, if L is partitionable, the maximal initialization words contain no loop. It follows that the potenti maximal initialization words are of size less than N, if N is the number of states of the transducer. Then, decide whether the language L is partitionable or not, we just have :

1 - to test if these words (that are in finite number) are initialization words, starting with t smallest ones, and avoiding the verification for the words that admit a left factor that has still recognized as an initialization word.

2 - To verify that L is equal to the union of the maximal components that have been found.

Thus, we can decide whether L is partitionable or not. And the theorem is proved. ◆

Example.
If we consider the example of figure 10, we see that the automaton defines a language that admits 2 maxim coherent components $w_1 = (n, c)$ and $w_2 = (s, c')$.
But the automaton of figure 11 is not partitionable, since it admits an infinity of maximal initialization word

4. Some properties of such a system

In this section we are interested in detecting some particular properties of the system.
We will first try to detect the possible lack of -utilization of some physical links of the system. A link will in a idle state if it is never logically connected. Formally:

Definition 7.
A link, defined by its component α in $\mu\,((\mathcal{D} \times \mathcal{N})^*)$ and a direction dir, is in a idle state iff :
$\forall$ nat $\in \mathcal{N}$, $[u(\text{dir}, \text{nat}) \in L \Rightarrow \mu(u) \neq \alpha]$

Proposition 7.

It is decidable whether a link is in a idle state.

Proof.
It suffices to test whether $\alpha \notin \cup_{\text{nat} \in \mathcal{N}}\ \mu(L(\text{dir}, \text{nat})^{-1})$, which is decidable.◆

We can also wonder if two points are logically linked together. We introduce three kinds of connectivity. order to simplify the notations we consider connectivities between the starting point and a given poi Nevertheless these results can be extended in a straightforward way to connectivity between two arbitrar choosen points of the net.

Definition 8.

1 - The point is *weakly connected* to the starting point iff there is a path u in L that leads to this point.

2 - The point is *strongly connected* to the starting point iff there is an infinity of paths u in L leading to it.

3 - The point is *completely connected* to the starting point iff every path leading to it belongs to L.

Proposition 8.

Let L be a rational language over $(\mathcal{D} \times \mathcal{N})^*$. Each of the 3 kinds of connectivity is decidable.

Proof.
Let us suppose that the considered point is given by its component α in $\mu\,((\mathcal{D} \times \mathcal{N})^*)$.
1 - One just has to test whether α belongs to $\mu(L)$ or not.
2 - One just has to test whether $\phi^{-1}(\alpha) \cap P(L)$ is finite or not.
3 - One has just to test whether $p_{\mathcal{D}}(\phi^{-1}(\alpha))$ is included in $p_{\mathcal{D}}(L)$ or not.◆

Finally, we can also look for paths of a given nature (the nature preserving connectivity).
More formally, let us suppose that a rational set N of logical links is given (a particular case is the case of nat*). Let us consider a point in the net given by its component α in μ $((\mathcal{D} \times \mathfrak{N})^*)$.
Is there a word u in L such that $\mu(u) = \alpha$ and $p_{\mathfrak{N}}(u) \in N$?

Proposition 9.
It is decidable whether a given point is nature preserving connected with the starting point.

Proof.
One just has to verify whether $\alpha \in \mu(L \cap p_{\mathfrak{N}}^{-1}(N))$ or not, which is decidable. In the particular case where $N = nat^*$, it enables to find paths of a given nature, i.e such that contain only link of a given logical nature.■

5. Perspectives

The title of the paper is intended to stress the fact that the properties we want to describe do not take into account any "meaning" for the nature of the logical links, or the nature of the processors that constitute the network.
In further work, we will investigate properties stemming from the nature of the processors and of the logical links. For example, a processor could be a finite automaton communicating with his neighbours, in the way described in [Beauquier…87]. The different types of logical connections would be then the description of the communicating states. Problems like local deadlock, arise then.
Also mechanisms that can realise these specifications have not been discussed here. Two possible mechanisms can be considered: an automaton that moves on the network, and establishes the different logical links, according to the rules given by the transducer. Another, is the broadcast mechanism, each node tells its neighbour, in wich state it is, and what type of link is established.

6. References

[Beauquier…87] J. Beauquier, "Systèmes distribués et automates finis", Actes du colloque C3, Angoulème, pp 27-44, 1987

[Berstel…79] J. Berstel, "Transductions and context free languages", Teubner Studienbucher Informatik, 1979.

[Ginsburg…66] S. Ginsburg, "The mathematical theory of context free languages" McGraw-Hill, 1966, New York.

[Karp…72] R.M. Karp, "Reducibility among combinatorial problems", in R.E. Miller and J.W. Tatcher (eds), Complexity of Computer Computations, Penum press, New York, pp. 85-103, 1972.

[Kohonen…88] T. Kohonen, "An introduction to neural computing", Neural Networks, vol. 1, pp 3-16, 1988.

[Parikh…66] R.J. Parikh, "On context-free languages", J. Assoc. Comput. Mach. 13, pp 570-581, 1966.

[Quinton…86] P. Quinton, "An introduction to systolic architectures", Future Parallel Computer, LNCS 272, pp 387-400, 1986.

Shuffle Equations, Parallel Transition Systems and Equational Petri Nets

Stephen L. Bloom[1] and Klaus Sutner
Stevens Institute of Technology
Hoboken, NJ 07030

1 Motivation

Consider how one might describe the sequences of atomic actions $[F]$ performable by a standard deter ministic flowchart scheme F. One can exhibit F as a finite directed graph whose vertices are labele with letters in a ranked alphabet Ω (if the vertex v has outdegree k, then the letter labeling v is i the set Ω_k). Then we say that the set $[F]$ is the set of all labels of paths in F from the 'begin verte to the exit. If one wants to refine this notion further, one can associate a labeled tree with F, the s called 'unfolding' of F, and define $[F]$ as this tree. To describe the tree in great detail, one can sho how to associate a tree of depth at most one with each node in the underlying graph of F. This as sociation takes the form of a system of equations. For example, suppose that F is the flowchart schem

```
1:  do ω; goto 2:
2:  if π is true do σ and goto 2:  else goto 3:
3:  exit.
```

Then we can introduce a 'variable' x_i for each instruction, and a corresponding equation (i):

$$
\begin{aligned}
x_1 &= \omega \cdot x_2 &(1\\
x_2 &= \pi_t \cdot \sigma \cdot x_2 + \pi_f \cdot x_3 &(2\\
x_3 &= \bot &(3
\end{aligned}
$$

Then, one solves this system of equations in an appropriate structure.

Our work began with the question of describing the sequences of atomic actions performable by flowchart algorithm which admits explicit nondeterminism and a forking type of parallelism. Th kind of flowchart scheme is made precise below in our definition of a parallel transition system; th corresponding systems of equations are the shuffle equations. We then saw that the same equation were determined by a subclass of free choice Petri nets, which we call the equational Petri nets. Thu shuffle equations, parallel transition systems and equational Petri nets are equivalent descriptions this class of algorithm scheme. For reasons of space, we concentrate here on the equational Petri net and state only two results concerning the parallel transition systems.

2 Preliminaries

We let N denote the set of nonnegative integers. For n in N, $[n]$ is the set {1,2,...,n}. A *multiset o* X is a function $\nu : X \rightarrow N$; the collection of all multisets on X will be denoted N^X. We identify subset of X with its characteristic function, so that, for example, the empty set $\emptyset$ is the multiset wit

[1]Partially supported by NSF grant CCR-8901693

constant value 0. For any x in $X, e_x : X \to N$ is the multiset with $e_x(y) = 1$ if $y = x$; 0 if $y \neq x$. If ν and ν' are multisets, $(\nu + \nu')(y) = \nu(y) + \nu'(y)$. We will also write $\nu - \nu'$ for the function defined by pointwise subtraction, even if the value of this function is negative.

A *word* of length n on a set A is a function $w : [n] \to A$, for some $n \geq 0$. An *infinite word* on A is a function $\{1, 2, \ldots\} \to A$. We denote the unique word of length 0 by ε. The familiar operation of concatenation of words w and u is denoted $w \cdot u$. The set of words on A of finite length is A^* ; the set of all finite and infinite words on A is A^∞. For $Y \subset A^\infty$, we write *pref* (Y) for the set of all finite words u such that $u \cdot v \in Y$, for some $v \in A^\infty$. A *ranked set* $\Omega = \{\Omega_k : k = 0, 1, ...\}$ is a collection of pairwise disjoint sets Ω_k. Elements in Ω_k are called function letters of rank k. For the remainder of the paper, Ω is a fixed ranked set such that Ω_0 is the singleton set $\{\bot\}$. We will make use of the *associated alphabet*

$$\Sigma := \bigcup_{k \geq 1} (\Omega_k \times [k]). \tag{4}$$

Thus, a letter in Σ is of the form (ω, j) for some $\omega \in \Omega_k$ and some $j \in [k]$. A *successor system* (V, ρ, σ) consists of a set V (of vertices or nodes), a "rank" function $\rho : V \to N$, and functions $\sigma_v : [\rho(v)] \to V$, for v in V; if $\sigma_v(i) = u$, we say that u is the *i-th successor* of the vertex v. To avoid subscripts, we will sometimes write $\sigma(v, i)$ instead of $\sigma_v(i)$. A *path* in a successor system from u to v is a sequence $u = u_0, u_1, \ldots, u_k = v$ of vertices such that u_{i+1} is a successor of u_i, for $i = 0, \ldots, k-1$. A *rooted successor system* is a successor system having a distinguished vertex r, called the root. An **Ω-labeled successor system** is a successor system equipped with a labeling function λ from the vertices to Ω which respects the ranking - i.e. $\rho(v) = n$ iff $\lambda(v) \in \Omega_n$.

For later use, we define a binary operation $\|$. The operation is first defined as a map from pairs of finite words in Σ^* to finite subsets of Σ^*.

$$u \| \varepsilon \;:=\; \varepsilon \| u = \{u\}, \text{for all } u \in \Sigma^*; \tag{5}$$

$$\sigma u \| \sigma' v \;:=\; \sigma \cdot (u \| \sigma' v) \cup \sigma' \cdot (\sigma u \| v). \tag{6}$$

If U and V are sets of finite words, $U \| V := \bigcup \{u \| v : u \in U, v \in V\}$. If Ω is a ranked set, a **Ω-coalgebra A** consists of a set A and a partial function

$$\omega_A : A \to A \times [n]$$

for each element ω in $\Omega_n, n \geq 1$. If $\omega \in \Omega_0, \omega_A$ is the empty function.

Example. The Ω-coalgebra $U\Sigma$ has Σ^* as its underlying set. The functions $\omega_{U\Sigma}$ are defined as follows. Let $s = \delta_1 \delta_2 \ldots \delta_n$ be a word in Σ^*. If $\omega \in \Omega_k, k > 0$, then

$$\omega_{U\Sigma} := \begin{cases} (\delta_2 \ldots \delta_n, j) & \text{if } n > 0 \text{ and } \delta_1 = (\omega, j), \\ \textit{undefined} & \text{otherwise} \end{cases}$$

3 Parallel systems

A *parallel successor system* is a rooted, $\Omega^\dagger$ -labeled successor system $\Re$, where $\Omega^\dagger$ is obtained from Ω by the addition of the two symbols $+$ and $\|$ to Ω_2. If $\Re$ is a parallel system, an *internal state* of $\Re$ is an element of $N^V \times \Sigma^*$. Recall that N^V is the collection of all multisets of vertices of $\Re$, and Σ is the associated alphabet $\bigcup_{k \geq 1} (\Omega_k \times [k])$. The meaning of the internal state (ν, s) is this: each vertex v with $\nu(v) > 0$ is allowed to "fire"; the word s is a record of the sequence of "externally observable" actions which took place in the course of reaching the current internal state from the initial state. The second components of the internal states will be used to define the set of *$\Re$-admissible words* in Σ^*.

Definition 1 *$\mathfrak{R}$ determines a "transition relation", denoted $\Rightarrow_{\mathfrak{R}}$, on the internal states. We defin the relation in two steps. First, for a vertex v of $\mathfrak{R}$, we define $\Rightarrow_{\mathfrak{R}}^{v}$ as follows:*

$$(\nu, s) \Rightarrow_{\mathfrak{R}}^{v} (\nu', s')$$

if $\nu(v) > 0$ and one of the following conditions hold.

1.1 *$\lambda(v) = +$, $s' = s$ and*
$\nu' = \nu - e_v + e_u$, where u is either $\sigma_v(1)$ or $\sigma_v(2)$.

1.2 *$\lambda(v) = \|$, $s' = s$ and*
$\nu' = \nu - e_v + e_u + e_w$, where $u = \sigma_v(1)$ and $w = \sigma_v(2)$.

1.3 *the label of v is $\omega \in \Omega_k, k > 0$, and for some $i \in [k]$, $s' = s \cdot (\omega, i)$ and*
$\nu' = \nu - e_v + e_u$, where $u = \sigma(v, i)$.

Lastly, we define

1.4 *$(\nu, s) \Rightarrow_{\mathfrak{R}} (\nu', s')$ if for some vertex v, $(\nu, s) \Rightarrow_{\mathfrak{R}}^{v} (\nu', s')$.*

The relation defined by the first two conditions (1.1) and (1.2) only is called the "internal transition" Thus, we may say $(\nu, s) \Rightarrow_{\mathfrak{R}}^{v} (\nu', s)$ via an *internal* transition, meaning that either (1.1) or (1.2) hold Similarly, if (1.3) holds, we say $(\nu, s) \Rightarrow_{\mathfrak{R}}^{v} (\nu', s \cdot (\omega, j))$ via a *visible* transition.

Definition 2 *The set $\Sigma(\mathfrak{R})$ of* **$\mathfrak{R}$-admissible words** *of a parallel system $\mathfrak{R}$ is defined as the set all words s in Σ^* such that $(e_r, \varepsilon) \Rightarrow_{\mathfrak{R}}^{*} (\nu, s)$, for some ν.*

Let $\mathbf{A} = (A, \omega_A : \omega \in \Omega)$ be an Ω-coalgebra. $\mathfrak{R}$ determines an A-indexed collection $\widehat{\mathfrak{R}}_A$ of sets finite and infinite sequences of elements in $A \times \mathbf{N}^V \times \Sigma^*$. In order to define which sequences belon to $\widehat{\mathfrak{R}}_A$, it is convenient to define three more binary relations.
Let $\mathfrak{R}$ be a fixed parallel system. Suppose that ν and ν' are multisets of vertices of $\mathfrak{R}$, that b and are elements of A, and that s and s' are words in Σ^*. Then

$$(b, \nu, s) \vdash_{\Omega}^{v} (c, \nu', s')$$

if $\nu(v) > 0, \lambda(v) = \omega \in \Omega_k, k > 0$, and for some j in $[k]$, $\omega_A(b) = (c, j)$, $\sigma(v, j) = u, s' = s \cdot (\omega, j)$ an $\nu' = \nu - e_v + e_u$.
We write

$$(b, \nu, s) \vdash_{\Omega} (c, \nu', s')$$

if $(b, \nu, s) \vdash_{\Omega}^{v} (c, \nu', s')$, for some v.

$$(b, \nu, s) \vdash_{in} (c, \nu', s')$$

if $c = b, s' = s$, and $(\nu, s) \Rightarrow_{\mathfrak{R}} (\nu', s)$ via an internal transition (recall 1.1 and 1.2). Lastly, define

$$(b, \nu, s) \vdash_{\mathfrak{R},A} (c, \nu', s')$$

if, for some multiset ν'', $(b, \nu, s) \vdash_{in}^{*} (b, \nu'', s)$ and $(b, \nu'', s) \vdash_{\Omega} (c, \nu', s')$.
We note that $c = b$ and $s' = s$ if $(b, \nu, s) \vdash_{in}^{*} (c, \nu', s')$. More importantly, if $(b, \nu, s) \vdash_{\mathfrak{R},A} (c, \nu', s'$ then $(\nu, s) \Rightarrow_{\mathfrak{R}} (\nu', s')$.

Remark. Whenever there is a vertex v in the current internal state which is labeled $+$ or $\|$, the $\mathfrak{R}$ may change its internal state without changing the "external state", i.e. the A-component. $(b, \nu, s) \vdash_{in}^{*} (b, \nu', s)$, then in state (ν, s), $\mathfrak{R}$ is capable of making a sequence of invisible intern changes of its state. On the other hand, if $(b, \nu, s) \vdash_\Omega (c, \nu', s')$, then in state (ν, s), $\Re$ may make an external, visible change of state. The "names" of the observable changes of state which occurred are recorded in the word s'.

Now we describe $\widehat{\Re}_A(a)$ for each element a in A. $\widehat{\Re}_A(a)$ is a set of sequences of elements in $A \times \mathbf{N}^V \times \Sigma^*$. A finite or infinite sequence $(\xi_0, \xi_1, \ldots)$ belongs to $\widehat{\Re}_A(a)$ if firstly, $\xi_0 = (a, e_r, \varepsilon)$ and $\xi_n \vdash_{\Re,A} \xi_{n+1}$ for each $n \geq 0$, whenever ξ_n and ξ_{n+1} are defined. Finally, $\widehat{\Re}_A$ is the function mapping $a \in A$ to the set of sequences $\widehat{\Re}_A(a)$.

For $\xi = (a, \nu, s)$ in $A \times \mathbf{N}^V \times \Sigma^*$, write the A–component of ξ as $pr_A(\xi) = a$. For a in A, we define $\Re_A(a)$ as the set of all sequences $(pr_A(\xi_1), pr_A(\xi_2), \ldots)$ for $(\xi_0, \xi_1, \ldots)$ in $\widehat{\Re}_A(a)$. Note that we are deliberately omitting the first element ξ_0.

Definition 3 *Let $\Re$ and $\Re'$ be parallel systems. $\Re$ is* **trace equivalent** *to $\Re'$ if for every coalgebra* **A**, *$\Re_A = \Re'_A$, i.e. the sets of sequences $\Re_A(a)$ and $\Re'_A(a)$ are identical, for each a in A.*

A fundamental question concerning parallel systems is this. Is there an algorithm to determine, given two parallel systems $\Re$ and $\Re'$, whether they are trace equivalent? We state a theorem which shows that this question reduces the the question of whether $\Re$ and $\Re'$ determine the same sets of sequences on one particular coalgebra. Recall the definition of the coalgebra $U\Sigma$ in Example 2 of the previous section.

Theorem 4 *Let $\Re$ and $\Re'$ be parallel systems. The following statements are equivalent. $\Re$ is trace equivalent to $\Re'$. $\Re_{U\Sigma} = \Re'_{U\Sigma}$. $\Sigma(\Re) = \Sigma(\Re')$.*

Our proof (omitted here) shows that the set $\Sigma(\Re)$ is determined by the following condition:

$s \in \Sigma(\Re) \iff$ there is a sequence $f = (f_1, \ldots, f_n)$ in $\Re_{U\Sigma}(s)$ where n is the length of s and $f_n = \varepsilon$.

Equations. The last condition above shows the importance of the set $\Sigma(\Re)$ of $\Re$-admissible words. We can give another description of this set by means of a system of equations. Suppose that $\Re = (V, \rho, \sigma, r, \lambda)$ is a fixed *finite* parallel successor system. Choose an enumeration of V, say $V = \{v_1, \ldots, v_m\}$. We consider each vertex as a "variable". For each variable v_i in V, we introduce a term t_i as follows. If $\lambda(v_i) = \omega \in \Omega_k$, write u_j for the $j-th$ successor vertex of v, $\sigma(v, j)$, for each j in $[k]$. Then the term t_i is defined as

$$t_i := (\omega, 1)u_1 + (\omega, 2)u_2 + \ldots + (\omega, k)u_k;$$

if $\lambda(v_i) = \perp$, the term t_i is just $\perp$. If $\lambda(v_i) = +$, then t_i is $u_1 + u_2$, where u_j is the $j-th$ successor of $v_i, j = 1, 2$. Lastly, if $\lambda(v_i)$ is $\|$, t_i is $u_1 \| u_2$, where again u_j is the $j-th$ successor of v_i. $\Re$ determines the system of fixed point equations

$$v_i = t_i, i = 1, \ldots, m. \tag{7}$$

Conversely, such a system of equations in turn determines a parallel system. The labeling of the vertex v_i can be deduced from the form of the corresponding term t_i, as can the values of the successor function. We take the root of the system to be v_1, the first variable. From now on, we identify a finite parallel system with a finite set of equations of the form 7. One can solve this system in an algebra $C\Sigma$ of closed subsets of the finite and infinite words (on the alphabet of those letters that label the nodes of a parallel system), in which $+$ is union and the operation $\|$ is the continuous extension of the operation given in Section 2, (5) and (6).

Consider a finite system of equations,

$$v_i = t_i, i = 1, \ldots, m \qquad (8$$

as above, where the set of variables on the left is $\{v_1, \ldots, v_m\}$. We will assume that the correspondin parallel system $\mathfrak{R}$ has root v_1 and that:

for each i, there is a path in $\mathfrak{R}$ from v_1 to v_i ;

for each i, there is a path in $\mathfrak{R}$ from v_i to a vertex labeled $\omega \in \Omega_k$, for some $k > 0$. every path c positive length from v to itself contains at least one vertex labeled $\omega \in \Omega_k$, for some $k > 0$.

Consider the m-tuple of terms $\tau := \langle t_1(v_1, \ldots, v_m), \ldots, t_m(v_1, \ldots, v_m)\rangle$. If we substitute t_i for simultaneously in each term $t_1, \ldots, t_m$, we get an m-tuple of terms τ^2; after m such substitution we get an m-tuple of terms $\tau^m = \langle h_1(v_1, \ldots, v_m), \ldots, h_m(v_1, \ldots, v_m)\rangle$. The meaning of the thre conditions above is the following. Identifying each term h_i with a finite tree in the usual way, *ever* path from the root of h_i to any leaf passes through some vertex labeled by a letter in Ω. In thi restricted situation, the set of equations (8) has a unique solution in $C\Sigma$, since the term τ^m is proper contraction map.

Theorem 5 *Let $(X_1, \ldots, X_m)$ be the unique solution of the system (8) of equations for the syste $\mathfrak{R}$. Then, letting $\mathfrak{R}_i$ denote the parallel system with root $v_i, i = 1, \ldots, m$, we have:*

$$\Sigma(\mathfrak{R}_i) = pref(X_i). \qquad (9$$

In the next section, we will consider several decision problems connected with systems of equations o the form (8).

4 Equational Petri Nets

In the remaining sections we consider the class of languages $\Sigma(\mathfrak{R})$ for parallel successor systems $\mathfrak{R}$ These languages also can be described in terms of a special class of Petri nets, which we call *th equational Petri nets.* We assume that the reader is familiar with the basic definitions concernin Petri nets and Petri net languages (see e.g. [6, 5, 4, 7, 8]). For our purposes, an *equational Petri n* is a quadruple $E = \langle P, T, \delta, \pi\rangle$ where P is a finite set of *places*, T is a finite set of *transitions*, and and δ are functions from the set of transitions to the set of places and multisets on P respectivel $\pi : T \to P$, $\delta : T \to \mathbf{N}^P$.

The place $\pi(t)$ is called the *source* of the transition t. In contrast to general Petri nets, every transitio is enabled by only one place, its source. A *marking* is a multiset on P. The *support* of a marking m the set $\{p \in \mathrm{P}: m(p) > 0\}$. For two markings m_1 and m_2, we write $m_1 \leq m_2$ if $\forall p \in P(m_1(p) \leq m_2(p)$ The *rank* of the marking m is the number of elements of m, i.e., $rk(m) := \Sigma_{p\in P} m(p)$. A *marke equational Petri net* is a pair $\langle E; m_0\rangle$ where E is an equational Petri net and m_0 is a marking, calle the initial marking.

We will abuse notation by identifying a place p with the multiset e_p introduced in section 1. Fo a transition t let $\Delta(t) := \delta(t) - \pi(t) \in \mathbf{Z}^P$ be the *yield* of t. A sequence of transitions is a *firin sequence*. For any marking m and firing sequence $w \in T^*$, we define $m \cdot w \in \mathbf{Z}^P$ by induction on th length of w:

$$m \cdot w := \begin{cases} m & \text{if } w = \varepsilon \\ (m + \Delta(t)) \cdot u & \text{if } w = t \cdot u \end{cases}$$

Definition 6 *A transition t is enabled at a marking m if $m \geq \pi(t)$, i.e., $m(\pi(t)) > 0$. A firing sequence $w = t \cdot u$, $u \in T^*$ is enabled at m if t is enabled at m and u is enabled at $m \cdot t$; $w \in T^\infty$ is enabled at m if for each finite prefix u of w, u is enabled at m. (By definition, the empty sequence is enabled at every marking.)*

We will use the following notation.

$$\begin{aligned} \mathcal{F}^\infty(m) &:= \{u \in T^\infty \mid u \text{ enabled at } m\}, \\ \mathcal{F}^*(m) &:= \{u \in T^* \mid u \text{ enabled at } m\}, \\ \mathcal{R}(m) &:= \{m + \Delta(w) \mid w \in \mathcal{F}^*(m)\}. \end{aligned}$$

$\mathcal{R}(m)$ is the set of markings *reachable* from m.

For any set Γ, a Γ-*labeled* equational Petri net is a triple $\langle E; \Gamma, h\rangle$ where E is an EPN, and where h is a function $h : T \to \Gamma$. We call h a *labeling* of T in Γ. In a $\Gamma \cup \{\varepsilon\}$-labeled net the labeling function has the format $h : T \to \Gamma \cup \{\varepsilon\}$. Similarly one defines labeled, marked EPNs. Define the finitary and infinitary *exhaustive language* of E with marking m_0 by

$$\begin{aligned} \mathcal{L}^*(m_0) &:= h[\mathcal{F}^*(m_0)] \subset \Gamma^*, \\ \mathcal{L}^\infty(m_0) &:= h[\mathcal{F}^\infty(m_0)] \subset \Gamma^\infty. \end{aligned}$$

For simplicity we will frequently write just $|m_0|$ for $\mathcal{L}^*(m_0)$ and $|m_0|_T$ for $\mathcal{F}^*(m_0)$. If the underlying net is not clear from context we will also write $\mathcal{F}^\infty(m_0; E)$, $\mathcal{L}^*(m_0; E)$ and so forth.

A language $L \subset \Gamma^*$ is an *equational Petri net language* (EPNL for short) iff for some Γ-labeled equational Petri net E and some marking m_0: $L = \mathcal{L}^*(m_0; E)$. Example 1 shows that EPNLs in general fail to be context free, even if the corresponding net is very simple.

It will be convenient to assume that $\delta(t) > 0$ for all transitions t in an equational Petri net. This property can be achieved by adding another place p_g to the net if necessary (for "garbage collection") and if $\delta(t) = 0$ for some transition t, we redefine δ at t by: $\delta(t) = p_g$ (see example 1). Furthermore, we will assume that there is at most one place that does not occur as the source of any transition. It is clear from the definitions just given that these conventions do not affect the firing sequences that are enabled in the net.

Example 1

For any alphabet Γ, a symbol $\gamma \in \Gamma$ let $\#_\gamma : \Gamma^* \to \mathbf{N}$ be the Parikh map: $\#_\gamma(x)$ is the number of occurrences of symbol γ in word $x \in \Gamma^*$. Define a net E_0 by $P = \{p_0, p_1, p_2, p_g\}$, $T = \{t_1, t_2, t_3\}$, $h(t_1) = a$, $h(t_2) = b$, $h(t_3) = c$, $\pi(t_i) = p_i$ and $\delta(t_1) = p_0 + p_1 + p_2$, $\delta(t_2) = \delta(t_3) = p_g$. Let $m_0 := p_0$ be the initial marking. Then $|m_0| = \{z \in \{a, b, c\}^* \mid \forall x \in Pref(z)(\#_a x \geq \#_b x, \#_c x)\}$. See figure 1 for a graphical representation of this net.

5 Equivalence of EPNLs and Languages of Admissible Words

There is a natural way to associate to a parallel successor system $\mathfrak{R} = \langle V, \rho, \sigma\rangle$ an $\Sigma \cup \{\varepsilon\}$-labeled equational Petri net E'. The net has as places the vertices of $\mathfrak{R}$ and its transitions are defined as follows. If the label of v is $\omega \in \Sigma_k$, $k > 0$, then there are k transitions t_i in the net labeled (ω, i), $i \in [k]$, such that $\pi(t_i) = v$ and $\delta(t_i) = \sigma_v(i)$. If there is a vertex v in $\mathfrak{R}$ labeled $\perp$ then the corresponding place has outdegree 0. If the label of v is $+$ then there are transitions t_i, $i = 1, 2$, labeled ε having source v such that $\delta(t_i) = \sigma_v(i)$, $i = 1, 2$. Lastly, if the label of v is '$\|$' then there is a transition t with source v labeled ε such that $\delta(t)$ is the multiset $\sigma_v(1) + \sigma_v(2)$. With these definitions it is quite straightforward to show that $\Sigma(\mathfrak{R}) = \mathcal{L}(r; E'_{\mathfrak{R}})$, where r is the root of $\mathfrak{R}$.

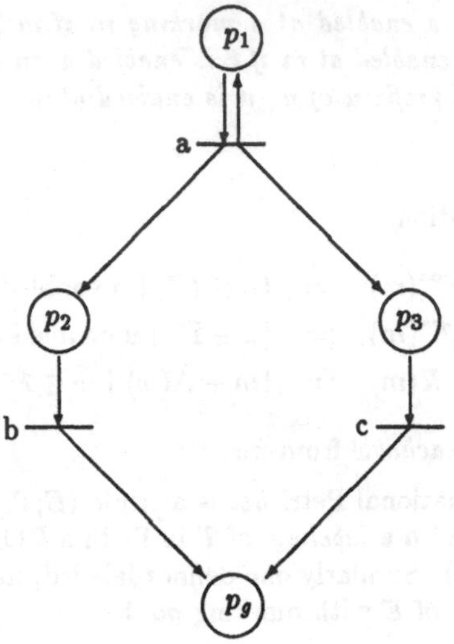

Figure 1: A primitive equational Petri net.

For a place p and a symbol $\sigma \in \Sigma$ define $eps\,(p,\omega) := \{m \in N^P \mid p \cdot w = m \wedge h(w) = \omega\}$. Thu $eps(p,\omega)$ is the set of all markings that can be reached from p by firing a transition labeled ω plus an number of ε-labeled transitions. If the parallel system satisfies condition 3, Section 2, then $eps(p,\omega$ must be finite for all places p in $E'_{\Re}$. This is the key to eliminating ε-labeled transitions from th net. Briefly, a new Γ-labeled net $E_{\Re}$ is obtained from $E'_{\Re}$ by deleting all the old transitions and, fc every p in P and every marking m in $eps(p,\omega)$, introducing a new transition t labeled ω such tha $\delta(t) = (p,m)$. Furthermore, one new place q with transitions t such that $h(t) = \omega$ and $\delta(t) = (q, m$ for all $m \in eps(r,\omega)$ is added to the net. Here $r \in V = P$ is the root of the successor system. The an easy induction shows that $\mathcal{L}(q; E_{\Re}) = \Sigma(\Re)$.

It is somewhat easier to describe the system of equations determined by a Γ-labeled EPN E; th variables appearing in the equations are the places, and for each place x and each transition t wit $\pi(t) = x$ we let the term $\tau(x,t)$ be defined by

$$\tau(x,t) := \omega(y_1 \| y_2 \| \cdots \| y_k),$$

where the label of t is ω and $\delta(t) = y_1 + \cdots + y_k$ (there may be repetitions among the y's). Th equation corresponding to the place x is then

$$x = \tau(x,t_1) + \cdots + \tau(x,t_m),$$

where $t_1, \ldots, t_m$ are all transitions with source x. (If x is a place of outdegree 0, the last equatio becomes $x = \perp$, or, equivalently in the algebra $C\Sigma$, $x = \{\varepsilon\}$.) The system of equations determine by E is the set of such equations, one for each place x in E. Note that these equations are already i normal form.

<u>Example 2</u>
Consider a parallel successor system $\Re$ defined by $V = \{r, v_1, v_2, u_1, u_2\}$, $\rho(r) = 1$, $\rho(v_i) = 2$, $\rho(u_i) =$ and lastly $\sigma_r(1) = v_1$, $\sigma_{v_1}(1) = r$, $\sigma_{v_1}(2) = v_2$ and $\sigma_{v_2}(i) = u_i$. The node-labels are given by $\lambda(r) =$ $\lambda(u_1) = b$ and $\lambda(u_2) = c$, $\{a,b,c\} = \Omega_1$, $\lambda(u_3) = \perp \in \Omega_0$; $\lambda(v_i) = \|$, $i = 1,2$. The correspondin system of equations is

$$r \;=\; a \cdot v_1,$$

$$\begin{aligned} v_1 &= \|(r, v_2), \\ v_2 &= \|(u_1, u_2), \\ u_1 &= b \cdot u_3, \\ u_2 &= c \cdot u_3, \\ u_3 &= \bot. \end{aligned}$$

Now define an equational Petri net $E_{\Re'}$ as follows. Let $P = V$ and $T = \{t_i \mid i \in [5]\}$. The maps π and δ as well as labeling h are defined in the following table.

i	$h(t_i)$	$\pi(t_i)$	$\delta(t_i)$
1	a	r	v_1
2	ε	v_1	$r + v_2$
3	ε	v_2	$u_1 + u_2$
4	b	u_1	u_3
5	c	u_2	u_3

Performing the ε-elimination procedure described above yields the following net $E_{\Re}$. There are 13 transitions $t_1, \ldots, t_{13}$; the functions π and δ as well as labeling h are again given by a table as follows. Note that $\mathcal{L}(r; E_{\Re}) = \mathcal{L}(m_0; E_0)$ where E_0 is the net from example 1.

i	$h(t_i)$	$\pi(t_i)$	$\delta(t_i)$
1	a	r	v_1
2	a	r	$r + v_2$
3	a	r	$r + u_1 + u_2$
4	a	v_1	$v_1 + v_2$
5	a	v_1	$r + 2v_2$
6	a	v_1	$r + v_2 + u_1 + u_2$
7	a	v_1	$r + 2u_1 + 2u_2$
8	b	v_1	$r + u_2 + u_3$
9	c	v_1	$r + u_1 + u_3$
10	b	v_2	$u_2 + u_3$
11	c	v_2	$u_1 + u_3$
12	b	u_1	u_3
13	c	u_2	u_3

The next three propositions indicate some major differences between general Petri nets and equational Petri nets. We omit their proofs. Let $\Gamma = \{\omega_1, \ldots, \omega_k\}$ and define the full Parikh map $\# : \Gamma^* \rightarrow N^k$ by $\#(x) := (\#_{\omega_1} x, \ldots, \#_{\omega_k} x)$.

Proposition 7 *In any Γ-labeled equational Petri net E with markings m_1, m_2 we have: $|m_1 + m_2| = |m_1| \,\|\, |m_2|$.*

It is demonstrated in [1] that the problem of testing equality of sets of reachable markings is undecidable for general Petri nets. However, this problem is decidable for equational Petri nets due to the simple periodic structure of their set of reachable markings.

Proposition 8 *In any equational Petri net, the set of reachable markings $\mathcal{R}(m)$ is semi-linear. Thus it is decidable whether two marked equational Petri nets have the same set of reachable markings.*

A similar argument shows that the commutative image of a EPNL is semi-linear. Hence many EPNLs are examples of non context free languages with semi-linear commutative image.

Proposition 9 *For every equational Petri net language L, the commutative image $\#(L)$ is semi-linear. Thus it is decidable if two equational Petri net languages are letter-equivalent.*

6 Regular and Context Free EPNLs

Clearly every Petri net language is context sensitive. As example 1 shows, even primitive equationa Petri nets may generate languages that fail to be context free. We will now give necessary an sufficient conditions for the set of firing sequences of an equational Petri net to be regular or contex free, respectively. A characterization of those Petri nets generating a regular exhaustive language ca be found in [9].

6.1 Firing Trees

The linear order inherent in a firing sequence obscures the concurrent nature of a Petri net. Therefor we introduce the notion of a *firing tree*: a firing tree is essentially a derivation tree in an appropriat rewriting system. Let E be an equational Petri net, m a marking and suppose $w \in |m|_T$. Assume som fixed linear ordering of the places in E. Define a forest $TR(m,w)$ consisting of exactly $rk(m)$ ordere trees whose nodes are labeled in P. $TR(m,w)$ is defined by induction on w as follows. - $TR(m,\varepsilon$ consists of $rk(m)$ isolated roots, the label of the i-th root is p where $\Sigma_{q<p} m(q) < i \leq \Sigma_{q \leq p} m(q)$. $TR(m,wt)$ is obtained from $TR(m,w)$ by finding the leftmost occurrence of a leaf z labeled $p = \pi(t$ and attaching new leaves $z_1,\ldots,z_k$ to z where $k = rk(\delta(t))$. The label of z_i is q where $\Sigma_{q'<q}\delta(t)(q')$ < $i \leq \Sigma_{q' \leq q}\delta(t)(q')$. We will write $\lambda(z)$ for this label. Now let w be a firing sequence w enabled at m and let p be a place. It is obvious from the definitions that
- the length of w is the number of interior nodes in $TR(m,w)$ and
- $(m \cdot w)(p)$ is the number of leaves in $TR(m,w)$ labeled p.

A place p in E is ***proper*** if there is a transition t enabled at p such that $\Delta(t)(p) = -1$, ***imprope*** otherwise. A set Q of places is ***unbounded*** (for the marking m_0) if $\forall k \geq 0 \exists m \in \mathcal{R}(m_0) \forall p$ $Q(m(p) \geq k)$. A set Q of places is ***proper unbounded*** iff Q is unbounded and all the places in Q ar proper. Lastly a net is ***k-unbounded*** iff $k = \max(|Q| \mid Q \subset P$ proper unbounded).

6.2 The Collapse of an Equational Petri Net

In what follows it will be convenient to think of an equational Petri net E as a bi-partite mult digraph. Thus for each transition t with source p there is an edge from p to t and edges from t t q (with multiplicity $\delta(t)(q)$) for all q in the support of $\delta(t)$. Note that the indegree of every vertex is at most one. The net E is ***primitive*** iff the corresponding graph contains no strongly connecte components (scc, for short) with more that two vertices. We now associate every equational Pet net E with a primitive net $C(E)$, the ***collapse*** of E. The nets E and $C(E)$ will have essentially th same sets of unbounded places. $C(E)$ is obtained from E by collapsing all the strongly connecte components C of E of size at least two into one place p_C and one transition t_C.

We will distinguish several types of strongly connected components as follows. Let C be a strongl connected component of E of size at least two.

C is ***inactive*** if $\forall t \in C(outdegree(t) \leq 1)$; otherwise C is ***active***. More specifically, C is ***external active*** if $\forall t \in C(outdegree_C(t) \leq 1) \wedge \exists t \in C(outdegree(t) \geq 2)$. C is ***internally active*** if $\exists t$ $C(outdegree_C(t) \geq 2)$.

Lastly define a collapse map κ from E onto $C(E)$ by

$$\kappa(z) := \begin{cases} z & \text{if the scc } C \text{ of } z \text{ is } \{z\}, \\ z_C & \text{otherwise.} \end{cases}$$

Note that $C(E)$ can be constructed from E in polynomial time using standard graph theory algorithm

The net E_0 from example 1 is primitive: it has only one non-trivial scc, namely $\{p_0, t_1\}$. This component is externally active. The place p_0 is improper, and $\{p_1, p_2, p_g\}$ is a proper unbounded set of places. As we will see shortly this implies that the language of the net is not context free. The net $E_{\mathfrak{R}}$ from example 2 also has only one active component, namely $\{r, v_1, t_1, \ldots, t_7\}$. All places are proper, and the set $\{v_2, u_1, u_2, u_3, u_3\}$ is proper unbounded.

Lemma 10 *Let E be a marked equational Petri net, $\mathcal{C}(E)$ its collapse. Then a set Q of places is unbounded in E iff $\kappa[Q]$ is unbounded in $\mathcal{C}(E)$.*

Proof. It clearly suffices to verify the following two claims. For all firing sequences w in E enabled at m_0 there exists a firing sequence w' in $\mathcal{C}(E)$ such that

$$\kappa(m_0 \cdot w) \leq \kappa(m_0) \cdot w'.$$

Conversely, for all firing sequences w' in $\mathcal{C}(E)$ enabled at $\kappa(m_0)$ there exists a firing sequence w in E such that

$$\kappa(m_0) \cdot w' \leq \kappa(m_0 \cdot w).$$

Since the proofs are similar, we give only the second argument. We proceed by induction on w'. Suppose $w' = u' \cdot t'$ where t' is some transition in $\mathcal{C}(E)$. By induction hypothesis $\kappa(m_0) \cdot u' \leq \kappa(m_0 \cdot u)$ for some firing sequence u in E. If t' is one of the transitions unaffected by the collapse (because the corresponding strongly connected component contains only t'), then we may set $w := u \cdot t'$. So assume $t' = t_C$. Note that at least one of the transitions in C is enabled at $m_0 \cdot u$. We distinguish several cases depending on the type of the strongly connected component C.

Case 1 C is inactive.
In this case $\Delta(t_C) = 0$. Pick a transition t in C which is enabled at $m_0 \cdot u$ and set $w := u \cdot t$. Note that $\kappa(m_0) \cdot u' \cdot t_C = \kappa(m_0 \cdot u') \leq \kappa(m_0 \cdot u) = \kappa(m_0 \cdot u \cdot t)$.

Case 2 C is active.
In this case $\Delta(t_C) > 0$ and possibly $\Delta(t_C)(p_C) = 1$ (if C is internally active). Pick a firing sequence v in C enabled at $m_0 \cdot u$ such that $\kappa(\Delta(v)) \geq \Delta(t_C)$. The existence of such a sequence follows from the fact that C is strongly connected and the definition of $\mathcal{C}(E)$: any edge (t_C, z) in $\mathcal{C}(E)$ comes from an edge (t, q) where t is a transition in C. Now set $w := u \cdot v$. Note that $\kappa(m_0) \cdot u' \cdot t_C = \kappa(m_0 \cdot u') + \Delta(t_C) \leq \kappa(m_0 \cdot u) + \kappa(\Delta(v)) = \kappa(m_0 \cdot u \cdot v)$. □

The following lemma establishes a normal form for the firing sequences that show that a set of places is unbounded.

Lemma 11 *Suppose that E is a primitive equational Petri net, $Q = \{q_1, \ldots, q_n\}$ a set of places in E and that m_0 is a marking. Then Q is unbounded for m_0 iff there exist repetition-free firing sequences u_i, v_i and transitions t_i, $i = 1, \ldots, n$, such that*

$$\forall k \geq 0, q \in Q(m_0 \cdot u_1 t_1^k v_1^k \ldots u_n t_n^k v_n^k(q) \geq k).$$

Proof. The sufficiency of our condition is obvious, to see necessity let $T_0 \subset T$ be the set of self-loop transitions in E with positive yield, and let P_0 be the corresponding set of places. Suppose Q is unbounded for m_0. Then for any $k \geq 0$ there must be a firing sequence w enabled at m_0 such that the firing tree $TR(m_0, w)$ contains at least k leaves labeled q for all q in Q. Choose $k := rk(m_0) \cdot r^{2|T|}$ where $r := \max(\delta(t) \mid t \in T)$ and let w be a minimal such firing sequence. Then at least one of the trees in $TR(m_0, w)$ must contain one of the transitions in T_0. Suppose $\delta(t) = \pi(p) + d$ where $d > 0$ is the top-most occurrence of any such transition. Then there is a repetition-free sequence u_1 of transitions in $T - T_0$ such that $u_1 t^* \subset |m_0|_T$. Let $\{q_1, \ldots, q_j\} \subset Q$ be all the places in Q that occur in at least one of the $rk(d)$ subtrees associated with the transition t. By the minimality of w, we have $j \geq 1$.

Hence there exist repetition-free firing sequences $v_1, \ldots, v_j$ such that $m_0 \cdot u_1 t_1^k v_1^k \ldots t_1^k v_j^k(q_i) \geq k$ fc $i = 1, \ldots, j$. Now delete all subtrees from $TR(m_0, w)$ that contain no leaves in $Q_1 := Q - \{q_1, \ldots, q_j$ Replace m_0 by $m_1 := m_0 \cdot v_1$ and Q by Q_1 and proceed by induction. □

Note that in general an equational Petri net has an exponential number of maximal unbounded set of places. Hence we cannot hope to enumerate them in polynomial time. However, the next theore shows that it is already NP-complete to determine whether a single given set Q of places is unbounde More precisely, consider the following decision problem.

Problem: Boundedness for EPNs
Instance: An equational Petri net E, a marking m and a set Q of places in E.
Question: Is Q an unbounded set of places for m in E?

Remark. The reachability problem for EPN's has been shown to be NP-complete in [3]. A relate boundedness problem has been considered in [2].

Theorem 12 *The problem of Boundedness for EPNs is NP-complete. The problem remains NP complete even if the net is required to be primitive.*

Proof. To see membership in NP first note that the primitive net $C(E)$, the collapse of an equation Petri net E, can be computed in polynomial time. Hence, by lemma 10, we may safely assume tha we are dealing with a primitive net. So let $Q = \{q_1, \ldots, q_n\}$ a set of places in E and m_0 a markin By lemma 11, the set Q is unbounded for m_0 iff there exist repetition-free firing sequences u_i, v_i an transitions t_i, $i = 1, \ldots, n$, such that $\forall k \geq 0,\ q \in Q(m_0 \cdot u_1 t_1^k v_1^k \ldots u_n t_n^k v_n^k(q) \geq k)$. The tot size of $u_1, \ldots, u_n, v_1, \ldots, v_n$ and $t_1, \ldots, t_n$ is polynomial in the size of E. Hence one can guess i non-deterministic polynomial time at these firing sequences and verify in deterministic polynomi time that they indeed have the desired properties.

To show hardness we will embed 3-Satisfiability (3SAT). An instance of 3SAT is a Boolean formu $\Phi = \phi_1 \wedge \phi_2 \wedge \ldots \wedge \phi_m$ in 3-conjunctive normal form using variables in $X = \{x_1, \ldots, x_n\}$. Suppos clause ϕ_i is $z_{i,1} \vee z_{i,2} \vee z_{i,3}$ where the $z_{i,j}$ are literals over X. Define a primitive equational Pet net E as follows. E has places $p_{i,j}$, $i = 1, \ldots, n$, $j = 0, 1, 2$ and q_k, $k = 1, \ldots, m$, the 'pockets There is a transition $t_{i,j}$ with yield $p_{i,j} - p_{i,0}$, $j = 1, 2$, $i \in [n]$. Furthermore, there is a transition s_i enabled at $p_{i,1}$ with yield $\Sigma_{x_i \in q_k} q_k$ and similarly $s_{i,2}$ is enabled at $p_{i,2}$ with yield $\Sigma_{\bar{x}_i \in q_k} q_k$, $i \in [n$ Hence the net consists of n basic components corresponding to the n boolean variables in Φ plus pockets corresponding to the clauses in Φ. Now consider the initial marking $m_0 := p_{1,0} + \cdots + p_{n,}$ In each component the first transition fired determines whether the corresponding variable is true false. The subsequent firings then allow the creation of an unbounded marking in those pockets whos corresponding clauses are satisfied by that particular choice of an assignment. It is easy to verify tha Φ is satisfiable iff $Q := \{q_1, \ldots, q_m\}$ is unbounded for the initial marking m_0. □

In contrast to the computational hardness of unboundedness for sets of places of arbitrary size we wi show that for a set Q of cardinality at most 2 one can determine in polynomial time whether Q unbounded. Hence there is a polynomial time algorithm to decide whether a net is 0-unbounded 1-unbounded.

For a strongly connected component C of an equational Petri net E let $D(C) := \{p \mid \exists t, q_1, q_2$ $C(\pi(t) = q_1 \wedge \delta(t)(q_2) > 0 \wedge \delta(t)(p) > 0)\}$. $D(C)$ is the collection of all places that can be "pumpe directly via firing sequences entirely within C. Clearly $D(C)$ can be determined in polynomial tim As a consequence of lemma 10 and lemma 11 we have the following two corollaries.

Lemma 13 *Let E be an equational Petri net, m_0 a marking and $Q = \{q\}$. Then Q is unbounded there exists an active strongly connected component C reachable from the support of m_0 such that is reachable from $D(C)$.*

Corollary 14 *Let E be an equational Petri net, m_0 a marking and $Q = \{q_1, q_2\}$. Then Q is unbounded iff there exist two active strongly connected components C_1 and C_2(not necessarily distinct) reachable from p_1 and p_2 respectively such that q_i is reachable from $D(C_i)$, $i = 1, 2$, and one of the following three conditions holds: $m_0 \geq p_1 + p_2$, $m_0 \geq p_1$* and *$p_2 \in C_1$, or $m_0 \geq p_2$* and *$p_1 \in C_2$.*

Theorem 15 *The set of all firing sequences enabled at m_0 in an equational Petri net E is regular iff E is 0-unbounded, i.e., E contains no proper unbounded places with respect to m_0.*

Proof. First suppose E is 0-unbounded. Let P_0 be the collection of all proper places in E. Then for some constant c we have: $\forall m \in \mathcal{R}(m_0), p \in P_0(m(p) \leq c)$. But for any improper place p, $m(p) > 0$ and $m' \in \mathcal{R}(m)$ implies $m'(p) > 0$: firing any transition enabled at p does not decrease $m(p)$. Now define the marking $m^\dagger$ by

$$m^\dagger(p) := \begin{cases} m(p) & \text{if } p \text{ proper,} \\ \min(m(p), 1) & \text{otherwise.} \end{cases}$$

Define a finite sink automaton M as follows: M has states $\{0, \ldots, c\}^{P_0} \times \{0,1\}^{P_0} \cup \{\bot\}$, where $m_0^\dagger$ is the initial state and $\bot$ is the sink. The transition function is defined by

$$\delta(m, t) := \begin{cases} (m \cdot t)^\dagger & \text{if } t \text{ is enabled at } m, \\ \bot & \text{otherwise.} \end{cases}$$

Clearly M accepts a firing sequence w iff w is enabled at m_0. For the opposite direction assume that E contains at least one proper unbounded place p_0. By lemma 11 there exist firing sequences u and w such that $\forall k \geq 0((m \cdot uw^k)(p_0) \geq k)$. Let t be a transition enabled at p such that $\Delta(t)(p) = -1$. Suppose for the sake of a contradiction that $|m_0|_T$ is regular. Note that for all $k \geq 0$: $uw^k t^k$ is enabled at m_0. By the pumping lemma there is some r sufficiently large such that $uw^r t^* \subset |m_0|$, contradiction. □

Theorem 16 *The set of all firing sequences enabled at m_0 in an equational Petri net E is context free (but not regular) iff E is 1-unbounded, i.e., any set of proper unbounded places with respect to m_0 has cardinality at most one and there exists at least one such set.*

Proof. First suppose E is 1-unbounded. Again let P_0 be the collection of all proper places in E. For any $c \geq 0$ call a marking m *c-critical* iff $\exists p(m(p) \geq c)$. As an immediate consequence of the 1-unboundedness of E we have for some sufficiently large c_0 and all m in $\mathcal{R}(m_0)$ there exists at most one proper place p such that $m(p) \geq c_0$. Thus for all $c \geq c_0$ and m in $\mathcal{R}(m_0)$ c-critical let $crit(m, c)$ denote that uniquely determined place.

<u>Claim</u> There exists a constant $c_1 \geq c_0$ such that for all $m \in \mathcal{R}(m_0)$ and $m' \in \mathcal{R}(m)$ c_0-critical we have: $crit(m, c_1) = crit(m', c_1)$. For suppose otherwise. Then there are proper places $p \neq q$ such that for any k we have m_k in $\mathcal{R}(m_0)$ and m'_k in $\mathcal{R}(m_k)$ such that $m_k(p) \geq k$ and $m'_k(q) \geq k$. But then for any $r \geq 0$ there are infinitely many pairs $i < j$ such that $m_i + rp \leq m_j$. Note that $m'_i + rp$ is in $\mathcal{R}(m_j)$. But then $\{p, q\}$ is unbounded: $(m'_i + rp)(p) \geq r$ and $(m'_i + rp)(q) \geq i$, contradiction.

Now we can define a counter automaton M that simulates the equational Petri net E as follows. Let c_1 be as in the claim. For any marking m define $m^\dagger$ by

$$m^\dagger(p) := \begin{cases} m(p) & \text{if } m(p) < c_1, p \text{ proper,} \\ min(m(p), 1) & \text{if } p \text{ improper,} \\ c_1 & \text{if } p = crit(m, c_1). \end{cases}$$

M has states $\{0, \ldots, c\}^P \cup \{\bot\}$, initial state $m_0^\dagger$ and every state other than $\bot$ is final. M works as follows: as long as the marking m of the net is non-critical m is represented by state $m^\dagger$ of M and the stack is empty. When the marking becomes c_1-critical for the first time M switches mode: from no on it uses its stack to represent $m(p)$, $p := crit(m, c_1)$, all other places are represented by the stat $m^\dagger$. By the claim all markings reachable in the future are not c_1-critical in any component other tha p. Hence M properly simulates E. We leave the details to the reader.

On the other hand assume that $\{p_1, p_2\}$ is proper unbounded, $p_1 \neq p_2$. By lemma 11 there exis firing sequences u_1, u_2, w_1 and w_2 such that $\forall k \geq 0((m_0 \cdot u_1 w_1^k u_2 w_2^k)(p_i) \geq k)$, $i = 1,2$. Let t_i be a transition enabled at p_i respectively such that $\Delta(t_i)(p_i) = -1$. As in the last argument for a $k \geq 0$ the firing sequence $u_1 w_1^k u_2 w_2^k t_1^k t_2^k$ is enabled at m_0. By the iteration theorem for context fre languages $u_1 w_1^* u_2 w_2^* t_1^* t_2^* \cap |m_0|$ fails to be context free and we are through. □

Combining theorem 15 and 16 and corollary 13 and 14 we have the following result.

Theorem 17 *There is a polynomial time algorithm to determine whether for a given equational Pet net E there exists a marking m_0 such that $|m_0|_T$ fails to be regular – or context free.*

6.3 The Equality Problem

We now turn to the problem of determining whether two parallel systems over the same ranke alphabet Ω have the same behavior. In terms of equational Petri nets the problem in its simplest for can be phrased as follows: given two equational Petri nets with the same set of transitions T, decid whether the same firing sequences are enabled in the two nets. We will show that this problem ca be answered in non-deterministic linear space. In fact, a slightly more general result holds.

Theorem 18 *Given two Γ-labeled marked equational Petri nets $\langle E_i; m_i; h_i, \Gamma\rangle$, $i = 1,2$, where h_2 injective, the problem to decide whether $\mathcal{L}^*(m_1; E_1) \subset \mathcal{L}^*(m_2; E_2)$ is in PSPACE.*

Proof. Let ω be an arbitrary symbol in Γ. For all positions p in P_1 and all markings m define

$$c_p := \max(i \leq \infty \mid \omega^i \in \mathcal{L}^\infty(p; E_1) \cup \mathcal{L}^*(p; E_1))$$

and

$$\mu_1^\omega(m) := \Sigma_{p \in P_1} c_p m(p).$$

Thus for a marking $m \mu_1^\omega(m)$ is the maximum number of transitions labeled ω that can be fired at m Define μ_2^ω similarly for the net E_2. For every symbol ω in Γ we will determine a bound on the lengt of the shortest firing sequence $w \in \mathcal{F}^*(E_1; m_1)$ – if such a sequence exists at all – that has followin two properties:

$$h_1(w) \in \mathcal{L}^*(m_2; E_2) \tag{10}$$

$$\mu_1^\omega(m_1 \cdot w) > \mu_2^\omega(m_2 \cdot h_2^{-1}(h_1(w))). \tag{11}$$

As $\mathcal{L}^*(m_1; E_1) \subset \mathcal{L}^*(m_2; E_2)$ iff no such ω and w exists this will prove our claim. We may safel assume that $\mathcal{L}^*(m_1; E_1)$ lies completely in the range of h_2; otherwise one can easily construct witness $x \in \mathcal{L}^*(m_1; E_1) - \mathcal{L}^*(m_2; E_2)$ of length at most n-1. Hence $h_2^{-1}(h_1(w))$ is defined for a $w \in |m_1|_T$; we will write $g(w) := h_2^{-1}(h_1(w))$. We will show that the desired bound is:

$$B := \gamma^{2n+2} \cdot \max(rk(m_1), rk(m_2)) + n. \tag{12}$$

Here $\gamma := \max(\max(rk(\delta_1(t)) \mid t \in T_1), \max(rk(\delta_2(t)) \mid t \in T_2))$ and $n := |P_1|$. We may safely assur that $\gamma \geq 2$, otherwise both nets are essentially finite automata and our claim follows from standar results on regular languages. Taking the bound B for granted, a PSPACE algorithm can be given as follows: non-deterministically generate a firing sequence w in T_1^* of length at most B and verify that $w \in |m_1|_{T_1}$ but $g(w) \notin |m_2|_{T_2}$. The markings that appear during the firing of w and $g(w)$ all have rank $O(\gamma^{cn})$. In binary notation their size is therefore linear in the size of the input $\langle E_1; m_1\rangle$ and $\langle E_2; m_2\rangle$. Hence the problem is solvable in non-deterministic linear space and thus in deterministic quadratic space by Savitch's theorem.

We now return to our main argument. A self-loop at a place p is a transition t enabled at p such that $\Delta(t)(p) \geq 0$. Clearly a self-loop at p labeled ω causes $c_p = \infty$. Let q_2 in P_2 be the place that enables a transition labeled ω in E_2. Note that by our assumption q_2 is uniquely determined and $\mu_2^\omega(m) = \infty$ or $\mu_2^\omega(m) = m(q_2)$ depending on whether q_2 has a self-loop labeled ω or not. Here a self-loop at place p is understood to be a transition t enabled at p such that $\Delta(t)(p) \geq 0$. Now suppose w is a minimal witness satisfying the above conditions 10 and 11.

Case 1 The place q_2 has no self-loop labeled ω in E_2.
First notice that we can restrict ourselves to the case where there is no cycle of transitions in E_1 all labeled by ω. For otherwise one can construct a witness v of length at most $n-1$ such that $\mu_1^\omega(m_1 \cdot v) = \infty > \mu_2^\omega(m_2 \cdot g(v))$ and there is nothing to show. Thus for any marking m we have $\mu_1^\omega(m) < \infty$. Now let β be the longest branch in $TR(m_1, w)$. By 12 the depth of $TR(m_1, w)$ is a least $n+2$, hence β has length at least $n+2$. Let $z_0, z_1, \dots, z_n$ be the $n+1$ bottom-most interior nodes on β (z_0 being the father of the leaf). Hence $\exists 0 \leq i < j \leq n(\lambda(z_i) = \lambda(z_j))$ by the pigeonhole principle. Let u be a firing sequence associated with the subtree of z_j minus the subtree of z_i, and w_2 a firing sequence associated with the subtree of z_i. One can rearrange w to a firing sequence $\bar{w} = w_1 u w_2 \in |m_1|$. By our choice of u the firing sequence $w_1 w_2$ is also enabled at m_1. But $|w_1 w_2| < |\bar{w}| = |w|$, hence $g(w_1 w_2)$ is enabled at m_2. Now $\mu_1^\omega(m_1 \cdot \bar{w}) = \mu_1^\omega(m_1 \cdot w) > \mu_2^\omega(m_2 \cdot g(w)) = \mu_2^\omega(m_2 \cdot g(\bar{w}))$ and $\mu_1^\omega(m_1 \cdot \bar{w}) = \mu_1^\omega(m_1 \cdot w_1 w_2) + d_1$, $\mu_2^\omega(m_2 \cdot g(\bar{w})) = \mu_2^\omega(m_2 \cdot g(w_1 w_2)) + d_2$ where $d_1 := \mu_1^\omega(\Delta_1(u))$ and $d_2 := \mu_2^\omega(\Delta_2(g(u)))$ are the respective gains in net E_1 and E_2 during the firing of u and $g(u)$. Note that $\Delta_1(u) \geq 0$, whence $d_1 \geq 0$. Again by the minimality of w we must have $\mu_1^\omega(m_1 \cdot w_1 w_2) \leq \mu_2^\omega(m_2 \cdot g(w_1 w_2))$. Thus $d_1 > d_2$. Also observe that $|u| \leq \gamma^n$ by our choice of z_j. By deleting repetitious labels on β above z_j one can construct a firing sequence $\bar{u}$ enabled at m_1 of length at most $n-1$ such that $\bar{u}u^* \subset |m_1|_T$. But we have just seen that $d_1 > d_2$, hence there exists a $k \leq rk(m_2)$ such that $g(\bar{u}u^k) \notin |m_2|_T$ and we are done.

Case 2 The place q_0 has a self-loop labeled ω in E_2.
Note that in this case we must have $\mu_1^\omega(m_1 \cdot w) > \mu_2^\omega(m_2 \cdot w) = 0$. Again let β be the longest branch in $TR(m_1, w)$. By 12 let $z_0, \dots, z_{2n}$ be the $2n+1$ bottom-most interior nodes on β (z_0 being the father of the leaf). By the pigeonhole principle we have

$$\exists 0 \leq i < j \leq r < s \leq 2n(\lambda(z_i) = \lambda(z_j) \wedge \lambda(z_r) = \lambda(z_s)).$$

One can rearrange w to a firing sequence $\bar{w} = w_1 v w_2 u w_3 \in |m_1|_T$ where u is a firing sequence associated with the subtree of z_j minus the subtree of z_i, v is a firing sequence associated with the subtree of z_s minus the subtree of z_r, and lastly w_3 a firing sequence associated with the subtree of z_i. As w is a minimal witness we must have $g(w_1 v w_2 w_3) \in |m_2|_T$. Hence the gain in the second net during the firing of $g(u)$ in position q_2 is not positive: $d_2 := \mu_2^\omega(\Delta_2(g(u))) = \mu_2^\omega(m_2 \cdot g(w)) - \mu_2^\omega(m_2 \cdot g(w_1 v w_2 w_3)) \leq 0$. If $d_2 < 0$ one can use the argument from case 1, so assume $d_2 = 0$. Then $\mu_2^\omega(m_2 \cdot g(w_1 v w_2 w_3)) = 0$ and by the minimality of $w \mu_1^\omega(m_1 \cdot w_1 v w_2 w_3) = 0$. It follows that $d_1 := \mu_1^\omega(\Delta_1(u)) \geq 1$. Now consider the firing sequence $w_1 w_2 u w_3$. Note that $\mu_1^\omega(m_1 \cdot w_1 w_2 u w_3) \geq 1$. Hence, again by our choice of w, $g(w_1 w_2 u w_3)$ is enabled at m_2 and $\mu_2^\omega(m_2 \cdot g(w_1 w_2 u w_3)) \geq 1$. Thus $d_2' := \mu_2^\omega(\Delta_2(g(v))) = \mu_2^\omega(m_2 \cdot g(w)) - \mu_2^\omega(m_2 \cdot g(w_1 w_2 u w_3)) < 0$ whereas $d_1' := \mu_1^\omega(\Delta_1(v)) \geq 0$. As in case 1 there is a firing sequence $\bar{v}$ of length at most $n-1$ such that $\bar{v}v^* \subset |m_1|_T$. However, for some $k \leq rk(m_2) + n \cdot B$, $g(\bar{v}v^k) \notin \mathcal{L}^*(E_2; m_2)$. This completes the proof. □

Corollary 19 *Given two marked EPNs* $\langle E_i; m_i \rangle$, $i = 1, 2$, *with free labelings* $h_i : T_i \to \Sigma$ *the Equalit Problem as well as the Subset Problem are in PSPACE. I.e., there are algorithms with polynomia space complexity to decide whether* $\mathcal{L}^*(m_1; E_1) = \mathcal{L}^*(m_2; E_2)$ *and whether* $\mathcal{L}^*(m_1; E_1) \subset \mathcal{L}^*(m_2; E_2)$

We conclude by stating some open problems. Corollary establishes an upper bound for the Equalit as well as the Subset Problem in equational Petri nets with free labelings. We are not aware of an lower bounds. Furthermore, it would be interesting to know whether these problems are decidable fo equational Petri nets that satisfy the disjoint labeling condition and ultimately for arbitrarily labele nets.

BIBLIOGRAPHY

[1] M. Hack. The equality problem for vector addition systems is undecidable. *Theoretical Compute Science*, 2:77 – 95, 1976.

[2] R.R. Howell, L.E. Rosier, and H. C. Yen. An $O(n^{1.5})$ algorithm to decide boundedness for conflic free vector replacement systems. Technical report, University of Texas at Austin, 1986.

[3] D.T. Huynh. Commutative grammars: the complexity of the uniform word problems. *Informatio and Control*, 57:21–39, 1983.

[4] M. Jantzen. Complexity of place and transition systems. In *Petri Nets: Central Models and the Properties*, volume 254 of *LNCS*. Springer-Verlag, 1986.

[5] M. Jantzen. Language theory of Petri nets. In *Petri Nets: Central Models and their Propertie* volume 254 of *LNCS*. Springer-Verlag, 1986.

[6] W. Reisig. *Petri Nets*, volume 4 of *EATCS*. Springer Verlag, 1986.

[7] G. Rozenberg and R. Verraedt. Subset languages of Petri nets part I: The relationship to strin languages and normal forms. *Theoretical Computer Science*, 26:301 – 326, 1983.

[8] G. Rozenberg and R. Verraedt. Subset languages of Petri nets part II: The closure propertie *Theoretical Computer Science*, 27:85 – 108, 1983.

[9] R.Valk and G.Vidal-Naquet. Petri nets and regular languages. *Journal of Computers and Syste Science*, 23:299 – 325, 1981.

Towards a Lambda-Calculus for Concurrent and Communicating Systems

Gérard Boudol

INRIA Sophia-Antipolis

06565-VALBONNE FRANCE

Abstract.

We introduce a calculus for concurrent and communicating processes, which is a direct and simple extension of the λ-calculus. The communication mechanism we use is that of Milner's calculus CCS: to communicate consists in synchronously sending and receiving a value through a shared port. Then the calculus is parameterized on a given set of port names, which are used in the two primitives for sending and receiving a value – as in the λ-calculus, a value can be any term. We use two parallel constructs: the first is interleaving, which does not allow communication between agents. The second, called cooperation, is a synchronizing construct which forces two agents to communicate on every port name. We show that the λ-calculus is a simple sub-calculus of ours: λ-abstraction is a particular case of reception (on a port named λ), and application is a particular case of cooperation.

1. Introduction.

The λ-calculus of Church formalizes in a very concise way the idea of functions being applied to arguments. Despite its simplicity, this calculus provides an astonishingly rich model for sequential evaluation, see [2]. A challenging problem that has emerged for some time is to devise a similar framework for concurrent and communicating processes, relying upon some "minimal" concepts for concurrency and communication. A natural claim is that such a formal model for processes should contain the λ-calculus as a simple sub-calculus – this would provide us with the full power of combinators. This note presents an attempt in this direction.

Regarding communication, our main source of inspiration is Milner's CCS [5]. Communication in CCS is a value passing act which two processes perform simultaneously: one of the two partners sends a value through a labelled port, while the other receives this value on a port

labelled by the same name, say α. Correspondingly there are two communication primitives in CCS, an output construct and an input construct. The output construct is $\bar{\alpha}e.p$, representing a process sending e on the port α, and then behaving as p. In this construct, e is an expression belonging to some outer language. The complementary input construct is $\alpha x.p$, representing a process receiving some value at the port α; here x is a bound variable, and receiving the value v yields a new process $p[x \mapsto v]$, that is p where v is substituted for x. Communication occurs when two concurrent processes perform matching send and receive actions. Therefore the *interaction law* may be stated, using $\|$ for parallel composition, as:

$$(\alpha x.p \parallel \bar{\alpha}e.q) \rightarrow (p[x \mapsto v] \parallel q)$$

where v is the result of evaluating e. In CCS such a transition is labelled by the communication action τ.

Let us discuss briefly how one could use CCS's ideas to find a generalization of the λ-calculus. Milner remarked (*cf.* [5] p.128) that one may compare a function's argument places with input ports of a process. Indeed the terms $\lambda x.p$ of Λ and $\alpha x.p$ of CCS behave quite similarly: both of them wait for a value to be substituted for x in p. This suggests that one could regard these two constructs as the same one – λ is thus a port name, the only one for the λ-calculus (*cf.* [5] p.49). Another obvious idea is that application of a function to its argument should be a special kind of communication (see again [5] p.128), or more precisely that *β-reduction should be the typical instance of an interaction law.* Then application appears as a parallel composition, where the argument is explicitly sent to the function. Regarding the sending primitive, we shall keep to the philosophy of the λ-calculus, where any term is a possible value. Then the sending construct is $\bar{\alpha}p.q$, where p is any agent. In fact we are only interested in the case where q is an idle (or terminated) process $\mathbb{1}$, which is like nil in CCS. Then $\bar{\alpha}p$ will be an abbreviation for $\bar{\alpha}p.\mathbb{1}$.

To work out the previous ideas, let us now introduce a first attempt – the calculus we actually propose will be a little bit more sophisticated. In order to build agents $\alpha x.p$ and $\bar{\alpha}p$ we need a denumerable set X of *variables* $x, y, z \dots$, and a non-empty set N of *port names*. We shall use α, $\beta \dots$ to range over port names. Then the syntax of the tentative calculus is given by the following grammar:

$$p ::= \mathbb{1} \mid x \mid \alpha x.p \mid \bar{\alpha}p \mid (p \parallel p)$$

where α is any port name. We shall use $p, q, r \dots$ to range over terms. As usual, the variable x is bound if it is in the scope of an αx, and some care is needed in defining substitution. For simplicity, we shall adopt Barendregt's variable convention ([2]): in any mathematical context where they occur, the terms $p_1, \dots, p_n$ are supposed to exhibit bound variables different from the free variables.

In this calculus, communication is given by an obvious adaptation of the interaction law of CCS, namely:

$$(\alpha x.p \parallel \bar{\alpha}q) \rightarrow (p[x \mapsto q] \parallel \mathbb{1})$$

The term $\mathbb{1}$ represents an idle process, and is a unit for parallel composition. Therefore the term we get after a communication, that is $(p[x \mapsto q] \parallel \mathbb{1})$, behaves like $p[x \mapsto q]$. Then the interaction law is similar to β-reduction, and, assuming that N contains a distinguished name λ, we can try to represent Λ as the subset of terms given by the grammar:

$$p ::= x \mid \lambda x.p \mid (p \parallel \bar{\lambda}p)$$

We could denote $(p \parallel \bar{\lambda} q)$ by (pq) (application), so that the previous rule is the β-rule, up to the simplification $(r \parallel \mathbb{1}) = r$. However this simple calculus fails to capture the λ-calculus. Let us see this point in more detail.

CCS also formalizes the natural idea that parallel composition is commutative and associative, so that processes need not be contiguous to communicate – unlike λ-terms where communication is sequential application. In other words, the following should hold in our tentative calculus:

$$(\cdots \parallel \alpha x.p \parallel \cdots \parallel \bar{\alpha} q \parallel \cdots) \rightarrow (\cdots \parallel p[x \mapsto q] \parallel \cdots \parallel \mathbb{1} \parallel \cdots)$$

Let us assume for a while that we have two rules stating that parallel composition is commutative and associative:

$$((p \parallel q) \parallel r) \rightarrow s \vdash (p \parallel (q \parallel r)) \rightarrow s$$

$$(p \parallel q) \rightarrow s \vdash (q \parallel p) \rightarrow s$$

These two rules introduce *conflicts*, arising from communication (technically speaking, we should say that associativity introduces overlapping redexes). As in CCS, there is a possibility that inputs at the same port may have different sources, and outputs at the same port different destinations. Then two communications are conflicting if they share the same destination, or the same source, the typical example being $((\alpha x.p \parallel \bar{\alpha} q) \parallel \bar{\alpha} r)$, and solving the conflict introduces non-determinism. To our view, the non-determinism arising from conflicting communications is a rather pleasant feature. But there is a negative consequence to the associativity (and commutativity) of parallel composition, namely that we lose the correspondence with λ-calculus application. For instance the term $((\lambda xy.x)u)v$ cannot be accurately represented by $p = ((\lambda x.\lambda y.x \parallel \bar{\lambda} u) \parallel \bar{\lambda} v)$ since we have $p \xrightarrow{*} ((v \parallel \mathbb{1}) \parallel \mathbb{1})$. One can recover from this failure by using the relabelling and restriction constructs of CCS, as shown by Thomsen in [7]. However, in Thomsen's CHOCS the λ-calculus is only caught up to observational equivalence; in particular the β-reduction is performed in two steps in his calculus, and thus is not an instance of the communication law.

We insist on obtaining the λ-calculus as a sub-calculus; more precisely, our goal is to find a direct generalization of the λ-calculus, that is a calculus where the operational semantics, once restricted to an appropriate subset of terms, gives an exact image of the β-reduction on Λ – recall: the β-reduction should be an instance of the interaction law. We do not want for instance to restrict the evaluation rules, by disallowing associativity of parallel composition for a particular kind of terms. Hence we must abandon the parallel composition of CCS – we do not expect to get CCS as a sub-calculus. One should observe that this parallel composition involves both the notion of concurrency and the notion of communication. Then our proposal is to split the into two constructors. The first is the usual *interleaving* construct $(p \mid q)$, which consists in juxtaposing p and q, without any communication wire between them. This operator represents *concurrency*. The second construct, denoted $(p \odot q)$ and called *cooperation*, consists in plugging together p and q – up to termination of one of them. This operator provides for *communication*, which can only occur within a $(p \odot q)$. On the other hand, a compound process $(p \mid q)$ can propose communications to its environment, and the interleaving operator is commutative and associative (and satisfies $(p \mid \mathbb{1}) = p$). Therefore the interaction law becomes:

$$\gamma: \quad ((\cdots \mid \alpha x.p \mid \cdots) \odot (\cdots \mid \bar{\alpha} q \mid \cdots)) \rightarrow ((\cdots \mid p[x \mapsto q] \mid \cdots) \odot (\cdots \mid \mathbb{1} \mid \cdots))$$

Like the operator considered by Milner in [5] (p.21), the operator $\odot$ is *not* associative. This allows us to represent λ-calculus application (pq) by a combination of cooperation and output, namely:

$$(pq) = (p \odot \bar{\lambda} q)$$

so that a particular case of the γ-rule above is:

$$(\lambda x.p \odot \bar{\lambda} q) \rightarrow (p[x \mapsto q] \odot \mathbb{1})$$

To ensure the correctness of this representation, the semantics of cooperation must be such that $(p \odot \mathbb{1})$ behaves like p. In other words, $\odot$ is not a *static* operator: $(p \odot q)$ cannot communicate with another process if p and q are not terminated, but it will be free to do so once p or q terminates. To capture the usual operational semantics of the λ-calculus, we must also introduce *structural rules*, which formalize the fact that reduction is compatible with the constructors. For instance there will be two rules allowing internal computations within a guarded process:

$p \rightarrow p' \ \vdash \ \bar{\alpha} p \rightarrow \bar{\alpha} p'$

$p \rightarrow p' \vdash \ \alpha x.p \rightarrow \alpha x.p' \qquad (\xi \text{ rule})$

These do not hold in CCS, where the transitions describe the behaviour of a reactive system, rather than an evaluation mechanism. Processes in our calculus could be qualified as *interactive* systems rather than reactive – in fact this is a matter of evaluation strategy.

One should observe that we still have conflicting communications, so that we can represent a *non-deterministic choice*, using the standard combinator $\mathsf{K} = \lambda x.\lambda y.x$, which chooses its first argument and deletes the second one:

$$(p \oplus q) = (\mathsf{K} \odot (\bar{\lambda} p \mid \bar{\lambda} q))$$

It is easy to see that we have:

$$(p \oplus q) \xrightarrow{*} (p \odot (\mathbb{1} \mid \mathbb{1})) \equiv p \qquad \text{and} \qquad (p \oplus q) \xrightarrow{*} (q \odot (\mathbb{1} \mid \mathbb{1})) \equiv q$$

where $\equiv$ is the syntactic equality, defined in the next section. An obvious consequence is that the Church-Rosser property no longer holds for reduction in our calculus – which therefore cannot be "included" in any way in the λ-calculus. Moreover it would be inconsistent to regard the associated conversion as establishing a notion of equality(‡). We shall adopt an intensional notion of equality, namely that of *observational equivalence* of Milner [5], which relies on the communication capabilities of a process. This has already been used by Abramsky in [1] to give a new semantics for λ-terms.

To conclude the informal presentation of our calculus, let us say a few words about the *binders*. In the λ-calculus, these are sequences $\lambda x_1. \ldots .\lambda x_k$ corresponding to application to a stream of arguments. Since in our calculus we may have interleaved arguments, it seems natural to correspondingly generalize the binders, allowing not only sequences of αx's, but also interleavings. Then we will have terms of the form $\langle \alpha_1 x_1 \mid \cdots \mid \alpha_k x_k \rangle .p$, meaning that p waits for k unordered values. This allows us for instance to represent non-deterministic choice as a simple variant of the combinator K, namely $\oplus =_{\text{def}} \langle \lambda x \mid \lambda y \rangle .x$. We will also see later how to represent some "parallel functions", not definable in the λ-calculus.

(‡) as a matter of fact, conversion is not regarded as the semantical equality for the λ-calculus, *cf.* [2] proposal 2.2.14.

2. The γ-calculus.

Given a denumerable set X of variables and a non-empty set N of port names, we first define the *binders*, which are the terms built according to the following grammar, where α is any port name, and x stands for any variable:

$$\rho ::= \varepsilon \mid \alpha x \mid (\rho \cdot \rho) \mid (\rho \mid \rho)$$

Intuitively αx represents a reception on the port α, $(\rho \cdot \rho')$ represents sequencing of such receptions while $(\rho \mid \rho')$ represents their interleaving. The term ε is an *empty* binder, therefore we shall consider binders up to the congruence $\doteq$ generated by the equations:

$$(\rho \cdot \varepsilon) = \rho = (\varepsilon \cdot \rho)$$
$$(\rho \mid \varepsilon) = \rho = (\varepsilon \mid \rho)$$

The congruence $\doteq$ defines the syntactic equality over binders. Any binder ρ will bind the variables belonging to the set $\mathsf{var}(\rho)$ defined as follows:

(i) $\mathsf{var}(\varepsilon) = \emptyset$

(ii) $\mathsf{var}(\alpha x) = \{x\}$

(iii) $\mathsf{var}(\rho \cdot \rho') = \mathsf{var}(\rho) \cup \mathsf{var}(\rho')$

(iv) $\mathsf{var}(\rho \mid \rho') = \mathsf{var}(\rho) \cup \mathsf{var}(\rho')$

The syntax of the γ-calculus is given by the following grammar, where α is any port name of N, ρ is any binder, and x stands for any variable:

$$p ::= \mathbb{1} \mid x \mid \bar{\alpha} p \mid \langle \rho \rangle . p \mid (p \odot p) \mid (p \mid p)$$

We denote by Γ the set of terms generated by this grammar, and we shall use $p, q, r \ldots$ to range over terms – which will be called *agents* or *processes*. For simplicity we shall denote $\langle \alpha x \rangle . p$ by $\alpha x . p$, and we shall omit some parentheses, using for instance $\langle \rho \mid \rho' \rangle . p$ instead of $\langle (\rho \mid \rho') \rangle . p$. A variable x is *bound* if it is in the scope of a binder. Then in substituting q for y in p, yielding $p[y \mapsto q]$, we might have to rename some bound variables of p. Although this is a standard matter (see [2], appendix C), it is worth to carefully define substitution. Here we adapt the definitions of [6]. The set $\mathsf{free}(p)$ of *free variables* of the term p is given by

(i) $\mathsf{free}(\mathbb{1}) = \emptyset$

(ii) $\mathsf{free}(x) = \{x\}$

(iii) $\mathsf{free}(\bar{\alpha} p) = \mathsf{free}(p)$

(iv) $\mathsf{free}(\langle \rho \rangle . p) = \mathsf{free}(p) - \mathsf{var}(\rho)$

(v) $\mathsf{free}(p \odot q) = \mathsf{free}(p) \cup \mathsf{free}(q) = \mathsf{free}(p \mid q)$

A term p is *closed* if $\mathsf{free}(p) = \emptyset$. A *substitution* is any mapping $\sigma: X \to \Gamma$. We use $\sigma, \sigma' \ldots$ to range over the set S of substitutions. The identity substitution is denoted ι. The *updating* operation on substitutions is defined as follows: let $x \in X$, $p \in \Gamma$ and $\sigma \in S$; then the new substitution $\sigma' = (x \mapsto p/\sigma)$ is given by:

$$\sigma'(y) = \begin{cases} p & \text{if } y = x \\ \sigma(y) & \text{otherwise} \end{cases}$$

For a binder ρ and a renaming, that is a substitution $\xi: X \to X$, the result $\rho[\xi]$ of applying ξ to ρ is defined in an obvious way, that is by structural induction starting from $(\alpha x)[\xi] = \alpha \xi(x)$. To define substitution on terms of Γ, we assume given for each pair V, W of finite subsets of X an injective mapping $\mathsf{new}_{V,W}: V \to X - W$; this assumption makes sense since X is infinite. Then the result $p[\sigma]$ of applying the substitution σ to the term p is defined by structural induction, the only non obvious case being $p = \langle\rho\rangle.q$ with $\mathsf{var}(\rho) \neq \emptyset$. In this case, let:

$$V = \mathsf{var}(\rho) = \{x_1, \ldots, x_n\}$$

$$W = \{v \mid \exists z \in \mathsf{free}(p)\ v \in \mathsf{free}(\sigma(z))\}$$

$$\mathsf{new}_{V,W}(x_i) = y_i \quad \text{for } 1 \leq i \leq n$$

$$\xi = [x_1 \mapsto y_1 / \cdots / x_n \mapsto y_n / \iota]$$

$$\sigma' = [x_1 \mapsto y_1 / \cdots / x_n \mapsto y_n / \sigma]$$

Then we define $p[\sigma]$ to be $\langle\rho[\xi]\rangle.q[\sigma']$. We shall denote $p[x \mapsto q/\iota]$ by $p[x \mapsto q]$ (similarly $\rho[x \mapsto y]$ denotes $\rho[x \mapsto y/\iota]$), and define the *composition* $\rho \bullet \sigma$ by $(\rho \bullet \sigma)(x) = \sigma(x)[\rho]$. As usual, we regard terms differing only on the name of bound variables as syntactically identical. Moreover we also regard $\mathbb{1}$ as a terminated, or idle agent, which should be cancelled from parallel combinations. Then our *syntactical equality* is the congruence $\equiv$ generated by the following equations:

$$(p \odot \mathbb{1}) = p = (\mathbb{1} \odot p)$$

$$(p \mid \mathbb{1}) = p = (\mathbb{1} \mid p)$$

$$\langle\varepsilon\rangle.p = p$$

$$\langle\rho\rangle.p = \langle\rho'\rangle.p \quad \text{if } \rho \doteq \rho'$$

$$\langle\rho\rangle.p = \langle\rho[x \mapsto y]\rangle.p[x \mapsto y] \quad \text{if } x \in \mathsf{var}(\rho) \text{ and } y \notin \mathsf{free}(p) \cup \mathsf{var}(\rho)$$

One could prove that $\equiv$ is *substitutive*, that is $p \equiv q \Rightarrow p[\sigma] \equiv q[\sigma]$ for all substitution σ (*cf.* [6]). We shall say that an agent p is *terminated*, or *idle*, in notation $p\dagger$, if $p \equiv \mathbb{1}$.

To define the semantics, that is the laws of reduction, we shall use Milner's technique of labelled transitions. This is the best way to formalize the idea that processes need not be contiguous to communicate. Let us introduce some technical definitions. The semantics is given by means of labelled transitions $p \xrightarrow{a} p'$ where the action a may be $\bar{\alpha}_p$, which means sending p at port α, or α_p, which means receiving p at port α, or the communication action τ. This could be formalized by saying that the set of actions is $A = (N \times \Gamma) \cup (\Gamma \times N) \cup \{\tau\}$ (if we regard α_p and $\bar{\alpha}_p$ as notations for (α, p) and (p, α) respectively). We shall say that a and b are *complementary* actions, in notation $a \frown b$, if $a = \alpha_p$ and $b = \bar{\alpha}_p$, or symmetrically $a = \bar{\alpha}_p$ and $b = \alpha_p$. To define the semantics of $\langle\rho\rangle.p$ we also need to specify the reception actions allowed by the binder ρ. To this end we introduce a transition relation $\rho \xrightarrow{a} \rho'$ between binders, where a has the form $\alpha_{x,p}$. This transition relation is the least one satisfying the following rules:

$$p \in \Gamma \ \vdash\ \alpha x \xrightarrow{\alpha_{x,p}} \varepsilon$$

$$\rho \xrightarrow{a} \rho' \ \vdash\ (\rho \cdot \rho'') \xrightarrow{a} (\rho' \cdot \rho'')$$

$$\rho \doteq \varepsilon \ \&\ \rho' \xrightarrow{a} \rho'' \ \vdash\ (\rho \cdot \rho') \xrightarrow{a} \rho''$$

$$\rho \xrightarrow{a} \rho' \ \vdash\ (\rho \mid \rho'') \xrightarrow{a} (\rho' \mid \rho'')$$

$$\rho \xrightarrow{a} \rho' \ \vdash\ (\rho'' \mid \rho) \xrightarrow{a} (\rho'' \mid \rho')$$

The transition relation $\rightarrow$ on agents is the least subset of $\Gamma \times A \times \Gamma$ satisfying the rules given below (where, as usual, we denote $(p, a, p') \in \rightarrow$ by $p \xrightarrow{a} p'$). The first two rules introduce the communication actions:

output R1: $\vdash \bar{\alpha}p \xrightarrow{\bar{\alpha}_p} \mathbb{1}$

input R2: $\rho \xrightarrow{\alpha_{x,q}} \rho' \vdash \langle\rho\rangle.p \xrightarrow{\alpha_q} \langle\rho'\rangle.p[x \mapsto q]$

One may observe that, due to R2, the *sort* of a process – that is the set of port names through which it may communicate – can evolve dynamically: if $p \xrightarrow{a} p'$ the sort of p' is not necessarily a subset of the sort of p. There is another rule concerning input, when the binder is empty:

input R3: $\rho \doteq \varepsilon \ \& \ p \xrightarrow{a} p' \vdash \langle\rho\rangle.p \xrightarrow{a} p'$

The interaction law is given by (the γ-rule):

communication R4 (γ): $p \xrightarrow{a} p'$, $q \xrightarrow{b} q'$ & $a \frown b \vdash (p \odot q) \xrightarrow{\tau} (p' \odot q')$

The following rules state that the transition relation $\xrightarrow{\tau}$ is compatible with all the constructors, and that $\xrightarrow{a}$ is compatible with interleaving for any $a \in A$:

output R5: $p \xrightarrow{\tau} p' \vdash \bar{\alpha}p \xrightarrow{\tau} \bar{\alpha}p'$

input R6: $p \xrightarrow{\tau} p' \vdash \langle\rho\rangle.p \xrightarrow{\tau} \langle\rho\rangle.p'$

cooperation (left) R7: $p \xrightarrow{\tau} p' \vdash (p \odot q) \xrightarrow{\tau} (p' \odot q)$

cooperation (right) R8: $q \xrightarrow{\tau} q' \vdash (p \odot q) \xrightarrow{\tau} (p \odot q')$

interleaving (left) R9: $p \xrightarrow{a} p' \vdash (p \mid q) \xrightarrow{a} (p' \mid q)$

interleaving (right) R10: $q \xrightarrow{b} q' \vdash (p \mid q) \xrightarrow{b} (p \mid q')$

Our last two rules formalize the fact that cooperation only holds up to termination of one partner:

cooperation (right unit) R11: $p \xrightarrow{a} p'$, $q\dagger \vdash (p \odot q) \xrightarrow{a} p'$

cooperation (left unit) R12: $q \xrightarrow{b} q'$, $p\dagger \vdash (p \odot q) \xrightarrow{b} q'$

One can readily see from these rules that if $p \not\equiv \mathbb{1} \not\equiv q$ then $(p \odot q)$ can only perform τ actions, while communication between p and q is forbidden within the construct $(p \mid q)$. We shall mostly denote $p \xrightarrow{\tau} p'$ by $p \rightarrow p'$, and by definition this is the *γ-reduction* between terms of Γ.

Our main purpose is to show that the γ-calculus contains the λ-calculus, up to syntactical equality. Then we first have to check that syntactical equality is consistent with the operational semantics. Formally speaking, this amounts to show that $\equiv$ is a bisimulation. Our notion of bisimulation is a slight extension of Park and Milner's one, in two respects: first we must regard the actions – made out of agents – up to bisimulation, second we must take into account the potential termination of agents. Moreover we wish to directly define bisimulation for non-closed terms; then two terms p and q are similar if all their instances $p[\sigma]$ and $q[\sigma]$ have similar behaviours. We shall use two notions of simulation: the first one, called strong, is relative to the transition relation $\rightarrow$. We will see the second (weak) one latter. Let $R \subseteq \Gamma \times \Gamma$ be a relation on terms; we define its extension $\widehat{R} \subseteq A \times A$ on actions as follows:

$$a \widehat{R} b \Leftrightarrow_{\text{def}} a = b \text{ or } \exists \alpha \in N \ \exists p, q. \ p R q \ \& \ a = \bar{\alpha}_p \ \& \ b = \bar{\alpha}_q$$

A relation $R \subseteq \Gamma \times \Gamma$ is

(i) a *strong simulation* if it satisfies

S1: $p\,R\,q \;\&\; p[\sigma] \xrightarrow{a} p' \;\Rightarrow\; \exists b.\, a\,\widehat{R}\,b\; \exists q'.\, p'\,R\,q' \;\&\; q[\sigma] \xrightarrow{b} q'$

S2: $p\,R\,q \;\&\; p\dagger \;\Rightarrow\; q\dagger$

(ii) a *strong bisimulation* if it is a symmetric strong simulation.

The first property defining a simulation is a refinement of the usual one: (instances of) strongly similar agents must perform *similar* actions. The second property states that strong simulation preserves the termination property. This property, that is $p \equiv \mathbb{1}$, should not be confused with the property of being a normal form, that is γ-irreducibility (p is γ-irreducible if $\{\, q \mid p \rightarrow q \,\} = \emptyset$): to our view a term such as $\alpha x.x$ is not terminated since it can perform some actions – namely α_p. One should note that every strong simulation is substitutive:

FACT. *If R is a strong simulation then $p\,R\,q \;\Rightarrow\; p[\sigma]\;R\;q[\sigma]$ for all substitution σ.*

This holds because $(p[\sigma])[\rho] = p[\rho \bullet \sigma]$ (*cf.* [6]).

PROPOSITION. *The congruence $\equiv$ is a strong simulation on Γ.*

The proof is straightforward: one proceeds by induction on the proof of $p[\sigma] \equiv q[\sigma]$ (using the fact that $\equiv$ is substitutive), and then by induction on the proof of the transition $p[\sigma] \xrightarrow{a} p'$ to show that $\exists q' \equiv p'\; q[\sigma] \xrightarrow{a} q'$. In the case of R4, one must show that:

$$p \xrightarrow{\alpha_q} p' \;\&\; q \equiv q' \;\Rightarrow\; \exists p''.\, p' \equiv p'' \;\&\; p \xrightarrow{\alpha_{q'}} p''$$

One must also prove that $\doteq$ satisfies:

$$\rho_0 \xrightarrow{a} \rho_0' \;\&\; \rho_0 \doteq \rho_1 \;\Rightarrow\; \exists \rho_1'.\, \rho_0' \doteq \rho_1' \;\&\; \rho_1 \xrightarrow{a} \rho_1'$$

The details are omitted □

This result allows us to define the transition relation $\rightarrow$ on $\Gamma/\!\equiv$. We shall abusively write transitions between simplified terms (obtained by cancelling $\mathbb{1}$ from parallel combinations), as for instance in a special case of the γ-rule: $(\alpha x.p \odot \bar{\alpha} q) \rightarrow p[x \mapsto q]$.

Now we can show that the γ-calculus contains the λ-calculus – which we assume to be well-known! The syntax of Λ is given by the following grammar:

$$M \;::=\; x \mid \lambda x.M \mid (MM)$$

The rules for β-reduction, denoted $M \rightarrow N$, are:

R′1 (β): $(\lambda x.M)N \rightarrow M[x \mapsto N]$

R′2: $M \rightarrow M' \;\vdash\; \lambda x.M \rightarrow \lambda x.M'$

R′3: $M \rightarrow M' \;\vdash\; (MN) \rightarrow (M'N)$

R′4: $N \rightarrow N' \;\vdash\; (MN) \rightarrow (MN')$

Assuming that $\lambda \in N$, we define the translation θ from Λ to Γ as follows:

$\theta(x) = x$

$\theta(\lambda x.M) = \langle \lambda x \rangle.\theta(M)$

$\theta(MN) = (\theta(M) \odot \bar{\lambda}\theta(N))$.

We assume that substitution is defined for λ-terms as it was defined for γ-terms, so that the translation preserves substitution, that is:

$$\forall M,N \in \Lambda \quad \theta(M[x \mapsto N]) = \theta(M)[x \mapsto \theta(N)]$$

PROPOSITION. *For all λ-terms M and N:*

(i) $M \rightarrow N \Rightarrow \exists p \equiv \theta(N).\ \theta(M) \rightarrow p$

(ii) $\theta(M) \rightarrow p \Rightarrow \exists N \in \Lambda.\ M \rightarrow N\ \&\ \theta(N) \equiv p$

PROOF: let $M, N \in \Lambda$ such that $M \rightarrow N$. We proceed by induction on the proof of this transition to show that $\exists p \equiv \theta(N)\ \theta(M) \rightarrow p$. If this transition is an instance of the β-rule, then we have $M = (\lambda x.P)Q$, $N = P[x \mapsto Q]$ and $\theta(M) = (\langle\lambda x\rangle.\theta(P) \odot \bar{\lambda}\theta(Q))$. Using R1 and R2 we have $\bar{\lambda}\theta(Q) \xrightarrow{a} \mathbb{1}$ with $a = \bar{\lambda}_{\theta(Q)}$ and $\langle\lambda x\rangle.\theta(P) \xrightarrow{b} \langle\varepsilon\rangle.\theta(P)[x \mapsto \theta(Q)]$ with $b = \lambda_{\theta(Q)}$. Therefore using the γ-rule (R4) we have $\theta(M) \rightarrow (\langle\varepsilon\rangle.\theta(P)[x \mapsto \theta(Q)] \odot \mathbb{1}) \equiv \theta(N)$. All the other cases are trivial.

Conversely let us assume that $\theta(M) \rightarrow p$. We proceed by induction on the structure of M to show that $\exists N \in \Lambda\ M \rightarrow N\ \&\ \theta(N) \equiv p$. We cannot have $M \in X$, since a variable cannot perform any action. If $M = \lambda x.P$ then $\theta(M) = \langle\lambda x\rangle.\theta(P)$, and the transition $\theta(M) \rightarrow p$ must be proved using R6 (since the action is τ). We easily conclude in this case using the induction hypothesis. If $M = (PQ)$ then $\theta(M) = (\theta(P) \odot \bar{\lambda}\theta(Q))$. The transition $\theta(M) \rightarrow p$ cannot be proved using R11 or R12 (since $\theta(P) \not\equiv \mathbb{1}$). If it is proved using R7 or R8 (and then R5) the result follows from the induction hypothesis. If it is proved using R4, we have $\theta(P) \xrightarrow{a} p'$ and $\bar{\lambda}\theta(Q) \xrightarrow{b} q'$ with $a \frown b$; hence $a \neq \tau \neq b$, and the second transition can only be proved by means of R1. This implies that $b = \bar{\lambda}_{\theta(Q)}$ (and $q' = \mathbb{1}$), and therefore $a = \lambda_{\theta(Q)}$. It is easy to prove that for all $P \in \Lambda$ if $\theta(P) \xrightarrow{\lambda_s} p'$ then $P = \lambda x.P'$ & $p' = \langle\varepsilon\rangle.\theta(P')[x \mapsto s]$. Consequently we have $M = ((\lambda x.P')Q)$, and $\theta(N) = \theta(P')[x \mapsto \theta(Q)] \equiv (\langle\varepsilon\rangle.\theta(P')[x \mapsto \theta(Q)] \odot \mathbb{1})$ □

This result allows us to regard the λ-terms as a special kind of γ-terms. To simplify the notations, we shall use (pq) as an abbreviation for $(p \odot \bar{\lambda}q)$ – recall also that $\lambda x.p$ is a notation for $\langle\lambda x\rangle.p$. We shall keep the usual notation for the standard combinators, *cf.* [2], chapter 6. For instance K is the (γ-)term $\lambda x.\lambda y.x$, or more simply $\lambda xy.x$.

We already saw that the γ-calculus is strictly more powerful than the λ-calculus: the term $\langle\lambda x \mid \lambda y\rangle.x$ (non-deterministic choice) does not have any image in Λ. Let us see another example, showing that we find in Γ "parallel functions" (we do not intend to precisely define them) which are not definable in Λ. It is known that K, which could be denoted also T, and $\mathsf{F} = \lambda xy.y$ may be regarded as the *truth values*. One can define in the λ-calculus a combinator representing the *left sequential disjunction*, namely $\mathsf{V} = \lambda xy.(x\mathsf{T})y$. This combinator is left-sequential since one cannot reduce $(\mathsf{V}M)\mathsf{T}$ into T without evaluating M. From Berry's sequentiality theorem (*cf.* [2]), one can show that there is no λ-term representing a *parallel disjunction* O, such that $(\mathsf{O}M)\mathsf{T}$ and $(\mathsf{O}\mathsf{T})M$ can be both reduced to T without evaluating M. On the other hand, this combinator is definable in the γ-calculus: this is just a parallel variant of V, namely

$$\mathsf{O} =_{\text{def}} \langle\lambda x \mid \lambda y\rangle.(x\mathsf{T})y$$

Then it is easy to see that:

$$\forall p \in \Gamma \quad (\mathsf{O}\,p)\mathsf{T} \xrightarrow{*} \mathsf{T} \qquad \text{and} \qquad (\mathsf{O}\,\mathsf{T})p \xrightarrow{*} \mathsf{T}$$

since the combinator **O** can choose what argument eventually needs to be evaluated first:

$$(\mathsf{O}\,p)q \xrightarrow{*} (p\,\mathsf{T})q \qquad \text{and} \qquad (\mathsf{O}\,p)q \xrightarrow{*} (q\,\mathsf{T})p$$

Obviously we also have:

$$(\mathsf{O}\,\mathsf{F})\mathsf{F} \xrightarrow{*} \mathsf{F}$$

Let us see another example, showing that we can retrieve in the γ-calculus some of CCS's ideas about concurrent processes. One of Milner's intentions in designing CCS was to formalize the idea that a process performs possibly infinite sequences of communications with its environment. One may wonder whether it is possible to describe in the γ-calculus a system made out of processes continuously exchanging messages. The answer is positive, thanks to the existence of endlessly reducible terms. In the λ-calculus, the typical example of such terms is $\Omega = \Delta\Delta$, where $\Delta = \lambda x.xx$ is the usual duplicator, for we have $\Omega \rightarrow \Omega$. Using this feature we can define a process which repeatedly accepts a message on a port α and then sends on port β a response elaborated using a "method" q. Let

$$\delta = \lambda y.\alpha x.(\bar{\beta}q \mid (y \odot \bar{\lambda}y)) \qquad \text{and} \qquad \omega = (\delta \odot \bar{\lambda}\delta)$$

These terms could be written more simply $\delta = \lambda y.\alpha x.(\bar{\beta}q \mid yy)$ and $\omega = (\delta\delta)$. Then we have:

$$\omega \rightarrow \alpha x.(\bar{\beta}q \mid \omega) \xrightarrow{\alpha_p} (\bar{\beta}q[x \mapsto p] \mid \omega) \xrightarrow{*} (\bar{\beta}r \mid \omega) \xrightarrow{\bar{\beta}_r} (\mathbb{1} \mid \omega)$$

We should say that evaluating ω repeatedly creates the communication channels α and β. More generally, recursion can be handled as in the λ-calculus, that is by means of fixed point combinators, like Turing's one $\Theta = (FF)$, where $F = \lambda x.\lambda y.y((xx)y)$, which is such that $\Theta M \xrightarrow{*} M(\Theta M)$.

Now let us return to the semantics of our calculus. We shall adopt Milner's observational equivalence [5] as our notion of equality. The observational equivalence is defined with respect to a transition relation where one abstracts from internal communications (i.e. τ actions). This transition relation $\Longrightarrow$ is the least subset of $\Gamma \times A \times \Gamma$ containing $\rightarrow$ and satisfying the following rules:

O1: $\vdash p \overset{\tau}{\Longrightarrow} p$

O2: $p \overset{a}{\Longrightarrow} p''\ ,\ p'' \overset{\tau}{\Longrightarrow} p' \vdash p \overset{a}{\Longrightarrow} p'$

O3: $p \overset{\tau}{\Longrightarrow} p''\ ,\ p'' \overset{a}{\Longrightarrow} p' \vdash p \overset{a}{\Longrightarrow} p'$

It should be clear that $p \overset{a}{\Longrightarrow} p'$ *iff* $a = \tau$ & $p' = p$ or

$$p(\xrightarrow{\tau})^* \xrightarrow{a} (\xrightarrow{\tau})^* p'$$

A relation $R \subseteq \Gamma \times \Gamma$ is

(i) a *weak*, or *observational simulation* if it satisfies

W1: $pRq\ \&\ p[\sigma] \overset{a}{\Longrightarrow} p' \Rightarrow \exists b.\ a\widehat{R}b\ \exists q'.\ p'Rq'\ \&\ q[\sigma] \overset{b}{\Longrightarrow} q'$

W2: $pRq\ \&\ p\dagger \Rightarrow \exists q'.\ q \overset{\tau}{\Longrightarrow} q'\ \&\ q'\dagger$

(ii) a *weak*, or *observational bisimulation* if it is a symmetric weak simulation.

Note that the we have:

$$pRq\ \&\ p[\sigma]\dagger \Rightarrow \exists q'.\ q[\sigma] \overset{\tau}{\Longrightarrow} q'\ \&\ q'\dagger$$

since $p[\sigma] \overset{\tau}{\Longrightarrow} p[\sigma]$, hence there exist q'' such that $q[\sigma] \overset{\tau}{\Longrightarrow} q''$ and $p[\sigma]\ R\, q''$, therefore there exists q' such that $q'\dagger$ and $q'' \overset{\tau}{\Longrightarrow} q'$, hence $q[\sigma] \overset{\tau}{\Longrightarrow} q'$. Consequently any observational simulation is substitutive:

FACT. *If R is an observational simulation then $p\,R\,q \Rightarrow p[\sigma]\ R\,q[\sigma]$ for all substitution σ.*

The following characterization of observational simulations is useful:

LEMMA. *A relation $R \subseteq \Gamma \times \Gamma$ is an observational simulation if and only if*

(i) R is substitutive: $p\,R\,q \Rightarrow \forall\sigma \in S\ p[\sigma]\,R\,q[\sigma]$

(ii) $p\,R\,q\ \&\ p \overset{a}{\longrightarrow} p' \Rightarrow \exists b.\ a\widehat{R}b\ \exists q'.\ p'\,R\,q'\ \&\ q \overset{b}{\Longrightarrow} q'$

(iii) $p\,R\,q\ \&\ p\dagger \Rightarrow \exists q'.\ q \overset{\tau}{\Longrightarrow} q'\ \&\ q'\dagger$

PROOF: it is clear that any observational simulation satisfies these properties. Let us assume that R satisfies (i)-(iii). We have to prove

$$p\,R\,q\ \&\ p[\sigma] \overset{a}{\Longrightarrow} p' \Rightarrow \exists b.\ a\widehat{R}b\ \exists q'.\ p'\,R\,q'\ \&\ q[\sigma] \overset{b}{\Longrightarrow} q'$$

Since R is substitutive, it is enough to prove this for $\sigma = \iota$ (note that $p[\iota] \equiv p$). We proceed by induction on the proof of $p \overset{a}{\Longrightarrow} p'$: if this transition is $p \overset{a}{\longrightarrow} p'$, then we conclude using (ii). The point is trivial if $p \overset{a}{\Longrightarrow} p'$ is an instance of O1, since $p' = p$, $a = \tau$ and $q \overset{\tau}{\Longrightarrow} q$ (by O1). The two other cases easily follow from the induction hypothesis ◻

The observational equivalence that we regard as our *semantic equality* is the coarsest weak bisimulation. Such a coarsest bisimulation exists, and is an equivalence, since we have:

DEFINITION and FACT. *Let us define:*

$$p \approx q \Leftrightarrow_{\mathrm{def}} \exists R \subseteq \Gamma \times \Gamma \text{ weak bisimulation such that } p\,R\,q$$

Then $\approx$ is a weak bisimulation. Moreover $\approx$ is an equivalence.

(the proof is omitted – the only point to check is that the composition of two weak simulations is a weak simulation).

A consequence of the previous lemma is that any strong simulation is also a weak one. Then for instance we have $\equiv \subseteq \approx$. This lemma also allows us to prove some algebraic properties of the operators with respect to $\approx$, as for example:

$$(p \odot q) \approx (q \odot p)$$
$$(p \odot \mathbb{1}) \approx p \approx (\mathbb{1} \odot p)$$
$$(p \mid (q \mid r)) \approx ((p \mid q) \mid r)$$
$$(p \mid q) \approx (q \mid p)$$
$$(p \mid \mathbb{1}) \approx p \approx (\mathbb{1} \mid p)$$

Note that the cooperation operator is not associative (up to $\approx$): for instance

$$((\lambda x.x \odot \bar{\lambda}p) \odot \bar{\lambda}q) \not\approx (\lambda x.x \odot (\bar{\lambda}p \odot \bar{\lambda}q))$$

since the first term – which is $(\mathsf{I}\,p)q$, where $\mathsf{I} = \lambda x.x$ is the usual identity combinator – can be reduced to (pq) (up to $\equiv$) while the second one, which could be written $\mathsf{I} \odot (\bar{\lambda}p \odot \bar{\lambda}q)$, cannot perform any computation.

3. Conclusion

To conclude this note, let us briefly discuss the relationships between the proposed γ-calculus and other established calculi.

An obvious question to investigate regards the relationship between CCS and the γ-calculus. We did not claim that CCS should appear as a subcalculus of our calculus. As a matter of fact, we suspect that most CCS primitives are not definable as γ-combinators. We conjecture that the observational equivalence $\approx$ is a congruence on Γ(‡). A consequence would be that we cannot define in the γ-calculus the CCS *sum* $(p+q)$, whose semantics is given by:

$$p \xrightarrow{a} p' \;\vdash\; (p+q) \xrightarrow{a} p'$$

$$q \xrightarrow{b} q' \;\vdash\; (p+q) \xrightarrow{b} q'$$

More precisely, if $\approx$ is a congruence then there is no γ-term r such that $(rp)q \approx (p+q)$ for all p and q, since $\approx$ is not a congruence with respect to CCS sum – for instance we have $(\mathsf{I}\,\mathbb{1}) \approx \mathbb{1}$ but $\mathsf{I} + (\mathsf{I}\,\mathbb{1}) \not\approx \mathsf{I} + \mathbb{1}$. To our view, non-definability of CCS sum could be a good point rather than a drawback, since there are some serious difficulties with $+$; the sum is a natural primitive for describing transition systems, but it is hard to devise a denotational interpretation of this construct. We saw that a non-deterministic combinator can be defined in the γ-calculus; this choice operator $\oplus$ is quite different from $+$, since it is an *internal* choice: $(p \oplus q)$ may evolve to one of p or q by internal communications. We also have, due to the structural rules:

$$p \rightarrow p' \;\Rightarrow\; (p \oplus q) \rightarrow (p' \oplus q) \qquad \text{and} \qquad q \rightarrow q' \;\Rightarrow\; (p \oplus q) \rightarrow (p \oplus q')$$

Note that in general $(p \oplus \mathbb{1}) \not\approx p$ – take for example $p = \mathsf{I}$.

A structural translation from CCS to the γ-calculus seems to be doomed to failure. One could however imagine to relate these two calculi in another way: CCS appears to be well-suited to describe concurrent and communicating "machines" – indeed Milner showed in [4] that his static operators (i.e. parallel composition, relabelling and restriction) have a natural interpretation as constructors of nets of agents. Then, with respect to the γ-calculus, CCS terms could play a rôle analogous to that of Turing machines with respect to the λ-calculus. From this point of view, one does not expect to translate directly the primitives of one calculus into the other: the correspondence is made on the ground of definability of abstract mathematical objects – like computable functions. For lack of an abstract notion of process, we leave this question for further investigation.

Regarding the relationship between the γ and λ calculi, let us examine briefly some λ-theories (*cf.* [2]) from the point of view of observational equivalence. First we should note that observational equivalence – which is consistent on Λ: $\mathsf{I} \not\approx \Omega$ – is not extensional. This means that the equation η (on λ-terms)

$$\lambda x.Mx = M \qquad x \text{ not free in } M$$

is not valid for $\approx$ – for instance $\lambda x.\Omega x \not\approx \Omega$, since the first term has a possible communication with its environment, while the second has no communication capability. For what regards

(‡) the main difficulty is to prove $p \approx q \;\Rightarrow\; r[x \mapsto p] \approx r[x \mapsto q]$. The same problem already arises in the λ-calculus, but here we cannot use the semantical method of Abramsky [1], since we do not have any model theory for the γ-calculus. Then we will have to extend the syntactical method of Lévy [3].

β-convertibility $=_\beta$ of λ-terms, one could prove that β-convertible λ-terms are observationally equivalent, that is

$$\forall M, N \in \Lambda \; M =_\beta N \;\Rightarrow\; \theta(M) \approx \theta(N)$$

This implies a restricted kind of η-conversion: if M is really a function, that is if there exists N such that $M \xrightarrow{*} \lambda z.N$, then we have $\lambda x.Mx \approx M$ (for x not free in M). The converse of the previous implication is not true: observational equivalence is strictly weaker than β-conversion. To see this, let us say that a γ-term p is *locked* if it has no communication capability, and can never terminate, that is if $p \stackrel{a}{\Longrightarrow} p' \;\Rightarrow\; a = \tau \;\&\; p' \not\equiv \mathbb{1}$. It is easy to see that any two locked terms are equivalent – note that if p is locked and $p \stackrel{a}{\Longrightarrow} p'$ then p' is locked as well. Then for instance we have $\Theta\Theta \approx \Omega$, where Θ is the Turing's fixed point combinator, while $\Theta\Theta$ and Ω are not convertible. One should note that the two terms Ω and $\Theta\Theta$ are *unsolvable* (*cf.* [2] for this notion). But an unsolvable λ-term is not necessarily locked, and observational equivalence does not equate all the unsolvables – for instance $\lambda x.\Omega$ is unsolvable, but not locked, and we saw that $\lambda x.\Omega \not\approx \Omega$. Then the observational equivalence of λ-terms is quite different from the usual semantics of the λ-calculus, which is based on the identification of unsolvable terms (*cf.* [2], proposal 2.2.14). However, Lévy showed in [3] that one can build a sound semantical theory of the λ-calculus without resorting to this identification – and it may be the case that this semantics is even "better" than the usual one, since Lévy's syntactical model provides some kind of "initial" interpretation. Moreover, Lévy's interpretation seems to be very close to the one introduced by Abramsky [1], using the idea of observational equivalence. One may expect that Abramsky's applicative bisimulation coincides with observational equivalence on λ-terms, as we defined it in this paper. A first step towards such a result would be to show that the "outermost evaluation" – that is evaluation performed without using the rules R5-R6 above – is correct with respect to observational equivalence. This task, involving the adaptation of Lévy's syntactical technique, is left for further work.

REFERENCES

[1] S. ABRAMSKY, *The Lazy Lambda-Calculus*, to appear in Declarative Programming, Ed. D. Turner, Addison Wesley (1988).

[2] H. P. BARENDREGT, *The Lambda Calculus*, Studies in Logic 103, North-Holland (1981).

[3] J.-J. LÉVY, *An Algebraic Interpretation of the λβK-Calculus; and an Application of a Labelled λ-Calculus*, Theoretical Comput. Sci. 2 (1976) 97-114.

[4] R. MILNER, *Flowgraphs and Flow Algebras*, JACM 26 (1979) 794-818.

[5] R. MILNER, *A Calculus of Communicating Systems*, Lecture Notes in Comput. Sci. 92 (1980) reprinted in Report ECS-LFCS-86-7, Edinburgh University.

[6] A. STOUGHTON, *Substitution Revisited*, Theoretical Comput. Sci. 59 (1988) 317-325.

[7] B. THOMSEN, *A Calculus of Higher Order Communicating Systems*, to appear in the Proceedings of POPL 89 (1989).

A Distributed, Net Oriented Semantics for Delta Prolog

Antonio Brogi & Roberto Gorrieri
Dipartimento di Informatica - Università di Pisa
Corso Italia,40 — I-56100 Pisa, Italy

ABSTRACT

A truly distributed operational semantics for Concurrent Logic Languages is defined here, differently from those semantics based on intrinsically sequential, interleaving models defined so far. Delta Prolog and the underlying Distributed Logic are taken as case studies, in order to semantically denote AND-parallelism and cooperation via explicit communication mechanism (local environment model of computation). A scheme for translating a Delta Prolog system into a 1-safe Petri net is given and properties of (perpetual) processes based on the notion of causality, e.g. fairness and deadlock, are addressed.

1. INTRODUCTION

The Horn Clause Language (HCL for short) is a subset of the First Order Logic which has been extensively studied and has generated the family of Logic Languages. The computation of a HCL program consists in proving a goal $G_1, \ldots, G_n$ by applying the clauses of the program. At each step of derivation there are two choices to be taken:

i) which goal G_i to reduce (*computation rule*)
ii) which clause C_j to apply to the selected goal (*search rule*)

This fact leads to implicit forms of potential parallelism. AND-parallelism exploits the parallelism in the choice of the goal to be reduced, i.e. all the goals can be derived in parallel (*parallel computation rule*). OR-parallelism is the parallelism in the clause selection phase, i.e. all the clauses whose head is unifiable with the goal are applied in parallel (*parallel search rule*). While models exploiting pure OR-parallelism do not present many problems because of the complete independence of the environments of the various branches of the OR-derivation, pure AND-parallelism presents several difficulties mainly due to the presence of shared logical variables.

As hardware VLSI technologies evolve, new highly parallel computer architectures become realizable which call for new languages for parallel programming. Logic languages are regarded as candidates for this purpose for their clean semantics and their implicit parallelism. Nevertheless, a parallel language requires some explicit mechanism for the control of concurrency: pure logic languages lack of synchronization mechanisms. This fact has led to two different approaches. The first one restricts the possible modes of the mechanism of unification by introducing an explicit distinction between producers and consumers of the values of a logical variable. The resulting model is a subcase of AND-parallelism, called STREAM-parallelism [7] based on an eager evaluation of structured data which can be treated as streams. Many concurrent logic languages based on STREAM-parallelism have been proposed, among which we mention

Concurrent Prolog [26], PARLOG [6] and Guarded Horn Clauses [27]. These languages introduce different types of constraints on the unification mechanism: Concurrent Prolog has *read only* annotations on variables, PARLOG provides *mode* declarations, while GHC solves the problem with the *input guards*. The other approach extends HCL by introducing explicit operators like fork and join, proposed in [18,9,12], and communication primitives for message-passing (send/receive), as in Delta Prolog [1,20,21,22].

We study the case of pure AND-parallelism where the cooperation between parallel goals is performed via explicit mechanisms of communication / synchronization. While this model of computation represents the correspondent of the message-passing model in the field of traditional concurrent languages (and therefore is naturally suitable for distributed descriptions), STREAM-parallelism corresponds to a shared memory model of computation. In Delta Prolog, the cooperation mechanism is based on the notion of event, originally introduced in [5,17]. These features of Delta Prolog make it a good candidate as a distributed logic language.

The presence of extra-logical primitives for the explicit control of concurrency decreases the logical flavour of concurrent logic languages. Phenomena such as synchronization, communication, deadlock and process creation may be more advantageously modeled using techniques stemming from imperative and functional concurrent languages, emphasizing *control* rather than *logic*. In fact, interleaving based operational semantics have been recently defined: in [25], Saraswat has provided operational semantics for several concurrent logic languages by means of transition systems defined in the Structural Operational Semantics style [23] (SOS for short). Also [14] has addressed this problem in a similar way for a subset of Theoretical Flat Concurrent Prolog by using techniques developed for TCSP in [5]. In a recent paper [3] an operational semantics has been given to Theoretical Concurrent Prolog resorting to techniques based on metric spaces.

A distributed, net oriented semantics for a concurrent logic language is presented here, to the best of our knowledge, for the first time. We model a concurrent system as a set of sequential processes, possibly located in different places, which cooperate in accomplishing a task: thus, neither a global state nor a global clock must be assumed. Behaviours of systems are represented through the causal relations among the events performed by subparts of their distributed states and then translated to a class of Petri nets [4,24], called 1-safe P/T nets. Petri nets are essentially nondeterministic automata enriched by giving states an additional structure of set in order to represent distributed states and by allowing transitions involving only some of the processes present in the actual state. The semantical description corresponds to an implementation where the goal is distributed on many processors, each containing a copy of the program.
In this paper we use a three step method. First, we describe the behaviour of sequential processes by means of two transition systems: one for programs and the other for sequential goals. The former describes the mechanisms of head unification and clause selection. The latter represents the evolution of sequential goals, possibly resorting to the program transition system when an atom goal needs a clause to unify with. Secondly, we represent the states of the concurrent system by sets of sequential processes and the evolution of the whole system is described in terms of process interactions. Finally, we map all our semantical structure in Petri Nets in quite a trivial manner. This constructive technique has been inspired by paper [11], where it was originally proposed, and [8,10,19], where it has been applied to concurrent languages.
As a matter of fact, we stipulate a sort of well-formedness of goals with respect to communications. More precisely, we define a condition of *closure*, which states that a goal must be "communication closed" at the top-level, i.e. unable to perform an event-communication with an external partner. Since we are considering programs and goals ready to be executed, we exclude possible open communications which have a sense

only during the composition of subsystems.

The classical operational semantics for logic languages denotes the set of solutions of a program as the set of computed answer substitutions. In our operational semantics, this success set together with the finite failures and the global deadlocks that may occur in a computation are observable from its final state. We could be dissatisfied for this *final state* semantics since it is not adequate w.r.t. infinite computations (perpetual processes) in the sense that relevant features of behaviours are not dealt with. A better description of a concurrent system should take into account all the actions performed (thus we know what the system is doing) and their causal relations, i.e. every computation is observed as the partial ordering of events it generates. This way, important properties, such as safety and liveness, can be observed.

Our semantical definitions do respect Delta Prolog specifications [1], nevertheless we do not compare our semantics with those of the authors of Delta Prolog since their operational model is interleaving while our one is distributed. It is worth observing that, with respect to the problem of dealing with distributed backtracking, our operational semantics is as far from implementation issues as the classical HCL operational semantics [28].

An outline of the paper follows. Section 2 contains the abstract syntax of Delta Prolog. The transition systems for programs and sequential goals are described in Section 3. In Section 4 we introduce a decomposition function in order to translate a goal into the set of its sequential processes. Then, we give a rewriting system for describing the interactions among sequential processes, and finally we consider closed systems. Section 5 contains the translation of the semantical structure into a 1-safe P/T Petri net, while several comments about semantics and properties of perpetual processes are addressed in Section 6.

2. PRELIMINARIES

We define the abstract syntax for a version of Delta Prolog [1,20,22]. The operators on goals resemble those of other concurrent languages such as CCS [17], TCSP [5] and CSP [15]. In Delta Prolog (DP for short) a clear distinction is made between sequential ";" and parallel composition "||" at the syntactical level. In HCL the meaning of the operator "," can not be immediately turned to either one. It is not a sequential composition operator since the success set does not depend on the order, if any, atoms in a goal have to be selected (i.e. on the computation rule) nor it is a parallel operator since atoms can not be independently derived in parallel (because of shared variables). Moreover, an explicit construct for exchanging information is introduced. Subgoals of a parallel goal can communicate by *event goals*, labelled by an event name representing the name of a communication channel.

The language *alphabet* is $\langle D, V, P\cup C\rangle$, where D is a family indexed on $\mathbf{N}$ (natural numbers) of data constructors (D_0 is the set of constants), V represents the set of variable symbols, P the set of all program-defined predicate symbols and C the set of (unary, postfixed) communication predicate symbols $\{?e, !e \mid e\in D_0\}$. In the following, let A represent a program-defined atomic formula and T a term in $T_{D(V)/\approx}$ (the free D-algebra on V modulo variance).

Definition 2.1. *(Delta Prolog Syntax)*

The abstract syntax of Delta Prolog is given by the following BNF-like grammar:

$P ::= A \leftarrow \mid A \leftarrow G \mid P \oplus P$

$G ::= A \mid G_e \mid G_c \mid G;G \mid G \| G$

$G_e ::= T?e \mid T!e$

$G_c ::= G_e;G \mid G_c \therefore G_c$

We denote by:

Pr the set of *programs*, ranged over by p (possibly indexed)

G the set of *goals*, ranged over by g (possibly indexed)

G_e the set of *event goals*, ranged over by g_e (possibly indexed)

G_c the set of *choice goals*, ranged over by g_c (possibly indexed)

A goal g is a *parallel goal* iff it has either form:

$g = g_1 \| g_2$ $\quad\quad$ $g = g';g''$ where g' is a parallel goal.

Otherwise g is a *sequential goal*. ♦

Let us give an informal overview of DP syntax and semantics. A program is a fact (A ←), a definite clause (A ← G) or the union of two programs (P ⊕ P): in other words a program is a set of facts and clauses. A goal g can be an atom A, an *event* goal g_e, a *choice* goal g_c, the sequential composition or the parallel composition of two goals. There are two types of events goals with the form, respectively: T?e and T!e, where T is a term (the message), e is the event name (the communication channel), "!" and "?" specify the communication mode (send/receive). In order to solve an event goal T?e, we have to solve simultaneously a complementary event goal T'!e. Moreover, both goals can solve if and only if T and T' unify and, after the unification is performed, they become ground [1]. As an example, the two event goals: 1!e and 2?e do not match as well as X!e and Y?e, while p(1,y)!e and p(x,2)?e do match by generating an exchange of values, i.e. the substitution {x/1,y/2}. Summing up, every communication is due to the simultaneous execution of exactly two complementary event goals, thus only symmetric, synchronous communication is allowed.
When dealing with a choice goal, only one goal is chosen for the rest of the derivation. If both goals could generate a step of derivation, one of them is nondeterministically selected. The operator of sequentialization ";" simply states that in $g_1;g_2$, g_1 has to be solved before g_2.
Finally, in order to solve a parallel goal $g_1 \| g_2$ we solve g_1 and g_2 in parallel. In contrast with the fact that Delta Prolog is based on a message-passing model of computation, it allows variable sharing by parallel goals. Two goals are independently solved in parallel generating local bindings. When both processes have terminated, the two environments are *joined*, i.e. we verify if the independently computed substitutions are compatible.
The target of our semantical definition is to model the peculiar aspects of control operators for concurrency. Therefore we do not deal with clause order or cut operator, i.e. we refer to HCL instead of Prolog as the support for Delta Prolog, as it is specified in the original proposal of Distributed Logic. It is worth observing that a complete definition of the transition system for Prolog with cut operator can be easily given [25].

3. The Transition Systems for Programs and Sequential Goals

We define the operational semantics in the Structural Operational Semantics style [23] as a hierarchy of abstract machines. The lower level machine, described as a transition system, defines the mechanisms of head unification and clause selection, i.e. the program dependent features. Then, an abstract machine for sequential goals is presented, again as a transition system possibly resorting to the former transition system when an atom goal needs a clause to unify with. Finally, a rewriting system representing the evolution of the whole system is introduced. A state is represented as a set of sequential processes, while the rewriting rules describe process interactions.

Definition 3.1. *(Labelled Transition System)*
A *labelled transition system* is a triple $\langle\Gamma, \Lambda, \rightarrow\rangle$, where Γ is a set of *configurations*, Λ is the set of *labels* and $\rightarrow \subseteq \Gamma \times \Lambda \times \Gamma$ is the *transition relation*. A configuration γ is called *stuck* iff
$\forall\lambda\in\Lambda, \forall\gamma'\in\Gamma: \langle\gamma, \lambda, \gamma'\rangle \notin \rightarrow$. ♦

We now define the sets of labels and configurations for programs and sequential goals. Since we give semantics to a pair ‹program, goal›, we should include the program into every configuration. However, since the program does never change during the evaluation of the goal, we do not insert it in all the configurations, but just consider it as an implicit parameter.

Definition 3.2. *(Labels and Configurations)*
We define the sets of *labels* and *configurations* for sequential goals and programs as follows.
Let $g \in G$ be a goal, $A \in$ Atom, $p \in Pr$, $\theta \in Sub$[1], $\square$ the empty goal, T a term and *failure* a special symbol.

$$\Lambda_G = \{\theta\} \cup \{T!e\} \cup \{T?e\} \qquad \Lambda_P = \{\theta\}$$

$$\Gamma_G = \langle g,\theta\rangle \cup \langle\square,\theta\rangle \cup \langle failure\rangle \qquad \Gamma_P = \Gamma_G \cup \langle A{:}p,\theta\rangle$$

These sets are ranged over by λ_G, λ_P, γ_G and γ_P, respectively, omitting the subscripts whenever unambiguous. ♦

A configuration $\langle g,\theta\rangle$ describes a goal g to be reduced and a substitution θ representing the history of the whole derivation from the initial goal to g. Every transition is labelled with the substitution computed in the last step of derivation or by an event goal. In order to describe the communication mechanism, we use labels of the form T!e, T?e. When a sequential goal communicates, it assumes the existence of a partner for the communication. The existence of such a partner will be checked only at composition time and thus we have to store the name, the communication mode (! or ?) and the term T of a not yet matched event goal. $\langle\square,\theta\rangle$ stands for the special stuck configuration "nothing to do", while $\langle A{:}p, \theta\rangle$ represents an atom goal A to be solved *via* program p. When goal A is unable to make further progress because there is no unifiable clause, the configuration ‹failure› is reached. Thus we can directly observe the case of failure and distinguish it from deadlock.

3.1. Transitions for Programs

The (possibly infinite) set of transitions is generated by means of inference rules defined through syntax driven rules in the SOS style.The problems of head unification and clause selection are program dependent [25]. We describe how a clause of the program is selected to unify its head with the goal. Note that a clause invocation fails if head unification fails, the activated goal fails only if all the clause invocations fail.

Definition 3.3.
The *program derivation relation* over configurations, written as $\gamma_P -\lambda_P\rightarrow_p \gamma'_P$, is defined as the least relation satisfying the following inference rules.

Head Unification)

$\theta = mgu(A_1, A_2)$ **implies** $\langle A_1{:}A_2\leftarrow g, \delta\rangle -\theta\rightarrow_p \langle g\theta, \delta\theta\rangle$
and $\langle A_1{:}A_2\leftarrow\ , \delta\rangle -\theta\rightarrow_p \langle\square, \delta\theta\rangle$

1 *Sub* denotes the set of substitutions.

$\neg \exists \, mgu(A_1, A_2)$ **implies** $\langle A_1{:}A_2{\leftarrow}g, \delta\rangle \text{-}\varepsilon\text{->}_p \langle failure\rangle$
and $\langle A_1{:}A_2{\leftarrow}, \delta\rangle \text{-}\varepsilon\text{->}_p \langle failure\rangle$

Clause Selection)

$\langle A{:}p, \delta\rangle \text{-}\theta\text{->}_p \langle g, \delta\theta\rangle$ **implies** $\langle A{:}(p \oplus p'), \delta\rangle \text{-}\theta\text{->}_p \langle g, \delta\theta\rangle$
and $\langle A{:}(p' \oplus p), \delta\rangle \text{-}\theta\text{->}_p \langle g, \delta\theta\rangle$

$\langle A{:}p, \delta\rangle \text{-}\varepsilon\text{->}_p \langle failure\rangle$ **and** $\langle A{:}p', \delta\rangle \text{-}\varepsilon\text{->}_p \langle failure\rangle$
implies $\langle A{:}(p \oplus p'), \delta\rangle \text{-}\varepsilon\text{->}_p \langle failure\rangle$
and $\langle A{:}(p' \oplus p), \delta\rangle \text{-}\varepsilon\text{->}_p \langle failure\rangle$ ♦

REMARKS

Head Unification) The first rule simply states that if there is a mgu for an atom and the head of a clause then the atom can be replaced by the body of the clause, instantiated with the generated substitution. If the head of a clause does not unify with the atom to be proved, the derivation of the atoms leads to the special configuration failure.

Clause Selection) This rule states that the choice of a clause unifying with the goal to be proved cuts off all the other clauses. With respect to the case of failure, the goal fails if and only if it is not possible to find a clause of the program which is head-unifiable with it, i.e. all the possible clause selections lead to failure.

3.2. Transitions for Sequential Goals

In the following definitions we use:

$\gamma_0 \text{—>} \gamma_1 / \ldots / \gamma_n$ **implies** $\gamma'_0 \text{—>} \gamma'_1 / \ldots / \gamma'_n$

as an abbreviation for the n rules:

$\gamma_0 \text{—>} \gamma_i$ **implies** $\gamma_0 \text{—>} \gamma_i$ for $i = 1,\ldots,n$.

Definition 3.4.

The *sequential goal derivation relation* over configurations, written as $\gamma_G = \lambda_G \Rightarrow \gamma'_G$, is defined as the least relation satisfying the following inference rules.

Goal Derivation)

$\langle A{:}p, \theta\rangle \text{-}\sigma\text{->}_p \langle g, \theta\sigma\rangle / \langle \square, \theta\sigma\rangle / \langle failure\rangle$
implies $\langle A, \theta\rangle = \sigma \Rightarrow \langle g, \theta\sigma\rangle / \langle \square, \theta\sigma\rangle / \langle failure\rangle$

Sequential Composition)

$\langle g, \delta\rangle = \lambda_G \Rightarrow \langle \square, \delta\theta\rangle / \langle g', \delta\theta\rangle / \langle failure\rangle$
implies $\langle g;g'', \delta\rangle = \lambda_G \Rightarrow \langle g''\theta, \delta\theta\rangle / \langle g';(g''\theta), \delta\theta\rangle / \langle failure\rangle$

Event Goal)

$\langle T!e, \delta\rangle = T!e \Rightarrow \langle \square, \delta\rangle$ **and** $\langle T?e, \delta\rangle = T?e \Rightarrow \langle \square, \delta\rangle$

Choice Goal)

$\langle g_c, \delta\rangle = \lambda_G \Rightarrow \langle \square, \delta\theta\rangle / \langle g, \delta\theta\rangle$ **implies** $\langle g_c \therefore g_c', \delta\rangle = \lambda_G \Rightarrow \langle \square, \delta\theta\rangle / \langle g, \delta\theta\rangle$
and $\langle g_c' \therefore g_c, \delta\rangle = \lambda_G \Rightarrow \langle \square, \delta\theta\rangle / \langle g, \delta\theta\rangle$ ♦

REMARKS

Goal Derivation) This rule simply states that an atom A has to be solved by means of the transition system for programs, and therefore clause selection and head unification are part of the next step. Note that p has to be a new, renamed copy of the program.

Sequential Composition) This rule states that the left component has to be solved first, and that the derived substitution has to be applied to the right component, too. On the other side, if the derivation leads to a failure, the whole goal fails.

Event Goal) According to [1], in order to solve an event goal T_1?e, we must simultaneously solve a complementary event goal T_2?e. Both goals solve if and only if T_1 and T_2 unify and then, after the value exchange, both T_1 and T_2 are ground terms. Therefore, we have to store all the relevant information in the label so that, when dealing with the process interaction, this condition can be checked.

Choice Goal) If a goal g_c can proceed somehow, the choice goal evolves by discarding the alternative and following the derivation of g_c.

Finally, note that all the configurations whose left component is a parallel goal are stuck in the transition system for sequential goals.

4. The Rewriting Derivation Relation for Parallel Goals

We now decompose a parallel goal into those subgoals representing its *sequential processes*, in the style of [11,10]. For instance, from goal g||g' we will obtain two subgoals g|id and id|g' (tag "|id" records that subterm g is in the left context of a parallel composition, symmetrically tag "id|"). More precisely, a sequential process is composed of a sequential goal (possibly the empty goal □ or failure), a computed substitution, the access path codifying its location in the abstract syntax tree of a parallel goal (i.e. a structure of tags id| and |id), and a continuation specifying what to do afterwards. Now, the evolution of a set of sequential processes I, included in a state, is represented by a rewriting rule of the form I–λ→ I'. The processes in I perform the action λ evolving to I', in other words, I causes I' through λ. The intended dynamic meaning of such a rewriting rule is that the set I occurring in the current state can be replaced, after showing an event (labelled by) λ, by the set of processes I'. We call these states *distributed*, since their components can be allocated in different places and can proceed on their own without requiring any centralized control, even when they synchronize. The description of parallel goal derivation relation is presented in Section 4.2.

Even if the variables of a parallel goal may be shared by its sub-goals, substitutions generated by the proofs of the sub-goals are to be considered independent. This *join* AND-parallelism allows us to describe computations with decentralized *loci* of control in a simple manner by checking the compatibility of independently computed substitutions at join time only.

Since we are interested in giving semantics to goals which are communication-closed at the top level, the rewriting system has to be constrained so that an event goal can not be performed by a sequential process if the complementary event goal is not ready to be executed. This aim is reached by introducing a rule which filters only the well-formed transitions and it will be described in Section 4.3.

4.1. Splitting a Delta Prolog parallel goal into a set of sequential processes

Definition 4.1. *(Sequential Processes)*

We give the syntax of *sequential processes*:

SP::= ‹ □, θ› / ‹g, θ› / ‹failure› /‹SP|id, g'› / ‹id|SP, g'›

where □ is the empty goal, g is a sequential goal, θ is a substitution, g' is a goal, failure is a special configuration and tags |id and id| record the context in which a sequential process is set. The set of sequential processes SP is ranged over by sp, and its subsets are named I, J (possibly indexed). ♦

Intuitively speaking, a sequential process is a sequential goal together with a substitution and an access path defining its location within the syntactical structure of the parallel goal. If the access path is not empty, the second component of the pair is a goal representing a continuation (what its ancestor has to do next). See also Example 4.1 below.
Now we describe how to map any pair ‹goal, substitution› into a (finite) set of sequential processes.

Definition 4.2. *(From goals to sets of sequential processes)*
Function **dec** : $\Gamma_G \longrightarrow fin(2^{SP})$ is defined by structural induction on goals.

Sequential Goal)

$dec(\langle\Box,\theta\rangle) = \{\langle\Box,\theta\rangle\}$		$dec(\langle failure\rangle) = \{\langle failure\rangle\}$
g is a sequential goal	**implies**	$dec(\langle g,\theta\rangle) = \{\langle g,\theta\rangle\}$

The following rules apply only to parallel goals.

Sequential Composition)

$\langle sp, g'\rangle \in dec(\langle g, \theta\rangle)$	**implies**	$\langle sp, g';g''\rangle \in dec(\langle g;g'', \theta\rangle)$
$\langle sp, \Box\rangle \in dec(\langle g, \theta\rangle)$	**implies**	$\langle sp, g'\rangle \in dec(\langle g;g', \theta\rangle)$

Parallel Composition)

$sp \in dec(\langle g, \theta\rangle)$	**implies**	$\langle sp|id, \Box\rangle \in dec(\langle g \| g', \theta\rangle)$
	and	$\langle id|sp, \Box\rangle \in dec(\langle g' \| g, \theta\rangle)$ ♦

The configurations for the empty goal, failure and for a sequential goal are singleton sets of sequential processes. The *Sequential Composition* rule simply states that the decomposition recursively splits the left component only and that the rôle of the right component is to contribute to the continuation. Finally, the sequential processes derived from a parallel goal are exactly those derived by its two components, augmented by tags |id or id|, and with the empty goal as continuation, since nothing has to be done after the parallel goal has been solved.

Example 4.1.
Consider the parallel goal $(((g_1\|g_2);g_3)\|g_4);g_5$, where all the g_i are sequential goals. Given a substitution θ, we have:

i) $dec(\langle g_i, \theta\rangle) = \{\langle g_i, \theta\rangle\}$ i= 1, 2,…,5; by *Sequential Goal* rule

ii) $dec(\langle g_1 \| g_2, \theta\rangle) = \{\langle\langle g_1, \theta\rangle|id, \Box\rangle, \langle id|\langle g_2, \theta\rangle, \Box\rangle\}$ by **i)** and *Parallel Composition* rule

iii) $dec(\langle(g_1 \| g_2);g_3, \theta\rangle) = \{\langle\langle g_1, \theta\rangle|id, g_3\rangle, \langle id|\langle g_2, \theta\rangle, g_3\rangle\}$ by **ii)** and *Sequential Composition* rule

iv) $dec(\langle(g_1 \| g_2);g_3) \| g_4, \theta\rangle) = \{\langle\langle\langle g_1, \theta\rangle|id, g_3\rangle|id, \Box\rangle, \langle\langle id|\langle g_2, \theta\rangle, g_3\rangle|id, \Box\rangle, \langle id|\langle g_4, \theta\rangle,\Box\rangle$
by **i)**, **iii)** and *Parallel Comp.* rule

v) $dec(\langle(((g_1\|g_2);g_3)\|g_4);g_5,\theta\rangle)=\{\langle\langle\langle g_1,\theta\rangle|id,g_3\rangle|id,g_5\rangle,\langle\langle id|\langle g_2,\theta\rangle,g_3\rangle|id,g_5\rangle, \langle\langle id|\langle g_4,\theta\rangle, \Box\rangle,g_5\rangle$
by **iv)** and *Sequential Comp.* rule ♦

Note that function dec is injective, but it is not surjective since there are sets of sequential processes which are not obtainable via dec: for instance, $\{\langle\langle g_1, \theta\rangle|id, \Box\rangle, \langle g_2, \theta\rangle\}$.
Now we characterize the set of sequential processes corresponding to a possible configuration reachable from a starting goal. Let us consider the empty goal □ and failure as legal Delta Prolog goals. We consider

$\square$ as a legal goal because we use it for representing a terminated component process waiting for the termination of its brother, and the motivations for the introduction of failure as a legal goal are analogous.

Definition 4.3. *(Complete sets of sequential processes)*
Given a set J of sequential processes, J is *complete* iff there is a goal g and a substitution θ such that J is dec(‹g,θ›) up to substitutions. ♦

The correspondence between dec(‹g,θ›) and J is up to substitutions because sequential processes generated by a parallel goal work in separate environments and thus their substitutions may be different.

4.2. Rewriting rules for Parallel Goals

Before giving the rewriting rules for parallel goals, we need an auxiliary definition. According to Delta Prolog specification, a parallel goal terminates if and only if all its sequential processes have succeeded and the computed substitutions are compatible.
Given a vector **X** of variables $\langle x_1,\ldots,x_n \rangle$ and a substitution θ, we say that $X\theta = \langle x_1\theta,\ldots,x_n\theta \rangle$ where $\forall x_i \notin Dom(\theta): x_i\theta = x_i$

Definition 4.4. *(Substitution Unifier)*
Given two substitutions θ_1 and θ_2, let $X_i = \{v_j \mid v_j \notin Dom(\theta_i) \wedge \exists\, v_k/t_k \in \theta_i : v_j \in Var(t_k)\}$ with i=1,2. Let X be the vector obtained from $\cup_{i=1,2} Dom(\theta_i) \cup X_i$.
θ_1 and θ_2 are **compatible** if and only if $\exists\, \sigma = mgu(X\theta_1, X\theta_2)$ and σ is called the **substitution unifier** (SU) of θ_1 and θ_2. ♦

Note that the definition above implies that: $\theta_1\sigma = \theta_2\sigma$, since the most general instance of $(X\theta_1, X\theta_2)$ is $X\theta_1\sigma = X\theta_2\sigma$.

Example 4.2.
Given $\theta_1 = \{x/2, y/f(s,z)\}$ and $\theta_2 = \{x/w, y/f(t,2)\}$, the sets X_1 and X_2 are $\{s,z\}$ and $\{w, t\}$, respectively. Thus $X = \langle x, y, z, s, w, t \rangle$, $X\theta_1 = \langle 2, f(s, z), z, s, w, t\rangle$ and $X\theta_2 = \langle w, f(t,2), z, s, w, t\rangle$. The more general instance of them is $\langle 2, f(t, 2), 2, t, 2, t\rangle$ and the unifying substitution σ, i.e. their SU is $\sigma = \{z/2,\ s/t,\ w/2\}$. ♦

Definition 4.5. *(Labels)*
The set Λ_D of *labels*, ranged over by λ_D, is the set $\{\theta\} \cup \{T!e\} \cup \{T?e\}$. ♦

Notation: the application of a substitution σ to a sequential process SP, denoted by SPσ, is defined as follows:

$\langle \square, \theta\rangle\sigma = \langle \square, \theta\sigma\rangle$ $\qquad$ $\langle g, \theta\rangle\sigma = \langle g\sigma, \theta\sigma\rangle$ $\qquad$ $\langle failure\rangle\sigma = \langle failure\rangle$

$\langle SP|id, g'\rangle\sigma = \langle SP\sigma|id, g'\rangle$ $\qquad$ $\langle id|SP, g'\rangle\sigma = \langle id|SP\sigma, g'\rangle$

and the set $\{SP\sigma \mid SP \in I\}$ is denoted by Iσ. We sometimes denote the set $\{\langle sp,g\rangle \mid sp \in I\}$ by the pair ‹I,g›. Besides, in the following definition, ɡ stands for a goal g or the empty goal $\square$.

Definition 4.6. *(Rewriting derivation relation for parallel goals)*
The *parallel goal derivation relation* $I_1 -\lambda_D\!\!\to I_2$ is defined as the least relation satisfying the following

axioms and inference rules.

Join)

θ_1 and θ_2 are *compatible* with *SU* σ **implies**

$\{sp_1, sp_2\}$ $-\sigma\text{->}$ $dec(\langle(g)\theta_1\sigma, \theta_1\sigma\rangle)$ **where** $sp_1 = \langle\langle\Box,\theta_1\rangle|id, g\rangle$ and $sp_2 = \langle id|\langle\Box,\theta_2\rangle, g\rangle$

$sp_1 = \langle\langle failure\rangle|id, g\rangle$ *and* $sp_2 = \langle id|\langle\Box,\theta_2\rangle, g\rangle$ **or**

$sp_1 = \langle id|\langle failure\rangle, g\rangle$ *and* $sp_2 = \langle\langle\Box,\theta_2\rangle|id, g\rangle$ **or**

$sp_1 = \langle\langle failure\rangle|id, g\rangle$ *and* $sp_2 = \langle id|\langle failure\rangle, g\rangle$ **implies** $\{sp_1, sp_2\}$ $-\varepsilon\text{->}$ $\{\langle failure\rangle\}$

Act)

$\langle g,\theta\rangle \Longrightarrow_{\lambda_G} \langle g',\theta'\rangle$ / $\langle\Box,\theta'\rangle$ / $\langle failure\rangle$

implies $\{\langle g,\theta\rangle\}$ $-\lambda_G\text{->}$ $dec(\langle g',\theta'\rangle)$ / $\{\langle\Box,\theta'\rangle\}$ / $\{\langle failure\rangle\}$

Async)

$I_1 -\lambda_D\text{->} I_2$ **implies** $\langle I_1|id, g\rangle -\lambda_D\text{->} \langle I_2|id, g\rangle$

and $\langle id|I_1, g\rangle -\lambda_D\text{->} \langle id|I_2, g\rangle$

Sync)

$I_1-\lambda_1\text{->}I'_1$ **and** $I_2 -\lambda_2\text{->}I'_2$ **and** $\lambda_1 = T_1!e$ and $\lambda_2 = T_2?e$

and $mgu(T_1, T_2) = \sigma$ **and** $T_i\sigma$ is ground, $i = 1,2$

implies $\langle I_1|id \cup id|I_2, g\rangle -\sigma\text{->} \langle(I'_1|id \cup id|I'_2)\sigma, g\rangle$

and $\langle I_2|id \cup id|I_1, g\rangle -\sigma\text{->} \langle(I'_2|id \cup id|I'_1)\sigma, g\rangle$

REMARKS

Join) As soon as both components of a parallel goal terminate, the continuation of the parallel goal is enabled and its substitution updated. If θ_1 and θ_2 are not compatible, then the set $\{sp_1, sp_2\}$ represents a deadlocked state. If one of the components (or both) has failed, then the whole system fails. Note that the failure of a parallel goal is detected at join time only.

Act) This is essentially an *import* rule for sequential goals. Everything a sequential goal can perform in the transition system for sequential goals can be also performed by the corresponding singleton set in the rewriting system for parallel goals.

Async) From the premise that a set J of sequential processes performs an action, we can infer that the same set can perform that action in any context.

Sync) The communication mechanism is handshake. Substitution σ is applied to the set of sequential processes in order to keep trace of the bindings generated by the communication.

4.3. Rewriting Rules for Systems

Given a Delta Prolog program p, an initial goal g and the empty substitution ε, we describe the system $S_{p,g}$ with a suitable rewriting system. The rewriting rules for system $S_{p,g}$ are those of the previous rewriting system without the rules for performing asynchronous event goals. In other words, we introduce a *closure* condition stating that a goal must be "communication closed" at the top-level, i.e. unable to perform an event-communication with an external partner. Since we are considering programs and goals ready to be executed, we exclude possible open communications at all.

Definition 4.7. *(Rewriting derivation relation for systems)*

The *system derivation relation* $I_1 =\lambda_D\Rightarrow I_2$ is defined as the least relation satisfying the following inference rule:

Communication-Closed)

$I_1 -\lambda_D\!\!\to I_2$	**and**	$\lambda_D \neq T!e$ and $\lambda_D \neq T?e$
	implies	$I_1 =\lambda_D\Rightarrow I_2$ ♦

Note that this filtering rewriting rule is very close to the rule for the restriction operator of CCS [17]. Instead of introducing a new rewriting system for filtering such derivations, we could have added a syntactic construct "·\" and denote with g\ the communication closed goal corresponding to a goal g and we could have imposed that the initial goal were a restricted goal.

Example 4.3.

Let us consider the following program p, where T is a ground term:

$$A \leftarrow (T!e \therefore T?f)\,;\, A \quad \oplus \quad (B \leftarrow T!f\,;\, B \;\; \oplus \;\; C \leftarrow T?e\,;\, C)$$

Let $A \parallel (B \parallel C)$ be the initial goal. The initial set of sequential processes is $dec(\langle A \parallel (B \parallel C), \varepsilon\rangle)$. From the definition of the dec function we have that:

$$dec(\langle A \parallel (B \parallel C), \varepsilon\rangle) = \{\langle\langle A, \varepsilon\rangle|id, \Box\rangle, \langle id|\langle\langle B, \varepsilon\rangle|id, \Box\rangle, \Box\rangle, \langle id|\langle id|\langle C, \varepsilon\rangle, \Box\rangle, \Box\rangle\}.$$

The first component of the set can make the following derivation:

$$\{\langle\langle A, \varepsilon\rangle|id, \Box\rangle\} =\varepsilon\Rightarrow \{\langle\langle (T!e \therefore T?f);A, \varepsilon\rangle|id, \Box\rangle\}$$

Below we give the complete deduction leading to that transition:

i) $\langle A : A \leftarrow (T!e \therefore T?f)\,;\, A, \varepsilon\rangle -\varepsilon\!\to_p \langle (T!e \therefore T?f)\,;\, A, \varepsilon\rangle$ — for *Head Unification*, since $\varepsilon = mgu(A, A)$

ii) $\langle A{:}p, \varepsilon\rangle -\varepsilon\!\to_p \langle (T!e \therefore T?f)\,;\, A, \varepsilon\rangle$ — by **i)** and *Clause Selection*

iii) $\langle A, \varepsilon\rangle =\varepsilon\Rightarrow \langle (T!e \therefore T?f)\,;\, A, \varepsilon\rangle$ — by **ii)** and *Goal Derivation*

iv) $\{\langle A, \varepsilon\rangle\} -\varepsilon\!\to \{\langle (T!e \therefore T?f)\,;\, A, \varepsilon\rangle\}$ — by **iii)** and *Act*

v) $\{\langle\langle A, \varepsilon\rangle|id, \Box\rangle\} -\varepsilon\!\to \{\langle\langle (T!e \therefore T?f)\,;\, A, \varepsilon\rangle|id, \Box\rangle\}$ — by **iv)** and *Async*

vi) $\{\langle\langle A, \varepsilon\rangle|id, \Box\rangle\} =\varepsilon\Rightarrow \{\langle\langle (T!e \therefore T?f)\,;\, A, \varepsilon\rangle|id, \Box\rangle\}$ — by **v)** and *Communication Closed*

In an absolutely analogous manner, it is possible to derive the following:

$$\{\langle id|\langle\langle B, \varepsilon\rangle|id, \Box\rangle, \Box\rangle\} =\varepsilon\Rightarrow \{\langle id|\langle\langle T!f;B, \varepsilon\rangle|id, \Box\rangle, \Box\rangle\} \text{ and}$$

$$\{\langle id|\langle id|\langle C, \varepsilon\rangle, \Box\rangle, \Box\rangle\} =\varepsilon\Rightarrow \{\langle id|\langle id|\langle T?e;C, \varepsilon\rangle, \Box\rangle, \Box\rangle\}.$$

The derivation corresponding to the communication between processes A and B follows:

a) $\langle T?f, \varepsilon\rangle =T?f\Rightarrow \langle\Box, \varepsilon\rangle$ — for *Event Goal*

b) $\langle (T!e \therefore T?f), \varepsilon\rangle, \Box\rangle =T?f\Rightarrow \langle\Box, \varepsilon\rangle$ — by **a)** and *Choice Goal*

c) $\langle (T!e \therefore T?f)\,;\, A, \varepsilon\rangle =T?f\Rightarrow \langle A, \varepsilon\rangle$ — by **b)** and *Sequential Composition*

d) $\{\langle\langle (T!e \therefore T?f)\,;\, A, \varepsilon\rangle|id, \Box\rangle\} -T?f\!\to \{\langle\langle A, \varepsilon\rangle|id, \Box\rangle\}$ — by **c)** and *Act*

Analogously, we derive:

e) $\{\langle id|\langle\langle T!f;B, \varepsilon\rangle|id, \Box\rangle, \Box\rangle\} -T!f\!\to \{\langle id|\langle\langle B, \varepsilon\rangle|id, \Box\rangle, \Box\rangle\}$.

Moreover, we have:

f) $\{\langle\langle (T!e \therefore T?f)\,;\, A, \varepsilon\rangle|id, \Box\rangle, \langle id|\langle\langle T!f;B, \varepsilon\rangle|id, \Box\rangle, \Box\rangle\} -\varepsilon\!\to \{\langle\langle A, \varepsilon\rangle|id, \Box\rangle, \langle id|\langle\langle B, \varepsilon\rangle|id, \Box\rangle\}, \Box\rangle\}$
by **d)**, **e)** and *Sync*

g) $\{\langle\langle (T!e \therefore T?f)\,;\, A, \varepsilon\rangle|id, \Box\rangle, \langle id|\langle\langle T!f;B, \varepsilon\rangle|id, \Box\rangle, \Box\rangle\} =\varepsilon\Rightarrow \{\langle\langle A, \varepsilon\rangle|id, \Box\rangle, \langle id|\langle\langle B, \varepsilon\rangle|id, \Box\rangle\}, \Box\rangle\}$
by **f)** and *Communication Closed*

Quite similarly, it is possible to derive the following:

$\{\langle\langle(T!e.\therefore.T?f) ; A, \varepsilon \rangle \text{lid}, \Box\rangle, \langle \text{idl}\langle\text{idl}\langle T?e;C, \varepsilon\rangle,\Box\rangle, \Box\rangle\} =\varepsilon\Rightarrow \{\langle\langle A, \varepsilon \rangle\text{lid}, \Box\rangle, \langle\text{idl}\langle\text{idl}\langle C, \varepsilon \rangle, \Box\rangle, \Box\rangle\}$ ♦

Our rewriting system is asynchronous. Actually, the derivations of a system are independent of those sequential processes which are concurrent with the rewritten ones, but inactive. In other words, each rule is context-independent.

Property 4.1. *(Asynchrony of system derivation relation)*
If $I_1 =\lambda_D\Rightarrow I_2$ is a derivation, there exists a set of sequential processes J such that $J \cap I_1 = \varnothing$ and $J \cup I_1$ is complete. Furthermore, for every such J we have also that $J \cap I_2 = \varnothing$ and that $J \cup I_2$ is a complete set of sequential processes.
Proof. Immediate by induction on the structure of the proof of the derivation. ♦

5. A Delta Prolog System as a 1-Safe P/T Net

Now we build a 1-safe P/T net given a Delta Prolog program p, an initial goal g to be proved and the empty substitution ε. In the following, we use standard definitions of Petri net theory [4,24]. Roughly, the places of the net are the sequential processes belonging to dec(‹g, ε›) and all those reachable from this starting set. The transitions are essentially the rewriting rules described in the previous section. The flow (causal) relation relates those sequential processes which are the left-hand side of a rewriting rule to the rewriting rule (transition) itself, and the rewriting rule to the new set of sequential processes composing the right-hand side of the rule.

Definition 5.1. *(From Delta Prolog systems to nets)*
The set of places S, the set of transitions T, the flow relation F and the initial marking M_0 of the P/T net $\Sigma_{p,g} = \langle S, T, F, M_0\rangle$ associated to system $S_{p,g}$ are defined as the least sets, relation and function satisfying the following inference rules:

- $sp \in dec(\langle g, \varepsilon\rangle)$ **implies**
  $sp \in S$ **and** $M_0(sp) = 1$
- $I_1 \subseteq S$ **and** $I_1 =\lambda_D\Rightarrow I_2$ **implies**
  $(I_1 =\lambda_D\Rightarrow I_2)\in T$ **and** $I_2 \subseteq S$ **and** $\forall\ sp \in I_2\backslash dec(\langle g, \varepsilon\rangle)\ M_0(sp) = 0$
- $(I_1 =\lambda_D\Rightarrow I_2)\in T$ **implies**
  $I_1\ F\ (I_1 =\lambda_D\Rightarrow I_2)$ **and** $(I_1 =\lambda_D\Rightarrow I_2)\ F\ I_2$ ♦

Note that $\Sigma_{p,g}$ is indeed a P/T net since ‹S, T, F› is a net (it satisfies the condition $S\cap T = \varnothing$) and M_0 is a marking. The condition which says that $\forall\ t \in T\ {}^{\bullet}t \neq \varnothing$ and $t^{\bullet} \neq \varnothing$ is satisfied (third rule).

A transition $t \in T$ is *enabled* by a marking *M* iff $\forall\ s \in {}^{\bullet}t,\ M(s) \geq 1$. Furthermore, the occurrence of a transition t *produces* the marking *M'* (*M* [t› *M'*):

$$M'(s) = \begin{cases} M(s) & \text{if } s \notin ({}^{\bullet}t \cup t^{\bullet}) \textit{ or } s \in ({}^{\bullet}t \cap t^{\bullet}) \\ M(s) - 1 & \text{if } s \in {}^{\bullet}t \setminus t^{\bullet} \\ M(s) + 1 & \text{if } s \in t^{\bullet} \setminus {}^{\bullet}t \end{cases}$$

$[M_0\rangle$ denotes the set of all the possible markings M reachable from M_0 by successive firing of enabled transitions. M a 1-safe marking iff $\forall s \in S, M(s) \leq 1$. The P/T net Σ is *1-safe* iff $\forall M \in [M_0\rangle, M$ is 1-safe. Finally, we are going to prove that $\Sigma_{p,g}$ is 1-safe.

Theorem 5.1.

Given a program p, a goal g and the substitution ε, the P/T net $\Sigma_{p,g}$ is 1-safe.

Proof. Since the P/T net $\Sigma_{p,g}$ is 1-safe if and only if $\forall M \in [M_0\rangle$ M is 1-safe, the proof is by induction on the set $[M_0\rangle$. At each step of induction we also prove that the set $P = \{s \in S \mid M(s) = 1\}$ is complete, in order to inherit the results of Property 4.1. By Definition 5.1, the initial marking M_0 is 1-safe and the set $P_0 = \{s \in S \mid M_0(s) = 1\} = \mathrm{dec}(g,\varepsilon)$ is complete. Let us suppose, by inductive hypothesis, that a reachable marking M is 1-safe and that its corresponding set P is complete. Let $t = I_1 =\!\lambda_D\!\Rightarrow I_2$ a M-enabled transition and $M\,[t\rangle\, M'$. We have to prove that the produced marking M' is 1-safe and that the corresponding set P' is complete. Let $J=P\backslash I_1$. By definition of produced marking, M' on the set of places $I_1\backslash I_2$ holds 0, on the set of places $I_2\backslash J \cup I_1\cap I_2$ holds 1, while for the set of places $J\cap I_2$ marking M' would hold 2. But, by Property 4.1, we have that $J\cap I_2=\emptyset$ - thus marking M' is 1-safe - and finally that $P' = J\cup I_2$ is complete. ♦

We eventually define the operational semantics of a Delta Prolog system $S_{p,g}$. The operational semantics can not be defined simply as the set of all the computed answer substitutions since occurrence sequences may be infinite or, though terminal (i.e. finite with no transition enabled by the last marking), may lead to deadlock. The possible last markings of a terminal occurrence sequence (o.s. for short) may be:

- the single place $\{\langle\square, \theta\rangle\}$, which corresponds to a correct answer substitution θ,
- the single place {‹failure›}, i.e. a failure occurs since an atom goal can not unify with any clause,
- a set I of places, which stands for a deadlock possibly due to (more than) one of the following reasons:
  - the non-existence of a compatible substitution (in case of termination of a parallel goal),
  - the inability to unify terms in a matching pair of event goals,
  - the possible partner for communication is terminated or failed.

Summing up, we can say that the operational semantics of a Delta Prolog system $S_{p,g}$ is:

$$
\begin{aligned}
[S_{p,g}]_o \quad = \quad & \{\theta \mid \xi \text{ is a terminal o.s. of } \Sigma_{p,g} \text{ and } \mathrm{last}(\xi) = \{\langle\square, \theta\rangle\}\} \\
\cup \quad & \{\mathrm{fail} \mid \xi \text{ is a terminal o.s. of } \Sigma_{p,g} \text{ and } \mathrm{last}(\xi) = \{\langle\mathrm{failure}\rangle\}\} \\
\cup \quad & \{\Delta \mid \xi \text{ is a terminal o.s. of } \Sigma_{p,g} \text{ and } |\mathrm{last}(\xi)| \geq 2\} \\
\cup \quad & \{\infty \mid \xi \text{ is an infinite o.s. of } \Sigma_{p,g}\}.
\end{aligned}
$$

6. Fairness and Deadlock

We could be dissatisfied for this *final state* semantics since it is not adequate w.r.t. infinite computations (perpetual processes) in the sense that relevant features of behaviours are not dealt with. A better description of system $S_{p,g}$ should take into account all the actions performed (thus we know what the system is doing) and their causal relations, i.e. every computation is observed as the partial ordering of events it generates. Moreover, important properties, such as safety and liveness properties, can be observed by considering the overall evolution of the system.

Let us consider the following Delta Prolog program taken from [21]. A counter object is programmed allowing commands for incrementing and checking the value of a counter:

counter(S) ← up(N)?mail; U is S+N; counter(U)
∴ equal_to(S)?mail; counter(S).

producer ← prod(X); X!mail; producer.

where prod is a predicate generating commands for the counter. An initial goal may be:

← counter(0) || producer || producer.

It is trivial to observe that the computations of this Delta Prolog system may not terminate while still having a precise meaning: an infinite sequence of communications and updatings of the counter value. The denotation of programs which do not compute answer substitutions is represented by the set of all the occurrence sequences (sequential behaviour) or by the set of processes [4,24] (concurrent behaviour)[2]. With these kinds of denotation, we are able to observe properties of perpetual processes like fairness and partial deadlock. The competition between two producers may be solved by the counter serving only one of them from a certain time onwards: this is a typical example of unfair computation. A situation of partial deadlock may arise if a producer sends an equal_to command to the counter process with an argument minor than the counter value: in this case the producer which has sent the message is deadlocked while the other producer and the counter may proceed.

A first attempt to express such properties in this framework may be the following. We look at an o.s. simply as the sequence of the sets of places with one token. Given a reachable marking M, let $P = \{s \in S \mid M(s) = 1\}$; an o.s. $\xi = M_0 t_1 M_1 t_2 M_2 t_3 M_3 \ldots$ are observed as the sequence $P_0 P_1 P_2 P_3 \ldots$.
Among the various notion of fairness (see [13] for a detailed description of this issue), we just consider a couple of them. The first is called *global* fairness in [11] and *communication* fairness in [13]. It states that a transition $t = I_1 \overset{\lambda_D}{\Longrightarrow} I_2$ which is always enabled will eventually be fired. This fact is expressed by the following formula:

$$\textit{not}\ (\exists\, i \in \mathbf{N}, \exists\, t \in T: {}^{\bullet}t \subseteq \cap_{j\geq i} P_j).$$

A second, more demanding notion of fairness (*local* fairness [11] or *process* fairness [13]) concerns the fact that a place which can always (infinitely often, respectively) be involved in a firing, possibly within different transitions, will eventually be consumed. We represent this fact as:

$$\textit{not}\ (\exists\, i \in \mathbf{N}, \exists\, s \in S: s \in \cap_{j\geq i} P_j \ \textit{and}\ \forall\, j\geq i\ \exists\, t \in T, {}^{\bullet}t \subseteq P_j \text{ and } s \in {}^{\bullet}t)$$

$$\textit{not}\ (\exists\, i \in \mathbf{N}, \exists\, s \in S: s \in \cap_{j\geq i} P_j \ \textit{and}\ \exists \text{ infinitely many } j, j\geq i, \exists\, t \in T, {}^{\bullet}t \subseteq P_j \text{ and } s \in {}^{\bullet}t).$$

We can also write formulae stating that an o.s. is free of some kind of deadlock. For instance, a sequence $P_0 P_1 P_2 P_3 \ldots$, satisfying the following formula:

$$\exists\, i \in \mathbf{N}, \exists\, s \in S: s \in \cap_{j\geq i} P_j \text{ and } \textit{goal}(s) = g_e \,/\, g_e; g \,/\, g_c \therefore g_c$$

shows a deadlock due to a missing matching event goal, where *goal*: S → G is the function that extracts from a place (sequential process) the sequential goal it represents. Note that a sequential process of this kind could have partners that may never perform the complementary event goal. Thus a sequence with a partial deadlock may be locally fair, as well.
Another manner to define the operational semantics of a DP system $S_{p,g}$ consists in considering its processes, i.e. the pairs $\pi = \langle N, \rho\rangle$, where N is an occurrence net and ρ: B∪E → S∪T a labelling of N.
For finite processes, we have that Max(N) may be $\{\langle\Box, \theta\rangle\}$, or {‹failure›}, or even a non-singleton set of

2 In order to better specify a system, it is necessary to consider structures which are less simplistic than sets, like event structures, synchronization trees or nondeterministic measurement systems. We will not be concerned with this aspect of the problem.

places. For infinite processes, Max(N) denotes the set of places that do not fire from a certain point onwards. If a transition t is such that $^{\bullet}t \subseteq$ Max(N), then ‹N, ρ› is not *globally* fair. On the contrary, nothing we can say about *local* fairness. Even if a place s belongs to Max(N), we do not know if there has always (infinitely often) been some transition t, which may be time-dependent, such that $s \in {}^{\bullet}t$.
This is a first possible approach to the problem of describing properties of perpetual processes. Further research would be mandatory to better investigate this subject.

CONCLUSIONS

In this paper a distributed, net oriented semantics for a Concurrent Logic Language is presented for the first time. We have studied a model of computation based on AND-parallelism together with an explicit mechanism of synchronization. Delta Prolog has been chosen as a case study.
In order to give a truly distributed semantics, we resorted to techniques developed for imperative and functional languages [8,10,11,19] and adapted them to the case of logic languages. This approach allows us to represent causal relations among processes in a very direct way and to address the problem of observing properties that are intrinsically based on causality.

Acknowledgments

We would like to thank Pierpaolo Degano for many valuable discussions and suggestions. Thanks also to Catuscia Palamidessi, Ugo Montanari and Franco Turini for their comments and encouragement.

REFERENCES

[1] J.N. Aparicio, J.C. Cunha, L.F. Monteiro, L.M. Pereira, "The Specification of Delta Prolog", *Draft manuscript*, June 1987.

[2] K.R. Apt, M.H. van Emden, "Contribution to the Theory of Logic Programming", *J. ACM* **29**,3, 1982, pp. 841-862.

[3] J.W. de Bakker, J.N. Kok, "Uniform Abstraction, Atomicity and Contractions in the Comparative Semantics of Concurrent Prolog", *Proc. of Int. Conf. on Fifth Generation Computer Systems*, Tokyo, 1988, pp 347-355, Vol. II.

[4] E. Best, R. Devillers, "Sequential and Concurrent Behaviour in Petri Net Theory", *Theoretical Computer Science* **55**, 1987, pp. 87-136.

[5] S.D. Brookes, C.A.R. Hoare, A.D. Roscoe, "A Theory of Communicating Sequential Processes", *J. ACM* **31**, 3, 1984, pp.560-599.

[6] K.L. Clark, S. Gregory, "PARLOG: A Parallel Logic Programming Language", *ACM Trans. on Prog. Lang. and Syst.* **8**, 1, Jan. 1986, pp. 1-49.

[7] J. Conery, D. Kibler, "Parallel Interpretations of Logic Programs", in *Proc. ACM Conf. on Functional Prog. Languages and Comp. Architectures*, 1981, pp. 163-170.

[8] P. Degano, R. De Nicola, U. Montanari, "A Distributed Operational Semantics for CCS based on Condition/Event Systems", Nota Interna B4-21, 1987. To appear in *Acta Informatica*.

[9] P. Degano, S. Diomedi, "A first order semantics of a connective suitable to express concurrency", in *Proc. 2nd Logic Programming Workshop*, Albufeira (Portugal), 1983, pp. 506-517.

[10] P. Degano, R. Gorrieri, S. Marchetti, "An Exercise in Concurrency: A CSP Process as a Condition/Event System", in *Proc. 8th European Workshop on Applications and Theory of Petri Nets*,

Zaragoza, 1987, pp 31-50. To appear in *Advances in Petri Nets 1988* (G. Rozenberg, ed.) *LNCS*, Springer 1988.

[11] P. Degano, U. Montanari, "Concurrent Histories: A Basis for Observing Distributed Systems", *J.CSS* **34**, 1987, pp. 442-461.

[12] M. Falaschi, G. Levi, C. Palamidessi "A Synchronization Logic: Axiomatics and Formal Semantics of Generalized Horn Clauses", *Information and Control*, **60**, 1-3 , pp.36-69.

[13] N. Francez, "Fairness", Text and Monographies in Computer Science (D. Gries, ed.), Springer, 1986.

[14] R. Gerth, M. Codish, Y. Lichtenstein, E.Y. Shapiro, "Full Abstract Denotational Semantics for Flat Concurrent Prolog", *Proc. Logic in Computer Science*,1988, pp 320-333.

[15] C.A.R. Hoare, "Communicating Sequential Processes", *C. ACM* **21**, 8, 1978, pp. 666-677.

[16] J.W. Lloyd, *Foundations of Logic Programming*, Springer 1984 (Second Edition 1987).

[17] R. Milner, "Notes on a Calculus for Communicating Systems", in *Control Flow and Data Flow: Concepts of Distributed Programming* (M. Broy,ed.), NATO ASI Series F, Vol. 14, Springer, 1984, pp. 205-228.

[18] L. Monteiro, "A proposal for distributed programming in logic", *Implementations of Prolog - Ellis Horwood, 1984*.

[19] E.-R. Olderog, "Operational Petri Net Semantics for CCSP", in *Advances in Petri Nets 1987*, (G. Rozenberg, ed.), *LNCS 266*, Springer 1987, pp. 196-223.

[20] L.M. Pereira, R. Nasr, "Delta Prolog: a distributed logic programming language", *Proc. of FGCS*, Tokyo, November 1984.

[21] L.M. Pereira, L.F. Monteiro, J. Cunha, J. Aparicio, M.C. Ferreira, *Delta Prolog User's Manual - Version 1.0*, June 1987.

[22] L.M. Pereira, L. Monteiro, J.C.. Cunha, J.N. Aparicio, "Delta Prolog: a distributed backtracking extension with events", *Proc. of 3rd International Conference on Logic Programming*, July 1986.

[23] G.D. Plotkin, "A Structural Approach to Operational Semantics", Technical Report DAIMI FN-19, CS Department, University of Aarhus, 1981.

[24] W. Reisig, *Petri Nets: An Introduction*, EATCS Monograph in Computer Science, Springer, 1985.

[25] V.A. Saraswat, "The Concurrent Language CP: Definition and Operational Semantics", *Proc. of the 14th Symposium on Principles of Programming Languages, ACM*, January 1987, pp. 49-62.

[26] E.Y. Shapiro, "Concurrent Prolog: a progress report", in *Fundamentals of Artificial Intelligence* (W. Bibel, Ph. Jorrand, eds.), *LNCS 232*, Springer, 1987, pp. 277-313.

[27] K. Ueda, *Guarded Horn Clauses*, Ph.D. Thesis, University of Tokyo, 1986, (also as Technical Report TR-103, ICOT 1986).

[28] M. van Emden, R.A. Kowalski, "The Semantics of Predicate Logic as a Programming Language", *J. ACM*, **23**, 1976, pp. 733-742.

Continuation Semantics for PROLOG with Cut

A. de Bruin

Faculty of Economics, Erasmus University
P.O.Box 1738, 3000 DR Rotterdam, the Netherlands

E.P. de Vink

Department of Mathematics and Computer Science, Free University
De Boelelaan 1081, 1081 HV Amsterdam, the Netherlands

ABSTRACT

We present a denotational continuation semantics for PROLOG with cut. First a uniform language $\mathscr{B}$ is studied, which captures the control flow aspects of PROLOG. The denotational semantics for $\mathscr{B}$ is proven equivalent to a transition system based operational semantics. The congruence proof relies on the representation of the operational semantics as a chain of approximations and on a convenient induction principle. Finally, we interpret the abstract language $\mathscr{B}$ such that we obtain equivalent denotational and operational models for PROLOG itself.

Section 1 Introduction

In the nice textbook of Lloyd on the foundations of logic programming [Ll] the cut, available in all PROLOG systems, is described as a controversial control facility. The cut, added to the Horn clause logic for efficiency reasons, affects the completeness of the refutation procedure. Therefore the standard declarative semantics using Herbrand models does not adequately capture the computational aspects of the PROLOG language. In the present paper we study the PROLOG cut operator in a sequential environment augmented with backtracking. Our aim is to provide a denotational semantics for PROLOG with cut and to prove this semantics equivalent to an operational one.

First of all we separate the "logic programming" details (such as most general unifiers and renaming indices) in PROLOG from the specification of the flow of control, (e.g. backtracking, the cut operator). This is achieved by extracting the uniform language $\mathscr{B}$ from PROLOG - uniform in the sense of [BKMOZ] - which contains only the latter issues. Fitting within the "Logic Programming without Logic" approach, ([Ba2]), our denotational model developed for the abstract backtracking language has enough flexibility for further elaboration to a non-uniform denotational model of PROLOG itself. Moreover, the equivalence of this denotational semantics and an operational semantics for PROLOG is a straightforward generalization for the congruence proof of $\mathscr{B}$.

Secondly, our denotational semantics uses continuations. This has several advantages over earlier semantics which (essentially) are based on a direct approach. (See [Br] for a discussion on the relative merits of continuations vs. direct semantics.) We arrive at a concise set of semantical equations in which there is no need for coding up the states using cut flags or special tokens (as in [JM], [DM], [Vi], [Bd]). Moreover, since operational semantics - being a goal stacking model - must contain (syntactical) continuations, congruence of the two semantics can be established much more elegantly.

Our final contribution can be found in the equivalence proof itself. The equivalence proof does not split - as usual - into $\mathcal{O} \subseteq \mathcal{D}$ and $\mathcal{D} \subseteq \mathcal{O}$. Rather, both the operational and denotational semantics are represented as least upperbounds of chains and we prove equality of the approximating elements. (See also [KR], [BM] where - although not made explicit - in the setting of complete metric spaces operational and denotational semantics can be represented as limits of Cauchy sequences.)

Related work on the comparative semantics of PROLOG with cut includes [JM], [DM], [Vi]. Jones & Mycroft present a direct Scott-Strachey style denotational semantics. They do not compare this semantics with an operational one. Instead, correctness of their semantics comes from its systematic construction. In [Vi] also a direct denotational model is developed and additionally proven correct w.r.t. a transition based operational meaning. The proof is rather involved, since the cut is modeled by a special token (as in [JM]). The semantics of Debray & Mishra is a mixture of a direct and continuation semantics. They (need to) have sequences of answers substitutions together with cut flags in their semantics. The denotational semantics is related to an operational one. However, it is not clear to us what makes their equivalence proof work. (In particular we do not understand the proof of theorem 4.1, case 5 in the appendix of [DM].) The semantics mentioned above all denote a program by a sequence of substitutions. In the present paper we only deliver the first one. This does not give rise to loss of generality, since our semantics allows extension to streams of substitutions, (as in [Vi], [Bd]). We have chosen not to do so for reasons of space and clarity of the presentation.

Acknowledgments. Our appreciation is due to Jaco de Bakker, Frank de Boer, Joost Kok, John-Jules Meyer and Jan Rutten, members of the Amsterdam Concurrency Group, who offered us a stimulating forum. We thank Aart Middeldorp and the referees for reading the manuscript. We are indebted to M279 for the hospitality the authors received during the preparation of this paper.

Section 2 Deterministic Transition Systems

In this section we introduce the notion of a transition system, ([Pℓ], [BMOZ]). For reasons of space we restrict ourselves to deterministic transition systems, which already suit our purposes. Collections of transition systems are turned into a cpo s.t. associating a valuation to a transition system becomes a continuous operation.

(2.1) DEFINITION A deterministic transition system T is a seven tuple $\langle C, I, F, \Omega, D, \alpha, S \rangle$ where the set of configurations C is the disjoint union of I, F and $\{\Omega\}$, I is a set of internal configurations, F is a set of final configurations, Ω is the undefined configuration, D is a domain of values, α: $F \to D$ is a valuation assigning a value to each final configuration and S is a deterministic step or transition relation, i.e. a partial function S: $C \to_{part} C$ with $dom(S) \subseteq I$.

Next we show how to extend the valuation α on final configurations to a valuation α_T on arbitrary configurations of a transition system T.

(2.2) DEFINITION Let $T = \langle C, I, F, \Omega, D, \alpha, S \rangle$ be a deterministic transition system. Denote by $D_\perp$ the flat cpo generated by D with least element $\perp$. We associate with T a mapping α_T: $C \to D_\perp$ defined as the least function in $C \to D_\perp$ s.t. $\alpha_T(\Omega) = \perp$, $\alpha_T(c) = \alpha(c)$ if $c \in F$, $\alpha_T(c) = \alpha_T(c')$ if $(c, c') \in S$ and $\alpha_T(c) = \perp$ otherwise.

Fix sets I and F of internal and final configurations, respectively. Fix an undefined configuration Ω, a domain of values D, a valuation function $\alpha : F \to D$ and put $C = I \cup F \cup \{\Omega\}$. Let $TS = \{ \langle C, I, F, \Omega, D, \alpha, S \rangle \mid S : C \to_{part} C$ with $dom(S) \subseteq I \}$ denote the collection of all deterministic transition systems with configurations in C, internal configurations in I, final configurations in F, undefined configuration Ω, domain of values D and valuation function α. In TS we identify a transition system with its transition relation. (In particular we may write $T(c)$ and $c \to_T c'$ rather than $S(c)$ or $(c, c') \in S$ for a transition system T with step relation S.)

We consider the set of configurations as a flat cpo with ordering $\leq_C$ and least element Ω. This induces an ordering $\leq_{TS}$ on TS as follows: $T_1 \leq_{TS} T_2 \Leftrightarrow dom(T_1) \subseteq dom(T_2)$ & $\forall c \in dom(T_1)$: $T_1(c) \leq_C T_2(c)$. We have that TS is a cpo when ordered by $\leq_{TS}$. (The nowhere defined transition system $\varnothing$ is the least element of TS; for a chain $\langle T_k \rangle_k$ in TS the transition system T with $dom(T) = \bigcup_k dom(T_k)$ and $T(c) = lub_k\, T_k(c)$ acts as least upperbound.) Moreover, the operation $\lambda T.\alpha_T$: $TS \to C \to D_\perp$ that assigns to a transition system the valuation it induces, is continuous with respect to $\leq_{TS}$. (See [Vi].)

REMARK Let $I_0 \subseteq I_1 \subseteq \cdots$ be an infinite sequence of subsets of internal configurations s.t. $I = \bigcup_k I_k$. Put $C_k = I_k \cup F \cup \{\Omega\}$. Then we can construct for each $T \in TS$ a chain of approximations $\langle T_k \rangle_k$ of T in TS, where T_k is defined as the smallest deterministic transition system s.t. $T_k(c) = T(c)$ if $c \in I_k$, $T(c) \in C_k$, and $T_k(c) = \Omega$ if $c \in I_k$ & $T(c)$ is defined but $T(c) \notin C_k$. Then it follows from the above that $T = lub_k\, T_k$ in TS. T_k is called the restriction of T to I_k since (by minimality of T_k) only configurations in I_k act as a left-hand side. Note also that only configurations in C_k act as a right-hand side.

Section 3 Operational Semantics of $\mathscr{B}$

In this section we introduce the abstract backtracking language $\mathscr{B}$ and present an operational semantics based on a deterministic transition system. $\mathscr{B}$ can be regarded as a uniform version of PROLOG with cut, in that it reflects the control structure of PROLOG. For a program $d|s$ in $\mathscr{B}$, the declaration d will induce a transition system $\to_d$ while the statement s induces (given a state) an initial configuration. The operational semantics then is the value of the final configuration (if it exists) of the maximal transition sequence w.r.t. $\to_d$ starting from the initial configuration w.r.t. s.

(3.1) DEFINITION Fix a set of actions *Action* and a set of procedure names *Proc*. We define the set of elementary statements $EStat = \{ a, \underline{fail}, !, s_1 \underline{or} s_2, x \mid a \in Action, s_i \in Stat, x \in Proc \}$, the set of statements $Stat = \{ e_1 ; .. ; e_r \mid r \in \mathbb{N}, e_i \in EStat \}$ and the set of declarations $Decl = \{ x_1 \leftarrow s_1 : .. : x_r \leftarrow s_r \mid r \in \mathbb{N}, x_i \in Proc, s_i \in Stat, i \neq j \Rightarrow x_i \neq x_j \}$. The backtracking language $\mathcal{B}$ is defined by $\mathcal{B} = \{ d | s \mid d \in Decl, s \in Stat \}$.

We let a range over *Action*, x over *Proc*, e over *EStat*, s over *Stat* and d over *Decl*. We write $x \leftarrow s \in d$ if $x \leftarrow s = x_i \leftarrow s_i$ (for some i) or if $s = \underline{fail}$ otherwise.

Next we give an operational semantics to our backtracking language $\mathcal{B}$. We associate with a declaration $d \in Decl$ a deterministic transition system $\rightarrow_d$. The internal configurations of $\rightarrow_d$ are stacks. Each frame on a stack represents an alternative for the execution of some initial goal, i.e. statement. As such a frame consists of a generalized statement and a local state. The state can be thought of holding the values of the variables for a particular alternative. The generalized statement is composed from ordinary statements supplied with additional information concerning the cut: Each component in a generalized statement corresponds with a (nested) procedure call. The left-most component is the body being evaluated at the moment, i.e. the most deeply nested one. Since executing a cut amounts to restoring the backtrack stack as it was at the moment of procedure entry, we attach to a statement a stack (or pointer), that constitutes (points to) the substack of the alternatives that should remain open after a cut in the statement is executed. We call this stack the dump stack of the statement, cf. [JM].

(3.2) DEFINITION Fix a set Σ of states. Define the set of generalized statements by $GStat = \{ \langle s_1, D_1 \rangle : .. : \langle s_r, D_r \rangle \mid r \in \mathbb{N}, s_i \in Stat, D_i \in Stack, i < j \Rightarrow D_i \geq_{ss} D_j \}$, γ denotes the empty generalized statement, the set of frames by $Frame = \{ [g, \sigma] \mid g \in GStat, \sigma \in \Sigma \}$ and the set of stacks by $Stack = \{ F_1 : .. : F_r \mid r \in \mathbb{N}, F_i = [\langle s_1, D_1 \rangle : .. : \langle s_q, D_q \rangle, \sigma] \in Frame$ s.t. $F_{i+1} : .. : F_r \geq_{ss} D_j \}$ (with $S \geq_{ss} S' \Leftrightarrow S'$ is a substack of S). Let $Conf = Stack \cup \Sigma \cup \{\Omega\}$ be the set of configurations.

Fix an action interpretation $I : Action \rightarrow \Sigma \rightarrow_{part} \Sigma$, that reflects the effect of the execution of an action on a state. (The language $\mathcal{B}$ gains flexibility if actions are allowed to succeed in one state, while failing in another. Hence we model failure as partiality.) Let *TS* be the collection of all deterministic transition system with configurations in *Conf*, internal configurations in *Stack*, final configurations in Σ, undefined configuration Ω, domain of values Σ, valuation $\alpha : \Sigma \rightarrow \Sigma$ with $\alpha(\sigma) = \sigma$. We distinguish $\delta \in \Sigma$ that will denote unsuccessful termination.

(3.3) DEFINITION Let $d \in Decl$. d induces a deterministic transition system in *TS* with as step relation the smallest subset of $Conf \times Conf$ s.t.

(i) $E \rightarrow_d \delta$

(ii) $[\gamma, \sigma] : S \rightarrow_d \sigma$

(iii) $[\langle \epsilon, D \rangle : g, \sigma] : S \rightarrow_d [g, \sigma] : S$

(iv) $[\langle a ; s, D \rangle : g, \sigma] : S \rightarrow_d [\langle s, D \rangle : g, \sigma'] : S$ if $\sigma' = I(a)(\sigma)$ exists

$[\langle a;s,D\rangle:g,\sigma]:S \rightarrow_d S$ otherwise

(v) $[\langle \underline{fail};s,D\rangle:g,\sigma]:S \rightarrow_d S$

(vi) $[\langle !;s,D\rangle:g,\sigma]:S \rightarrow_d [\langle s,D\rangle:g,\sigma]:D$

(vii) $[\langle x';s,D\rangle:g,\sigma]:S \rightarrow_d [\langle s',S\rangle:\langle s,D\rangle:g,\sigma]:S$ if $x' \leftarrow s' \in d$

(viii) $[\langle (s_1 \ \underline{or} \ s_2);s,D\rangle:g,\sigma]:S \rightarrow F_1:F_2:S$ where $F_i = [\langle s_i;s,D\rangle:g,\sigma]$ $(i=1,2)$

We comment briefly on each of the above transitions (more precisely transition schemes): (i) The empty stack, denoted by E, has no alternatives left to be tried. Hence the computation terminates unsuccessfully yielding δ. (ii) If the top frame contains the empty generalized statement, denoted by γ, the computation terminates successfully. The local state σ of the frame is delivered as result. (iii) If the left-most component of a generalized statement has become empty (as is the case when a procedure call or the initial statement has terminated), i.e. has format $\langle \epsilon,D\rangle$, the statement-dump stack pair is deleted from the frame. The computation continues with the remaining generalized statement. (iv) In case an action a in the top frame has become active, the action interpretation I is consulted for the effect of a in σ. If $I(a)(\sigma)$ is defined the state is transformed accordingly. If $I(a)(\sigma)$ is not defined the frame fails and is popped of the stack. (v) Execution of *fail* amounts to failure of the current alternative. Hence the top frame is popped of the backtrack stack. Control is transferred to the new top frame. (vi) The transition concerning the cut represents removal of alternatives; the top frame continues its execution. Since the dump stack D is a substack of the backtrack stack S, replacing the backtrack stack by the current dump stack indeed amounts - in general - to deletion of frames, i.e. of alternatives. (Note that the right-hand stack is well-formed by definition of *GStat*.) (vii) A call initiates body replacement. The body is looked up in the declaration d and becomes the active component of the generalized statement in the top frame. This component has its own dump stack, which is (a pointer to) the backtrack stack at call time. (viii) Execution of an alternative composition yield two new frames: an active frame corresponding to the left component of the or-construct and a suspended frame corresponding to the right component.

(3.4) DEFINITION The operational semantics $\mathcal{O}: \mathcal{B} \rightarrow \Sigma \rightarrow \Sigma_\perp$ for the backtracking language $\mathcal{B}$ is defined by $\mathcal{O}(d|s)(\sigma) = \alpha_d([\langle s,\mathrm{E}\rangle,\sigma])$ where $\alpha_d : Conf \rightarrow \Sigma_\perp$ is the valuation associated with the deterministic transition system induced by d.

Section 4 Denotational Semantics for $\mathcal{B}$

By now a standard approach has been established for defining a denotational semantics of a sequential procedural language. Cf. [MS], [St], [Ba1]. We show that a semantics of $\mathcal{B}$ in this section and PROLOG in section 6 can also be given along these lines. Standard semantics uses environments and continuations.

Environments are needed because the denotation $[\![s]\!]_\Delta$ of a statement s depends amongst others on the meaning of the procedure names occurring in s. Therefore the function $[\![\bullet]\!]_\Delta$ takes an environment $\eta \in Env$ as a parameter which defines the meaning of all procedure names.

The flow of control will be described using continuations. For languages like PASCAL, where flow of control is not very intricate, a denotation $[\![s]\!]_\delta$ needs only one continuation as a parameter. Languages containing backtrack constructs, like SNOBOL4, are best described using two continuations, cf. [Te]. In order to capture the effects of the cut operator yet another continuation will be needed. (As is observed independently by M. Felleisen in [Wi], p. 273.) In order to explain how these continuations will be used we introduce them one after another. First we shall discuss the PASCAL-subset of $\mathcal{B}$, i.e. $\mathcal{B}$ without <u>or</u> , <u>fail</u> and cut ! . Thereafter we shall examine the SNOBOL4-subset of $\mathcal{B}$, introducing the <u>or</u> and <u>fail</u> constructs, and finally we shall explain how all three continuations are used in describing full $\mathcal{B}$.

In order to understand the essence of continuation semantics, consider a substatement s that is part of a statement s' (in the PASCAL-fragment of $\mathcal{B}$). The denotation $[\![s]\!]_\delta$ will be a function that will, in the end, deliver an answer in $\Sigma_\perp$. This answer is not the result of executing s alone, but the result of evaluating the whole statement s' of which s is a substatement. Therefore the result does not only depend on an environment η and an initial state σ, it also depends on a denotation ξ of the remainder of the statement, to be executed once evaluation of s has terminated. This leads to the following functionality of $[\![\bullet]\!]_\delta$: $Env \to Cont \to \Sigma \to \Sigma_\perp$. Here $Cont = \Sigma \to \Sigma_\perp$ since the future ξ of a computation will in the end yield an answer, but this answer depends on an intermediate state, viz. the result of evaluating s alone. A typical clause in our semantics up till now, describing the composition operator ";", will be $[\![e;s]\!]_\delta \eta\xi\sigma = [\![e]\!]_e \eta\{[\![s]\!]_\delta \eta\xi\}\sigma$, which says that the answer obtained by executing $e;s$ before ξ will be equal to the answer resulting from execution of e before { execution of s before ξ }.

The next stage is to introduce backtracking in the language by adding the constructs <u>or</u> and <u>fail</u> (and by allowing actions to fail). Describing the flow of control is more complicated now. The problem is that the notion "future of the computation" is not that obvious any more. Evaluation of a statement s can terminate for two reasons now. The first one, successful termination, is similar to the situation we had before. In this case the future of the computation is realized by executing the remainder of the statement textually following s. But now it is also possible that evaluation of s terminates in failure, e.g. by executing a <u>fail</u> statement. Now the rest of the computation is determined by backtracking to the alternatives built up through execution of <u>or</u> - statements in earlier stages of the computation. Such a doubly edged future can best be captured by two continuations, a success continuation $\xi \in SCont$ and a failure continuation $\phi \in FCont$. So now $[\![\bullet]\!]_\delta$ has a new functionality: $[\![\bullet]\!]_\delta : Env \to SCont \to FCont \to \Sigma \to \Sigma_\perp$. The meaning $[\![s]\!]_\delta \eta\xi\phi\sigma$ of s will depend on ξ, denoting the rest of the statement following s, and on ϕ, which is a denotation of the stack of alternatives built up in the past. $FCont$ is best understood by investigating the meaning of the <u>or</u> construct: $[\![s_1 \ \underline{or}\ s_2]\!]_\delta \eta\xi\phi\sigma = [\![s_1]\!]_\delta \eta\xi\phi'\sigma$. This says that executing s_1 <u>or</u> s_2 amounts to executing s_1 with a new failure continuation ϕ' describing what will happen if s_1 terminates in failure. In that case s_2 should be executed, and only if this also ends in failure the computation should proceed as indicated by the original failure continuation ϕ. Hence we have that ϕ' equals $[\![s_2]\!]_\delta \eta\xi\phi\sigma$. Combining all this we obtain that $[\![s_1 \ \underline{or}\ s_2]\!]_\delta \eta\xi\phi\sigma = [\![s_1]\!]_\delta \eta\xi\{[\![s_2]\!]_\delta \eta\xi\phi\sigma\}\sigma$. Apparently we have $FCont = \Sigma_\perp$. As far as the structure of $SCont$ is concerned, it must be realized that the answer obtained from evaluation of the rest of the statement s' does not only depend on the intermediate state resulting from the evaluation of s but also on the alternatives built up by executing s' up to and including s. For it can very well happen that evaluation of the rest of the statement will terminate in failure. We therefore have $SCont = FCont \to \Sigma \to \Sigma_\perp$. We notice that the meaning of the <u>fail</u> statement is straightforward. The answer is the one provided by the failure continuation: $[\![\, \underline{fail} \,]\!]_e \eta\xi\phi\sigma = \phi$. This is also the case if an action a does not succeed in a state σ, i.e. $[\![\, a \,]\!]_e \eta\xi\phi\sigma = \phi$ if $I(a)(\sigma)$ is undefined, (where I is the fixed action interpretation). If a does succeed the state is transformed according to I and the failure continuation and new state are passed to the success continuation ξ. So $[\![\, a \,]\!]_e \eta\xi\phi\sigma = \xi\phi\sigma'$ if $\sigma' = I(a)(\sigma)$ exists.

The only construct of full $\mathscr{B}$ that we did not take into account up to now is the cut operator **!**. This statement resembles the dummy statement because it does not affect the state. There is a side effect however, since a number of alternatives is thrown away. To be more precise, evaluation of **!** discards the alternatives which have been generated since the procedure body in which the **!** occurs has been entered. For our semantics this means that evaluation of **!** amounts to applying the success continuation to the original state (this is the dummy statement aspect), but also to a new failure continuation. This new failure continuation ϕ' is in fact an old one, namely the failure continuation which was in effect on entry of the procedure body in which the **!** occurs. A natural way to obtain this old continuation, which we will call the cut continuation $\chi \in CCont$ in the sequel, is to provide it as an argument of the meaning function $[\![\, \bullet \,]\!]_s$. We finally arrive at the functionality $[\![\, \bullet \,]\!]_s : Stat \to Env \to SCont \to FCont \to CCont \to \Sigma \to \Sigma_\perp$, with $SCont = FCont \to CCont \to \Sigma \to \Sigma_\perp$ and $FCont = CCont = \Sigma_\perp$. The denotation of **!** can now be given by $[\![\, ! \,]\!]_e \eta\xi\phi\chi\sigma = \xi\chi\chi\sigma$. On entry of a procedure body a new cut continuation has been established. The meaning of a procedure call is straightforward. We have $[\![\, x \,]\!]_e \eta\xi\phi\chi\sigma = \eta x \xi\phi\chi\sigma$, i.e. the arguments ξ, ϕ, χ and σ are passed to the meaning ηx of x in the environment η. The real work is performed in the definition of the environment η, which should be derived from the declaration d in the program. We want η to be a fixed point such that ηx, the meaning of the procedure name x is given by $\eta x \xi\phi\chi\sigma = [\![\, s \,]\!]_s \eta\{\lambda\bar{\phi}\bar{\chi}.\xi\bar{\phi}\chi\}\phi\phi\sigma$ if $x \leftarrow s \in d$. Two effects can be noticed here. First of all a new cut continuation, viz. the failure continuation ϕ, is "passed", secondly on (successful) termination of s the old cut continuation should be restored and this is captured by passing $\lambda\bar{\phi}\bar{\chi}.\xi\bar{\phi}\chi$ instead of ξ to the body s.

We now give the denotational semantics of $\mathscr{B}$. We first give the domains: the set of failure continuations $FCont = \Sigma_\perp$, the set of cut continuations $CCont = \Sigma_\perp$, the set of success continuations $SCont = FCont \to CCont \to \Sigma \to \Sigma_\perp$ and the set of environments $Env = Proc \to SCont \to FCont \to CCont \to \Sigma \to \Sigma_\perp$. We denote by σ, ϕ, χ, ξ and η typical elements of Σ, $FCont$, $CCont$, $SCont$ and Env, respectively.

(4.1) DEFINITION

(i) $[\![\, \bullet \,]\!]_e : EStat \to Env \to SCont \to FCont \to CCont \to \Sigma \to \Sigma_\perp$

$[\![\, a \,]\!]_e \eta\xi\phi\chi\sigma = \xi\phi\chi\sigma'$ if $\sigma' = I(a)(\sigma)$ exists

$[\![\, a \,]\!]_e \eta\xi\phi\chi\sigma = \phi$ otherwise

$[\![\, \underline{fail} \,]\!]_e \eta\xi\phi\chi\sigma = \phi$

$[\![\, ! \,]\!]_e \eta\xi\phi\chi\sigma = \xi\chi\chi\sigma$

$[\![s_1 \underline{or} s_2]\!]_e \eta\xi\phi\varkappa\sigma = [\![s_1]\!]_s \eta\xi\{[\![s_2]\!]_s \eta\xi\phi\varkappa\sigma\}\varkappa\sigma$

$[\![x]\!]_e \eta\xi\phi\varkappa\sigma = \eta x \xi\phi\varkappa\sigma$

(ii) $[\![\bullet]\!]_s : Stat \to Env \to SCont \to FCont \to CCont \to \Sigma \to \Sigma_\perp$

$[\![\epsilon]\!]_s \eta\xi\phi\varkappa\sigma = \xi\phi\varkappa\sigma$

$[\![e;s]\!]_s \eta\xi\phi\varkappa\sigma = [\![e]\!]_e \eta\{[\![s]\!]_s \eta\xi\}\phi\varkappa\sigma$

(iii) $\Phi : Decl \to Env \to Env$

$\Phi d\eta x\xi\phi\varkappa\sigma = [\![s]\!]_s \eta\{\lambda\bar{\phi}\bar{\varkappa}.\xi\bar{\phi}\varkappa\}\phi\phi\sigma$ if $x \leftarrow s \in d$

(iv) $[\![\bullet]\!]_{\mathscr{B}} : \mathscr{B} \to \Sigma \to \Sigma_\perp$

$[\![d|s]\!]_{\mathscr{B}}\sigma = [\![s]\!]_s \eta_d \xi_0\phi_0\varkappa_0\sigma$

where η_d is the least fixed point of $\Phi(d)$, $\xi_0 = \lambda\phi\varkappa\sigma.\sigma$ and $\phi_0 = \varkappa_0 = \delta$.

REMARK The least fixed point η_d defined in 4.1(iv) can be obtained as the least upperbound of a chain of iterations $\langle \eta_{d,i} \rangle_i$, with $\eta_{d,i}$ defined by $\eta_{d,0} = \lambda x\xi\phi\varkappa\sigma.\perp$ and $\eta_{d,i+1} = \Phi(d)(\eta_{d,i})$. From the continuity of $[\![\bullet]\!]_s$ we have $[\![s]\!]_s \eta_d = lub_i [\![s]\!]_s \eta_{d,i}$.

We conclude this section with some comment on the similarity of the operational semantics from the previous section and the denotational semantics of this one. There is a natural correspondence between components of a configuration and the parameters of the denotation of a statement. We compare the answer resulting from evaluation of an elementary statement e and the value obtained from a configuration in which e is about to be executed: $[\![e]\!]_e \eta\xi\phi\varkappa\sigma$ vs. $[\langle e;s,D\rangle : g,\sigma]:S$. Here ξ is a denotation of the statements to be executed once e has terminated successfully. So ξ corresponds to the statement s followed by the statements in the generalized statement g. The failure continuation ϕ is the denotational counterpart of the backtrack stack S, the cut continuation $\varkappa$ corresponds to the dump stack D. It is to be expected that if the correspondence is set up as above, the resulting answers should be the same. This will be formalized in the next section and is pivotal to the equivalence proof given there.

Section 5 Equivalence of $\mathscr{O}$ and $\mathscr{D}$

In this section we prove the equivalence of the operational and denotational semantics, thus justifying the definition of the latter one.

(5.1) THEOREM For all $d|s \in \mathscr{B}$: $[\![d|s]\!]_{\mathscr{B}} = \mathscr{O}(d|s)$.

In order to prove theorem 5.1 we use the cpo-structure on the collection of transition systems *TS* and the continuity of the statement evaluator $[\![\bullet]\!]_s$. According to the remark at the end of section 2 and the remark following definition 4.1 we have that both the operational and denotational semantics can be represented as the limit of a chain, i.e. $\mathscr{O}(d|s)(\sigma) = lub_i\ \alpha_{d,i}([\langle s,\mathrm{E}\rangle,\sigma])$ and $[\![d|s]\!]_{\mathscr{B}}\sigma = lub_i [\![s]\!]_s \eta_{d,i}\xi_0\phi_0\varkappa_0\sigma$.

However, in order to compare $\alpha_{d,i}([\langle s,\mathrm{E}\rangle,\sigma])$ and $[\![s]\!]_s \eta_{d,i}\xi_0\phi_0\varkappa_0\sigma$ we need a stronger result. An intercedent is needed between $\alpha_{d,i}$ and $[\![\bullet]\!]_s$. We define a (denotational) function

$[\![\bullet]\!]_{\mathscr{C}}$ on (operational) configurations with parameters d and i and show that for all configurations C we have that the value $\alpha_{d,i}(C)$ equals $[\![C]\!]_{\mathscr{C}} di$. The desired result then follows from $[\![s]\!]_{\mathscr{A}} \eta_{d,i} \xi_0 \phi_0 \chi_0 \sigma = [\![[\langle s, \mathrm{E} \rangle, \sigma]]\!]_{\mathscr{C}} di$ which can be checked by routine. Also, perhaps more surprisingly, the equality of $[\![\bullet]\!]_{\mathscr{C}} di$ and $\alpha_{d,i}$ will be easy to check once the appropriate tool is available.

Having outlined the strategy for the equivalence proof we continue with the definition of the intermediate function $[\![\bullet]\!]_{\mathscr{C}}$. First we have to specify the subsets of configurations $Stack_i$. The environment $\eta_{d,i}$, being the i-th iteration of the bottom-environment, yields the right answer in x provided the call of x leads to at most $i-1$ nested inner calls. This depth of nesting can be controlled in our operational semantics as well. Each component $\langle s_i, D_i \rangle$ in a generalized statement g corresponds to a (nested) procedure call. The depth of nesting in $g = \langle s_1, D_1 \rangle : \ldots : \langle s_r, D_r \rangle$ therefore equals r. (Note that, although $g = \langle s_1, D_1 \rangle : \ldots : \langle s_r, D_r \rangle$ we do not require $D_j \in Stack_i$, this is the case if $[g, \sigma]:S \in Stack_i$ since $[g, \sigma]:S \in Stack$ implies that D_j is a substack of S.)

(5.2) DEFINITION Let $GStat_i = \{ g \in GStat \mid \|g\| \leq i \}$ where $\|g\| = r$ for $g = \langle s_1, D_1 \rangle : \ldots : \langle s_r, D_r \rangle$, $Frame_i = \{ [g, \sigma] \in Frame \mid g \in GStat_i \}$ and $Stack_i = \{ F_1 : \ldots : F_q \in Stack \mid F_j \in Frame_i \}$.

Next we define the intermediate function $[\![\bullet]\!]_{\mathscr{C}}$. Given a stack S the definition of $[\![\bullet]\!]_{\mathscr{C}}$ can only be elaborated further if $S \in Stack_i$. Otherwise the value $\perp$ is returned. Intuitively $\perp$ expresses uncertainty about the value of the configuration. So $\perp$ will be delivered if the elaboration asks for a chain of nested calls of length exceeding i.

(5.3) DEFINITION

(i) $[\![\bullet]\!]_{\mathscr{C}} : Conf \to Decl \to \mathbb{N} \to \Sigma_\perp$

$[\![S]\!]_{\mathscr{C}} di = [\![S]\!]_{\mathscr{S}} di$, $[\![\sigma]\!]_{\mathscr{C}} di = \sigma$, $[\![\Omega]\!]_{\mathscr{C}} di = \perp$

(ii) $[\![\bullet]\!]_{\mathscr{S}} : Stack \to Decl \to \mathbb{N} \to \Sigma_\perp$

$[\![\mathrm{E}]\!]_{\mathscr{S}} di = \delta$, $[\![F:S]\!]_{\mathscr{S}} di = [\![F]\!]_{\mathscr{F}} di \{ [\![S]\!]_{\mathscr{S}} di \}$ if $F:S \in Stack_i$,

$[\![S]\!]_{\mathscr{S}} di = \perp$ if $S \notin Stack_i$

(iii) $[\![\bullet]\!]_{\mathscr{F}} : Frame \to Decl \to \mathbb{N} \to FCont \to \Sigma_\perp$

$[\![[g, \sigma]]\!]_{\mathscr{F}} di\phi = [\![g]\!]_{\mathscr{G}} di\phi\sigma$

(iv) $[\![\bullet]\!]_{\mathscr{G}} : GStat \to Decl \to \mathbb{N} \to FCont \to \Sigma \to \Sigma_\perp$

$[\![\gamma]\!]_{\mathscr{G}} di\phi\sigma = \sigma$, $[\![\langle s, D \rangle : g]\!]_{\mathscr{G}} di\phi\sigma = [\![s]\!]_{\mathscr{A}} \eta_{d,i,g} \{ \lambda\bar{\phi}\bar{x}. [\![g]\!]_{\mathscr{G}} di\bar{\phi} \} \phi \{ [\![D]\!]_{\mathscr{S}} di \} \sigma$

where $\eta_{d,i,g} = \eta_{d,j}$ with $j = i \dot{-} (\|g\|+1)$ where $\|\langle s_1, D_1 \rangle : \ldots : \langle s_r, D_r \rangle\| = r$

$[\![\langle s, D \rangle : g]\!]_{\mathscr{G}} di$ should yield the right answer only if this can be obtained with less than i nested calls. Now g is responsible for a nesting depth $\|g\|$. So the whole generalized statement $\langle s, D \rangle : g$ has a chain of $\|g\|+1$ nested calls already. This means that $\eta_{d,i,g}$ should allow less then $i \dot{-} (\|g\|+1)$ calls. (Here $\dot{-}$ denotes the monus, i.e. subtraction in $\mathbb{N}$.)

The desired property of the function $\llbracket \bullet \rrbracket_{\mathscr{C}}$ is stated in lemma 5.5. First we establish Noetherianity of the restrictions of $\rightarrow_d$ to $Stack_i$, i.e. the absence of infinite transition sequences w.r.t. $\rightarrow_{d,i}$. This supplies us with an induction principle that we shall use in the proof of 5.5.

(5.4) LEMMA Let $d \in Decl$, $i \in \mathbb{N}$ and $\rightarrow_{d,i}$ be the restriction of $\rightarrow_d$ to $Stack_i$. Then we have that $\rightarrow_{d,i}$ is Noetherian.

PROOF Omitted. □

We proceed with the proof of the equality $\llbracket \bullet \rrbracket_{\mathscr{C}} di = \alpha_{d,i}$ (*) . First we notice that this holds for final configurations $\sigma \in \Sigma$, for the undefined configurations Ω, and for internal configurations that admit no transition, i.e. stacks not in $Stack_i$.

We shall prove that (*) is also satisfied by internal configurations that do admit a transition, i.e. stacks in $Stack_i$. For this we observe that given the above it suffices to prove: if $C \rightarrow_{d,i} C'$ and (*) holds for C' then (*) holds for C too, by virtue of the Noetherianity of the transition system $\rightarrow_{d,i}$. (This is the principle of Noetherian induction, although in our - deterministic - case it specializes to induction on the length of the maximal transition sequence (which is finite) out of a configuration. See e.g. [Hu].) By definition of the valuation $\alpha_{d,i}$ we have $\alpha_{d,i}(C) = \alpha_{d,i}(C')$ provided $C \rightarrow_{d,i} C'$. So we only need to show: if $C \rightarrow_{d,i} C'$ then $\llbracket C \rrbracket_{\mathscr{C}} di = \llbracket C' \rrbracket_{\mathscr{C}} di$.

(5.5) LEMMA For all $d \in Decl$ and $i \in \mathbb{N}$ we have $\llbracket \bullet \rrbracket_{\mathscr{C}} di = \alpha_{d,i}$.

PROOF Let $d \in Decl$, $i \in \mathbb{N}$ and $C, C' \in Conf$ s.t. $C \rightarrow_{d,i} C'$. Note $C \in Stack_i$. It suffices to show by structural induction on C: $\llbracket C \rrbracket_{\mathscr{S}} di = \llbracket C' \rrbracket_{\mathscr{C}} di$.

We only treat case (vi) of definition 3.3: $C = [\langle x';s, D\rangle:g, \sigma]:S$. Say $x' \leftarrow s' \in d$. We distinguish two subcases: Subcase (a): $\|\langle x';s, D\rangle:g\| = i$. Then we have $C' = \Omega$ and $\|g\| = i-1$.

$$\begin{aligned}
&\llbracket [\langle x';s, D\rangle:g, \sigma]:S \rrbracket_{\mathscr{S}} di \\
&= \llbracket x' \rrbracket_{e} \eta_{d,i,g} \{ \llbracket s \rrbracket_{s} \eta_{d,i,g} \{\lambda\phi x. \llbracket g \rrbracket_{\mathscr{G}} di\phi\} \} \{\llbracket S \rrbracket_{\mathscr{S}} di\} \{\llbracket D \rrbracket_{\mathscr{S}} di\} \sigma \\
&= \eta_{d,i,g} x' \{ \llbracket s \rrbracket_{s} \eta_{d,i,g} \{\lambda\phi x. \llbracket g \rrbracket_{\mathscr{G}} di\phi\} \} \{\llbracket S \rrbracket_{\mathscr{S}} di\} \{\llbracket D \rrbracket_{\mathscr{S}} di\} \sigma \\
&= \eta_{d,0} x' \{ \llbracket s \rrbracket_{s} \eta_{d,i,g} \{\lambda\phi x. \llbracket g \rrbracket_{\mathscr{G}} di\phi\} \} \{\llbracket S \rrbracket_{\mathscr{S}} di\} \{\llbracket D \rrbracket_{\mathscr{S}} di\} \sigma \\
&= \bot \\
&= \llbracket \Omega \rrbracket_{\mathscr{C}} di.
\end{aligned}$$

Subcase (b): $\|\langle x';s, D\rangle:g\| < i$. Then we have $C' = [\langle s', S\rangle:\langle s, D\rangle:g, \sigma]:S$ and $\|g\| < i-1$.

$$\begin{aligned}
&\llbracket [\langle x';s, D\rangle:g, \sigma]:S \rrbracket_{\mathscr{S}} di \\
&= \eta_{d,i,g} x' \{ \llbracket s \rrbracket_{s} \eta_{d,i,g} \{\lambda\phi x. \llbracket g \rrbracket_{\mathscr{G}} di\phi\} \} \{\llbracket S \rrbracket_{\mathscr{S}} di\} \{\llbracket D \rrbracket_{\mathscr{S}} di\} \sigma \\
&= \Phi d \eta_{d,i-1,g} x' \{ \llbracket s \rrbracket_{s} \eta_{d,i,g} \{\lambda\phi x. \llbracket g \rrbracket_{\mathscr{G}} di\phi\} \} \{\llbracket S \rrbracket_{\mathscr{S}} di\} \{\llbracket D \rrbracket_{\mathscr{S}} di\} \sigma \\
&= \llbracket s' \rrbracket_{s} \eta_{d,i-1,g} \bar{\xi} \{\llbracket S \rrbracket_{\mathscr{S}} di\} \{\llbracket S \rrbracket_{\mathscr{S}} di\} \sigma \\
&\text{where } \bar{\xi} = \lambda\bar{\phi}\bar{x}. \llbracket s \rrbracket_{s} \eta_{d,i,g} \{\lambda\phi x. \llbracket g \rrbracket_{\mathscr{G}} di\phi\} \bar{\phi} \{\llbracket D \rrbracket_{\mathscr{C}} di\} \\
&= \llbracket s' \rrbracket_{s} \eta_{d,i,\langle s, D\rangle:g} \{\lambda\bar{\phi}\bar{x}. \llbracket \langle s, D\rangle:g \rrbracket_{\mathscr{G}} di\bar{\phi}\} \{\llbracket S \rrbracket_{\mathscr{S}} di\} \{\llbracket S \rrbracket_{\mathscr{S}} di\} \sigma \\
&= \llbracket [\langle s', S\rangle:\langle s, D\rangle:g, \sigma]:S \rrbracket_{\mathscr{S}} di
\end{aligned}$$

The other cases are similar, (easier) and omitted here. □

Finally we give the congruence proof of the operational and denotational semantics for $\mathscr{B}$. In the next section we shall modify both this operational and denotational semantics in order to give meaning to PROLOG with cut.

PROOF (of theorem 5.1) Let $\sigma \in \Sigma$. $[\![d|s]\!]_{\mathscr{B}}\sigma = [\![s]\!]_{\mathscr{A}} \eta_d \xi_0 \phi_0 \varkappa_0 \sigma$ (by definition) $=$ $lub_i\, [\![s]\!]_{\mathscr{A}} \eta_{d,i} \xi_0 \phi_0 \varkappa_0 \sigma$ (by continuity of $[\![\bullet]\!]_{\mathscr{A}}$) $= lub_i\, [\![[\langle s,\mathrm{E}\rangle,\sigma]]\!]_{\mathscr{S}} di$ (straightforward) $= lub_i\, \alpha_{d,i}([\langle s,\mathrm{E}\rangle,\sigma])$ (by the lemma) $= \alpha_d([\langle s,\mathrm{E}\rangle,\sigma])$ (by continuity of $\lambda T.\alpha_T$) $=$ $\mathscr{O}(d|s)(\sigma)$ (by definition). □

Section 6 Interpretation of $\mathscr{B}$ into PROLOG

At the moment PROLOG is probably the most important programming language featuring backtracking. It can be viewed as Horn clause logic with a left-most depth-first computation rule. Nevertheless PROLOG contains execution oriented constructs, e.g. the cut, that makes the standard declarative semantics, that associates to a set of clauses its least Herbrand model ([AE], [EK]), less satisfactorily. Although dating from the early seventies it has lasted until 1984 before a denotational semantics for PROLOG was presented, viz. [JM], that gave account to the behavioural aspects of the language. More recently other (denotational) semantics based on several approaches have appeared, e.g. [DM], [Vi], [Bd]. (See also [Fi], [Fr], [DF], [AB], [BW].)

Our work on the backtracking language $\mathscr{B}$ in the previous sections makes yet another semantics easily available: we can interpret the abstract or uniform statements, declarations and states such that: a set of PROLOG clauses can be regarded as a declaration, a PROLOG goal corresponds with a statement in the abstract language, while a substitution can be viewed as a state. (After all this is not surprising since we designed $\mathscr{B}$ as an abstraction of PROLOG.)

This can be done similarly for the operational semantics. Moreover, the interpretation or de-uniformization is done in such a way that the equivalence proof remains valid (after adaptation to minor technicalities). Having factorized the work for a PROLOG semantics in a control flow component (the abstract language $\mathscr{B}$) and a logical component (the interpretation of $\mathscr{B}$ towards PROLOG) we obtain presently a congruence proof for the denotational and operational semantics almost for free. Stated otherwise, we have an instance of the "Algorithm = Logic + Control" paradigm ([Kw]) at the meta level. (In fact, several semantics of logic programming languages can be considered as generalizations of established models for imperative languages w.r.t. the control; the extensions made are concerned with the particular logic component. Cf. [MR], [GCLS], [Kk], [Ba2]. See in particular [BK] for a related approach in the setting of Concurrent Prolog.)

Unfortunately there is a price to pay for our two pass approach, albeit just a syntactical one. Since we restrict procedure names in $\mathscr{B}$ to have just one procedure body, we can consider clauses with pairwise different head predicates only. We feel free to do so, because this is by no means a computational restriction in the presence of the explicit or-construct and actions interpreted as unifications. (One can use a so called homogeneous form for clauses, as in e.g. [EY], and "or" together clauses with the same head predicate.)

Next we define our variant of the PROLOG language. (Note the similarity with the definition of the language $\mathscr{B}$ in section 3.)

(6.1) DEFINITION Let $\mathscr{F}$ be a collection of function symbols, $\mathscr{V}$ a collection of variables and $\mathscr{R}$ a collection of predicate letters. Let *Term* denote the collection of terms generated by $\mathscr{F}$ over $\mathscr{V}$. Define the set of atomic goals $AGoal = \{ t_1 = t_2, \underline{fail}, \; ! \;, G_1 \; \underline{or} \; G_2, R(t_1,..,t_k) \mid t_i \in Term, G_i \in Goal, R \in \mathscr{R}$ of arity $k \}$, the set of goals $Goal = \{ A_1\&..\&A_r \mid r \in \mathbb{N}, A_i \in AGoal \}$, $\underline{true}$ is the empty goal, the set of PROLOG programs $Prog = \{ A_1 \leftarrow G_1 :..: A_r \leftarrow G_r \mid r \in \mathbb{N}, A_i = R_i(\vec{t_i}) \in AGoal, \; i \neq j \Rightarrow R_i \neq R_j, \; G_i \in Goal \}$. Define PROLOG $= \{ P \mid G \mid P \in Prog, G \in Goal \}$.

We next develop an operational semantics for PROLOG along the lines of section 3. In order to obtain a most general answer substitution (i.e. to avoid clashes of logical variables) one is only allowed to resolve an atom against a program clause provided that the variables of the clause are fresh w.r.t. the computation so far. We can achieve this by having infinite supply of copies of the class of variables and tagging every goal with an index that it should be renamed with. (This is in fact structure sharing.) In a global counter we keep track of the number of the first class of variables not used yet.

(6.2) DEFINITION Let *Term'* be the set of terms generated by $\mathscr{F}$ over $\mathscr{V} \times \mathbb{N}$ and Σ be the collection of substitutions over *Term'*, i.e. $\Sigma = \{ \sigma : Term' \rightarrow Term' \mid \sigma$ homomorphic $\}$. The set *GGoal* of generalized goals is defined by $GGoal = \{ \langle G_1, D_1, m_1 \rangle : .. : \langle G_r, D_r, m_r \rangle \mid r \in \mathbb{N}, G_i \in Goal, D_i \in Stack, i \leq j \Rightarrow D_i \geq_{ss} D_j, m_i \in \mathbb{N} \}$, the set of frames $Frame = \{ [g, \sigma, n] \mid g \in GGoal, \; \sigma \in \Sigma, \; n \in \mathbb{N} \}$, the set of stacks $Stack = \{ F_1 :..: F_r \mid r \in \mathbb{N}, F_i = [\langle G_1, D_1, m_1 \rangle : .. : \langle G_r, D_r, m_r \rangle, \sigma, n] \in Frame$ s.t. $F_{i+1} : .. : F_r \geq_{ss} D_j \}$ and the set of configurations $Conf = Stack \cup \Sigma \cup \{\Omega\}$.

The transition system underlying the operational semantics is a straightforward modification of definition 3.3.

Execution of actions $t_1 = t_2$ and procedure calls $R(t_1,..,t_k)$ involve unification. We use a black box unification algorithm *mgu* that yields a most general unifier for two atoms or terms if one exists, and is undefined otherwise. (Cf. [JM], [Fr].) So the effect of the execution of an action $t_1 = t_2$ in state σ is the update $\sigma\theta$, i.e. composition of substitutions, of σ w.r.t. the most general unifier θ of t_1 and t_2 in state σ (and appropriately renamed).

Slightly more deviating is procedure handling, since one has to unify first the call and the head of the particular clause successfully before body replacement can take place. (Stretching a point one may consider PROLOG as a form of conditional rewriting. See also [BW], [EY].) A call is operationally described as follows. Consider a call, i.e. atom, $R(t_1,..,t_k)$. First the concerning procedure definition, i.e. clause, is looked up in the declaration, i.e. PROLOG program. Say this is $R(\bar{t}_1,..,\bar{t}_k) \leftarrow \bar{G}$. Next we try to unify $R(t_1,..,t_k)$ and $R(\bar{t}_1,..,\bar{t}_k)$ (considering renaming and the current substitution). If this is possible, i.e. a most general unifier exists, we replace the call by the procedure body, i.e. body of the program clause, extended with dump stack and renaming index, and change the state and global counter according to the side effect, i.e. the result of *mgu*, initiated by the call. We refer the reader to the nice tutorial of [Le] for a discussion on unification in logic programming vs. parameter passing and value return in imperative languages.

(6.3) DEFINITION Let $P \in Prog$. P induces a deterministic transition system $\rightarrow_P$ with as transition relation the smallest subset of $Conf \times Conf$ s.t.

(i) $E \rightarrow_P \delta$

(ii) $[\, \gamma, \sigma, n \,]{:}S \rightarrow_P \sigma$

(iii) $[\, \langle \underline{true}, D, m \rangle{:}g, \sigma, n \,]{:}S \rightarrow_P [\, g, \sigma, n \,]{:}S$

(iv) $[\, \langle t_1 = t_2 \,\&\, G, D, m \rangle{:}g, \sigma, n \,]{:}S \rightarrow_P [\, \langle G, D, m \rangle{:}g, \sigma\theta, n \,]{:}S$
if $\theta = mgu(t_1^{(m)}\sigma, t_2^{(m)}\sigma)$ exists
$[\, \langle t_1 = t_2 \,\&\, G, D, m \rangle{:}g, \sigma, n \,]{:}S \rightarrow_P S$ otherwise

(v) $[\, \langle \underline{fail} \,\&\, G, D, m \rangle{:}g, \sigma, n \,]{:}S \rightarrow_P S$

(vi) $[\, \langle ! \,\&\, G, D, m \rangle{:}g, \sigma, n \,]{:}S \rightarrow_P [\, \langle G, D, m \rangle{:}g, \sigma, n \,]{:}D$

(vii) $[\, \langle R(t_1,..,t_k) \& G, D, m \rangle{:}g, \sigma, n \,]{:}S \rightarrow_P [\, \langle \bar{G}, S, n \rangle{:}\langle G, D, m \rangle{:}g, \sigma, n+1 \,]{:}S$
if $R(\bar{t}_1,..,\bar{t}_k) \leftarrow \bar{G} \in P$ and $\theta = mgu(R(t_1^{(m)},..,t_k^{(m)})\sigma, R(\bar{t}_1^{(n)},..,\bar{t}_k^{(n)}))$ exists
$[\, \langle R(t_1,..,t_k) \& G, D, m \rangle{:}g, \sigma, n \,]{:}S \rightarrow_P S$ otherwise

(viii) $[\, \langle (G_1 \, \underline{or} \, G_2) \,\&\, G, D, m \rangle{:}g, \sigma, n \,]{:}S \rightarrow F_1{:}F_2{:}S$
where $F_i = [\, \langle G_i \& G, D, m \rangle{:}g, \sigma, n \,]$

In the above definition we denote by $t^{(m)}$ the term in $Term'$ obtained by renaming in t variables in $\mathscr{V}$ into the corresponding variables in $\mathscr{V} \times \{m\}$. We use suffix notation for the application and composition of substitutions.

The operational semantics is defined similar to definition 3.4. Here, in the context of logic programming, we choose to fix the start state, viz. the identity substitution σ_{id}. The renaming index is set to 1 having used 0 for the top-level goal already.

(6.4) DEFINITION The operational PROLOG-semantics $\mathscr{O}$: PROLOG $\rightarrow \Sigma_\perp$ is defined by $\mathscr{O}(P|G) = \alpha_P([\, \langle G, E, 0 \rangle, \sigma_{id}, 1 \,])$ where $\alpha_P : Conf \rightarrow \Sigma_\perp$ is the valuation associated with the transition system induced by P.

Having discussed already the idiosyncrasies of PROLOG w.r.t. unification-action and call, it is clear how to adapt the denotational semantics of $\mathscr{B}$ in order to obtain a denotational semantics for PROLOG.

First we redefine the functionality of environments and success continuations. Define $Atom = \{\, R(t_1,..,t_k) \mid R \in \mathscr{R}$ *of arity* k, $t_i \in Term \,\}$. (*Atom* is the PROLOG-counterpart of *Proc*.) Let $Env = Atom \rightarrow \mathbb{N} \rightarrow SCont \rightarrow FCont \rightarrow CCont \rightarrow \Sigma \rightarrow \mathbb{N} \rightarrow \Sigma_\perp$ and $SCont = FCont \rightarrow CCont \rightarrow \Sigma \rightarrow \mathbb{N} \rightarrow \Sigma_\perp$. We take *FCont* and *CCont* as defined previously (with $\Sigma_\perp$ implicitly changed).

(6.5) DEFINITION

(i) $[\![\bullet]\!]_{\mathscr{A}} : AGoal \rightarrow Env \rightarrow \mathbb{N} \rightarrow SCont \rightarrow FCont \rightarrow CCont \rightarrow \Sigma \rightarrow \mathbb{N} \rightarrow \Sigma_\perp$
$[\![t_1 = t_2]\!]_{\mathscr{A}} \eta m \xi \phi \chi \sigma n = \xi \phi \chi \{\sigma\theta\} n$ if $\theta = mgu(t_1^{(m)}\sigma, t_2^{(m)}\sigma)$ exists
$[\![t_1 = t_2]\!]_{\mathscr{A}} \eta m \xi \phi \chi \sigma n = \phi$ otherwise
$[\![\underline{fail}]\!]_{\mathscr{A}} \eta m \xi \phi \chi \sigma n = \phi$

$[\![\; ! \;]\!]_{\mathscr{A}} \eta m \xi \phi \chi \sigma n = \xi \chi \chi \sigma n$

$[\![G_1 \; \underline{or} \; G_2]\!]_{\mathscr{A}} \eta m \xi \phi \chi \sigma n = [\![G_1]\!]_{\mathscr{G}} \eta m \xi \{ [\![G_2]\!]_{\mathscr{G}} \eta m \xi \phi \chi \sigma n \} \chi \sigma n$

$[\![R(\vec{t})]\!]_{\mathscr{A}} \eta m \xi \phi \chi \sigma n = \eta \{ R(\vec{t}) \} m \xi \phi \chi \sigma n$

(ii) $[\![\bullet]\!]_{\mathscr{G}} : Goal \rightarrow Env \rightarrow \mathbb{N} \rightarrow SCont \rightarrow FCont \rightarrow CCont \rightarrow \Sigma \rightarrow \mathbb{N} \rightarrow \Sigma_\perp$

$[\![\underline{true}]\!]_{\mathscr{G}} \eta m \xi \phi \chi \sigma n = \xi \phi \chi \sigma n$

$[\![A \; \& \; G]\!]_{\mathscr{G}} \eta m \xi \phi \chi \sigma n = [\![A]\!]_{\mathscr{A}} \eta m \{ [\![G]\!]_{\mathscr{G}} \eta m \xi \} \phi \chi \sigma n$

(iii) $\Phi : Prog \rightarrow Env \rightarrow Env$

$\Phi P \eta \{ R(\vec{t}) \} m \xi \phi \chi \sigma n = [\![G_0]\!]_{\mathscr{G}} \eta n \{ \lambda \bar{\phi} \bar{\chi} . \xi \bar{\phi} \chi \} \phi \phi \{ \sigma \theta \} \{ n+1 \}$

if $R(\vec{t_0}) \leftarrow G_0 \in P$ and $\theta = mgu(R(\vec{t}^{(m)}) \sigma , R(\vec{t_0}^{(n)}))$ exists

$\Phi P \eta \{ R(\vec{t}) \} m \xi \phi \chi \sigma n = \phi$ otherwise

(iv) $[\![\bullet]\!]_{\mathscr{Prolog}}$: PROLOG $\rightarrow \Sigma_\perp$

$[\![P | G]\!]_{\mathscr{Prolog}} = [\![G]\!]_{\mathscr{G}} \eta_P 0 \xi_0 \phi_0 \chi_0 \sigma_{id} 1$

where η_P is the least fixed point of $\Phi(P)$, $\xi_0 = \lambda \phi \chi \sigma n . \sigma$ and $\phi_0 = \chi_0 = \delta$

It is a matter of routine to obtain the equivalence of the operational and denotational semantics for PROLOG along the lines of section 5.

Section 7 References

[AE]. K.R. Apt and M.H. van Emden, "Contributions to the Theory of Logic Programming," *Journal of the ACM* **29**, pp. 841-862 (1982).

[AB]. B. Arbab and D.M. Berry, "Operational and Denotational Semantics of Prolog," *Journal of Logic Programming* **4**, pp. 309-329 (1987).

[Bd]. M. Badinet, "Proving Termination Properties of PROLOG Programs: A Semantic Approach," pp. 336-347 in *Proc. LICS'88*, Edinburgh (1988).

[BW]. J.C.M. Baeten and W.P. Weijland, "Semantics for Prolog via Term Rewrite Systems," pp. 3-14 in *Proc. 1st International Workshop on Conditional Term Rewriting Systems*, ed. S. Kaplan and J.-P. Jouannaud, LNCS 308, Springer (1987).

[Ba1]. J.W. de Bakker, *Mathematical Theory of Program Correctness*, Prentice Hall International, London (1980).

[Ba2]. J.W. de Bakker, "Comparative Semantics for Flow of Control in Logic Programming without Logic," Report CS-R88.., Centre for Mathematics and Computer Science, Amsterdam, to appear (1988).

[BK]. J.W. de Bakker and J.N. Kok, "Uniform Abstraction, Atomicity and Contractions in the Comparative Semantics of Concurrent Prolog," in *Proc. FGCS'88*, Tokyo (1988).

[BKMOZ]. J.W. de Bakker, J.N. Kok, J.-J.Ch. Meyer, E.-R. Olderog, and J.I. Zucker, "Contrasting Themes in the Semantics of Imperative Concurrency," pp. 51-121 in *Current Trends in Concurrency: Overviews and Tutorials*, ed. J.W. de Bakker, W.P de Roever & G. Rozenberg, LNCS 224, Springer (1986).

[BM]. J.W. de Bakker and J.-J.Ch. Meyer, "Metric Semantics for Concurrency," *BIT* **28**, pp. 504-529 (1988).

[BMOZ]. J.W. de Bakker, J.-J.Ch. Meyer, E.-R. Olderog, and J.I. Zucker, "Transition Systems, Metric Spaces and Ready Sets in the Semantics for Uniform Concurrency," *Journal of Computer System Sciences* **36**, pp. 158-224 (1988).

[Br]. A. de Bruin, *Experiments with Continuation Semantics: Jumps, Backtracking, Dynamic Networks*, Dissertation, Free University, Amsterdam (1986).

[DM]. S.K. Debray and P. Mishra, "Denotational and Operational Semantics for Prolog," *Journal of Logic Programming* **5**, pp. 61-91 (1988).

[DF]. P. Deransart and G. Ferrand, "An Operational Formal Definition of PROLOG," Rapport de Recherche 598, INRIA, Rocquencourt (1986).

[EK]. M.H. van Emden and R.A. Kowalski, "The Semantics of Predicate Logic as a Programming Language," *Journal of the ACM* **23**(4), pp. 773-742 (1976).

[EY]. M.H. van Emden and K. Yukawa, "Logic Programming with Equations," *Journal of Logic Programming* **4**, pp. 265-288 (1987).

[Fi]. M. Fitting, "A Deterministic PROLOG Fixpoint Semantics," *Journal of Logic Programming* **2**, pp. 111-118 (1985).

[Fr]. G. Frandsen, "A Denotational Semantics for Logic Programming," DAIMI PB-201, Aarhus University, Aarhus (1985).

[GCLS]. R. Gerth, M. Codish, Y. Lichtenstein, and E. Shapiro, "Fully Abstract Denotational Semantics for Concurrent Prolog," pp. 320-335 in *Proc. LICS'88*, Edinburgh (1988).

[Hu]. G. Huet, "Confluent Reductions: Abstract Properties and Applications to Term Rewriting Systems," *Journal of the ACM* **27**(4), pp. 797-821 (1980).

[JM]. N.D. Jones and A. Mycroft, "Stepwise Development of Operational and Denotational Semantics for Prolog," pp. 281-288 in *Proc. SLP'84*, Atlantic City (1984).

[Kk]. J.N. Kok, "A Compositional Semantics for Concurrent Prolog," pp. 373-388 in *Proc. STACS'88*, ed. R. Cori & M. Wirsing, LNCS 294, Springer (1988).

[KR]. J.N. Kok and J.J.M.M. Rutten, "Contractions in Comparing Concurrency Semantics," pp. 317-332 in *Proc. ICALP'88*, ed. T. Lepistö & A. Salomaa, LNCS 317, Springer (1988).

[Kw]. R. Kowalski, "Algorithm = Logic + Control," *Communications of the ACM* **22**(7), pp. 424-436 (1979).

[Le]. G. Levi, "Logic Programming: the Foundations, the Approach and the Role of Concurrency," pp. 396-441 in *Current Trends in Concurrency: Overviews and Tutorials*, ed. J.W. de Bakker, W.P de Roever & G. Rozenberg, LNCS 224, Springer (1986).

[Lℓ]. J.W. Lloyd, *Foundations of Logic Programming*, Springer, Berlin (1984).

[MR]. A. Martelli and G. Rossi, "On the Semantics of Logic Programming Languages," pp. 327-334 in *Proc. ICLP'86*, ed. E. Shapiro, LNCS 225, Springer (1986).

[MS]. R. Milne and C. Strachey, *A Theory of Programming Language Semantics*, Chapman & Hall, London and Wiley, New York, 2 Volumes (1976).

[Pℓ]. G.D. Plotkin, "A Structural Approach to Operational Semantics," DAIMI FN-19, Aarhus University, Aarhus (1981).

[St]. J.E. Stoy, *Denotational Semantics - The Scott-Strachey Approach to Programming Language Theory*, MIT Press, Cambridge (1977).

[Te]. R.D. Tennent, "Mathematical Semantics of SNOBOL4," pp. 95-107 in *Proc. POPL'73*, Boston (1973).

[Vi]. E.P. de Vink, "Equivalence of an Operational and a Denotational Semantics for a Prolog-like Language with Cut," Report IR-151, Free University, Amsterdam (1988).

[Wi]. *Formal Description of Programming Concepts - III*, ed. M. Wirsing, North-Holland, Amsterdam (1987).

LABELED TREES AND RELATIONS ON GENERATING FUNCTIONS

M.P. DELEST, J.M. FEDOU

Bordeaux I University
LABRI[+]
Laboratory of Computer Science

Abstract. *In this paper, we give a combinatorial interpretation for a property on generating functions gived by R. Stanley. Our proof is based upon the study of special kind of labeled trees and forests.*

Résumé. *Nous donnons ici une interprétation combinatoire d'une propriété des fonctions génératrices citée par R. Stanley. Notre preuve utilise une classe particulière d'arbres et de forêts étiquettées.*

INTRODUCTION

Trees are presents in many subjects of theorical computer science. There are the natural representation of a lot of objects such as programs, arithmetic expressions, words or algebraic expressions in language theory.

The structure of tree is very efficient in the study of complexity of algorithm because they constitute a dynamic data structure which allows to organize informations and to measure precisely the complexity of algorithm (see for example [4][6][11]). Trees are also present in other fields as shown in the paper of Viennot [14].

In some purely combinatorics subjects, they are the nice objects for understanding formulas. For example labeled trees, as defined in [2], are usefull in many subjects, see for instance Cori and Vauquelin [3] for the construction of a bijection between planar graphs and well labeled trees, or Moon [9] for identities on labelled forests. Moreover, the bijections are the basis of new theories in combinatorics (see for example [7],[13]) and trees are very studied in this context.

In this paper, we introduce a special kind of set of labeled trees, the k-shaped forests, in order to give a bijective proof and a generalization of a result about generating functions

---

+ Unité de Recherche Associée au Centre National de la Recherche Scientifique n°726.

* Post Mail : 351 Cours de la Libération, 33405 TALENCE Cedex, FRANCE.

* Electronic Mail : maylis@geocub.greco-prog.fr

* This work was supported by the "PRC de Mathématiques et Informatique".

given by R. Stanley in [10] (see proposition 19 at the end of this paper). In fact, we study the generating functions

$$F_k(X) = \sum_{n>0} N_k(n)\, X^n,$$

where

$$N_k(n) = \Sigma(f_{n_1}+f_{n_2}+...+f_{n_k})(f_{n_2}+f_{n_3}+...+f_{n_{k+1}})...(f_{n_s}+f_{n_{s+1}}+...+f_{n_{s+k-1}}).$$

R. Stanley gives the result for k=2 and 3, the proof only for k=2, saying that the formula for k=3 appears after an enormous amount of cancelation and tedious computation. Moreover he says that he does not known a simpler alternative method. We give it in paragraph 3 and in some way we generalize his result.

After some definitions and notations, in section 2 and 3, we bring back the problem of the determination of $F_k(X)$ to the enumeration of k-shaped forests according to the number of sons of each vertex.

This problem of enumeration is then solved, in section 4, in the particular cases where k=2, 3 and 4. We solve a linear system of equations obtained by the mean of the k-shaped forests. The readers are refered to [4][5][14] for examples of similar methods.

1. DEFINITIONS AND NOTATIONS

A *rooted tree* is a connected graph G without cycle [1] with a distinguished vertex r called *root*. If (f,s) belongs to G, f (resp. s) is called *father* (resp. *son*) of s (resp. f). Here, we will consider trees which have sometimes a loop on the root. A set of trees is called a forest . A *labeled* tree is a tree with an integer, called label, associated to each vertex. The *entering degree* deg(i) of a vertex labelled i is the number of sons of this vertex. The *difference degree* $\delta(i)$ of a vertex labeled i is the difference between the label of the father of this vertex and i. The difference degree of a loop is equal to 0. It is easy to generalize these definitions to the forests. We note ε the empty forest.

Let K be an half ring and Y be $\{y_1, y_2, ..., y_k\}$. We denote by K[Y] the ring of the formal power series over Y with coefficients in K. Let $f(x)=\Sigma_{n>0} f_n x^n$ and $g(x)=\Sigma_{n>0} g_n x^n$ be two formal power series from K[{x}], the product of Hadamard of f and g is defined by

$$f*g(x) = \Sigma_{n>0}\, f_n g_n x^n.$$

Generaly speaking, we will denote by $(*f)^p$ the product of Hadamard p times of the serie f that is $(*f)^p = \Sigma_{n>0} f_n{}^p x^n$. By convention, we note

$$A = (*f)^0 = \Sigma_{n\geq 1} x^n = x/(1-x).$$

We call parameter an application from $\mathbb{N}$ into $\mathbb{N}$. The two degrees deg and δ are parameters.

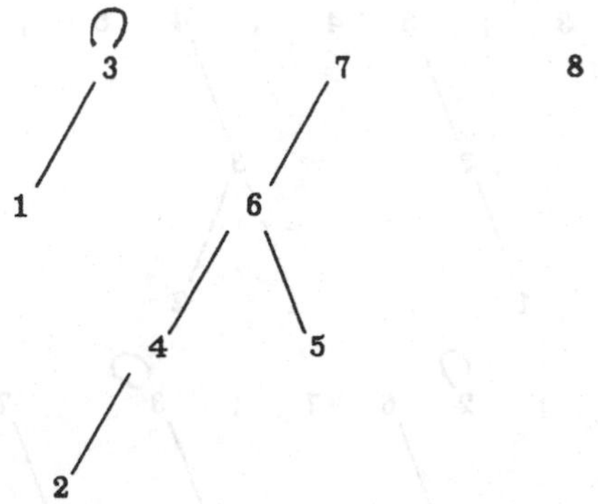

Figure 1. A forest of $\mathcal{S}_{3,6}$.

2. K-SHAPED FORESTS

We introduce a special kind of labeled forest, in order to "explain" products such as $(y_1+...+y_k)...(y_s+...+y_{s+k-1})$.

Definition 1. *Let $\mathcal{A}$ be a forest having* t *vertices labeled from* 1 *to* t. *The forest $\mathcal{A}$ is* k-*labeled when, for every* i *in* [1..t]

$$0 \leq \delta(i) < k .$$

Definition 2. *A* k-*labeled forest $\mathcal{A}$ is said* k-*shaped if $\mathcal{A}$ has* s+k-1 *vertices with* s≥0 *such that*

i) every root in $\mathcal{A}$ labeled with l *in* [1..s] *has a loop,*

ii) the k-1 *last vertices labeled from* s+1 *to* s+k-1 *are roots without loop.*

Notations.

$\mathcal{S}_{k,s}$ is the set of k-shaped forests having s+k-1 vertices.

$\mathcal{S}_{k,0}$ is the unique forest made with k-1 isolated vertices, we denote it by e.

$\mathcal{S}_k$ is the set of k-shaped forests.

See for example the figure 1, the forest $\mathcal{A}$ belongs to $\mathcal{S}_{3,6}$.

Remark 3. Let $\mathcal{A}$ be in $\mathcal{S}_{k,s}$, we obtain k forests in $\mathcal{S}_{k,s+1}$ by the mean of the following algorithm:

```
begin
  for every vertex in 𝒜  do label(vertex):=label(vertex)+1;
  Let 1 be the new vertex; then 1 is
      (i) either a loop ,
      (ii)either a son of one of the vertices numbered 2 to k.
end
```

Thus, the number of k-shaped forests having n+k-1 vertices is k^n.

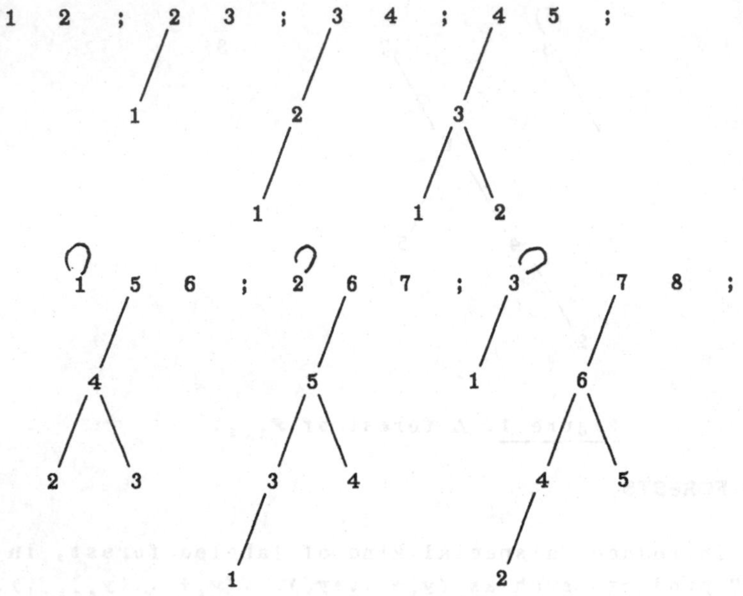

<u>Figure 2</u>. An example of the algorithm of remark 3

The figure 2 shows the construction of the forest of figure 1 using this algorithm. In this example, we have k=3.

Remark 4. Let P be the product

$$(y_1+y_2+y_3)(y_2+y_3+y_4)(y_3+y_4+y_5)(y_4+y_5+y_6)(y_5+y_6+y_7)(y_6+y_7+y_8).$$

In some way, we can say that the forest of figure 1 "represents" the term $y_3{}^2y_4y_6{}^2y_7$ in the development of P. In fact, on this forest every relation "j is father of i" means "we choose y_j in the ith factor $(y_i+y_{i+1}+y_{i+2})$".

Let p be a parameter.

Definition 5. *The weight according to* p *of a forest* $\mathcal{A}$ *from* $\mathcal{F}_{k,s}$ *is the monomial defined by*

$$\pi_p(\mathcal{A})(y_1,\ldots,y_{s+k-1}) = \prod_{1\leq i\leq s+k-1} y_i{}^{p(i)}.$$

For example, if $\mathcal{A}$ is the forest of figure 1, the weights according to the parameters entering degree and difference degree are repectively

$$\pi_{deg}(\mathcal{A})=y_3{}^2y_4y_6{}^2y_7,$$

and

$$\pi_\delta(\mathcal{A})=y_1{}^2y_2{}^2y_4{}^2y_5y_6\ .$$

Notation. For every forest $\mathcal{A}$ from $\mathcal{S}_{k,s}$, let φ be the map from $\mathcal{S}_k$ into $\mathbb{Z}[Y]$ defined by

$$\varphi(\mathcal{A}) = \prod_{i=1}^{s+k-1} y_{deg(i)}.$$

Define the formal power serie $\Phi(\mathcal{S}_k)$ of $\mathbb{Z}[Y]$ by

$$\Phi(\mathcal{S}_k) = \sum_{\mathcal{A}\in\mathcal{S}_k} \varphi(\mathcal{A}).$$

$\Phi(\mathcal{S}_k)$ is the generating function of the k-shaped forest according to the number of sons of the vertices that is

$$\Phi(\mathcal{S}_k) = \sum_{i_0,\ldots,i_{k-1}} n_{i_0\ldots i_k} Y_0^{i_0}\ldots Y_k^{i_k},$$

where $n_{i_0\ldots i_k}$ is the number of k-shaped forests such that, for each j in [0..k], there is i_j vertices which have j sons.

Definition 6. *Let* $f(x) = \Sigma_{n\geq 1} f_n x^n$ *be a formal power serie of* $K[x]$ *and* $\mathcal{A}$ *be an element of* $\mathcal{S}_{k,s}$. *We denote by* Δ_f *the map from* $\mathcal{S}_k$ *into* $\mathbb{Z}[x]$ *defined by*

$$\Delta_f(\mathcal{A}) = \prod_{1\leq i\leq s+k-1} (*f)^{deg(i)}.$$

Property 7. Δ_f *and* Π_{deg} *are morphisms over the forests.*

We obtain $\Delta_f(\mathcal{A})$ substituting formally to each y_i the value $(*f)$ in $\Pi_{deg}(\mathcal{A})$ or the value $(*f)^i$ in $\varphi(\mathcal{A})$. Moreover, let $\{u,\mathcal{A}'\}$ be a forest where u is a tree and $\mathcal{A}'$ a forest, applying the previous definitions to the labeled trees gives the following equalities which prove the property 7

$$\Delta_f(\{u,\mathcal{A}'\}) = \Delta_f(u)\ \Delta_f(\mathcal{A}')$$

and

$$\Pi_{deg}(\{u,\mathcal{A}\}) = \Pi_{deg}(u)\ \Pi_{deg}(\mathcal{A}').$$

Examples: In $\mathcal{S}_k$, we have $\Delta_f(\varepsilon)=A^{k-1}$. If $\mathcal{A}$ is the forest of figure 1, we have $\Delta_f(\mathcal{A}) = A^4 f^2 (*f)^2$ and $\varphi(\mathcal{A}) = Y_0{}^4 Y_1{}^2 Y_2{}^2$.

Definition 8. *Let $\mathcal{A}_1$ and $\mathcal{A}_2$ be two forests from $\mathcal{F}_{k,s}$. The forests $\mathcal{A}_1$ and $\mathcal{A}_2$ are equivalents when, if*

$$\Pi_{deg}(\mathcal{A}_1) = y_1^{n_1} \ldots y_{s+k-1}^{n_{s+k-1}}$$

then there exists a permutation σ of $[1..s+k-1]$ *such that*

$$\Pi_{deg}(\mathcal{A}_2) = y_{\sigma(1)}^{n_1} \ldots y_{\sigma(s+k-1)}^{n_{s+k-1}} .$$

The property 7 gives immediatly the following proposition.

Proposition 9. *Let $\mathcal{A}_1$ and $\mathcal{A}_2$ be two equivalent forests from $\mathcal{F}_{K,S}$. We have the following properties*

(i) $\varphi(\mathcal{A}_1) = \varphi(\mathcal{A}_2)$,

(ii) *if* $f \in K[x]$, $\Delta_f(\mathcal{A}_1) = \Delta_f(\mathcal{A}_2)$.

3. INTERPRETATION OF F_k IN TERM OF k-SHAPED FORESTS

In this paragraph, we give an expression for $F_k(X)$ using $\mathcal{F}_k$ that is

Theorem 10. *The function $F_k(x)$ is deduced from the enumerating function $\Phi(\mathcal{F}_k)$ of the k-shaped forests substituting formaly each Y_i by $(*f)^i$.*

First, we give some precision about F_k. Define, for every integers k,s and n

$$N_{k,s}(n) = \Sigma(f_{n_1}+f_{n_2}+\ldots+f_{n_k})(f_{n_2}+f_{n_3}+\ldots+f_{n_{k+1}})\ldots(f_{n_s}+f_{n_{s+1}}+\ldots+f_{n_{s+k-1}}),$$

where the sum is taken over all ordered partitions of $n=n_1+n_2+\ldots n_{s+k-1}$ with $n_i \geq 1$ and

$$F_{k,s}(x) = \Sigma_{n>1} N_{k,s}(n) x^n .$$

Note that we have

$$N_k(n) = \Sigma_{s>0} N_{k,s}(n), \quad F_k(x) = \Sigma_{s>0} F_{k,s},$$

and also

$$F_{k,s}(x) = \Sigma \, P_{k,s}(f_{p_1}, \ldots, f_{p_{s+k-1}}) x^n ,$$

where the sum is over all ordered partitions of n with $n_i \geq 1$.

Let $P_{k,s}(y_1,\ldots,y_{s+k-1})$ be the polynomial of $\mathbb{Z}[Y]$ equal to 1 if s=0 and otherwise given by

$$P_{k,s}(y_1,\ldots,y_{s+k-1}) = (y_1+\ldots y_k)(y_2+\ldots+y_{k+1})\ldots(y_s+\ldots+y_{s+k-1}).$$

In order to prove the theorem 10 we prove the following

Lemma 11.

$$P_{k,s}(y_1,\ldots y_{s+k-1}) = \sum_{\mathcal{A}\in\mathcal{F}_{k,s}} \pi_{deg}(\mathcal{A})(y_1,y_2,\ldots,y_{s+k-1}).$$

If s=0, the result is immediate. Otherwise, an induction shows that

$$P_{k,s}(y_1,\ldots,y_{s+k-1}) = \sum_{i=1}^{k} y_i P_{k,s-1}(y_2,\ldots,y_{s+k-1})$$

$$= \sum_{i=1}^{k} \sum_{\mathcal{A}'\in\mathcal{F}_{k,s-1}} y_i \pi_{deg}(\mathcal{A}')(y_2,\ldots,y_{s+k-1}),$$

Using the algorithm of construction of $\mathcal{F}_{k,s}$, each forest $\mathcal{A}$ of $\mathcal{F}_{k,s}$ comes from a forest $\mathcal{A}'$ of $\mathcal{F}_{k,s-1}$. If the vertex labeled 1 is a loop in $\mathcal{A}$,

$$\Pi_{deg}(\mathcal{A})(y_1,\ldots,y_{s+k-1})=y_1\Pi_{deg}(\mathcal{A}')(y_2,\ldots,y_{s+k-1}).$$

Else, the vertex labeled 1 is a son of a vertex i in [2..k] and

$$\Pi_{deg}(\mathcal{A})(y_1,\ldots,y_{s+k-1})=y_i\Pi_{deg}(\mathcal{A}')(y_2,\ldots,y_{s+k-1}).$$

Thus the lemma 11 is proved. We deduce

$$F_{k,s}(x) = \sum_{n\geq 1} \sum_{p_1+\ldots+p_{s+k-1}=n} \sum_{\mathcal{A}\in\mathcal{F}_{k,s}} f_{p_1}^{deg(1)} \ldots f_{p_{s+k-1}}^{deg(s+k-1)} x^n,$$

and thus by factorisation, we obtain

$$F_{k,s}(x) = \sum_{\mathcal{A}\in\mathcal{F}_{k,s}} \prod_{i=1}^{s+k-1} \left(\sum_{p_i\geq 1} f_{p_i}^{deg(i)} x^{p_i} \right).$$

Consequently we have

$$F_{k,s}(x) = \sum_{\mathcal{A}\in\mathcal{F}_{k,s}} \Delta_f(\mathcal{A}).$$

Using the notation $F_k(x) = \Sigma_{s>0} F_{k,s}(x)$, we get

$$F_k(x) = \sum_{\mathcal{A} \in \mathcal{S}_k} \Delta_f(\mathcal{A})$$

and theorem 10 follows from property 7.

4. RELATIONS BETWEEN SUBSETS OF $\mathcal{S}_k$

In this section, we part the set $\mathcal{S}_k$ into particular subsets which basic properties allow us to decompose the formal power serie $\Phi(\mathcal{S}_k)$ as sum of formal power series $P_{i_1 i_2 \ldots i_p}$ whose are given by

$$P_{i_1 i_2 \ldots i_p} = \sum_{\mathcal{A} \in E_{i_1 \ldots i_p}} \varphi(\mathcal{A}) .$$

Then, we provide relations between these formal power series and we obtain two types of relations:

- lemmas 12 to 15 will allow us to express $P_{i_1 \ldots i_p \ldots i_{p+q}}$ with respect to $P_{i_p \ldots i_{p+q}}$,
- lemma 16 will allows us to write $P_{i_1 \ldots i_{p+q}}$ as $P_{j_1 \ldots j_{p+q}}$ in order to apply lemmas 12 to 15.

Most of these lemmas are based upon the relation of equivalence (definition 8) of particular types of forests and are deduced by the application of proposition 9.

<u>Notations</u>: For every positive integer p, and for every p-uple in $[0..k-1]^p$, we denote by

. $E_{i_1 \ldots i_p}$ the set of forests belonging to $\mathcal{S}_k$ having at least p+k-1 vertices such that $\forall\, j \in [1..p],\ \delta(j)=i_j$,

. $e_{i_1 \ldots i_p}$ the forest of $\mathcal{S}_k$ with p+k-1 vertices such that

$$\forall\, j \in [1..p],\ \delta(j)=i_j .$$

By convention, let E (resp. e) be $\mathcal{S}_k$ (resp. $\mathcal{S}_{k,0}$).

<u>Example</u>: the forest $\mathcal{A}$ of the example 1 is exactly e_{220211} and belongs to the sets E, E_2, E_{22}, E_{220}, ...

Lemma 12. *For every* p-*uple* $(i_1, i_2, .., i_p)$ *of* $[0..k-1]^p$,

$$E_{i_1 \ldots i_p} = e_{i_1 \ldots i_p} + \sum_{h=0}^{k-1} E_{i_1 \ldots i_p h}, \qquad (1)$$

and

$$P_{i_1 \ldots i_p} = \varphi(e_{i_1 \ldots i_p}) + \sum_{h=0}^{k-1} P_{i_1 \ldots i_p h}. \qquad (2)$$

Indeed, a forest of $E_{i_1 \ldots i_p}$ is

- either the forest $e_{i_1 \ldots i_p}$,
- or a forest having at least p+k vertices, such that the half degree of the vertex p+1 is in [0..k-1].

Examples: $E = e + \sum_{i=0}^{k-1} E_i$, $\Phi(\mathcal{S}_k) = y_0^{k-1} + \sum_{i=0}^{k-1} P_i$,

$$E_i = e_i + \sum_{j=0}^{k-1} E_{ij}, \quad P_i = y_1\, y_0^{k-1} + \sum_{j=0}^{k-1} P_{ij}.$$

Lemma 13. $$P_{p(p-1)\ldots 10} = y_0^p\, y_{p+1}\, \Phi(\mathcal{S}_k) \qquad (4)$$

Proof: An element $\mathcal{A}$ of $E_{p(p-1)\ldots 10}$ (see figure 3) is a forest $\{u, \mathcal{A}'\}$ where u is the labeled tree whose root labeled p+1 has a loop and p sons labeled from 1 to p and where $\mathcal{A}'$ is any forest of $\mathcal{S}_k$ where each label l of a vertex has been changed in (l+p) . Thus we have the announced result using the property 7.

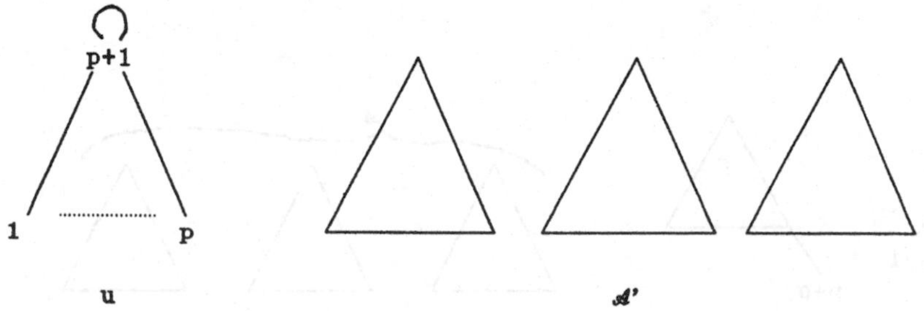

Figure 3. An element of $E_{p(p-1)\ldots 10}$.

In the same way, we fix j supplementary vertices and we get

Lemma 16. $P_{p(p-1)\ldots 10 i_1 \ldots i_j} = y_0^{\,p} y_{p+1} P_{i_1 \ldots i_j}$. (5)

<u>Example</u>: $P_0 = y_1 \Phi(\mathscr{S}_k)$, $P_{10} = y_0 y_2 \Phi(\mathscr{S}_k)$, $P_{210} = y_0^{\,2} y_3 \Phi(\mathscr{S}_k)$ ...
and also $P_{01} = y_1 P_1$, $P_{1023} = y_0 y_2 P_{23} \ldots$

Lemma 15. *Let* $i_1 = \alpha+p-1, i_2 = \alpha+p-2, \ldots, i_p = \alpha$ *and*
$$i_{p+1} \neq \alpha-1, i_{p+2} \neq \alpha-2, \ldots, i_{p+\alpha} \neq 0 \qquad (6)$$
then
$$P_{i_1 i_2 \ldots i_{\alpha+p}} = y_0^{\,p-1} y_p \, P_{i_{p+1} \ldots i_{\alpha+p}}. \qquad (7)$$

Let $\mathscr{A}$ be a forest from $P_{i_1 i_2 \ldots i_{\alpha+p}}$ satisfying (6).

Then the vertex labeled $p+\alpha$ has exactly p sons labeled from 1 to p as shown in figure 4. Indeed, the condition $i_{p+q} \neq \alpha-q$ in (6) means that the vertex labeled $p+\alpha$ has no son other than those labeled from 1 to p. Thus $\mathscr{A}$ is equivalent to the forest obtained by displacing the vertices labeled from 1 to p so that the vertex p is a loop having p-1 sons labeled from 1 to p-1 as shown in figure 4. Therefore, we can use the lemma 14 which proves lemma 15.

<u>Example</u>: $P_{3233} = y_0 y_2 P_{33}$.

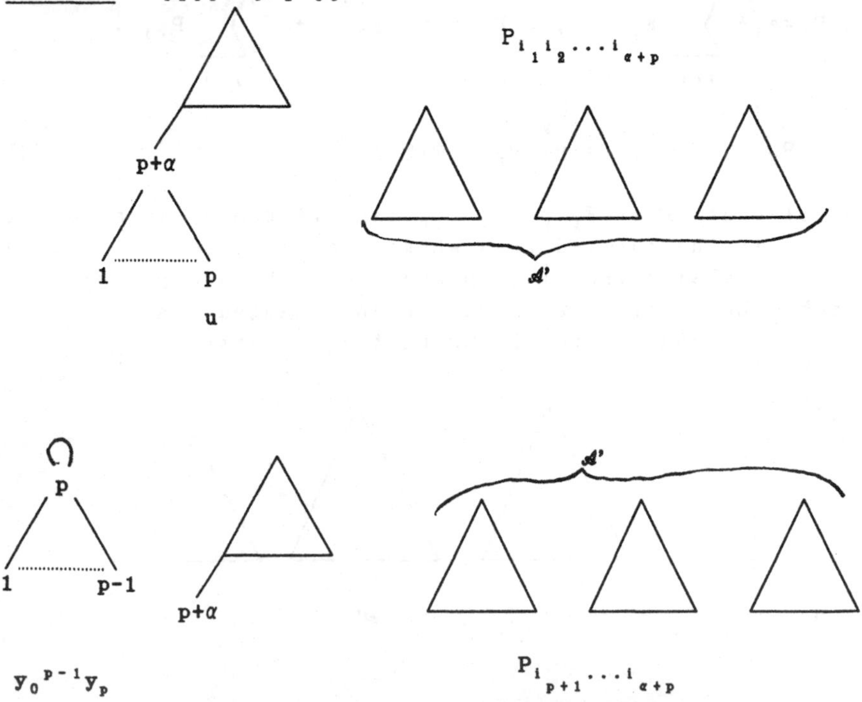

<u>Figure 4.</u> An example for lemma 16.

Lemma 16. *If the forest $\mathcal{A}$ of $\mathcal{S}_k$ is such that, for somme positive integers $\alpha,\beta,\gamma,\delta$,* father$(\alpha)=\alpha+\beta+\gamma+\delta$ *and* father$(\alpha+\beta)=\alpha+\beta+\gamma$, *then, $\mathcal{A}$ is equivalent to the forest obtained by the permutation of the subtrees raising from the vertices α and $\alpha+\beta$.*

The proof is immediate because this transformation does not affect the entering degree. Thus we have the following

Corollary 17. *If* $i_\alpha=\beta+\gamma+\delta$ *and* $i_{\alpha+\beta}=\gamma$ *then*

$$e_{i_1..i_\alpha..i_{\alpha+\beta}..i_p} = e_{i_1..i'_\alpha..i'_{\alpha+\beta}..i_p},$$
$$E_{i_1..i_\alpha..i_{\alpha+\beta}..i_p} = E_{i_1..i'_\alpha..i'_{\alpha+\beta}..i_p},$$
$$P_{i_1..i_\alpha..i_{\alpha+\beta}..i_p} = P_{i_1..i'_\alpha..i'_{\alpha+\beta}..i_p}, \tag{8}$$

where $i'_\alpha=\beta+\gamma$ *and* $i'_{\alpha+\beta}=\gamma+\delta$.

<u>Example</u>: $P_{20}=P_{11}$

5. ENUMERATION OF THE K-SHAPED FORESTS

In this section, we enumerate the k-shaped forests according to the distribution of each vertices in the forests.

The lemmas of the section 4 allow us to write out a linear system of equations satisfied by $\Phi(\mathcal{S}_k)$. However, the number of equations increases exponentialy with k. Thus it was inefficient to write in Macsyma [12] an algorithm of generation for such a general system. Thus, we just solve explicitly the systems for k=2, 3 and 4.

Theorem 18. *The generating function for k-shaped forests are*

for k=2,
$$\Phi(\mathcal{S}_2) = \frac{y_0}{(1-y_1)^2 - y_0y_2},$$

for k=3,
$$\Phi(\mathcal{S}_3) = \frac{y_0^2}{(1-y_1)^2(1-y_1-y_1^2)-2y_0y_2(1-y_1^2)-y_0^2(y_2^2+y_3)},$$

and for k=4,
$$\Phi(\mathcal{S}_4) = \frac{y_0^3\ (1 + y_1^2 - y_0y_2)}{Q_0 + Q_1y_0 + Q_2y_0^2 + Q_3y_0^3 + Q_4y_0^4},$$

where
$$Q_0= (1-y_1)^2(1-y_1-y_1^2-y_1^3),$$
$$Q_1= -2y_0y_2(1-y_1)(2+y_1-y_1^2-3Y_1^3-3Y_1^4-2Y_1^5),$$
$$Q_2= 2Y_3(Y_1-1)(1+3Y_1+Y_1^2+Y_1^3) - Y_2^2(1-4Y_1^3-6Y_1^4),$$
$$Q_3= -(Y_2^3+4Y_2Y_3+Y_4) + 2Y_1Y_2(Y_3-Y_2^2) -Y_1^2\ (Y_4+4Y_2Y_3+4Y_2^3),$$
$$Q_4= (Y_2+Y_3)^2 + Y_2Y_4 - 2Y_3^2.$$

We write systems of linear equations by using first the lemma 12, then by applying lemmas 13 to 16.

For k=2, we have

$$\Phi(\mathcal{P}_2) = Y_0 + P_0 + P_1,$$
$$P_1 = Y_0 Y_1 + P_{10} + P_{11},$$
$$P_0 = Y_1 \Phi(\mathcal{P}_2),$$
$$P_{10} = Y_0 Y_2 \Phi(\mathcal{P}_2),$$
$$P_{11} = Y_1 P_1.$$

Solving it, gives the result of theorem 18. We give in the annex the system for k=3 and 4. They have been solved using the symbolic manipulation system MACSYMA from MIT [12]. We deduce from the theorem 19 the expression of $F_k(X)$ for k=2, 3 and 4 using the theorem 11. Note that in the case k=3, we find the result of R. Stanley [10] that is

Proposition 19. *Define*

$$N(n) = \Sigma (f_{n_1} + f_{n_2} + f_{n_3})(f_{n_2} + f_{n_3} + f_{n_4}) \ldots (f_{n_s} + f_{n_{s+1}} + f_{n_{s+2}}),$$

where f *is any function* $f: \mathbb{N} \to \mathbb{C}$ *and where the sum is over all ordered partitions* $n_1 + n_2 + \ldots + n_{s+2} = n$ *of* n $(n_i \geq 1)$. *By convention, a summand with* s=0 *is* 1. *Define*

$$F(X) = \sum_{n=1}^{\infty} N(n) X^n, \qquad f = \sum_{n=1}^{\infty} f_n X^n, \qquad A = X/(1-X).$$

Let $*$ *denote Hadamard product. Then*

$$F(X) = \frac{A^2}{(1-f)^2(1-f-f^2) - 2A(f*f)(1-f^2) - A^2((f*f)^2 + (f*f*f))}.$$

However, we did not find a general formula for $F_k(X)$, or, at least a recursive relation on these functions. Yet, the time of computation was very slight, and we think that it is possible to compute the result for k=5.

REFERENCES

[1] C. BERGE, *Graphes et hypergraphes*, Dunod, Paris (1970).

[2] N.G. de BRUIJN and B.J.M. MORSELT, A note on plane trees, J. Comb. Th. 2 (1967), 27-34.

[3] R. CORI and B. VAUQUELIN, Planar maps are well labeled trees, Canadian J. of Math. 33 (1981) 1023-1042.

[4] M.P. DELEST, Utilisation des langages algébriques et du calcul formel pour le codage et l'énumération des polyominos, Thèse d'Etat, Université de Bordeaux I, 1987.

[5] J.M. FEDOU, Enumération de certains polyominos selon les paramètres périmètre et aire, mémoire de D.E.A, Bordeaux I, 1987.

[6] P. FLAJOLET, Analyse d'algorithmes de manipulation d'arbres et de fichiers, Cahiers du B.U.R.O. 34-35(1981), 1-209.

[7] A. JOYAL, Une théorie combinatoire des séries formelles, Adv. in Math., 42 (1981), 1-82.

[8] D. KNUTH, *The art of computer programming*, Vol. 1, Addison Wesley Reading, 1968.

[9] J.W. MOON, A note on an identity and labelled forests, Caribb. J. Math., 3, (2) 59-65.

[10] R.P. STANLEY, Generating function, Studies in Comb, 100-148, 17, Math. Assoc. America Washington DC 78.

[11] J.M. STEYEART, Complexité et structure des algorithmes, Thèse Université Paris VII, 1984.

[12] SYMBOLICS INC., Macsyma Reference manual version ten, third printing, december 1984.

[13] G. VIENNOT, Heap of pieces: Basic definitions and combinatorial lemmas, in *Combinatoire Enumérative*, UQAM 1985, Montréal, G. Labelle et P. Leroux ed., Lecture Notes in Mathematics, n°1234, Springer Verlag, 321-350, 1986.

[14] G. VIENNOT, Trees, Rivers, RNAs and many other things, preprint , Bordeaux I, 1987.

ANNEXE

For k=3,

$$\Phi(\mathcal{S}_3) = y_0{}^2 + P_0 + P_1 + P_2 ,$$
$$P_1 = y_0{}^2 y_1 + P_{10} + P_{11} + P_{12},$$
$$P_2 = y_0{}^2 y_0 + P_{20} + P_{21} + P_{22},$$
$$P_{20} = y_0{}^2 y_1{}^2 + P_{200} + P_{201} + P_{202},$$
$$P_{21} = y_0{}^3 y_2 + P_{210} + P_{211} + P_{212},$$
$$P_{22} = y_0{}^2 y_1{}^2 + P_{220} + P_{221} + P_{222},$$
$$P_0 = y_1 \ \Phi(\mathcal{S}_3),$$
$$P_{10} = y_0 \ y_2 \Phi(\mathcal{S}_3),$$
$$P_{11} = y_0 \ P_1,$$
$$P_{12} = y_0 \ P_2,$$
$$P_{200} = P_{110} = y_1 \ P_{10},$$
$$P_{201} = P_{111} = y_1 \ P_{11},$$
$$P_{202} = P_{112} = y_1 \ P_{12},$$
$$P_{210} = y_0{}^2 y_3 \ \Phi(\mathcal{S}_3),$$
$$P_{211} = y_0 y_2 \ P_1 .$$

For k=4,

$$\Phi(\mathcal{S}_4) = y_0{}^3 + P_0 + P_1 + P_2 + P_3,$$
$$P_0 = y_1 \ \Phi(\mathcal{S}_4),$$
$$P_1 = y_0{}^3 y_1 + P_{10} + P_{11} + P_{12} + P_{13},$$
$$P_2 = y_0{}^3 y_0 + P_{20} + P_{21} + P_{22} + P_{23},$$
$$P_3 = y_0{}^3 y_0 + P_{30} + P_{31} + P_{32} + P_{33},$$
$$P_{10} = y_0 \ y_2 \Phi(\mathcal{S}_4),$$
$$P_{11} = y_1 \ P_1,$$

$$
\begin{aligned}
P_{12} &= y_1 P_2, \\
P_{13} &= y_1 P_3, \\
P_{20} &= y_1 P_1, \\
P_{21} &= y_0^4 y_2 + P_{210} + P_{211} + P_{212} + P_{213}, \\
P_{22} &= y_0^3 y_1^2 + P_{220} + P_{221} + P_{222} + P_{223}, \\
P_{23} &= y_0^3 y_1^2 + P_{230} + P_{231} + P_{232} + P_{233}, \\
P_{30} &= P_{12}, \\
P_{31} &= P_{22}, \\
P_{32} &= y_0^4 y_2 + P_{320} + P_{321} + P_{322} + P_{323}, \\
P_{33} &= y_0^3 y_1^2 + P_{330} + P_{331} + P_{332} + P_{333}, \\
P_{210} &= y_0^2 y_3\ \Phi(\mathcal{S}_4), \\
P_{211} &= P_{220} = y_0 y_2\ P_1, \\
P_{212} &= y_0 y_2\ P_2, \\
P_{213} &= y_0 y_2\ P_3, \\
P_{221} &= P_{320} = y_1\ P_{21}, \\
P_{222} &= P_{330} = y_1\ P_{22}, \\
P_{223} &= y_1\ P_{23}, \\
P_{230} &= y_0 y_2\ P_2, \\
P_{231} &= y_1\ P_{31}, \\
P_{232} &= y_1\ P_{32}, \\
P_{233} &= y_1\ P_{33}, \\
P_{321} &= y_0^5 y_3 + P_{3210} + P_{3211} + P_{3212} + P_{3213}, \\
P_{322} &= P_{331} = y_0^4 y_1 y_2 + P_{3220} + P_{3221} + P_{3222} + P_{3223}, \\
P_{323} &= y_0^4 y_1 y_2 + P_{3230} + P_{3231} + P_{3232} + P_{3233}, \\
P_{332} &= y_0^4 y_1 y_2 + P_{3320} + P_{3321} + P_{3322} + P_{3323}, \\
P_{333} &= y_0^3 y_1^3 + P_{3330} + P_{3331} + P_{3332} + P_{3333}, \\
P_{3210} &= y_0^3 y_4\ \Phi(\mathcal{S}_4), \\
P_{3211} &= P_{3220} = y_0^2 y_3\ P_1, \\
P_{3212} &= P_{3230} = y_0^2 y_3\ P_2, \\
P_{3213} &= y_0^2 y_3\ P_3, \\
P_{3221} &= P_{3320} = y_0 y_2\ P_{21}, \\
P_{3222} &= P_{3330} = y_0 y_2\ P_{22}, \\
P_{3223} &= y_0 y_2\ P_{23}, \\
P_{3231} &= y_0 y_2\ P_{31}, \\
P_{3232} &= y_0 y_2\ P_{32}, \\
P_{3233} &= y_0 y_2\ P_{33}, \\
P_{3321} &= y_1\ P_{321}, \\
P_{3322} &= y_1\ P_{322}, \\
P_{3323} &= y_1\ P_{323}, \\
P_{3331} &= y_1\ P_{331}, \\
P_{3332} &= y_1\ P_{332}, \\
P_{3333} &= y_1\ P_{333}.
\end{aligned}
$$

PROOFS OF DECLARATIVE PROPERTIES OF LOGIC PROGRAMS

Pierre DERANSART
INRIA
Domaine de Voluceau
B.P. 105 - Rocquencourt
78153 LE CHESNAY Cédex
Tel. : (33-1) 39 63 55 36
uucp: deransar@minos.inria.fr

Abstract

In this paper we shall consider proofs of declarative properties of Logic Programs, i.e. properties associated with the logical semantics of pure Logic Programs, in particular what is called the partial correctness of a logic program with respect to a specification. A specification consists of a logical formula associated with each predicate and establishing a relation between its arguments. A definite clause program is partially correct iff every possible answer substitution satisfies the specification.

This paper generalizes known results in logic programming in two ways : first it considers any kind of specification, second its results can be applied to extensions of logic programming such as functions or constraints.

In this paper we present two proof methods adapted from the Attribute Grammar field to the field of Logic Programming. Both are proven sound and complete. The first one consists of defining a specification stronger than the original one, which furthermore is inductive (fix-point induction).

The second method is a refinement of the first one : with every predicate, we associate a finite set of formulas (we call this an annotation), together with implications between formulas. The proofs become more modular and tractable, but the user has to verify the consistency of his proof, which is a decidable property. This method is particularly suitable for proving the validity of specifications which are not inductive.

Introduction

The problem of proving the correctness of a definite Horn clause program (clauses with exactly one positive literal or logic programs), with respect to a given specification, saying what the answer substitutions (if any) should be, although essential, has been rarely considered [Cla 79, Hog 84, Dev 87, DrM87, SS 86]. Most published works present some variant of the inductive assertion method (fix-point induction) which is illustrated with examples. No adaptation of the presentation to the field of logic programming is always made, no other original method is presented and no proof of soundness and completeness of the method are always given.

In this paper we shall consider the partial correctness of a logic program with respect to a specification. A specification consists of a logical formula associated with each predicate and establishing a relation between its arguments. A definite clause program is partially correct iff every possible answer substitution satisfies the specification.

Partial correctness does not depend on any interpretation strategy ; it does not refer to any operational semantics of logic programs but only to their constructive (proof-trees) or logical semantics [DF 88]. Specifications are properties of the proof-tree roots. They are used to detect properties of goals before any execution or testing of a logic program.

The proof methods presented in this paper are part of a general methodology we are developing for logic programming and have turned out to be of practical use for software development in PROLOG style.

Their purpose is not to be fully automatizable as in [KS 86, Fri 88] but to give the user some better ability to deal with and understand logic programs.

Partial correctness does not say anything about the existence of answer substitutions (completeness) or effective form of the goals during or after computations ("run-time properties" in [DrM 87] STP in [DF 88]), or whether any solution can be reached using some computation rule (termination, [FGK 85]) ...etc... All together these problems are part of the total correctness of a definite clause program. They have been considered in other works [also in DM 89] but for their solution many problems use partial correctness properties (or valid specifications), and, as such, the partial correctness proof methods are the basic elements of any general validation system for logic programming.

Sometimes it is argued that logic programs are axioms, hence they do not need to be verified. However, this is not the case. In particular programmers frequently write general axioms which are used in a restricted manner. This is mainly due to the high level of expression permitted by the language.

Consider for example the program "concatenate" using difference-lists given by the unique clause :

concatenate (L1- L2, L2-L3, L1-L3) ←.

In its logical semantics [AvE 82, Llo 84] (we use in place the denotation, i.e the set of all atomic logical consequences [Cla 79, Fer 85]) there are not only difference-lists : for example concatenate([ ]-[1], [1]-[1, 2], [ ]-[1, 2]) is a logical consequence also. But it is sufficient (for partial correctness purposes only) from the programmers point of view to know that concatenate is correct w.r.t. the fact that if the two first arguments are difference-lists representing lists l1 and l2, then (if the goal succeeds) the third argument is a difference-list representing the concatenation of l1 and l2.

Proof of partial correctness are thus required to certify the validity of the written axioms in the specific model the user has in mind. In practice, however, one does not expect to write axioms for any kind of model but more frequently axioms of a specific model.

The purpose of this paper is to rephrase the proof methods developed in [CD 88] for Attribute Grammars and Logic Programming without unnecessary references to Attribute Grammars.Two proof methods are presented. Both are proven, sound and complete (provided the specification language is large enough [Coo 78]). The first one consists of defining a specification stronger than the original one, which furthermore is inductive (fix-point induction).

Our second method is a refinement of the first one : with every predicate we associate a finite set of formulas (we call this an annotation), together with implications between formulas. The proofs become more modular and tractable, but its consistency has to be proven, this is a decidable property. This method is particularly suitable for proving the validity of specifications which are not inductive.

The proof methods presented here are a direct adaptation of the proof methods described in [Der 83, Cou 84, CD 88] for attribute grammars to the case of logic programming. In [CD 88] the adaptation is sketched using the results of [DM 85] : a definite clause program can be thought of as an attribute grammar. In this paper the methods are defined directly in the logic programming framework without unnecessary references to the attribute grammar field, and the formulation of the second method is more general than in [CD 88] such that many applications concerning the proof of other properties like "run time" or safe use of the negation can thus be considered.

The paper is organized as follows :

Section 1 recalls basic definitions of logical languages used to describe the specifications, section 2 defines the notion of validity of a specification for a definite clause program. Section 3 and 4 present and study the two proof methods.

It is important to remark that even if the semantics of logic programs is usually defined on non sorted algebras of terms, the specifications may be expressed in any logical language in which terms are interpreted. For that reason we use many sorted languages to describe the specifications of the programs. Obviously presentation using homogeneous algebras could be also made.

1 - Basic definitions and notations

(1.1) Sort, signatures, terms

Context free grammars and logical languages will be defined in the algebraic style of [CD 88]. Some definitions are also useful to describe logic programs.

Let S be a finite set of sorts. A S-sorted signature F is a finite set of function symbols with two mappings : the arity α (some word in S* representing the sorts of the arguments in the same order), the sort σ of the function symbol. The length of $\alpha(f)$ is called the rank of f and denoted $\rho(f)$. If $\alpha(f) = \varepsilon$ (the empty word) then f is a constant symbol. The pair $< \alpha(f), \sigma(f) >$ is the profile of f. A constant of sort s has profile $< \varepsilon, s >$.

A heterogeneous F-algebra is an object A :

$$A = < \{A_s\}_{s \in S}, \{f_A\}_{f \in F} >$$

where $\{A_s\}$ is a family of non empty sets indexed by S (the carriers) and each f_A a mapping :

$$A_{s1} \times ... \times A_{sn} \rightarrow A_s \text{ if f has profile } < s1...sn, s >.$$

Let V be a S-sorted set of variables (each v in V has arity ε, sort $\sigma(v)$ in S). The free F-algebra T generated by V, also denoted T(F, V) is identified as usual as the set of the well-formed terms, "well typed" with respect to sorts and arities. Terms will also be identified with trees in a well known manner. T(F) denotes the set of all terms without variables, i.e. the ground terms, $T(F)_s$ denotes the set of the ground terms of sort s.

A term t in $T(F)_s$ is considered as denoting a value t_A in A_s for a F-algebra A. For any F-algebra A and a S-sorted set of variables, an assignment of values in A_s to variables V_s, for all s in S, is an S-indexed family of functions.

$$\nu = \{ \nu_s : V_s \rightarrow A_s \}_{s \in S}$$

It is well-known that this assignment can be extended into a unique homomorphism

$$\nu' = \{ \nu'_s : T(F, V)_s \rightarrow A_s\}_{s \in S}$$

In T assignments are called substitutions. For any assignment ν in T and term t in T(F, V), νt is called an instance of t.

(1.2) Grammars

Proof-trees of a logic program can be thought of as abstract syntax trees with associated atoms. These abstract syntax trees can be represented by abstract context free grammars. An abstract context free grammar is the pair < N, P > where N is a finite set (the non-terminal alphabet) and P a N-sorted signature (for more details see [DM 85]).

(1.3) Many sorted logical languages

The specification will be given in some logical language together with an interpretation that we define as follows. Let S be a finite set of sorts containing the sort bool of the boolean values true, false. Let V be a sorted set of variables, F a S-signature and R a finite set of many sorted relation symbols (i.e. a set of symbols, each of them having an arity and, implicitely, the sort bool).

A logical language L over V, F, R is the set of formulas written with V, F, R and logical connectives like $\forall$, $\exists$, $\Rightarrow$, $\wedge$, $\vee$, not, ... We denote by free (φ) the possibly empty set of the free variables of the formula φ of L (free (φ) $\subseteq$ V), by AND A (resp OR A) the conjunction (resp. the disjunction) of formulas (AND Ø = true, OR Ø = false), and by $\varphi\,[t_1/v_1,\dots, t_n/v_n]$ the result of the substitution of t_i for each free occurrence of v_i (some renaming of variables may be necessary). We do not restrict a priori the logical language to be first order.

Let $\mathcal{C}$(L) denote a class of L-structures, i.e. objects of the form :

$$D = < \{D_s\}_{s \in S}\ ,\ \{f_D\}_{f \in F},\ \{r_D\}_{r \in R} >$$

where $< \{D_s\}_{s \in S}\ ,\ \{f_D\}_{f \in F} >$ is a heterogeneous F-algebra and for each r in R, r_D is a total mapping

$$D_{s1} \times \dots \times D_{sn} \rightarrow \{\underline{true}, \underline{false}\} = \underline{bool} \quad \text{if } \alpha(r) = s1 \dots sn.$$

The notion of validity is defined in the usual way. For every assignment ν, every D in $\mathcal{C}$(L), every φ in L, one assumes that (D, ν) $\models \varphi$ either holds or does not hold. We say φ is valid in D, and write D $\models \varphi$, iff (D, ν) $\models \varphi$ for every assignment ν.

2 - Definite Clauses Programs, specifications

(2.1) Definition : Definite Clause Program (DCP).

A DCP is a triple P = <PRED, FUNC, CLAUS> where PRED is a finite set of predicate symbols, FUNC a finite set of function symbols disjoint of PRED, CLAUS a finite set of clauses defined as usual [Cla 79, Llo 87] with PRED and TERM = T(FUNC, V). Complete syntax can be seen in examples (2.2) and (2.3). A clause is called a fact if it is restricted to an atomic formula. An atomic formula is built as usual with PRED and TERM.

(2.2) Example PRED = {plus} ρ(plus) = 3
FUNC = {zero, s} ρ(zero) = 0, ρ(s) = 1
CLAUS = {c1 : plus (zero, X, X) $\leftarrow$,
c2 : plus (s(X), Y, s(Z)) $\leftarrow$ plus (X, Y, Z)}

variables begin with uppercase letters.

(2.3) Example "List" terms are represented in Edinburgh syntax

PRED = {perm, extract} ρ(perm) = 2, ρ(extract) = 3

FUNC = {[],[_|_]} ρ([]) = 0, ρ([_|_]) = 2

CLAUS = {c1 : perm ([], []) ←,

c2 : perm ([A|L], [B|M]) ← perm(N, M), extract([A|L], B, N),

c3 : extract ([A|L], A, L),

c4 : extract ([A|L], B, [A|M]) ← extract (L, B, M)}

(2.4) Definition : denotation of a DCP P : DEN(P)

The denotation of a DCP is the set of all its atomic logical consequences :

DEN(P) = {a | P ⊢ a}

We do not give any more details on the notions of models of P (structures in which the clauses are valid formulas) and of logical consequences (all atoms of DEN(P) are valid in the models of P), since we won't make use in this paper of the logical semantics of a logic program, but rather of its constructive semantics that we shall now define. Other details can be found in [Cla 79, Ave 82, Llo 87, Fer 85].

(2.5) Definition : proof-tree [Cla 79, DM 85]

A proof-tree is an ordered labeled tree whose labels are atomic formulae (possibly including variables). The set of the proof-trees of a given DCP P = < PRED, FUNC, CLAUS > is defined as follows :

1 - If A ← is an instance of a fact of CLAUS (instances built with TERM), then the tree consisting of one vertice with label A is a proof-tree.

2 - If $T_1,..., T_q$ for some q > 0 are proof-trees with roots labeled $B_1,..., B_q$ and if A ← $B_1,..., B_q$. is an instance of a clause of CLAUS, then the tree consisting of the root labeled with A and the subtrees $T_1,..., T_q$ is a proof-tree.

By a partial proof-tree we mean any finite tree constructed by "pasting together" instances of clauses. Thus a proof-tree is a partial proof-tree all of whose leaves are instances of facts. We denote by PTR(P) the set of all root labels of proof-trees of P, in short the proof-tree roots of P. Note that every instance of a proof-tree is a proof-tree.

(2.6) Proposition [Cla 79, Fer 85, DF 86a] - Constructive semantics

Given a DCP P : DEN(P) = PTR(P).

Thus, instead of the logical semantics of a logic program, one can deal with its constructive semantics. As pointed out in [DM 85, DM 88], proof-trees can be thought as syntax trees (terms of a clauses-algebra) "decorated" by atoms as specified in the proof-tree definition. Thus inductive proof methods as defined in [CD 88] may be applied to logic programs. This will be done in the next section. All the definitions are adapted from [CD 88] to the logic programming case.

(2.7) Definition : Specification **S** on (L, $\mathbb{D}$) of a logic program P.

A specification of a logic program P is a family of formulas $\mathbf{S} = \{ \mathbf{S}^p \}_{p \in PRED}$ of a logical language L over V, F, R such that V contains the variables used in P and F contains FUNC, together with a L-structure $\mathbb{D}$. For every p of PRED, we denote by varg(p) = { p1, ..., pρ(p) } the set of variable names denoting any possible term in place of the 1st, ... or $\rho(p)^{th}$ argument of p. Thus we impose free($\mathbf{S}^p$) $\subseteq$ varg(p). Variables in a specification also begins with an uppercase letter.

(2.8) Definition : valid specification **S** for the DCP P.

A specification **S** on (L, $\mathbb{D}$) is valid for the DCP P (or P is correct w.r.t **S**) iff

$$\forall p(t_1, ..., t_n) \in DEN(P), \quad \mathbb{D} \models \mathbf{S}^p[t_1/p_1, ..., t_n/p_n], \quad n = \rho(p).$$

In the following the notation will be abbreviated into $\mathbf{S}^p[t_1, ..., t_n]$ if no confusion may arise.

This means that every atom of the denotation satisfies the specification, hence every atom in any proof-tree. It also means that every answer substitution (if any) satisfies the specification. It corresponds to a notion of partial correctness referring to the declarative (i.e. constructive or logical) semantics since nothing is specified about the existence of proof-trees (the denotation can be empty), the way to obtain them or the kind of resulting answer substitution for a given goal.

(2.9) Example : specification for (2.2) :

L_1 = V_1 contains varg(plus) = {plus1, plus2, plus3}
F_1 = {zero, s, + }
R_1 = { = }

$\mathbb{D}_1$ = $\mathbb{N}$ the natural integers with usual meaning, zero is interpreted as 0, s as the increment by 1, + as the addition. (i.e. $\mathbb{N}$ is the L1-structure $\mathbb{D}$1)

$\mathbf{S}_1$ = { $\mathbf{S}_1^{plus}$ }, $\mathbf{S}_1^{plus}$: plus3 = plus1 + plus2

The validity of $\mathbf{S}_1$ (which is proved in the next section) means that the program "plus" in (2.2) specifies the addition on $\mathbb{N}$, or that every n-uple of values corresponding to the interpreted arguments of the elements of the denotation satisfy the specification plus3 = plus1+ plus2 .

L_2 = V_2 as in L_1 , F_2 = { zero, s}
R_2 = { ground } ρ(ground) = 1.

$\mathbb{D}_2$ contains the term algebra T(F_2, V_2), and ground(t) is true iff t is a ground term.

$\mathbf{S}_2$ = { $\mathbf{S}_2^{plus}$ }, $\mathbf{S}_2^{plus}$: ground(plus3) $\Rightarrow$ [ground(plus1) $\wedge$ ground(plus2)]

$\mathbf{S}_2$ is a valid specification (it can be observed on every proof-tree and will be proven in the next section).

(2.10) Example : specification for (2.3). This example uses a many sorted L_3 structure.

L_3 = V_3 contains varg(perm) and varg(extract).

F_3 = { [], [_|_], nil, . , append } [] and nil are constants, the other operators have rank 2.

R3 = { is-a-list, permut } ρ(is-a-list) = 1, ρ(permut) = 2.

D_3 contains two domains : list the usual domain of the lists of unspecified arguments, any the domain of the unspecified arguments.

the profiles are :

for functions

[] ∈ < ε, list >, is interpreted as nil .

[_|_] ∈ < any list, list > , is interpreted as . (cons of LISP)

nil ∈ < ε, list > (as usual)

. ∈ < any list, list > (as usually : cons of LISP)

append ∈ < list list, list > (lists concatenation).

for relations

permut ∈ <list list, bool > it is the relation defining the pairs of permuted lists.

S_3 = { Sperm : permut (perm1, perm2) ,

Sextract : ∃ L1, L2 (extract1 = append (L1, extract2.L2) ∧ extract3 = append (L1, L2) }

3 - Inductive proof method

(3.1) Definition : Inductive specification S of a DCP P.

A specification S on (L, D) of a DCP P is inductive iff for every c in CLAUS,

$$c: r_0(t01, ..., t0n_0) \Leftarrow r_1(t11, ..., t1n_1), ..., r_m(tm1, ..., tmn_m) ,$$

$$D \models (\mathop{AND}_{1 \le k \le m} S^{rk} [tk1, ..., tkn_k] \Rightarrow S^{r0} [t01, ..., t0n_0])$$

i.e. a specification is inductive iff in every clause, if the specification holds for the atoms of the body, it holds for the head.

(3.2) Proposition :

If S of P is inductive then S is valid for P .

Proof : by an easy induction on the size of the proof-trees, if $PTR_n(P)$ denotes the set of all roots of the proof-tree of size ≤ n, S holds in $PTR_{n+1}(P)$ (by definition (2.5) and (3.1) and the notion of validity), thus in $DEN(P) = \bigcup_{n \ge 0} PTR_n(P)$. QED.

(3.3) Definition : strongest (weakest) specification of P

Let S and S' be two specifications of P on (L, D). One says that S is weaker than S' (or S' stronger than S), and denotes it by D |= (S'⇒ S) iff ∀p ∈ PRED, D |= ($S'^p \Rightarrow S^p$) .

Given a program P, we consider an L-structure $\mathbb{D}$ such that $\mathbb{D}$ contains an interpretation of FUNC. Then we consider L' the language of all relations on $\mathbb{D}$. Obviously $L \subseteq L'$. Note that L' may not be first order. We denote S_P the following specification on (L', $\mathbb{D}$) defined as :

$$\forall p, t_1, \ldots, t_n \quad (\mathbb{D} \models S_P{}^p [t_1, \ldots, t_n]) \text{ iff } p(t_1, \ldots, t_n) \in DEN(P).$$

We denote $S_{\underline{true}}$ the specification such that $Sp_{\underline{true}}$: <u>true</u> for all p in PRED (i.e no specification).

(3.4) <u>Proposition</u> : Given a DCP P, S_P and $S_{\underline{true}}$ are respectively the <u>strongest</u> and the <u>weakest</u> valid specification for P and S_P is inductive i.e. all valid specifications S satisfy :

$\mathbb{D} \models (S_P \Rightarrow S \Rightarrow S_{\underline{true}})$, and S_P is inductive.

<u>Proof</u> : it is easy to observe that S_P on (L', $\mathbb{D}$) is inductive, hence valid and that every valid specification S on (L, $\mathbb{D}$) – thus on (L', $\mathbb{D}$) – satisfies $\mathbb{D} \models S_P \Rightarrow S$. Obviously $\mathbb{D} \models S \Rightarrow S_{\underline{true}}$.

(3.5) <u>Theorem</u> (soundness and completeness of the inductive proof method)

A specification S on (L, $\mathbb{D}$) is valid for P <u>iff</u> it is weaker than some inductive specification S' on (L', $\mathbb{D}$) :

i.e. 1) $\exists$ S' inductive
2) $\mathbb{D} \models S' \Rightarrow S$.

<u>Proof</u> : Soundness is trivial since if S' is inductive, it is valid by proposition (3.2) and if $\mathbb{D} \models S' \Rightarrow S$ and S' true, then S is also true.

Completeness is stated in proposition (3.4) with $S' = S_P$. Note that one does not have the completeness if one restricts the specification language like first order logic (Proof of this claim will be given in an extended version of this paper).

(3.6) <u>Example</u> (2.2): the specification S_1 (2.9) is inductive.

Following the definition (3.1) it is sufficient to prove :

$$\mathbb{N} \models S_1[\text{zero}, X, X] \text{ and}$$

$$\mathbb{N} \models S_1[X, Y, Z] \Rightarrow S_1[s(X), Y, s(Y)]$$

thus:

$$\mathbb{N} \models 0 + X = X \text{ and}$$

$$\mathbb{N} \models (X + Y = Z \Rightarrow X+1 + Y = Z+1)$$

which are valid formulas on $\mathbb{N}$.

(3.7) Example (2.2): the specification S_2 (2.9) is inductive.

In the same way it is easy to show that the following formulas are valid on $\mathbb{D}_2$:

$$\text{ground}(0) \wedge \text{ground}(X) \Rightarrow \text{ground}(X)$$

$$[\text{ground}(Z) \Rightarrow (\text{ground}(X) \wedge \text{ground}(Y))] \Rightarrow [\text{ground}(s(Z)) \Rightarrow (\text{ground}(s(X)) \wedge \text{ground}(Y)]$$

(3.8) Example (2.3): the specification S_3 (2.10) is inductive.

It is easy to show that the following formulas are valid on $\mathbb{D}_3$: (some replacements are already made in the formulas and universal quantifications on the variables are implicit)

in c1 : permut (nil, nil)

in c2 : permut (N, M) $\wedge$ $\exists$ L1, L2 (A.L = append(L1, B.L2) $\wedge$ N = append (L1, L2))
$\Rightarrow$ permut (A.L, B.M)

in c3 : $\exists$ L1, L2 (A.L = append(L1, A.L2) $\wedge$ L = append (L1, L2))
(L1 and L2 are lists) take L1 = nil and L2 = L.

in c4 : $\exists$ L1, L2 L = append(L1, B.L2) $\wedge$ M = append (L1, L2))
$\Rightarrow$ $\exists$ L'1, L'2 (A.L = append(L'1, B.L'2) $\wedge$ A.M = append (L'1, L'2))

take L'1 = A.L1 and L'2 = L2.

(3.9) Example:

We achieve this illustration by a proof of the claim in the introduction concerning the program concatenate :

$S^{\text{concatenate}}$ = repr(concatenate3) = append(repr(concatenate1), repr(concatenate2))

defined on (L_3, $\mathbb{D}_3$) enriched with two operators: "-" of profile <list list, diflist> and "repr" of profile <diflist, list>, defining the list represented by a difference list.

The claim is obvious as
repr(L1-L3) = append(repr(L1- L2), repr(L2 -L3))

(3.10) Remark that as noticed in [CD88] this proof method can be viewed as a fix point induction on logic programs. It seems very easy to use, at least on simple programs, which are sometimes very hard to understand. The ability of the programmer to use this method may improve his ability to understand and thus handle axioms of logic programs.

4 - Proof method with annotations

The practical usability of the proof method of theorem (3.5) suffers from its theoretical simplicity : the inductive specifications **S**' to be found to prove the validity of some given specification **S** will need complex formulas $\mathbf{S}'^p$ since we associate only one for each p in PRED. It is also shown in [CD 88] that **S**' may be exponential in the size of the DCP (to show this result we can use the DCP's transformation into attribute grammars as in [DM 85]). The proof method with annotations is introduced in order to reduce the complexity of the proofs : the manipulated formulas are shorter, but the user has to provide the organization of the proof i.e. how the annotations are deducible from the others. These indications are local to the clauses and described by the so called logical dependency scheme. It remains to certify the consistency of the proof, i.e. that a conclusion is never used to prove itself. Fortunately this last property is decidable and can be verified automatically, using the Knuth algorithm [Knu 68] or its improvements [DJL 86].

(4.1) Definition : annotations of a DCP

Given a DCP P, an annotation is a mapping Δ assigning to every p in PRED a finite set of formulas or assertions $\Delta(p)$ built as in definition (2.7). It will be assumed that assertions are defined on (L, $\mathbb{D}$).

The set $\Delta(p)$ is partitioned into two subsets $I\Delta(p)$ (the set of the inherited assertions of p) and $S\Delta(p)$ (the set of the synthesized assertions of p).

The specification $\mathbf{S}_\Delta$ associated with Δ is the family of formulas :

$$\{\ \mathbf{S}^p_\Delta :\ \text{AND}\ I\Delta(p)\ \Rightarrow\ \text{AND}\ S\Delta(p)\ \}_{p\,\in\,\text{PRED}}$$

(4.2) Definition : validity of an annotation Δ for a DCP P.

An annotation Δ is valid for a DCP P iff for all p in PRED in every proof-tree T of root $p(t_1, ..., t_{np})$: if $\mathbb{D} \models$ AND $I\Delta(p)\ [t_1, ..., t_{np}]$ ($np = \rho(p)$) then every label $q(u_1, ..., u_{nq})$ ($nq = \rho(q)$) in the proof-tree T satisfies : $\mathbb{D} \models$ AND $\Delta(q)\ [u_1, ..., u_{nq}]$.

In other words, an annotation is valid for P if in every proof-tree whose root satisfies the inherited assertions, all the assertions are valid at every node in the proof-tree, hence the synthesized assertions of the root.

(4.3) Proposition : if an annotation Δ for the DCP P is valid for P, then $\mathbf{S}_\Delta$ is valid for P.

Proof : It follows from the definition of $\mathbf{S}_\Delta$, the definitions of validity of an annotation (4.2) and of a specification (2.8).

Note that $\mathbf{S}_\Delta$ can be valid but not inductive (see example (4.14)).

We shall give sufficient conditions insuring the validity of an annotation and reformulate the proof method with annotations. This formulation is slightly different from that given in [CD 88]. The introduction of the proof-tree grammar is a way of providing a syntaxic formulation of the organization of the proof.

(4.4) <u>Definition</u> : <u>Proof-tree grammar</u> (PG)

Given a DCP P = < PRED, FUNC, CLAUS >, we denote Gp and call it the <u>proof-tree grammar of P</u>, the abstract context free grammar < PRED, RULE > such that RULE is in bijection with CLAUS and r of RULE has profile $< r_1 r_2 \ldots r_m, r_0 >$ iff the corresponding clause in CLAUS is $c : r_0(\ldots) \leftarrow r_1(\ldots), \ldots, r_m(\ldots)$.

Clearly a (syntax) tree in Gp can be associated to every proof-tree of P. But not every tree in Gp is associated a proof-tree of P.

(4.5) <u>Definition</u> : <u>Logical dependency scheme for Δ</u> (LDS_Δ).

Given a DCP P and an annotation Δ for P, a logical dependency scheme for Δ is $LDS_\Delta = < Gp, \Delta, D >$ where Gp = < PRED, RULE > is the proof-tree grammar of P and D a family of binary relations defined as follows.

We denote for every rule r in RULE of profile $< r_1 r_2 \ldots r_m, r_0 >$
$W_{\underline{hyp}}(r)$ (resp. $W_{\underline{conc}}(r)$) the sets of the hypothetic (resp. conclusive) assertions which are :
$W_{\underline{hyp}}(r) = \{\varphi_k \mid k = 0, \varphi \in I\Delta(r_0) \text{ or } k > 0, \varphi \in S\Delta(r_k)\}$
$W_{\underline{conc}}(r) = \{\varphi_k \mid k = 0, \varphi \in S\Delta(r_0) \text{ or } k > 0, \varphi \in I\Delta(r_k)\}$

where φ_k is φ in which the free variables <u>free</u> (φ) = $\{p_1, \ldots, p_n\}$ have been renamed into <u>free</u> (φ_k) = $\{p_{k1}, \ldots, p_{kn}\}$.

The renaming of the free variables is necessary to take into account the different instances of the same predicate (if $r_i = r_j = pr$ in a clause for some different i and j) and thus different instances of the same formula associated with pr, but this will not be explicit anymore by using practically the method.

$$D = \{D(r)\}_{r \in RULE}, D(r) \subseteq W_{\underline{hyp}}(r) \times W_{\underline{conc}}(r).$$

From now on we will use the same name for the relations D(r) and their graph. For a complete formal treatment of the distinction see for example [CD 88]. We denote by <u>hyp</u> (φ) the set of all formulas ψ such that (ψ, φ) ∈ D(r) and by <u>assoc</u> (φ) = $p(t_1, \ldots, t_n)$ the atom to which the formula is associated by Δ in the clause c corresponding to r.

(4.6) <u>Example</u> : annotation for example (2.2) and specification S_2 (2.9).

$\Delta(plus) = I\Delta(plus) \cup S\Delta(plus)$
$I\Delta(plus) = \{\varphi : \text{ground (Plus3)}\}$
$S\Delta(plus) = \{\psi : \text{ground (Plus1)}, \delta : \text{ground (Plus2)}\}$
$S_\Delta plus = S_2 plus$

G_{plus} : PRED = {plus}
RULE = $\{r_1 \in <\varepsilon, plus>, r_2 \in <plus, plus>\}$

D : $D(r_1) = \{\varphi_0 \rightarrow \delta_0\}$
$D(r_2) = \{\varphi_0 \rightarrow \varphi_1, \psi_1 \rightarrow \psi_0, \delta_1 \rightarrow \delta_0\}$ (see the scheme below)

$W_{\underline{hyp}}(r_1) = \{\varphi_0\}, W_{\underline{conc}}(r_1) = \{\psi_0, \delta_0\}$
$W_{\underline{hyp}}(r_2) = \{\varphi_0, \psi_1, \delta_1\}, W_{\underline{conc}}(r_2) = \{\varphi_1, \psi_0, \delta_0\}$

Note that in r_2 for example :
φ_0 = ground (Plus03)
φ_1 = ground (Plus13)...

In order to simplify the presentation of D we will use schemes as in [CD 88] representing the rules in RULE and the LDS of Δ. Elements of $W_{\underline{conc}}$, will be underlined. Inherited (synthesized) assertions are written on the left (right) hand side of the predicate name. Indices are implicit : 0 for the root, 1... to n following the left to right order for the sons.

r 1 :

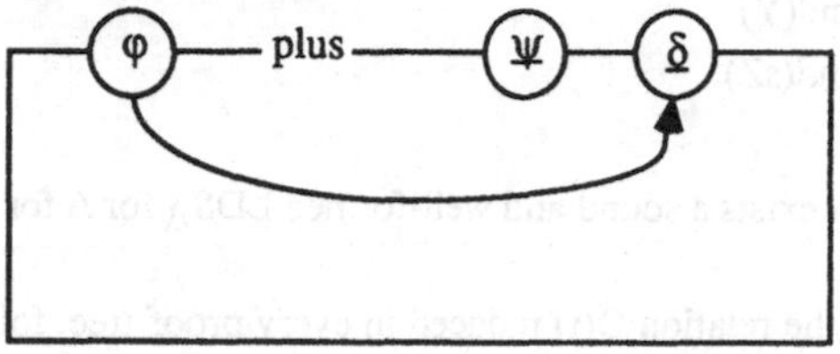

r 2 :

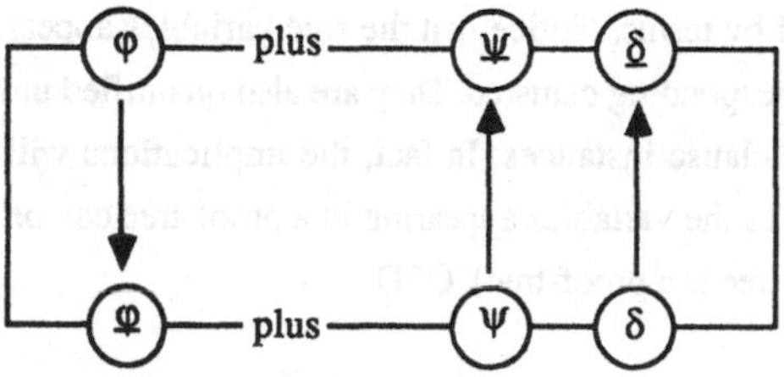

(4.7) <u>Definition</u> : <u>Purely-synthesized LDS, well-formed LDS</u>.

A LDS for Δ is <u>purely-synthesized</u> iff IΔ = Ø, i.e. there are no inherited assertions.

A LDS for Δ for P is <u>well-formed</u> iff in every tree t of Gp the relation of the induced dependencies D(t) is a partial order (i.e. there is no cycle in its graph).

To understand the idea of well-formedness of the LDS it is sufficient to understand that the relations D(r) describe dependencies between instances of formulas inside the rules r. Every tree t of Gp is built with instances of rules r in RULE, in which the local dependency relation D(r) defines dependencies between instances of the formulas attached to the instances of the non-terminals in the rule r. Thus the dependencies in the whole tree t define a new dependency relation D(t) between instances of formulas in the tree. A complete treatment of this question can be found in [CD 88]. We recall here only some important results [see Knu 68, DJL 88 for a survey on this question] :

(4.8) Proposition - The well-formedness property of an LDS in decidable.
- The well-formedness test is intrinsically exponential.
- Some non trivial subclasses of LDS can be decided in polynomial time.
- A purely-synthesized LDS is (trivially) well-formed.

(4.9) Definition : Soundness of a LDS for Δ.

A LDS for Δ < G_p, Δ, D > is sound iff for every r in RULE and every
φ in W_{conc}(c) with assoc (φ) = $q(u_1, ..., u_{nq})$ the following holds :

$$D \models AND\{ \psi[t_1, ..., t_{np}] \mid \psi \in hyp(\varphi) \text{ and } assoc(\psi) = p(t_1, ..., t_{np}) \} \Rightarrow \varphi[u_1, ..., u_{nq}]$$

(Note that the variable q_i (p_i) in a formula φ (ψ) is replaced by the corresponding term u_i (t_i)).

(4.10) Example : The LDS given in example (4.6) is sound. In fact it is easy to verify that the following holds in T :

in r1 : ground(X) ⇒ ground(X)
ground(0)
in r2: ground(sX) ⇒ ground(X)
ground(Y) ⇒ ground(Y)
ground(Z) ⇒ ground(sZ)

(4.11) Theorem : Δ is valid for P if it exists a sound and well-formed LDS_Δ for Δ for P.

Sketch of the proof by induction on the relation D(t) induced in every proof-tree, following the scheme given in [Der 83] or the proof given in [CD 88, theorem (4.4.1)]. The only difference comes from the lack of attribute definitions replaced by terms. Notice that the free variables appearing in the formulas (4.9) are the free variables of the corresponding clause c. They are also quantified universally. Hence the results, as a proof-tree is built with clause instances. In fact, the implications will hold also in every instance of a clause in the proof-tree as the variables appearing in a proof-tree can be viewed universally quantified (every instance of a proof-tree is a proof-tree). QED.

(4.12) Theorem (soundness and completeness of the annotation method for proving the validity of specifications) : We use the notations of (3.3) and (3.5).

A specification **S** on (L, D) is valid iff it is weaker than the specification S_Δ of an annotation Δ on (L', D) with a sound and well-formed LDS (L' as in 3.3).
i.e. :

1) there exists a sound and well-formed LDS_Δ.
2) $D \models S_\Delta \Rightarrow S$.

Proof (soundness) follows from theorem (4.11).
(completeness) follows from the fact that S_P on (L', D) is a purely synthesized (thus well-formed) sound annotation.

We complete this presentation by giving some examples.

(4.13) <u>Example</u> (4.10) continued.

The LDS is sound and well-formed, thus $S_{\Delta}^{plus} = S_2^{plus}$ is a valid specification.

(4.14) <u>Example</u> We borrow from [KH 81] an example given here in a logic programming style : it computes multiples of 4.

c1 : fourmultiple (K) ← p(0, H, H, K).

c2 : p(F, F, H, H) ←

c3 : p(F, sG, H, sK) ← p(sF, G, sH, K)

$\mathcal{S}^{fourmultiple}$: ∃ N, N ≥ 0 ∧ Fourmultiple1 = 4∗N

L, $\mathbb{D} = \mathbb{D}_1$ as in (2.9) enriched with ∗, ≥ 0 etc...

The following annotation Δ is considered in [KH 81] :

IΔ (fourmultiple) = Ø, SΔ (fourmultiple) = { $\mathcal{S}^{fourmultiple}$} = {δ}

IΔ(p) = {β} SΔ(p) = {α, γ}

α : ∃N, N ≥ 0 ∧ P2 = P1 + 2 ∗ N

β : P3 = P2 + 2 ∗ P1

γ : P4 = 2 ∗ P2 + P1

The assertions can be easily understood if we observe that such a program describes the construction of a "path" of length 4∗N and that p1, p2, p3 and p4 are lengths at different steps of the path as shown in the following figure :

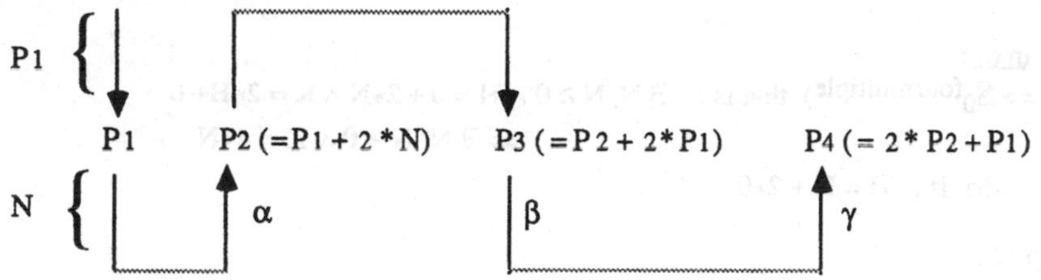

The LDS for Δ is the following :

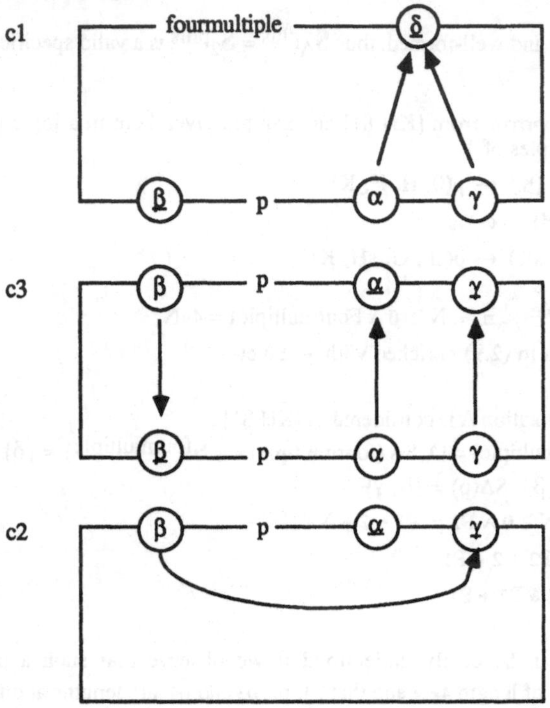

The LDS is sound and well-formed. For example it is easy to verify that the following fact holds in D_1 :

in c1 :

$(\alpha_1 \wedge \gamma_1 \Rightarrow S_0^{fourmultiple})$ that is : $\exists N, N \geq 0 \wedge H = 0 + 2*N \wedge K = 2*H + 0 \Rightarrow \exists N, N \geq 0 \wedge K = 4*N$

(β_1) that is : $H = H + 2*0$

in c2 :

$(\beta_0 \Rightarrow \gamma_0)$ that is : $H = F + 2*F \Rightarrow H = 2*F + F$

(α_0) that is : $\exists N, N \geq 0 \wedge F = F + 2*N$ (with $N = 0$)

etc...

Note that as S_Δ is inductive, this kind of proof modularization can be viewed as a way to simplify the presentation of the proof of S_Δ.

Now we consider on the same program a non inductive valid specification ξ defined on L_2, D_2 (2.9) :

$\xi^{fourmultiple}$: ground (Fourmultiple1)

ξ^p : [ground(P1) ∧ ground(P3)] ⇒ [ground(P2) ∧ ground(P4)]

The specification clearly is valid but not inductive since the following does not hold with D_2 (term-algebra) in c1

$$D_2 \not\models \xi P_1 \Rightarrow \xi\text{fourmultiple}_0$$

i.e.

$$D_2 \not\models [\ (\text{ground}(0) \wedge \text{ground}(H)) \Rightarrow (\text{ground}(H) \wedge \text{ground}(K))\ ] \Rightarrow \text{ground}(K)$$

But it is easy to show that the following LDS is sound and well-formed :

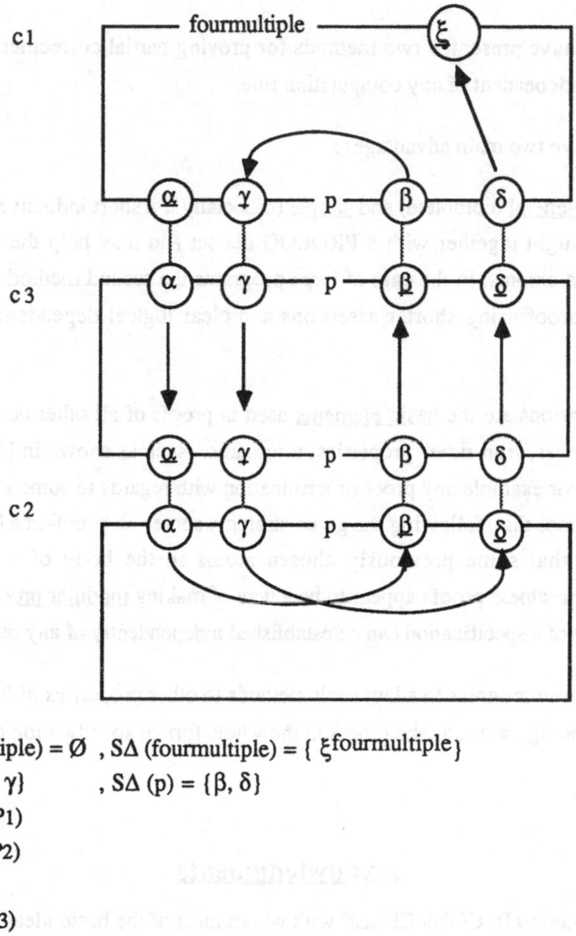

$I\Delta$ (fourmultiple) = Ø , $S\Delta$ (fourmultiple) = { ξfourmultiple }

$I\Delta$ (p) = {α, γ} , $S\Delta$ (p) = {β, δ}

α : ground(P1)

β : ground(P2)

γ : ground(P3)

δ : ground(P4)

Dotted lines help to observe that the LDS is well-formed (without circularities). The proofs are trivial.

Note that the corresponding inductive specification is $(\alpha \Rightarrow \beta) \wedge (\gamma \Rightarrow \delta)$. It is shown in [CD 88] how the inductive specification can be inferred from LDS_Δ.

Notice that this kind of proof of correctness corresponds to some kind of mode verification. It can be automatized for a class of programs identified in [DM 84] and experimentally studied in [Dra 87] (the class of simple logic programs). As shown in [DM 84] this leads to an algorithm of automatic (ground) modes computation for simple logic programs which can compute (ground) modes which are not inductive.

Conclusion

In this paper we have presented two methods for proving partial correctness of logic programs. Both are modular and independent of any computation rule.

These methods have two main advantages :

1) They are very general (complete) and simple (especially if a short inductive assertion is proven). As such they can be taught together with a PROLOG dialect and may help the user to detect useful properties of the written axioms. In the case of large programs the second method may help to simplify the presentation of a proof using shorter assertions and clear logical dependency schemes between assertions.

2) Valid specifications are the basic elements used in proofs of all other desirable logic program properties as completeness, "run-time" properties, termination such as shown in [DF 88] or safe use of the negation [Llo 87]. For example any proof of termination with regards to some kind of used goals and some strategy will suppose that, following the given strategy, some sub-proof-tree has been successfully constructed and thus that some previously chosen atoms in the body of a clause satisfy their specifications. Thus correctness proofs appear to be a way of making modular proofs of other properties also. In fact the validity of a specification can be established independently of any other property.

Work is at present in progress to adapt such methods to other properties of logic programs. These methods are currently being used to make proofs in the whole formal specification of standard PROLOG [DR 88].

Aknowledgments

We are indebted to B. COURCELLE with whom most of the basic ideas have been developed and to G. FERRAND and J.P. DELAHAYE who helped to clarify this text.

References

[AvE 82] K.R. Apt, M.H. Van Emden : Contributions to the theory of Logic Programming. JACM V29, N° 3, July 1982 pp 841-862.

[CD 88] B. Courcelle, P. Deransart : Proof of partial Correctness for ·Attribute Grammars with application to Recursive Procedure and Logic Programming. Information and Computation 78, 1, July 1988 (First publication INRIA RR 322 - July 1984).

[Cla 79] K.L. Clark : Predicate Logic as a Computational Formalism. Res. Mon. 79/59 TOC. Imperial College, December 1979.

[Coo 78] S.A. Cook : Soundness and Completeness of an Axiom System for Programs Verification. SIAM Journal. Comput. V7, n° 1, February 1978.

[Cou 84] B. Courcelle : Attribute Grammars : Definitions, Analysis of Dependencies, Proof Methods. In Methods and Tools for Compiler Construction, CEC-INRIA Course (B. Lorho ed.). Cambridge University Press 1984.

[Der 83] P. Deransart : Logical Attribute Grammars. Information Processing 83, pp 463-469, R.E.A. Mason ed. North Holland, 1983.

[Dev 87] Y. Deville : A Methodology for Logic Program Construction.PhD Thesis, Institut d'Informatique, Facultés Universitaires de Namur (Belgique), February 1987.

[DF 87] P. Deransart, G. Ferrand : Programmation en Logique avec Négation : Présentation Formelle. Publication du laboratoire d'Informatique, University of Orléans, RR 87-3 (June 1987).

[DF 88] P. Deransart, G. Ferrand : Logic Programming, Methodology and Teaching. K. Fuchi, L. Kott editors, French Japan Symposium, North Holland, pp 133-147, August 1988.

[DF 88] P. Deransart, G. Ferrand : On the Semantics of Logic Programming with Negation. RR 88-1, LIFO, University of Orléans, January 1988.

[DJL 88] P. Deransart, M. Jourdan, B. Lorho : Attribute Grammars : Definitions, Systems and Bibliography, LNCS 323, Springer Verlag, August 1988.

[DM 84] P. Deransart, J. Maluszynski : Modelling Data Dependencies in Logic Programs by Attribute Schemata. INRIA, RR 323, July 1984.

[DM 85] P. Deransart, J. Maluszynski : Relating Logic Programs and Attribute Grammars. J. of Logic Programming 1985, 2 pp 119-155. INRIA, RR 393, April 1985.

[DM 89] P. Deransart, J. Maluszynski : A Grammatical View of Logic Programming. PLILP'88, Orléans, France, May 16-18, 1988, LNCS 348, Springer Verlag, 1989.

[DR 88] P. Deransart, G. Richard : The Formal Specification of PROLOG standard. Draft 3, December 1987. (Draft 1 published as BSI note PS 198, April 1987, actually ISO-WG17 document, August 1988).

[Dra 87] W. Drabent, J. Maluszynski : Do Logic Programs Resemble Programs in Conventional Languages. SLP87, San Francisco, August 31 -September 4 1987.

[DrM 87] W. Drabent, J. Maluszynski : Inductive Assertion Method for Logic Programs. CFLP 87, Pisa, Italy, March 23-27 1987 (also : Proving Run-Time Properties of Logic Programs. University of Linköping. IDA R-86-23 Logprog, July 1986).

[Fer 85] G. Ferrand : Error Diagnosis in Logic Programming, an Adaptation of E. Y. Shapiro's Methods. INRIA, RR 375, March 1985. J. of Logic Programming Vol. 4, 1987, pp 177-198 (French version : University of Orléans, RR n° 1, August 1984).

[FGK 85] N. Francez, O. Grumberg, S. Katz, A. Pnuelli : Proving Termination of Prolog Programs. In "Logics of Programs, 1985", R. Parikh Ed., LNCS 193, pp 89-105, 1985.

[Fri 88] L. Fribourg : Equivalence-Preserving Transformations of Inductive Properties of Prolog Programs. ICLP'88, Seattle, August 1988.

[Hog 84] C.J. Hogger : Introduction to Logic Programming. APIC Studies in Data Processing n° 21, Academic Press, 1984.

[KH 81] T. Katayama, Y. Hoshino : Verification of Attribute Grammars. 8th ACM POPL Conference. Williamsburg, VA pp 177-186, January 1981.

[Knu 68] D.E. Knuth : Semantics of Context Free Languages. Mathematical Systems Theory 2, 2, pp 127-145, June 1968.

[KS 86] T. Kanamori, H. Seki : Verification of Prolog Programs using an Extension of Execution. In (Shapiro E., ed.), 3rd ICLP, LNCS 225, pp 475-489, Springer Verlag, 1986.

[Llo 87] J. W. Lloyd : Foundations of Logic Programming. Springer Verlag, Berlin, 1987.

[SS 86] L. Sterling, E. Y. Shapiro : The Art of Prolog. MIT Press, 1986.

THE REACHABILITY PROBLEM FOR GROUND TRS AND SOME EXTENSIONS

A. DERUYVER and R. GILLERON
LIFL UA 369 CNRS
Universite des sciences et techniques de LILLE FLANDRES ARTOIS
U.F.R. d' I.E.E.A. Bat. M3 59655 VILLENEUVE D'ASCQ CEDEX FRANCE

ABSTRACT

The reachability problem for term rewriting systems (TRS) is the problem of deciding, for a given TRS S and two terms M and N, whether M can reduce to N by applying the rules of S.
We show in this paper by some new methods based on algebraical tools of tree automata, the decidability of this problem for ground TRS's and, for every ground TRS S, we built a decision algorithm. In the order to obtain it, we compile the system S and the compiled algorithm works in a real time (as a fonction of the size of M and N).
We establish too some new results for ground TRS modulo different sets of equations : modulo commutativity of an operator σ, the reachability problem is shown decidable with technics of finite tree automata; modulo associativity, the problem is undecidable; modulo commutativity and associativity, it is decidable with complexity of reachability problem for vector addition systems.

INTRODUCTION

The reachability problem for term rewriting systems (TRS) is the problem of deciding, for given TRS S and two terms M and N, whether M can reduce to N by applying the rules of S. It is well-known that this problem is undecidable for general TRS's .In a first part we study this problem for more simple systems, more specifically in the case of ground term rewriting systems.
A TRS is said to be ground if its set of rewriting rules R={ li->ri | i∈ I} (where I is finite) is such that li and ri are ground terms(no variable occurs in these terms). The decidability of the reachability problem for ground TRS was studied by Dauchet M. [4],[5] as a consequence of decidability of confluence for ground TRS. Oyamaguchi [15] and Togushi-Noguchi have shown this result too for ground TRS and in the same way for quasi-ground TRS.We take again this study with two innovator aspects:
- the modulary aspect of the decision algorithm which use all algebraical tools of tree automata, that permits to clearly describe it.
- the exchange between time and space aspect which have permitted to obtain some time complexities more and more reduced.
Therefore we have proceeded in three steps:
1- We begin with the TRS S not modified which gives the answer to the problem with a time complexity not bounded.

supported by "Greco programmation" and "PRC mathematique et informatique"

2- We transform the system S in a GTT (ground tree transducer) which simulates it, we will call this system, S'. Then the decision algorithm will have a quadratic time complexity. The memory space of S' will be in O((number of rules of S)^2).
3- Then, we obtain, after a compilation of S' which could be realised in an exponential time (reduction of nondeterminism), a real time decision algorithm (linear complexity). The necessary memory space, after the compilation of S', will be in O(exp(number of rules of S)).
If we make a comparison with the result of Oyamaguchi M.[15] we can have the next figure:

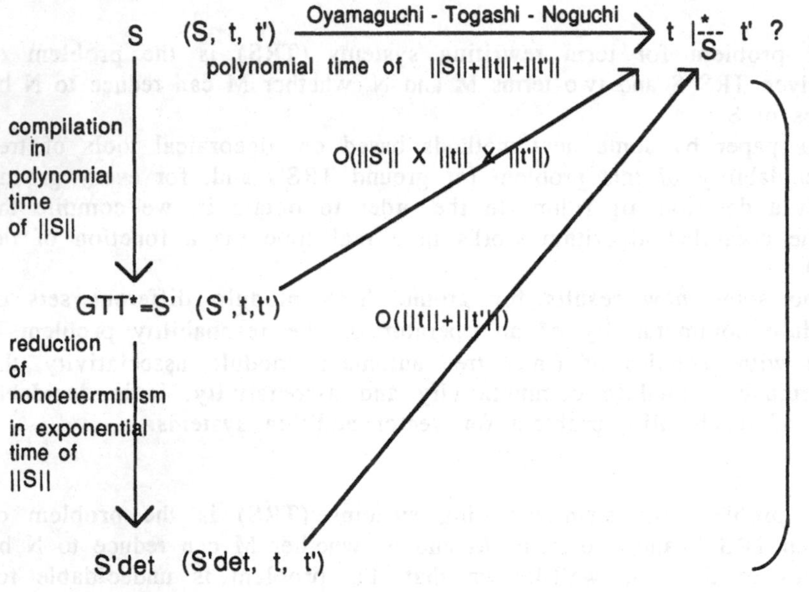

S = rewriting system
S'= our system S after compilation
S'det = our system S' after the reduction of nondeterminism
t,t' = the given trees
||S|| = size of the rewriting system S
||t|| = size of the tree t

A program, which is called VALERIANN, written in PROLOG realizes at the present time this algorithm.(on SUN machine)
In a second part, we consider the case of a ground TRS R_G modulo different sets of equations in the next three cases:

E_C :commutativity of an operator σ

E_A :associativity of an operator σ

E_{AC} :associativity and commutativity of an operator σ

R_C, R_A, R_{AC} denote the TRS obtained by orientation of equations into rules.We look at the two next problems:

For a TRS S equal to $R_G \cup R_C, R_G \cup R_A, R_G \cup R_{AC}$ and with conditions on the configuration of terms (i) if F is a recognizable forest, is the class of F modulo S recognizable ?(ii)decidability of the reachability problem
We have different results for each case:

For RGC, we have a positive answer for (i) and henceforth for (ii)
For RGA, we have a negative answer for the problems (i) and (ii)
For RGAC, we have a negative answer for (i) and a positive answer for (ii) with the complexity of the reachability problem for vector addition systems.

I-PRELIMINARIES

Let us recall some classical definitions and some usefull results:

1- tree automata and recognizable forests.

Let Σ be a finite ranked alphabet.
T_Σ is the set of terms (or trees) over Σ.

Definition1: A frontier-to-root (bottom-up) tree automaton is a quadruplet M=(Σ , Q , F, R) where

* Σ is a finite ranked alphabet.
* Q is a finite set of states.
* F is the set of final states, with F Q
* R is a finite set of transition rules, these rules have the next configuration: $c(q_{i1}[x1], \ldots, q_{in}[xn])$ --> $q[c(x1, \ldots, xn)]$
  if n=0, the rule is c' --> q[c']

We can dually define root-to-frontier(top-down) tree automata.
For more development see Gecseg F. & Steinby M.[7].

Definition2: A forest F is said to be recognizable if and only if there is a frontier-to-root tree automaton which accepts it.

properties: the class REC of recognizable forests is closed under union, intersection, and complementary.

2-algorithm of decision on tree automata

notation:
we note ||m|| the number of rules of the automaton m and $|m_q|$ the number of states of the automaton m.
we note m the automaton which accepts the complementary of the language accepted by m

a-Decision of the emptiness (M= ∅)

Let M an automaton. The time complexity to answer to the next problem:
Is the language which is accepted by M empty ?
is:
* linear, for word languages, if we have direct access to rules and if we use a naive algorithm.
* in $O(||M|| \times |M_q|)$, for tree languages.

b-Intersection of two automata M and M', and decision of the emptiness of this intersection. (M ∩M' ≠ ∅)

* for word languages, the time complexity to answer to this problem is in $O(||M|| \times ||M'||)$.
* for tree languages, the time complexity is more important, it is in $O(||M|| \times ||M'|| \times |M_q| \times |M_{q'}|)$.

c- Equivalence of M and M' (M=M')

$$M=M' \Leftrightarrow M \cap \underline{M}' = \emptyset \quad \text{and} \quad \underline{M} \cap M' = \emptyset$$

*deterministic case:
we can transform M in $\underline{M}$ by exchanging final states and the other states. Then we return to the same case than b-.

* nondeterministic case:
the time complexity contains the time of reduction of nondeterminism which is exponential.

3- Ground TRS and GTT.

*A tree rewriting system (TRS) S on T_Σ is a set of directed rewriting rules $R=\{li \text{-->} ri \mid i \in I\}$. Here, we only consider finite TRS (where I is finite), For more development see Huet G. & Oppen D.[8].
|--- is the extension of --> according to tree substitutions.
The reduction relation |-*- on T_Σ is the reflexive and transitive closure of -->.
S is a ground TRS if and only if no variable occurs in rules.

*A ground tree transducer on T_Σ (a GTT in short) is the relation T or (G,D) associated with two tree automata G and D and defined as follows:

$$t \overset{T}{\rightarrow\leftarrow} t' \text{ iff there exist } u \in T_\Sigma \cup Eg \cup Ed \text{ such that } t \underset{G}{\rightarrow} u \underset{D}{\leftarrow} t'.$$

where Σ is a finite ranked alphabet.
Eg and Ed are sets of states.
In order to produce actual pairs of terms, the set Eg and Ed are supposed non disjoint. $Eg \cap Ed$ is called the interface.

*Dauchet M. and Tison S., and Dauchet , Heuillart, Lescanne and Tison have proved the next results:

Proposition1: There is an algorithm which associates to each ground TRS S a GTT Ts such that S = Ts where:

$$S= \{ (t,t') \mid t \overset{*}{\underset{S}{\vdash}} t'\} \quad \text{and} \quad Ts=\{(t,t') \mid t \underset{G}{\longrightarrow} u \underset{D}{\longleftarrow} t'\}.$$

Proposition2: The confluence of ground TRS is decidable.

Proposition3: The reachability problem for ground TRS is decidable.

Proposition4:
If F is recognizable then $[F]_S= \{ t' \mid \exists\, t \in F,\ t \overset{*}{\underset{S}{\vdash}} t' \}$ is recognizable.

II- COMPILATION OF A GROUND TRS S AND DECISION ALGORITHM FOR THE REACHABILITY PROBLEM

We will construct systems S' and S" , from the ground rewriting system S, so as to reduce more and more the time of answer to the reachability problem .

To do that, we use the next tools: Automata, Recognizable forest and ground tree transducer.

1-Creation of the system S'.

All along of the different steps, we will use the same example, so as to easily follow the different transformations which are realized.

Let us write the next ground rewriting system:

```
Σ ={ b1, q , q1' , p1 , b , q', p,    a  ,   c  }
                        |   |   |    / \    /|\
```

rules =	1- b(b1) -> b1	2- a(b1, q) -> q	3- q1' -> q'(q1')
	4- q(q1') -> q1'	5- q1' -> a (q1', q1')	6- b(q'(q'1)) -> c(p1, p(p1), p1)
	7-p1 -> p(p1)	8- a(b1, a(q, b1)) -> b(q1')	

First step:

In this part, we have to construct a GTT, from the system S, its frontier-to-root automaton will accept left hand sides of rules of S, and its root-to-frontier automaton will generate right hand sides of rules of S. Its interface states will make the connexion between left hand sides and right hand sides. for example we built for the rule 8 a frontier-to-root automaton which accepts the left hand side, where the terminal state is i8, and the other states are e14, e15, e16, e17.

Consider again our last system, then we will have the next rules:

	frontier-to-root automaton G	root-to-frontier automaton D
1-	b1 -> e1 b(e1) -> i1	i1-> b1
2-	b1-> e2 q-> e3 a(e2, e3) -> i2	i2 -> q
3-	q1'-> i3	i3 -> q'(e4) e4 -> q1'
4-	q1' -> e5 q'(e5) -> i4	i4 -> q1'
5-	q1' -> i5	i5 -> a(e6, e6) e6 -> q1'
6-	q1'-> e7 q'(e7) -> e8 b(e8) -> i6	i6 -> c(e9, e10, e11) e9 -> p1 e12 -> p1 e11 -> p1 e10 -> p(e12)
7-	p1 -> i7	i7 -> p(e13) e13 -> p1
8-	q -> e14 b1 -> e15 b1 -> e17 a(e14, e15) -> e16 a(e17, e16) -> i8	i8 -> b(e18) e18 -> q1'

Interface states are: I={ i1,i2,i3,i4,i5,i6,i7,i8}

<u>Second step:</u>
Creation of the GTT, G*, which simulates the ground rewriting system S.
The principle is:

" it's not good generating, to nible"

To do that , we create some ε-transitions, with the next induction rules:

$$\frac{\begin{array}{c} e \to f(e1, \ldots, en) \\ e1 \to e1', \ldots, en \to en' \\ f(e1', \ldots, en') \to e' \end{array}}{e \to e'}$$

The algorithm is :

1- we take a rule of the root-to-frontier automaton D

2- We examine, if we can find the right hand-side of this rule in the left hand-side of one rule of the frontier-to-root automaton G.

* if it is the case, we create an ε-transition with in left hand-side, the left hand side of the rule of D which is choosen (a state of D), and in right hand side, the state in which we arrive when we apply the rule of G which was found.Then we choose the next rule of D and we start again in 2.

* if it is not the case, we choose a new rule of D and we start again in 2.

Such a transformation can be illustrated with the diagram of the figure 2.This operation is realized in a polynomial time of n where n=||G||x||D|| .

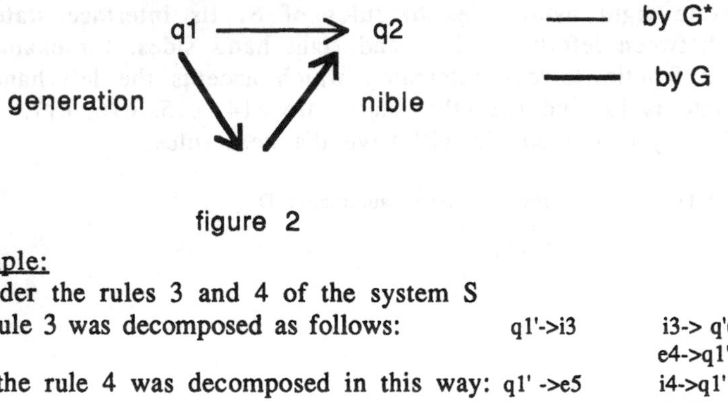

figure 2

<u>Example:</u>
Consider the rules 3 and 4 of the system S
The rule 3 was decomposed as follows: q1'->i3 i3-> q'(e4) e4->q1'

And the rule 4 was decomposed in this way: q1' ->e5 i4->q1' q'(e5) -> i4

Consider the state e4 ,we get: e4 -> q1' and q1' ->e5, q1' ->i3
so we get e4 -> e5 and e4 -> i3

Consider now the state i3 we get: i3 -> q'(e4)
and by the last step we get e4 ->e5 so i3-> q'(e5)
And we find q'(e5) -> i4 in the decomposition of the rule 4
So we deduce the next ε-transition i3 -> i4

So, instead of doing the next rewritings:

i3 -> q'(e4) -> q'(q1') -> q'(e5)-> i4

the GTT, G*, will directly pass by i3 to i4
So we have constructed in two steps, a GTT, denoted by G*, which simulates the system S, we call G* , the system S'. The answer to our problem will be given with S' in a quadratic time.

2-Creation of the system S".

In first time, we modify again the system S', so as to construct a frontier-to-root automaton. This one will accept a forest, which symbolizes all transformations that we can realize with the system S.

We can depict a tree which belongs to this forest like that:

A=

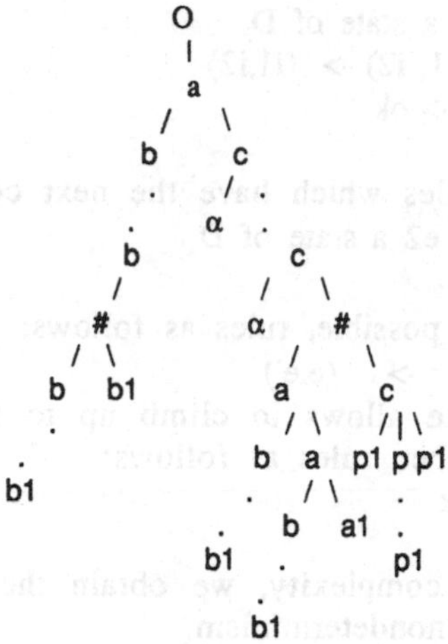

Inside this tree, we can bring to light, two trees t and t', with two morphisms φ and φ' .

by φ, we get:

t=

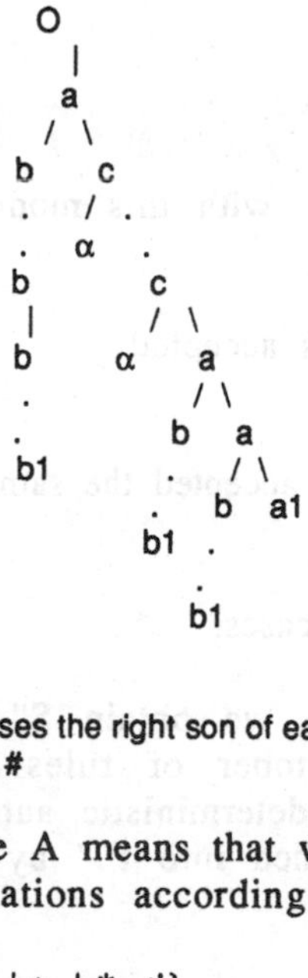

φ erases the right son of each node #

by φ', we get:

t'=

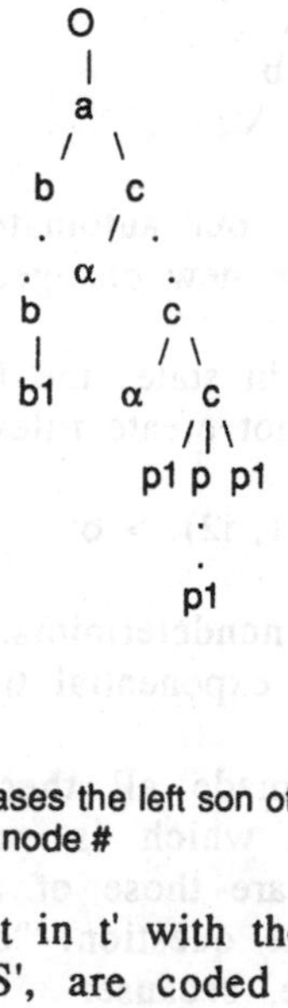

φ' erases the left son of each node #

Like that, the tree A means that we can transform t in t' with the system S'.
So, all transformations according to the system S', are coded in a recognizable forest F.

with F={ t#t' | t |-*- t'}
S

To create a frontier-to-root automaton which will accept this forest, we proceed in three steps:
1- We keep nible rules of the sytem S'
2- We reverse generation rules of the system S' so as to convert them in bottom-up rules (by reversing the arrows)
3- For interface states we add next rules:

if i1=i2 with i1 a state of G and i2 a state of D

#(i1, i2) -> ok and #(i1, i2) -> (i1,i2)

and (i1,i2) -> ok

and then, for the other pairs of states, rules which have the next configuration:

if e1$\neq$ e2 with e1 a state of G and e2 a state of D

#(e1, e2) -> (e1,e2)

for all letters 'a' of Σ, we add, when it is possible, rules as follows:

a((e1,e1'), (e2,e2'), . . . , (en,en')) -> (e,e')

Finally, when we know that the 'ok' state allows to climb up to the root of the tree, we add, for all letter 'a' of the alphabet, rules as follows:

a(ok, ok, . . . , ok) -> ok

but, now, in order to improve the time complexity, we obtain the automaton S'', by transforming F and by reducing the nondeterminism.
1- Suppression of a hidden difficulty:
We bring down, into F, # nodes, the lower as possible, so that descendant letters of the # node would be always different.

ie: The next tree t:

```
t=     #        will be replace by        b
     /   \                              /   \
    b     b                            #     #
  /  \  /  \                          / \   / \
```

in order that our automaton could accept F with this modification. we must bring it some new changes:

a- We keep in states the last letter which is accepted.
b- We will not create rules as follows:

#(i1, i2) -> ok   if i1 and i2 had accepted the same letter.

2- Reduction of nondeterminism.
We do that in an exponential time, in the worst cases.

When we have made all these different steps, we obtain S''.This one have a number of rules which is running to exp(number of rules of S) (not very readable), which are those of a frontier-to-root deterministic automaton S''.
The answer to the question: "can t be transformed into t'?" by the system S, is made in real time, because:

$$t \mathrel{\mathop{|\!\!-\!*\!-}\limits_{S}} t' \quad \Leftrightarrow \quad t\#t' \text{ is accepted by } S''$$

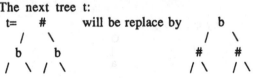

Remark:
But the reduction of the nondeterminism stands some very important problems, all at once of memory space and of time of answer.
To avoid this problem, we can consider another method. We will see this method in the next paragraph.

3-Resolution of the reachability problem by using the system S' (G*)

Let us take G and D, which are automata of the GTT G* (see p6), and let us take Mt and Mt', which are automata which accept t and t'.
We call D_{inv}, the automaton obtained from D by reversing its arrows. So D_{inv} is a frontier-to-root automaton.
To solve our problem we can study two cases:

a- When G and D_{inv} are nondeterministic

We can answer to $t \mathrel{|\!-\!*\!-}_{S} t'$? by using F

In fact $t \mathrel{|\!-\!*\!-} t' \Leftrightarrow \varphi{-}1(t) \cap \varphi'{-}1(t') \cap F$

φ and φ' are morphisms which are defined above (p7). These one are independant of t, t' and S.
So we have a complexity equal to K(S) x || Mt || x || Mt'|| by omitting the access time.
Besides, we can proceed in the same way to express the set of all transformations of t, that we will call S(t), because:

$S(t)= \varphi '(\varphi{-}1(t) \cap F)$

The creation of the automaton which accepts S(t) is made with the next algorithm:
1-We make the intersection between the automaton Mt and the frontier-to-root automaton G of the GTT G*, but this thing by keeping all rules which accept t.
2-We search inside this automaton, rules which conduct to a couple of states (q,i) where i is an interface state of the GTT and q is any state, and we add all rules of the root-to-frontier automaton D of the GTT which start from this interface state i (this by reversing the arrows so as to always have a frontier-to-root automaton). Such an algorithm is realized with a time complexity in $O((||Mt|| \times ||G|| \times |Mt_q| \times |G_q|)+||D||)$.We will call this new automaton M_{st}
to answer to $t \mathrel{|\!-\!*\!-} t'$, we make the intersection between the automata Mt' and M_{st}. So as to know if $S(t) \cap t' \neq \emptyset$
The answer is given after a time in $O(||M_{st}|| \times ||Mt'|| \times |M_{stq}| \times |Mt'_q|)$

b- When G and D_{inv} are deterministic

As G is deterministic, it can accept the tree t, likewise for $D_{inv}$and t'.So we can, by recognition of t by G (resp of t' by D_{inv}), mark all subtrees of t which could be accepted by G (resp subtrees of t' accepted by D_{inv}).Our aim, is to have two new automata which will accept all at once t(or t') and trees which have the next configuration:

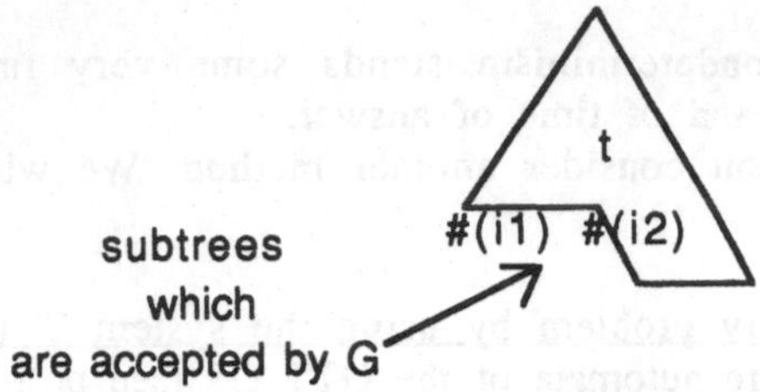

figure 3

where i1 and i2 are final states of the subtrees accepted by G (in fact, they are interface states), here, #(i1) and #(i2) replace these subtrees, they are leafs of the tree.

This operation is called, the "marking" operation.
Here,the algorithm used:

```
marking(x,l,y,e)
x: node
l: list of sons of the node x
y: state in which we arrive when we have accepted the node x (with rules of the automaton
Mt or Mt')
e: state in which we arrive when we have accepted the node x (with rules of the automaton G
or Dinv)
begin
if it exists a rule of G (or Dinv) accepting the node x with the list l then
         - We keep the state e of G (or of Dinv) in which we arrive
           after having made the recognition.
         - We search if this state is an interface state :
         if yes then we add the next rule
                       #(e) -> state(y) in front of the list of rules of
                    the automaton Mt (or Mt').
              else nothing
         endif
else We keep a fictitious state 'p' so as to continue the
         exploration of the tree
           (remark: fathers of the node x ,couldn't be accepted by
         G(or Dinv))
endif
end

study-node(x,l,y,e)
begin
if the letter x is a leaf then  marking(x,l,y,e)
else if the letter x is a node then
                        for each son fi of x  do
                               -Take the rules of Mt (or Mt') which conducts
                           to this son :
                          < xi,li>-> state(fi)
                              -study-node(xi,li,fi,ei)
                              -keep each ei in the list l'
                        end
                          marking(x,l',y,e)

    endif
endif
end
```

```
main program
begin
    Take the rule which accepts the root node of t or t'
     <x,l>->state(y)              x: root node
                                  l:list of sons
     study-node(x,l,y,e)
/ *exploration of the tree t (or t') with "marking" operation*/
end
```

The two automata obtained, after having applied this algorithm on t and t', are called Mt_m and Mt_m'.
we can remark that we make one and only one "marking" operation for each node of the considered tree (ie: for each rule of the associated automaton).
Besides, the "marking" operation of a node is made in a linear time, so we can deduce that the creation of automata Mt_m and Mt_m' is made with a time complexity in O(||Mt||+||Mt'||).
Now, we only have to compare these automata so as to find a tree common to the forest accepted by Mt_m and Mt_m',this tree will have the next configuration:

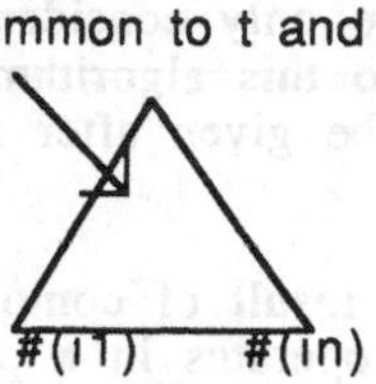

figure 4

i1, ..., in are interface states which represent all transformations made when we go from t to t'.
this operation is made in a very short time by using the next algorithm:

```
main program
 begin
     each automaton have only one final state, so we search the rules
     which conduct to these states:
   i.e.:        y-> state(ft)           ft and ft' are final states of Mtm and

              y1->state(ft')        Mtm'

        compare-node(y,y1,ft,ft') /* comparison of nodes y and y1*/
 end

compare-node(y,y1,e,e')
  begin
    if y=<x,L>   and  y1=<x1,L1>  then
          if x=x1 then consider each list
                           L=e1,e2,...,en and
                           L1= e1',e2',...,en'
                              (ei and ei' are states)
                           for each couple of states (ei,ei') do
                                 -search rules which conduct to these two
                            states
                              i.e.: yi->state(ei)
                                    yi'->state(ei')
                                -compare-node(yi,yi',ei,ei')
                           endfor
                 else fail
        endif
```

```
else if y=#(e1) and y1=<x1,L1> then
                                  - take the rule so that y=<x,L>
                                       (this rule always exists)
                                  - compare-node(<x,L>,<x1,L1>,e,e')
    else if y=<x,L> and y1=#(e1') then
                                  -take the rule so that
                                  y1=<x1,L1>
                                       (this rule always exists)
                                  -compare-node(<x,L>,<x1,L1>,e,e')
    else if y=#(e1) and y1=#(e1) then OK
         else take rules so that y=<x,L> and y1=<x1,L1>
                   compare-node(<x,L>,<x1,L1>,e,e')
         endif
       endif
     endif
   endif
end
```

We can see that in this case too, we only consider one and only one time, each rule of automata Mt_m and Mt_m'. So this algorithm is executed in a linear time, and the answer to our problem will be given after a time complexity in the order of $\|Mt_m\|+\|Mt_m'\|$.

We can compare this result with the result of complexity obtained in the paper of Oyamaguchi M.[15]. Their algorithm operates in a polynomial time of n where $n=\|M_t\|+\|M_t'\|+\|S\|$, where $\|S\|$ is the size of the given rewriting system. In our case, the complexity is began linear and the size of S is not consider. This fact can be explained because we have made a first operation of compilation on our rewriting system (this operation is made only once). So after this operation, we can ask as many questions as we want without making it again, that is why we earn much time.

Remark:

If G and D_{inv} are nondeterminist, the reduction of the nondeterminism on them is more realizable than on S" (the automaton which accepts all transformations that we can make with S), because, they have a smaller number of rules than S", so the time of execution of this operation is reduced.

III.Some extensions of ground TRS

Notation: Σ is a finite ranked alphabet

σ is a letter of arity 2 and $\sigma \notin \Sigma$

$\Delta=\Sigma \cup \{\sigma\}$

T_Σ, T_Δ are the set of terms (trees) over Σ,Δ

X is a set of variables

$T_\sigma(X)$ the set of terms over σ indexed by X

$T_\sigma(\Sigma)=\{t=t_\sigma(t_1,\ldots,t_n)/t_\sigma \in T_\sigma(X), \forall i, t_i \in T_\Sigma\}$

Let $R=\{l_i \to r_i / l_i, r_i \in T_\Sigma\}$ be a ground TRS on T_Σ and $R_\sigma=\{l_i \to r_i / l_i, r_i \in T_\sigma(\Sigma), l_i \notin T_\Sigma\}$ be a ground TRS on T_Δ, the condition $l_i \notin T_\Sigma$ is necessary because we consider terms in $T_\sigma(\Sigma)$ and recognizable forests included in $T_\sigma(\Sigma)$.

Let $R_G=R \cup R_\sigma$

Let $E_C=\{\sigma(x,y) = \sigma(y,x)\}$ and $R_C=\{\sigma(x,y) \to \sigma(y,x)\}$

Let $E_A=\{\sigma(\sigma(x,y),z) = \sigma(x,\sigma(y,z))\}$

and $R_A=\{\sigma(\sigma(x,y),z) \to \sigma(x,\sigma(y,z))\,;\,\sigma(x,\sigma(y,z)) \to \sigma(\sigma(x,y),z)\}$

and $E_{AC}=E_A \cup E_C$ and $R_{AC}=R_A \cup R_C$.

1.Commutativity

$R_{GC}=R_G\cup R_C=R\cup R_\sigma\cup R_C$ is the union of two ground TRS R and R_σ and of R_C TRS associated with commutativity of the operator σ

Example:

$R_{GC}=\{1{:}f(a)\to a\ ;2{:}\ g(a,a)\ \to b;3{:}\ \sigma(a,b)\ \to b;4{:}\ \sigma(x,y)\to\ \sigma(y,x)\ \}$

$R=\{1;2\},R_\sigma=\{3\},R_G=\{1;2;3\},R_C=\{4\}$

1.1:Recognizability of $[F]_{R_{GC}}$

Lemma 1.1:There exists a TRS S_σ verifying: $\forall t,t'\in T_\sigma(\Sigma)$

$$(t \underset{R_{GC}}{\text{|--*--}} t') \Leftrightarrow (\exists\, t_1,t_2\in T_\sigma(\Sigma),\ t \underset{R}{\text{|-*-}} t_1 \underset{S_\sigma\cup R_C}{\text{|---*---}} t_2 \underset{R}{\text{|-*-}} t')$$

Proof : - Construction of S_σ

We add to R_σ new rules to simulate rewritings by R on terms of T_Σ which appear in rules of R_σ.

$$R_\sigma=\{l_i=l_{i\sigma}.(l_{i1},\dots,l_{in_i})\to r_i=r_{i\sigma}.(r_{i1},\dots,r_{ip_i})/i\in I,n_i>1;l_{i\sigma},r_{i\sigma}\in T_\sigma(X)\ \forall i,j,k,l_{ij},r_{ik}\in T_\Sigma\ \}$$

Let $G=\bigcup_{i\in I}\{l_{i1},\dots,l_{in_i}\}$ and $D=\bigcup_{i\in I}\{r_{i1},\dots,r_{ip_i}\}$

Let $S=\{\ (r,l)\in D\times G\ /\ r \underset{R}{\text{|-*-}} l\ \}$

DxG is finite and for every r of D the set $[r]_R=\{t'/\ r \underset{R}{\text{|-*-}} t'\}$ is recognizable ([4]) so we can construct S.

Let $R'=\{\ r\to l\ /\ (r,l)\in S\}$ and $S_\sigma=R_\sigma\cup R'$

- $\Leftrightarrow$

$\Leftarrow$ is obvious and $\Rightarrow$ is proved by induction on the number n of utilisations of rules of R_σ .It is based on the two results:

-Each rule of R can commute with each rule of R_C.

-Each rule of R' simulate the rewriting of a term of D in a term of G by R

So we obtain the decomposition of lemma 1.1, moreover we use a rule of R' only to transform a term of D in a term of G.

Lemma 1.2:There exists a TRS V_σ verifying: $\forall\ t,t'\in T_\sigma(\Sigma)$,

$$(t \underset{R_{GC}}{\text{|--*--}} t')\Leftrightarrow(\exists t_1,t_2,t_3,t_4\in T_\sigma(\Sigma)\ ,\ (\ t \underset{R}{\text{|-*-}} t_1 \underset{R_C}{\text{|-*-}} t_2 \underset{V_\sigma}{\text{|-*-}} t_3 \underset{R_C}{\text{|-*-}} t_4 \underset{R}{\text{|-*-}} t'\ )$$

Proof : To obtain V_σ,we just add to S all the rules obtained by using commutativity of the operator σ on left-hand-side of rules of R_σ.

Proposition.1.3: F is a forest included in $T_\sigma(\Sigma)$

(F recognizable) $\Rightarrow$ ($[F]_{R_{GC}}$ recognizable)

Proof : $[F]_{R_{GC}}=\{t'\ /\ \exists\ t\in F\ ,\ t \underset{R_{GC}}{\text{|-*-}} t'\}$

For every recognizable forest F and every ground TRS S, the forest $[F]_S$ is recognizable([4]).For every recognizable forest F, the forest $[F]_{R_C}$ is recognizable (obvious with the bottom-up automaton recognizing F).So with the decomposition of lemma 1.2 we have $[F]_{R_{GC}}$ recognizable (R and V_σ are ground TRS).

1.2 :Reachability problem for R_{GC} in $T_\sigma(\Sigma)$

Proposition.1.4: For every t and t' in $T_\sigma(\Sigma)$, we can decide whether t can reduce to t' by applying rules of R_{GC} i.e the reachability problem is decidable for R_{GC} in $T_\sigma(\Sigma)$.

Proof : For every t in $T_\sigma(\Sigma)$,we have $[t]_{R_{GC}}$ recognizable (consequence of proposition 1.3) so we can decide if t' is in $[t]_{R_{GC}}$.

2.Associativity.

$R_{GA}=R_G \cup R_A=R \cup R_\sigma \cup R_A$ is the union of two ground TRS R and R_σ and of R_A TRS associated with associativity of the operator σ.

Example: R_{GA}={1: f(a, a) →a;2: σ(σ(a, b), a) → σ(b, b) ;3: σ(σ(x, y), z) → σ(x, σ(y, z)) ;4: σ(x, σ(y, z)) → σ(σ(x, y), z) }
R={1} ; R_σ={2} ; R_G={1;2} ; R_A={3;4}

2.1:Recognizability of $[F]_{R_{GA}}$

Example: Let $R=R_\sigma=\emptyset$ and so $R_{GA}=R_A$ and F be the recognizable forest generated by the regular grammar { A → σ(a, σ(A, b)) ; A → σ(a, b) }
Then $[F]_{R_{GA}}=[F]_{R_A}=\{ t\in T_{\{\sigma;a;b\}} / \phi(t)=a^n b^n, n>0 \}$ where φ(t) denotes the frontier of the term t and $[F]_{R_{GA}}$ is not recognizable so in general F recognizable does not imply $[F]_{R_{GA}}$ recognizable

2.2:Reachability problem for R_{GA} in $T_\sigma(\Sigma)$

Proposition 2.1: The reachability problem for R_{GA} in $T_\sigma(\Sigma)$ is undecidable
Proof: Let Γ be a finite alphabet and R_w be a word rewriting system on Γ* ,let $\Delta=\Gamma\cup\{\sigma\}$ be a finite ranked alphabet (all letters of Δ are of arity 0 except σ which arity is 2).
Let $f : \Gamma^* \to T_\Delta$
$m \mapsto f(m)$ defined by (if |m|=1 then f(m)=m)
and(if |m|>1 and $m=a_1a_2...a_n$ then $f(m)=\sigma(a_1, \sigma(a_2, \sigma(a_3, ... \sigma(a_{n-1}, a_n))))$)

So we can associate to R_w={ l → r / l,r ∈ Γ* } a TRS denoted R_G defined by
R_G={ f(l) → f(r)/ l→r ∈ R_w }and thus we can prove $(t \overset{*}{\underset{R_{GA}}{\vdash}} t') \Leftrightarrow (\Phi(t) \overset{*}{\underset{R_w}{\vdash}} \Phi(t'))$
The reachability problem for R_w in Γ* is known undecidable so the reachability problem for R_{GA} in $T_\sigma(\Sigma)$ is undecidable.

3.Associativity and commutativity.

$R_{GAC}=R_G\cup R_{AC}=(R\cup R_\sigma)\cup(R_A\cup R_C)$ is the union of the ground TRS R_G and of R_{AC} TRS associated with commutativity and associativity of the operator σ .R_G is itself the union of the ground TRS R on T_Σ and of the TRS R_σ on T_Δ(with $\Delta=\Sigma\cup\{\sigma\}$ and conditions on the configuration of rules of R_σ,see III Notations).

3.1: Recognizability of $[F]_{R_{GAC}}$

Example: With $R=R_\sigma=\emptyset$ and so $R_{GAC}=R_{AC}$
with the forest F of the example of the section III.2.1 we have $[F]_{R_{GAC}}=[F]_{R_{AC}}=\{ t\in T_{\{\sigma,a,b\}} / |\Phi(t)|_a=|\Phi(t)|_b \}$ (where Φ(t) is the frontier of the term t and $|\Phi(t)|_a$ the number of occurences of a in the word Φ(t)) which is not recognizable .So generally F recognizable does not imply $[F]_{R_{GAC}}$ recognizable .

3.2: Reachability problem for R_{GAC} in $T_\sigma(\Sigma)$

Lemma 3.1: There exists a TRS S_σ such that : ∀t,t' ∈ $T_\sigma(\Sigma)$
$(t \overset{*}{\underset{R_{GAC}}{\vdash}} t') \Leftrightarrow (\exists t_1,t_2\in T_\sigma(\Sigma) , t \overset{*}{\underset{R}{\vdash}} t_1 \overset{*}{\underset{S_\sigma\cup R_{AC}}{\vdash}} t_2 \overset{*}{\underset{R}{\vdash}} t')$
Proof: Similar to lemma 1.1.
With the notations of section III.1.1,let $M=G\cup D=\{u_1,...,u_m\}$ be the set of all subterms of T_Σ which appear as subterms of rules of S_σ.
Example: S_σ={ σ(f(a, a), c) → σ(f(a, a), d) ;2: σ(a, f(a, a)) → c ;3: σ(c, d) → c }
then M={ a , f(a, a) , c , d }

Let $X=\{x_1,...,x_m\}$ be an alphabet one to one with M
on X^* we define the relation ($m \equiv m'$)$\Leftrightarrow$($\forall x \in X$, $|m|_x = |m'|_x$)
Let $f: \quad T_\sigma(\Sigma) \quad \rightarrow \quad X^*/_\equiv$

$t=t_\sigma.(t_1,...,t_n) \mapsto f(t)=x_1^{y_1}...x_m^{y_m}$ where yi is the number of occurences of the term u_i (which belongs to M) in $\{t_1,...,t_n\}$

Thus to each tree t of $T_\sigma(\Sigma)$, we can associate f(t) in $X^*/_\equiv$ and g(t) the list (or multiset) of terms of $\{t_1,...,t_n\}$ which are not in M (g(t) is the list of subterms of t which cannot be transformed by S_σ).

Example: $S_\sigma=\{1,2,3\}$;$M=\{a, f(a, a), c, d\}$;$X=\{x,y,z,t\}$
with $t= \sigma(\sigma(\sigma(f(a, a), c), b), \sigma(c, b))$
We get $f(t)=yz^2$ and $g(t)=(b,b)$

Moreover to each rule $l_i \rightarrow r_i$ of S_σ, we can associate the rule $f(l_i) \rightarrow f(r_i)$ on $X^*/_\equiv$ and thus to S_σ is associated a TRS S_X on $X^*/_\equiv$.

Example: With S_σ,M,X defined in the previous example
we get $S_X=\{ 1_X : yz \rightarrow yt ; 2_X : xy \rightarrow z ; 3_X : zt \rightarrow z \}$

Lemma 3.2: $\forall\ t_1,t_2 \in T_\sigma(\Sigma)$
$(t_1 \ \overset{*}{\underset{S_\sigma \cup R_{AC}}{\vdash\!\!-\!\!-\!\!-}}\ t_2) \Leftrightarrow (f(t_1) \overset{*}{\underset{S_X}{\vdash\!\!-}} f(t_2)$ and $g(t_1)$ and $g(t_2)$ contain exactly the same terms)

Proof: -S_X is a TRS on $X^*/_\equiv$ and by definition of $X^*/_\equiv$ the rewritings are made modulo commutativity and associativity so each rule of S_X simulates commutativity,associativity and one rule of S_σ

-the trees of $g(t_1)$ cannot be rewritten by S_σ so we must have the second condition.

Example: With S_σ,M,X,t of the previous example we have
$t \underset{\{1\}}{\vdash\!\!-\!\!-\!\!-} \sigma(\sigma(\sigma(f(a, a), d), b), \sigma(c, b)) \overset{*}{\underset{R_{AC}\cup\{3\}}{\vdash\!\!-\!\!-\!\!-}} t'= \sigma(\sigma(f(a, a), c), \sigma(b, b))$
and $f(t)=yz^2$, $f(t')=yz$, $yz^2 \underset{1_X}{\vdash\!\!-\!\!-\!\!-} yzt \underset{3_X}{\vdash\!\!-\!\!-\!\!-} yz$
$g(t)=g(t')=(b,b)$.

Lemma 3.3: The reachability problem for S_X in $X^*/_\equiv$ is decidable.

Proof: To the TRS S_X on $X^*/_\equiv$,we can associate the Petri net PS_X defined as follow:
-Set of places $P=\{p_1,...,p_m\}$,p_i is associated with x_i of X
-Set of transitions $T=\{t_1,...,t_n\}$,t_i is associated with the rule $l_i \rightarrow r_i$.
-Pré and Post are defined by :
if $x_1^{l_{i1}}...x_m^{l_{im}} \rightarrow x_1^{r_{i1}}...x_m^{r_{im}}$ is the rule $l_i \rightarrow r_i$ of S_X then for the transition t_i we have $Pré(p_j,t_i)=l_{ij}$ and $Post(p_j,t_i)=r_{ij}$

Moreover to each $m=x_1^{y_1}...x_m^{y_m}$ of $X^*/_\equiv$ we associate the vector v(m) of N^m such that $v(m)(i)=y_i$.

We can dually associate to a Petri net a TRS on $P^*/_\equiv$ so the reachability problem for S_X in $X^*/_\equiv$ is equivalent to the reachability problem for Petri net indeed decide if m can reduce to m' by applying rules of S_X is decide if the vector v(m') is reachable for the Petri net SP_X with the initial marking v(m).The reachability problem in Petri nets is decidable (Kosaraju[11],Mayr[13]) and so the reachability problem for S_X in $X^*/_\equiv$ is decidable.

Example: With S_σ,M,$X=\{x,y,z,t\}$,$S_X=\{1_X:yz \rightarrow yt;2_X:xy \rightarrow z;3_X:zt \rightarrow z\}$ and t of the previous example we have $P=\{p,q,r,s\}$, $T=\{t ,t',t''\}$ and

$$Pre=\begin{pmatrix}010\\110\\101\\001\end{pmatrix};Post=\begin{pmatrix}000\\100\\010\\101\end{pmatrix};\text{Initial marking } v(m)=\begin{pmatrix}0\\1\\2\\0\end{pmatrix}$$

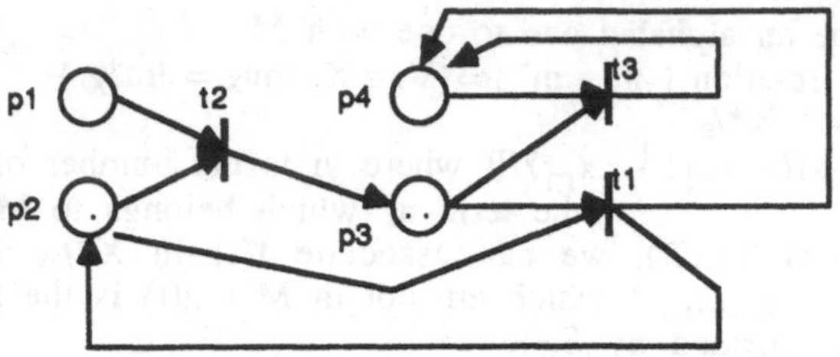

Petri net SPX with the initial marking v(m)

Proposition 3.4: The reachability problem for R_{GAC} in $T_\sigma(\Sigma)$.is decidable
Proof:-If t and t' belong to T_Σ then R_{GAC}=R and the reachability problem for the ground TRS R is decidable

-If t belongs to T_Σ and t' does not belong to T_Σ,t cannot be rewrited in t' because of the condition $l_i \notin T_\Sigma$ for rules of R_σ(we forbid the generation of σ from terms of T_Σ).

-If $t=t_\sigma.(t_1,...,t_n) \in T_\sigma(\Sigma)$ and $t'=t'_\sigma.(t'_1,...,t'_p) \in T_\sigma(\Sigma)$ .We use the decomposition of lemma 3.1 so we first rewrite the terms t_i by the ground TRS R on T_Σ and so each term t_i can produce terms of $M=G\cup D$ or not so we consider $F(t)=\{t_\sigma.(u_1,...,u_n)/$ if $[t_i]_R \cap M=\emptyset$ then $u_i=t_i$;if $[t_i]_R \cap M=\{m_1,...,m_j\}$ then $u_i=t_i$ or $u_i=m_1$ or ... or $u_i=m_j$ }.M is a finite set of terms and for each t_i,the forest $[t_i]_R$ is recognizable so we can build the finite set F(t) for every t of $T_\sigma(\Sigma)$.Dually, using decomposition of lemma 3.1,we consider
$F^{-1}(t')=\{t'_\sigma.(u'_1,...,u'_p)$ / if $[t'_i]^{-1}_R \cap M = \emptyset$ then $u'_i=t'_i$;if $[t'_i]^{-1}_R \cap M=\{m'_1,...,m'_k\}$ then $u'_i=t'_i$ or $u'_i=m'_1$ or...or $u'_i=m'_k\}$ with $[t']^{-1}_R=\{t/$ t|-*- t' for the TRS R} which is recognizable so we can build the finite set $F^{-1}(t')$ for every t' of $T_\sigma(\Sigma)$.We are now ready to show

Lemma3.5:($t \mathrel{\underset{R_{GAC}}{|\text{--*--}}} t'$)$\Leftrightarrow$(($\exists T \in F(t)$, $\exists T' \in F^{-1}(t')$, $f(T) \mathrel{\underset{S_X}{|\text{-*--}}} f(T')$) and
(there exists a one to one correspondance h between g(T) and g(T') such that we have $g(T) \ni u \mathrel{\underset{R}{|\text{-*-}}} h(u) \in g(T')$))

Proof: We use in this proof the results of lemma 3.1, lemma 3.2 and the construction of the finite sets F(t) and $F^{-1}(t')$.$\Leftarrow$ is without difficulty using these results.For $\Rightarrow$ we have to examine the rewriting of t in t' by R_{GAC} using the decomposition of lemma 3.1 and build T of F(t) and T' of $F^{-1}(t')$ verifying the two properties.

$$t= t_\sigma.(t_1,...,t_n) \mathrel{\underset{R}{|\text{-*-}}} t_\sigma.(v_1,...,v_n)$$

$$t_\sigma.(v_1,...,v_n) \mathrel{\underset{S_\sigma \cup R_{AC}}{|\text{----*----}}} t'_\sigma.(v'_1,...,v'_p)$$

$$t'_\sigma.(v'_1,...,v'_p) \mathrel{\underset{R}{|\text{-*-}}} t'_\sigma.(t'_1,...,t'_p)=t'$$

This construction is not difficult,we just have to look at every possible cases for the rewritings by R: t_i|-*-$v_i \in M$; t_i |-*-$v_i \notin M$ and then $t_i \in M$ or $t_i \notin M$;and dually $M \ni v'_i$|-*-t'_i ; $v'_i \notin M$,v'_i|-*-t'_i and then $t'_i \in M$ or $t'_i \notin M$.

Moreover, F(t) and $F^{-1}(t')$ are finite sets, for every (t,t') of $F(t) x F^{-1}(t')$, we can decide if the properties of lemma 3.5 are satisfied or not (lemma 3.4 and decidability of the reachability problem for the ground TRS R) and so the reachability problem for R_{GAC} in $T_\sigma(\Sigma)$ is decidable.

CONCLUSION

These works could permit to obtain some algebraical methods to realise the compilation of TRS, so as to have an execution of these sorts of systems in a real time.
Besides, these researches show the difficulty to have some good classes and make us researching some partial algorithms of decision of the reachability problem based on our methods for these classes.

BIBLIOGRAPHY

[1] BRAINERS : Tree-generating regular systems, Info and control (1969)

[2].G.W.BRAMS : Reseaux de Petri:theorie et pratique,tomes 1&2, Masson,Paris (1983)

[3] CHEW : An improved algorithm for computing with equations,21st FOCS (1980)

[4] DAUCHET, HEUILLARD, LESCANNE, TISON : The confluence of ground term tewriting systems is decidable,2nd LICS (1987)

[5] DAUCHET & TISON : Tree automata and decidability in ground term rewriting systems FCT' 85 (LNCS n° 199)

[6] N.DERSHOWITZ,J.HSIANG,N.JOSEPHSON and D.PLAISTED : Associative-commutative rewriting,Proc.10th IJCAI, LNCS 202 (1983)

[7] GECSEG F. & STEINBY M : tree automata,Akadémiai Kiado, Budapest (1984)

[8] HUET.G & OPPEN.D.: Equations and rewrite rules:a survey,in formal languages :perspective and open problems,Ed.Book R.,Academic Press(1980)

[9] J.P.JOUANNAUD : Church-Rosser computations with equational term rewriting systems,Proc.4th Conf on Automata, Algebra and programming,LNCS 159 (1983)

[10] C.KIRCHNER : Methodes et outils de conception systematique d'algorithmes d'unification dans les théories équationnelles,These d'etat de l'universite de Nancy I (1985)

[11] S.R.KOSARAJU : Decidability of reachability in vector addition systems, Proc.14th Ann.Symp.on Theory of Computing, 267-281.(1982)

[12] KOZEN : Complexity of finitely presented algebra, 9th ACM th. comp. (1977)

[13] E.W.MAYR : An algorithm for the general Petri net reachability problem, Siam J Comput.13 441.- 460

[14] NELSON & OPPEN : Fast decision algorithms based on congruence closure, JACM 27 (1980)

[15] OYAMAGUCHI M.: The reachability problem for quasi-ground Term Rewriting Systems, Journal of Information Processing , vol 9 , n°4 (1986)

[16] PLAISTED D. & BACHMAIR L. : Associative path ordering, Proc. 1st conference on Rewriting Techniques and Applications, LNCS 202 (1985)

[17] RAOULT J.C. : Finiteness results on rewriting systems, RAIRO, IT, vol 15 (1985)

Order-Sorted Completion: The Many-Sorted Way (Extended Abstract)

Harald Ganzinger*
Fachbereich Informatik, Universität Dortmund
D-4600 Dortmund 50, W. Germany
e-mail: hg@informatik.uni-dortmund.de

Order-sorted specifications can be transformed into equivalent many-sorted ones by using injections to implement subsort relations. In this paper we improve a result of Goguen/Jouannaud/Meseguer about the relation between order-sorted and many-sorted rewriting. We then apply recent techniques in completion of many-sorted conditional equations to systems obtained from translating order-sorted conditional equations. Emphasis will be on ways to overcome some of the problems with non-sort-decreasing rules.

1 Introduction

1.1 Operational Semantics for Order-Sorted Specifications

Many-sorted equational logic is a major candidate to serve as a basis for algebraic specifications and logic programming. Order-sorted equational logic originated with [Gog78] and was further elaborated — in different variations — in [Gogo86], [GM87], [SNGM87], [KKM88], and [Poi88], among others. In [GJM85], an operational semantics for order-sorted specifications based on a translation scheme into many-sorted specifications is introduced. The main idea of this translation is to add auxiliary injection operators $i_{s \subset s'}$ to implement subsort containments $s \subset s'$. This provides the ability to use standard many-sorted concepts of rewriting, completion, and theorem proving in implementations of specification languages based on order-sorted logic.

In [GKK87], a completion procedure specifically tailored to *unconditional* order-sorted equations and based on order-sorted rewriting is proposed as an alternative. The main motivation for this approach is that order-sorted rewriting can be more efficient than naive many-sorted rewriting with the translated rules [KKM88].

The purpose of this paper is not to enter a discussion about efficiency of rewrite relations — we believe that both approaches are of interest in their own right — but to improve both previous approaches. There are two directions of improvement which this paper wants to contribute to.

The first problem left open by the approaches of [GJM85] and [GKK87] is the handling of non-sort-decreasing rules. This problem had been overlooked in [GJM85] as most of the results in this paper which relate order-sorted to many-sorted rewriting are only valid for sort-decreasing rules. Hence, standard Knuth-Bendix completion when applied to the many-sorted translation of order-sorted equations fails whenever a non-sort-decreasing rule is encountered. In this case, the operationally awkward injectivity axiom for the injections has to be taken into consideration — a problem which standard Knuth-Bendix completion is not prepared to handle. For a related reason, the order-sorted completion procedure of [GKK87] fails when a non-sort-decreasing rule is generated. Order-sorted replacement of equals by equals is not a complete proof method for order-sorted deduction in this case [SNGM87].

The second problem is to complete order-sorted specifications with *conditional* equations. Fortunately, the state-of-the-art in completion of many-sorted conditional equations which has originated with [Kap84] has been substantially advanced during the last year, mainly by the work of Rusinowitch and Kounalis [KR88], [Rus87], and by this author [Gan87], [Gan88], [BG88]. (Other related relevant work is described in [ZR84], [ZR85] and [KaR87].) In particular our own approach, based on the method of proof orderings

*This work is partially supported by the ESPRIT-project PROSPECTRA, ref#390.

[Bac87], [BDH86], appears to be well-engineered towards practical use due to its power in simplifying and eliminating equations during completion. This advance in technology is the main reason of why this paper again picks up the translation idea of [GJM85] to develop order-sorted completion the many-sorted way.

As one can immediately see, both problem areas are closely related. To handle non-sort-decreasing rules requires to consider the effect on the equational theory of conditional equations such as the injectivity axiom.

1.2 Summary of Main Results

With regard to the relation between order-sorted rewriting and many-sorted rewriting we improve results by Goguen, Jouannaud and Meseguer. In particular we show that one step of order-sorted rewriting with a set of rules R is one step of many-sorted rewriting with the lowest parses $\lambda(R)$ of R modulo the axioms LP of the lowest parse. This result also holds in the case of non-sort-decreasing rules. It could be made the basis of employing the concepts of completion modulo equations for order-sorted specifications — an idea which will, however, not be investigated any further in this paper. A second result is that to any sort-decreasing canonical system of order-sorted *conditional* rewrite rules there exists an equivalent, canonical many-sorted system (consisting of LP and the lowest parses $\lambda(R_S)$ of the sort specializations R_S of the given order-sorted rules R). This result is not completely obvious as the critical pairs lemma is not true for conditional rewrite rules in general [DOS88]. (A related result in [GJM85] which, by the way, is also only true for non-sort-decreasing rules, employs the composed relation $\rightarrow_{\lambda(R_S)} \circ \rightarrow_{LP}{}^{nf}$ in the many-sorted world.) Hence, many-sorted unfailing completion [HR87], [Bac87] will generate the reduced version of this system. The proofs are contained in the full version [Gan88b] of this paper and omitted from this extended abstract.

We show that in many practical cases non-sort-decreasing rules can be replaced by sort-decreasing ones without changing the initial algebra. These replacements are rules with extra variables in the condition and in the right side. Fortunately they belong to the class of what we call *quasi-reductive rules*. Quasi-reductive rules are a generalization of reductive conditional rewrite rules and the associated rewrite process is similarly efficient.

We outline an unfailing completion procedure for conditional equations that can handle both nonreductive equations such as injectivity axioms and quasi-reductive equations as they are introduced during replacing non-sort-decreasing rules. It is an extension of the one which has been presented and proved correct in detail in [Gan87] and [Gan88]. The correctness of the extensions (unfailing completion, quasi-reductive equations) is shown in [BG88] and [BG89]. We demonstrate by means of examples that these techniques perform successfully on practical examples of order-sorted specifications.

2 Basic Notions and Notations

We will only introduce the syntactic aspects of order-sorted logic and refer to [GM87] and [SNGM87] for the two main variants of semantics for order-sorted specifications. In this paper, notions and notations mainly follow [SNGM87] and [GKK87].

Every *variable* x comes with a sort s^x which is a *sort symbol.* For every sort symbol there exist infinitely many variables having this sort.

A *subsort declaration* is an expresion of form $s < s'$, where s and s' are sort symbols. A *function or operator declaration* has the form $f : s_1 \ldots s_n \rightarrow s_0$, where n is the arity of f and s_i are sort symbols. An *order-sorted* signature, usually denoted Σ^{os}, is a set of sort symbols, subsort and function declarations. A *many-sorted signature*, usually denoted Σ, is a particular case of an order-sorted signature with an empty set of subsort declarations. In a many-sorted signature we do not allow more than one declaration for any function symbol. By S and Ω we denote the set of sorts and operators, respectively, of a many-sorted signature Σ.

The *subsort order* $s \leq_{\Sigma^{os}} s'$ is the least quasi-order on the sort symbols of Σ^{os} generated by the subsort declarations. Throughout this paper we restrict attention to signatures for which $<_{\Sigma^{os}}$ is a partial order. If the signature is clear from the context, we will omit the subscript Σ^{os} in $<_{\Sigma^{os}}$. $\leq_{\Sigma^{os}}$ extends to tuples of sort symbols of the same length by $(\underline{s}_1, \ldots, \underline{s}_n) \leq_{\Sigma^{os}} (s_1, \ldots, s_n)$, iff $\underline{s}_i \leq_{\Sigma^{os}} s_i$, for $1 \leq i \leq n$. We write $s <_b s'$, if $s < s'$ and if there does not exist any s'' such that $s < s'' < s'$.

Given a set of variables X, a Σ^{os}*-term of sort* s *in* $T_{\Sigma^{os}}(X)_s$ is either a variable x such that $s^x \leq s$, or has the form $f(t_1, \ldots, t_n)$, where $f : s_1 \ldots s_n \rightarrow s_0$ is a function declaration in Σ^{os} such that $s_0 \leq s$ and

$t_i \in T_{\Sigma^{os}}(X)_{s_i}$ is a term of sort s_i, for $1 \leq i \leq n$.

Many-sorted terms are defined like order-sorted ones. The main difference is that the sort of a many-sorted term t is uniquely determined and will be denoted s^t in this paper. Given a many-sorted signature Σ, by $T_\Sigma(X)_s$ we denote the set of many-sorted terms of sort s over the variables in X. For both order-sorted and many-sorted terms t we use the notation $t : s$ to indicate that t is a term of sort s.

A Σ^{os}*-equation* is a pair of Σ^{os}-terms u and v of the same sort written as $u \doteq v$. A *conditional equation over* Σ^{os} is a formula of form $C \Rightarrow u \doteq v$, where $u \doteq v$ is a Σ^{os}-equation and C is a finite conjunction of Σ^{os}-equations $u_1 \doteq v_1 \wedge \ldots \wedge u_n \doteq v_n$, $n \geq 0$. If E is a set of (conditional) Σ^{os}-equations, by $E^=$ we denote the set

$$E^= = \{C \Rightarrow u \doteq v \mid C \Rightarrow u \doteq v \in E \text{ or } C \Rightarrow v \doteq u \in E\}.$$

An *order-sorted (equational) specification* consists of an order-sorted signature Σ^{os} and a set E of conditional Σ^{os}-equations.

If O is a syntactic Σ^{os}-object, i.e. a term or (conditional) equation, by $var(O)$ we denote the set of variables occurring in O.

A Σ^{os}*-substitution* is a function σ from Σ^{os}-terms to Σ^{os}-terms such that

1. if u is a term of sort s, then $\sigma(u)$ is a term of sort s,
2. $\sigma(f(t_1, \ldots, t_n)) = f(\sigma(t_1), \ldots, \sigma(t_n))$,
3. $dom(\sigma) := \{x \mid \sigma(x) \neq x\}$, the *domain* of σ is finite.

We will also use the notation $u\sigma$ for $\sigma(u)$.

An order-sorted signature Σ^{os} is called *pre-regular*, if for any Σ^{os}-function symbol f and every string $s_1 \ldots s_n$ of Σ^{os}-sorts the set $\{\bar{s} \mid (f : \bar{s}_1 \ldots \bar{s}_n \rightarrow \bar{s}) \in \Sigma^{os},\ s_1 \ldots s_n \leq_{\Sigma^{os}} \bar{s}_1 \ldots \bar{s}_n\}$ is either empty or has a minimum element wrt. the subsort order of Σ^{os}. Σ^{os} is called *regular*, if for any string $s_1 \ldots s_n$ of Σ^{os}-sorts such that there exists a function declaration $(f : \bar{s}_1 \ldots \bar{s}_n \rightarrow \bar{s}) \in \Sigma^{os}$ with $s_1 \ldots s_n \leq_{\Sigma^{os}} \bar{s}_1 \ldots \bar{s}_n$, then there exists a least $(\underline{s}_1, \ldots, \underline{s}_n, \underline{s})$ such that $(f : \underline{s}_1 \ldots \underline{s}_n \rightarrow \underline{s}) \in \Sigma^{os}$ and $s_1 \ldots s_n \leq_{\Sigma^{os}} \underline{s}_1 \ldots \underline{s}_n$. Regularity of a signature implies pre-regularity.

Pre-regularity is required for the existence of initial algebras in the semantics of [SNGM87]. For the semantics of [GM87], regularity is a sufficient condition for the existence of initial algebras. Although the notion of order-sorted deduction which is used in this paper corresponds to the semantics of [GM87], pre-regulary will already be a sufficient condition for the syntactic properites on which our approach of order-sorted completion is based.

An order-sorted signature is called *coherent*, if each equivalence class of sorts under the equivalence closure of $\leq_{\Sigma^{os}}$ has a maximal element.

From now on we will assume order-sorted signatures to be pre-regular and coherent.

3 Translation of Order-Sorted Specifications

3.1 Many-Sorted Representations of Order-Sorted Terms

Definition 3.1 *Let* Σ^{os} *be an order-sorted signature. Its* **translation** *into a corresponding many-sorted signature* Σ *is defined as follows:*

1. *The sorts in* S *are the sorts in* S^{os}.
2. *If* $f : s_1 \ldots s_n \rightarrow s_0$ *is an operator declaration in* Σ^{os}, *then* $f_{s_1 \ldots s_n \rightarrow s_0} : s_1 \ldots s_n \rightarrow s_0 \in \Omega$.
3. *If* $s <_b s'$ *in* Σ^{os}, *then* $i_{s \subset s'} \in \Omega$.

The translation disambiguates overloaded function symbols and introduces injections $i_{s \subset s'}$ to represent the containment of s in s'. Injections along chains of subsort relations can be represented by terms in elementary injections. As there may be more than one way of going from an s to an s' we have to order these paths if we want a unique representation. For that purpose we assume the partial subsort order to be extended to an arbitrary but fixed total order $<_t$ on S. The same notation is used for the *lexicographic* extension of $<_t$ to sequences of sorts. (This is different to the extension of $\leq_{\Sigma^{os}}$ to tuples which we have used to define the regularity properties. The latter was defined component-wise.)

Now we define composite injections to proceed along minimal paths in the subsort graph.

Definition 3.2 *Let $s \leq s'$. Let furthermore $s_n s_{n-1} \dots s_0$, $n \geq 0$, be a sequence of sorts minimal wrt. $<_t$ such that $s_0 = s$, $s_n = s'$, and $s_i <_b s_{i+1}$, for $0 \leq i < n$. Then the* **minimal composite injection**

$$i_{s_{n-1} \subset s_n} \circ i_{s_{n-2} \subset s_{n-1}} \circ \dots \circ i_{s_0 \subset s_1}$$

from s to s' will be denoted as $I_{s \subset s'}$. (If $s = s'$, $I_{s \subset s'}$ is the identity for which there is no explicit operator symbol in Σ. Hence, terms $I_{s \subset s'}(t)$ and t are not distinguished in this case.)

We can now go on and define mappings between order-sorted terms and their many-sorted representations as terms over Σ. Any many-sorted term t in Σ represents one unique order-sorted term $\omega(t)$ which is obtained by deleting injections and by collapsing the disambiguated operator symbols into the original overloaded symbol.

Definition 3.3 *The mapping $\omega(\_) : T_\Sigma(X) \rightarrow T_{\Sigma^{os}}(X)$ is inductively defined as follows:*

1. *If $x : s$ is a variable, then $\omega(x : s) = x : s$.*
2. *$\omega(f_{s_1 \dots s_n \rightarrow s_0}(t_1, \dots, t_n)) = f(\omega(t_1), \dots, \omega(t_n))$, for non-injections $f_{s_1 \dots s_n \rightarrow s_0} \in \Omega$.*
3. *$\omega(i_{s \subset s'}(t)) = \omega(t)$, for injections $i_{s \subset s'} \in \Omega$.*

In the reverse direction, $\lambda(\_)$ will compute the lowest parse of an order-sorted term.

Definition 3.4 *The* **lowest parse** *$\lambda(\_) : T_{\Sigma^{os}}(X) \rightarrow T_\Sigma(X)$ is inductively defined as follows:*

1. *$\lambda(x : s) = x : s$, for variables $x : s \in X$.*
2. *$\lambda(f(t_1, \dots, t_n)) = f_{s_1 \dots s_n \rightarrow s_0}(t'_1, \dots, t'_n)$, where $f : s_1 \dots s_n \rightarrow s_0$ is the operator declaration in Σ^{os} for which $s_0 s_1 \dots s_n$ is minimal wrt. $<_t$ such that $\lambda(t_i) \in T_\Sigma(X)_{s'_i}$ and $s'_i \leq s_i$, $1 \leq i \leq n$, and where $t'_i = I_{s'_i \subset s_i}(\lambda(t_i))$, $1 \leq i \leq n$.*

As we are putting the codomain s_0 of f at the beginning of the sort sequence $s_0 s_1 \dots s_n$ when looking for a minimal declaration for f, $\lambda(t)$ will always have a lowest possible sort.

Proposition 3.5 *1. $\omega(\lambda(t)) = t$*

2. $s^{\lambda(\omega(t))} \leq s^t$

3.2 Computation of Minimal Parses by Rewriting

In the preceeding section we have defined a function $\lambda(\_)$ which produces a many-sorted representation of an order-sorted term. The representation always is a term of a lowest possible sort and, hence, unique for pre-regular signatures. On the other hand, there are usually many different many-sorted terms that represent the same order-sorted term via $\omega(\_)$. In this section we describe a canonical set of rewrite rules over Σ which, for any given many-sorted term, computes the lowest parse of the order-sorted term $\omega(t)$ it represents.

The set of rules consists of rules for computing the minimal path (wrt. $<_t$) between any two sorts $s < s'$ and of rules which represent the inheritance axioms for overloaded function symbols on subsort hierarchies.

Axioms (CI) for composite injections

$$i_{s' \subset s''}(I_{s \subset s'}(x)) = I_{s \subset s''}(x),$$

for $s < s' < s''$, if $I_{s \subset s''} \neq i_{s' \subset s''} \circ I_{s \subset s'}$.

Axioms (INH) for inheritance

$$f_{s_1 \dots s_n \rightarrow s_0}(I_{\bar{s}_1 \subset s_1}(x_1), \dots, I_{\bar{s}_n \subset s_n}(x_n)) = I_{s'_0 \subset s_0}(f_{s'_1 \dots s'_n \rightarrow s'_0}(I_{\bar{s}_1 \subset s'_1}(x_1), \dots, I_{\bar{s}_n \subset s'_n}(x_n))),$$

if $f : s_1 \dots s_n \rightarrow s_0$ and $f : s'_1 \dots s'_n \rightarrow s'_0$ are operator declarations, $s'_0 s'_1 \dots s'_n <_t s_0 s_1 \dots s_n$, $s'_0 \leq s_0$, and if $\bar{s}_i$ are maximal[1] sorts such that $\bar{s}_i \leq s_i$ and $\bar{s}_i \leq s'_i$.

[1]To select maximal sorts $\bar{s}_i$ is not really required. However, (INH)-axioms for maximal $\bar{s}_i$ subsume (INH)-axioms for non-maximal ones.

We will now prove that the equation system $LP = CI \cup INH$, oriented from left to right, forms a canonical system of rewrite rules. First we will define a precedence on Σ-operators such that the induced recursive path ordering proves the termination of the system.

Definition 3.6 *By $>_I$ we denote the following partial order on Ω:*

1. $i_{s_1 \subset s_2} > i_{s'_1 \subset s'_2}$, *iff* $s'_2 < s_2$ *or if* $s'_2 = s_2$ *and* $s'_1 < s_1$, *for injections* $i_{s_1 \subset s_2}$ *and* $i_{s'_1 \subset s'_2}$.

2. $f_{s_1 \ldots s_n \to s_0} > i_{s \subset s'}$, *for any order-sorted operator* f *and any injection* $i_{s \subset s'}$.

3. $f_{s_1 \ldots s_n \to s_0} > f_{s'_1 \ldots s'_n \to s'_0}$, *iff* $s'_0 s'_1 \ldots s'_n <_t s_0 s_1 \ldots s_n$, *for any two declarations* $f : s'_1 \ldots s'_n \to s'_0$ *and* $f : s_1 \ldots s_n \to s_0$ *of the same order-sorted operator symbol* f.

By $>_I$ we denote the recursive path ordering on $T_\Sigma(X)$ induced by $>_I$.

Proposition 3.7 *Orienting the equations in LP from left to right into rules $L \to R$, we have $L >_I R$, for any of these rules.*

The confluence of the system will be proved using the following proposition:

Proposition 3.8 *If*

$$J(x) = i_{s_{n-1} \subset s_n} \circ i_{s_{n-2} \subset s_{n-1}} \circ \ldots \circ i_{s_0 \subset s_1}(x)$$

is some composite injection from s_0 to s_n, $n \geq 1$, then $J(x) \to^*_{CI} I_{s_0 \subset s_n}(x)$.

As a consequence we have $J_1(x) \downarrow_{CI} J_2(x)$, for any two composite injections J_1 and J_2 from s to s'.

Lemma 3.9 *For any two Σ-terms $t : s$ and $t' : s'$ such that $s \leq s_0$, and $s' \leq s_0$, we have $\omega(t) = \omega(t')$, iff $I_{s \subset s_0}(t) \equiv_{LP} I_{s' \subset s_0}(t')$ such that the $\equiv_{LP}$-proof only involves intermediate terms smaller than $I_{s \subset s_0}(t)$ or $I_{s' \subset s_0}(t')$ with respect to $>_I$.*

Proposition 3.10 *The set of rules LP is locally confluent, hence confluent by 3.7.*

From 3.7 and 3.10 it follows that LP is canonical. We will now prove that the LP-normalforms of terms t represent the lowest parse of the corresponding order-sorted terms $\omega(t)$. More precisely,

Lemma 3.11 *Let $t \in T_\Sigma(X)_s$ and $s^{\lambda(\omega(t))} =: s' \leq s'' \leq s$.*

1. $I_{s' \subset s''}(\lambda(\omega(t)))$ *is irreducible under* $\to_{LP}$.

2. $t \geq_I I_{s' \subset s''}(\lambda(\omega(t)))$.

In particular, from 1, 3.9, 3.7, and 3.10 we have that $t \downarrow_{LP}= I_{s' \subset s}(\lambda(\omega(t)))$.

Altogether we have shown that two terms $t_1, t_2 \in T_\Sigma(X)$ are representations of the same order-sorted terms, iff they are equivalent under LP. Moreover, the equivalence can be decided by rewriting the appropriately injected terms to their $\to_{LP}$-normalforms.

3.3 Order-Sorted Deduction and Rewriting

The notion of order-sorted deduction here is the one for the variant of order-sorted logic in [GM87]. Order-sorted deduction is described by the following set of inference rules, cf. e.g. [GKK87]:

Definition 3.12 (Order-Sorted Deduction) *Let E be a set of order-sorted equations over Σ^{os}.*

Reflexivity

$$E \vdash_X t \doteq t\ ,$$

for any $t \in T_{\Sigma^{os}}(X)$.

Symmetry

$$\frac{E \vdash_X t \dot{=} t'}{E \vdash_X t' \dot{=} t}$$

Transitivity

$$\frac{E \vdash_X t \dot{=} t',\ E \vdash_X t' \dot{=} t''}{E \vdash_X t \dot{=} t''}$$

Congruence

$$\frac{E \vdash_Y \theta(x) \dot{=} \theta'(x),\ \forall x \in X}{E \vdash_Y \theta(t) \dot{=} \theta'(t)},$$

for $\theta, \theta' : X \to T_{\Sigma^{os}}(Y)$, $t \in T_{\Sigma^{os}}(X)$

Substitutivity

$$\frac{E \vdash_Y \theta(t_i) \dot{=} \theta(t_i'),\ 1 \le i \le n}{E \vdash_Y \theta(t) \dot{=} \theta(t')},$$

for $\eta = t_1 \dot{=} t_1' \wedge \ldots \wedge t_n \dot{=} t_n' \Rightarrow t \dot{=} t' \in E$ *and* $\theta : var(\eta) \to T_{\Sigma^{os}}(Y)$ *a substitution.*

Clearly, $E \vdash_X t \dot{=} t'$, iff $t \equiv_E^X t'$, where $\equiv_E^X = \cup_{n \in N} \equiv_n^X$, with $\equiv_0^X = \emptyset$ and $t \equiv_n^X t'$, iff $t \equiv_{n-1}^X t'$ or if there exist u_j, u_j' such that $u_j \equiv_{n-1}^X u_j'$ and $t \dot{=} t'$ can be derived from $u_j \dot{=} u_j'$ using one of the above inference rules.

We will now extend our notion of lowest parses $\lambda(\_)$ to unconditional equations. Let $t_1 \dot{=} t_2$ be an order-sorted equation, and assume that $s_i = s^{\lambda(t_i)}$.

$$\lambda(t_1 \dot{=} t_2) \;=\; I_{s_1 \subset s}(\lambda(t_1)) \dot{=} I_{s_2 \subset s}(\lambda(t_2)),$$

where s is some minimal supersort of both t_1 and t_2, i.e. $t_1, t_2 \in T_{\Sigma^{os}}(X)_s$. (Due to the coherence of Σ^{os} such an s exists. There may be more than one choice for s. This, however is irrelevant in our context.) In particular, if $s_2 \le s_1$, the left side of $\lambda(t_1 \dot{=} t_2)$ will not have an injection as top symbol.

Let now

$$E^{\#} = CI \cup INH \cup IN \cup \lambda(E),$$

where

$$\lambda(E) = \{\ldots \lambda(t_i \dot{=} t_i') \ldots \Rightarrow \lambda(t \dot{=} t') \ldots t_i \dot{=} t_i' \ldots \Rightarrow t \dot{=} t' \in E\}$$

are the minimal parses of the equations in E and where

$$IN = \{i_{s \subset s'}(x) \dot{=} i_{s \subset s'}(y) \Rightarrow x \dot{=} y \mid i_{s \subset s'} \in \Omega\}$$

is the set of injectivity axioms for the injections in Σ.

The following is the proof-theoretic equivalent of the satisfaction theorem in [GJM85]:

Theorem 3.13 *For* $t_1, t_2 \in T_\Sigma(X)_s$, $t_1 \equiv_{E^{\#}}^X t_2$, *iff* $\omega(t_1) \equiv_E^X \omega(t_2)$.

This theorem proves the equivalence of order-sorted deduction in E with standard many-sorted equational logic in $E^{\#}$.

We now go on and compare order-sorted rewriting to many-sorted rewriting.

Definition 3.14 *An* **order-sorted conditional rewrite rule** *is an order-sorted conditional equation* $C \Rightarrow l \dot{=} r$ *satisfying* $(var(C) \cup var(r)) \subset var(l)$ *and denoted* $C \Rightarrow l \dot{\to} r$.

Definition 3.15 *A term* $t \in T_{\Sigma^{os}}(X)$ **rewrites** *to* t' *with a rewrite rule* $\rho = C \Rightarrow l \dot{\to} r$ *in* R *at occurrence* o, *which is denoted* $t \to_{R^X} t' = t[o \leftarrow \sigma(r)]$ *whenever*

1. *σ is a substitution* $\sigma : var(\rho) \to T_{\Sigma^{os}}(X)$ *such that* $t/o = l\sigma$,
2. *there is a sort s such that, for x a variable of sort s,* $t[o \leftarrow x]$ *is a well-formed term and* $l\sigma, r\sigma \in T_{\Sigma^{os}}(X)_s$,
3. *for any* $u \dot{=} v \in C$ *there exists a term w such that* $u\sigma \to_{R^X}^* w$ *and* $v\sigma \to_{R^X}^* w$, *with* $\to_{R^X}^*$ *the reflexive and transitive closure of* $\to_{R^X}$.

If $X = var(t)$, we will also write $\rightarrow_R$ and $\rightarrow_R^*$ for $\rightarrow_{R^X}$ and $\rightarrow_{R^X}^*$, respectively. The smallest fixpoint of this recursion defines $\rightarrow_{R^X}$.

Many-sorted rewriting is defined like order-sorted rewriting in signatures with an empty set of subsort relations. In this case, the second condition of the previous definition becomes trivial.

Theorem 3.16 *$u \rightarrow_{\lambda(R)/LP} v$ iff $\omega(u) \rightarrow_R \omega(v)$.*

This theorem proves that order-sorted rewriting is equivalent to rewriting the many-sorted representations of terms modulo the axioms of the lowest parse LP, using the lowest parses of the order-sorted rules as rewrite rules. If $\rightarrow_R$ is canonical, $\lambda(R)/LP$ is canonical, too. However, rewriting modulo LP does not seem to be very efficient. Fortunately, forming the closure R_S of R by all *specializations* of the rules, will make $\lambda(R_S) \cup LP$ canonical, provided R is sort-decreasing and canonical.

To formally introduce the notion of specialization it is useful to define the notion of a sort assignment. A sort assignment is a map $\alpha : \overline{X} \rightarrow S$, where $\overline{X}$ is the set of names of variables in X. Hence, a sorted set of variables is a pair $(\overline{X}, \alpha)$, denoted X_α. Sort assignments inherit the subsort ordering such that $\alpha \leq \alpha'$, iff $\alpha(x) \leq \alpha'(x)$, for any $x \in \overline{X}$. A specialization is a substitution $\rho : X_\alpha \rightarrow X_{\alpha'}$, where $\alpha' \leq \alpha$, sending $x : \alpha(x)$ to $x : \alpha'(x)$. To *specialize* a order-sorted term or formula ϕ means to apply a specialization to ϕ. If Φ is a set of order-sorted terms of formulas, by Φ_S we denote the set of all specializations of terms or formulas in Φ.

Definition 3.17 *An order-sorted rule $C \Rightarrow s \dot{\rightarrow} t$ is called* **sort-decreasing,** *iff for any specialization ρ, $s\rho \dot{\rightarrow} t\rho$ has a lowest parse such that $s^{\lambda(s\rho)} \geq s^{\lambda(t\rho)}$. A many-sorted rule $C \Rightarrow s \dot{\rightarrow} t$ is called* **sort-decreasing,** *iff the left side s does not carry an injection as its top symbol. A set of rules is sort-decreasing, if each of its members is sort-decreasing.*

An immediate consequence of 3.16 is the following corollary.

Corollary 3.18 *$u \rightarrow_{\lambda(R_S)/LP} v$ iff $\omega(u) \rightarrow_R \omega(v)$.*

Confluence and termination of R carry over to $R_\# = \lambda(R_S) \cup LP$ as we shall see in the next theorem. Let us first make a few remarks about reduction orderings. If we are given a reduction ordering $>$ on $T_{\Sigma^{ms}}(X)$, it can be extended to a reduction ordering $>_{ms}$ on $T_\Sigma(X)$ which is compatible with $\equiv_{LP}$, simply by defining $t >_{ms} t'$, iff $\omega(t) > \omega(t')$. In addition, the transitive closure $\tilde{>}$ of $(>_{ms} \cup >_I)$, where $>_I$ is the recursive path ordering that we have introduced to order the LP axioms, also is a reduction ordering on $T_\Sigma(X)$ which is compatible with LP. This ordering can be used to order $\lambda(R_S) \cup LP$.

Theorem 3.19 *Let R be a set of order-sorted rules.*

1. *R is sort-decreasing, iff $R_\#$ is sort-decreasing.*

2. *Let R be sort-decreasing. R is canonical iff $R_\#$ is canonical with a reduction ordering that is compatible with $\equiv_{LP}$.*

3. *If $R_\#$ is canonical and sort-decreasing, then, $\downarrow_{R_\#} = \equiv_{E^\#}$, i.e. for any two terms $u, v \in T_\Sigma(X)_s$ we have $u \equiv_{E^\#}^X v$, iff $u \downarrow_{R_\#} v$.*

In the case of unconditional rewrite rules R, $R_\#$ is unconditional, too. Unfailing many-sorted completion [HR87], [Bac87] will generate any reduced variant of $R_\#$.

3.4 Elimination of Non-Sort-Decreasing Rules

Theorem 3.19 requires order-sorted rules to be sort-decreasing for the construction of an equivalent canonical system of many-sorted rules. Likewise, order-sorted completion as proposed in [GKK87] requires rules to be sort-decreasing and fails, if non-sort-decreasing rules are generated. We shall see in section 5.1 that translating into many-sorted specifications and applying conditional equation completion (to deal with the injectivity axiom of injections) is successful in simple cases of non sort-decreasing rules. In many interesting cases, like in the subsequent example, the completion procedure which we will describe in section 4 below will not terminate.

Example 3.20

```
sort nzNat < nat, nat < int, nzNat < nzInt, nzInt < int
op
    0 :  nat
    s :  nat -> nzNat
    + :  int*int    -> int, nat*nat      -> nat,  nat*nzNat -> nat
         nzNat*nat -> nat, nzNat*nzNat -> nzNat
    - :  nat -> int, nzNat -> nzInt, int -> int, nzInt -> nzInt
    * :  int*int -> int, nat*nat -> nat
    square :  int*int -> nat
var i:int, j:int, n:nat
axioms
    -(0) = 0
    -(-i) = i
    i+0 = i
    0+i = i
    k+s(m) = s(k+m)
    (-s(k)) + s(m) = (-k) + m
    i + (-j) = -((-i)+j)
    i*0 = 0
    0*i = 0
    i*s(n) = i*n + i
    i*(-j) = -(i * j)
    (-i)*j = -(i * j)
    square(i) = i*i
```

The last axiom, when oriented from left to right, is clearly not sort-decreasing. A specification with the same initial algebra would be the one in which this equation is replaced by

$$i * i \doteq n \Rightarrow square(i) \doteq n,$$

with n a variable of sort *nat*. This equation, when oriented from left to right, is sort-decreasing. However, it has the extra variable n in its condition and right side. The lowest parse of this equation would be

$$i * i \doteq i_{nat \subset int}(n) \Rightarrow square(i) \doteq n.$$

Equations of this kind are usually not admitted as rewrite rules. In fact, we plan to associate a specific operational semantics with it. It should be specifying the replacement of a $square(i)$ by any n which can be obtained from normalizing $i * i$ and type checking the result by matching $i_{nat \subset int}(n)$ with the normalform. If the normalform is unique, this process of finding the substitution for i and n at rewrite-time is completely backtrack-free. Unfortunately, this idea of deterministic oriented goal solving is not a complete goal solving method in general. Fortunately, an adequately designed completion procedure can make it become complete.

Our idea of how one can replace non-sort-decreasing equations by sort-decreasing ones should be obvious, not requiring any further formalization. However, we should be saying something about when this replacement preserves the initial algebra of a specification. We assume to be given a set E of order-sorted equations, as well as its many-sorted equivalent $E^\#$.

Definition 3.21 *Let C be a set of unconditional equations, let $t \in T_\Sigma(X)_s$ be a term of sort s, $var(C) \subset X$, and $s' \leq s$. We say that t* **is of type s' in context C**, *if for any ground substitution σ of the variables in X such that $C\sigma \subset \equiv^{\emptyset}_{E^\#}$ there exists a term $u_\sigma \in T_\Sigma(X)_{s'}$ such that $t\sigma \equiv^{\emptyset}_{E^\#} I_{s' \subset s}(u_\sigma)$.*

In our above example we have $i * i$ of type *nat* in the empty context as for any ground substitution $i * i$ is equal to $i_{nat \subset int}((i_{nzNat \subset nat} \circ s)^{i*i}(0))$.

Proposition 3.22 *Let $C \Rightarrow I_{s^l \subset s}(l) \doteq I_{s^r \subset s}(r)$ be a conditional equation and let $I_{s^r \subset s}(r)$ be of type s_l in context C. Then, replacing $C \Rightarrow I_{s^l \subset s}(l) \doteq I_{s^r \subset s}(r)$ by*

$$C \wedge I_{s^r \subset s}(r) \doteq I_{s^l \subset s}(x) \Rightarrow l \doteq x,$$

with x a new variable of sort s^l, preserves $\equiv^{\emptyset}_{E^\#}$, and hence the initial algebra of the specification.

One half of the proof of this proposition is that paramodulation of $C \Rightarrow I_{s^l C_s}(l) \dot{=} I_{s^r C_s}(r)$ on the condition of the injectivity axioms for the injections will generate the replacement $C \wedge I_{s^r C_s}(r) \dot{=} I_{s^l C_s}(x) \Rightarrow l \dot{=} x$. We shall see that during completion superpositions of this kind will be performed anyway. Yet, completion very often will not terminate because of other superpositions on the originally given equation. We believe that sufficiently powerful *ground completion procedures*, once they have been developed for conditional equations, can solve this problem in cases where the previous proposition applies. In other words, ground completion of the *original* specification can be expected to terminate in this case.

4 Completion of Many-Sorted Conditional Equations

In this section, we will assume to be given a fixed many-sorted signature Σ. Equations, terms, substitutions, etc. will be taken over this signature, unless specified otherwise. Furthermore, we assume a reduction ordering $>$ to be given on $T_\Sigma(X)$. $>_{st}$ will denote the transitive closure of $> \cup\, st$, with st the strict subterm ordering. $>_{st}$ is noetherian and stable under substitutions.

4.1 Annotated Equations and Reductive Rewriting

We do not put any restrictions on the syntactic form of conditional equations. In particular, conditions and right sides may have extra variables. To compensate for this permissiveness, the application of equations at rewrite-time will be restricted. Completion will guarantee that this restricted application is complete. Formally, application restrictions can be modelled by considering a given set E of equations as a *generator* for rewrite rules.[2] In particular, the set E^r of *reductive* instances of the equations in E is of interest:

$$E^r = \{C\sigma \Rightarrow s\sigma \dot{\rightarrow} t\sigma \mid C \Rightarrow s \dot{=} t \in E^=, s\sigma > t\sigma, s\sigma > u\sigma, s\sigma > v\sigma, \text{ for any } u \dot{=} v \in C\}$$

In the general case, $\rightarrow_{E^r}$ can be quite inefficient and require (restricted) paramodulation to solve conditions of equations in E. Furthermore, the computed solutions have to be tested for reductivity. To increase efficiency of rewriting, it can be useful to further restrict application of equations at rewrite-time.

We will annotate equations to specify in which way their use at rewrite-time should be restricted. For the purposes of this paper, an equation can be annotated as *operational* or *nonoperational*. The intuitive meaning is that a nonoperational equation should not contribute at all to the equational theory. Injectivity axioms, for example, should be irrelevant at rewrite-time.

In *operational* equations $C \Rightarrow s \dot{=} t$, condition equations $u \dot{=} v \in C$ will be annotated as either *oriented* or *unoriented*. We will use the notation $u \stackrel{\rightarrow}{=} v$ to indicate the annotation "oriented". For a oriented condition oriented goal solving is wanted. Altogether:

Definition 4.1 *Let E be a set of annotated equations. E is viewed to generate the set E^a of rewrite rules $C\sigma \Rightarrow s\sigma \dot{\rightarrow} t\sigma$ such that*

1. *$C \Rightarrow s \dot{=} t \in E^=$ is annotated as operational and $C\sigma \Rightarrow s\sigma \dot{\rightarrow} t\sigma \in E^r$, i.e. the instance is reductive,*
2. *if $u \stackrel{\rightarrow}{=} v \in C$, then $v\sigma\sigma'$ is $\rightarrow_{E^a}$-irreducible for any $\rightarrow_{E^a}$-irreducible substitution σ'.*

Clearly, $\rightarrow_{E^a} \subset \rightarrow_{E^r} \subset \rightarrow_E$, where the subset inclusions are proper, hence $\stackrel{*}{\leftrightarrow}_{E^a} \neq \equiv_E$ in general. E is called *complete*, iff $\stackrel{*}{\leftrightarrow}_{E^a} = \equiv_E$ and if $\rightarrow_{E^a}$ is canonical. A completion procedure attempts to complete E in this sense.

In many practical cases, a final system E obtained by completion will have additional properties which make $\rightarrow_{E^a}$ to be efficiently computable. For example, if

$$i * i \stackrel{\rightarrow}{=} i_{nat \subset int}(n) \Rightarrow square(i) \dot{=} n.$$

with the condition annotated as oriented, is element of a complete E, $square(i)$ needs only be rewritten for those instances of i for which the $\rightarrow_{E^a}$-normalform of $i * i$ is of form $i_{nat \subset int}(n)$. Moreover, if $i * i$ is smaller in the reduction order than $square(i)$, the replacement n will also always be smaller than $square(i)$, making any application of the equation reductive. No reductivity tests are required at rewrite-time. Equations which have this property will be called *quasi-reductive* below. Note that let-expressions with patterns in functional programming languages such as MIRANDA are another example of equations with oriented conditions, cf. definition of quicksort below.

[2] In [BG88] we develop a more general concept of application restrictions based on a notion of relevant substitutions.

4.2 Quasi-Reductive Equations

To simplify the formal treatment in this section, we can assume that operational equations have oriented conditions only. (If an equation has an unoriented condition $u \stackrel{\downarrow}{=} v$, we can replace the latter by the two oriented conditions $u \equiv x$ and $v \equiv x$, where x is a new variable.)

In the classical case of unoriented conditions, the class of reductive equations [Kap84], [JW86], allows for efficient rewriting [Kap87]. In particular, conditions of equations are easily proved or disproved, and no goal solving is required. Moreover, there are no reductivity tests required at rewrite-time, as any instance of a reductive equation is reductive.

In the case of oriented goal solving there exists a similarly efficient class of equations. Oriented goal solving $u\sigma \rightarrow_E^* v\sigma$ reduces to normalizing $u\sigma$ and, then, matching $v\sigma$ with the normal form, if any of the variables of u is already bound by the matching of the redex, or by the solution of some other condition equation. To formalize this idea, we will have to look at how variables are bound within an equation. We call a conditional equation $u_1 \dot{=} v_1 \wedge \ldots \wedge u_n \dot{=} v_n \Rightarrow s \dot{=} t$ (with oriented conditions) *deterministic*, if, after appropriately changing the orientation of the consequent and choosing the order of the condition equations, the following holds true:

$$var(u_i) \subset var(s) \cup \bigcup_{1 \le j < i} (var(u_j) \cup var(v_j)),$$

and

$$var(t) \subset var(s) \cup \bigcup_{j=1}^{n} (var(u_j) \cup var(v_j)).$$

To arrive at a concept for avoiding reductivity proofs at rewrite-time, let us now assume the existence of some enrichment $\Sigma' \supseteq \Sigma$ of the signature such that the given reduction ordering on $T_\Sigma(X)$ can be extended to a reduction ordering on $T_{\Sigma'}(X)$.

Definition 4.2 *A deterministic equation* $u_1 \dot{=} v_1 \wedge \ldots \wedge u_n \dot{=} v_n \Rightarrow s \dot{=} t$, $n > 0$, *is called* **quasi-reductive**, *if there exists a sequence* $h_i(\xi)$ *of terms in* $T_{\Sigma'}(X)$, $\xi \in X$, *such that* $s > h_1(u_1)$, $h_i(v_i) \ge h_{i+1}(u_{i+1})$, $1 \le i < n$, *and* $h_n(v_n) \ge t$. *An unconditional equation* $s \dot{=} t$ *is quasi-reductive, if* $s > t$.

The equation

$$i * i \dot{=} i_{nat \subset int}(n) \Rightarrow square(i) \dot{=} n$$

becomes quasi-reductive under a recursive path ordering, if $square > *$ in precedence. Choosing, $h_1(\xi) = \xi$, the inequalities $square(i) > i * i$ and $i_{nat \subset int}(n) \ge n$ are obvious. Also quasi-reductive is

$$split(x, l) \dot{=} (l_1, l_2) \Rightarrow sort(cons(x, l)) \dot{=} append(sort(l_1), cons(x, sort(l_2))),$$

with $h_1(\xi) = f(\xi, x)$, where f is a new auxiliary function symbol. The termination proofs can be given by an appropriately chosen polynomial interpretation. It has to be verified that $f(sort(cons(x, l)), x) > f(split(x, l), x)$ and $f((l_1, l_2), x) \ge append(sort(l_1), cons(x, sort(l_2)))$.

Quasi-reductivity is a proper generalization of reductivity:

Proposition 4.3 *If the equation* $u_1 \dot{=} u_{n+1} \wedge \ldots \wedge u_n \dot{=} u_{2n} \Rightarrow s \dot{=} t$ *is reductive, then the equation*

$$u_1 \dot{=} x_1 \wedge u_{n+1} \dot{=} x_1 \wedge \ldots \wedge u_n \dot{=} x_n \wedge u_{2n} \dot{=} x_n \Rightarrow s \dot{=} t,$$

is quasi-reductive, if the x_i *are new, pairwise distinct variables.*

Lemma 4.4 *Let* E *be finite and* $u_1 \dot{=} v_1 \wedge \ldots \wedge u_n \dot{=} v_n \Rightarrow s \dot{=} t \in E$ *a quasi-reductive equation.*

1. *If* σ *is a substitution such that* $u_i\sigma \rightarrow_E^* v_i\sigma$, $1 \le i \le n$, *then,* $s\sigma > t\sigma$.
2. *If* $N' \rightarrow_{E^a} N''$ *is decidable for all terms* N' *such that* $N >_{st} N'$, *then the applicability of the equation* $u_1 \dot{=} v_1 \wedge \ldots \wedge u_n \dot{=} v_n \Rightarrow s \dot{=} t$ *in* N *is decidable.*

Corollary 4.5 *Let* E *be a set of annotated equations in which any operational equation is quasi-reductive. Then,* $\rightarrow_{E^a}$ *is decidable.*

For confluent $\rightarrow_{E^a}$, the applicability of a quasi-reductive equation can be decided by matching the left side and, then, for $1 \le i \le n$, matching the v_i against the normal forms of the substituted u_i to obtain another part of the substitution. As quasi-reductive equations are deterministic, each variable in u_i is bound at the time when the i-th condition is to be checked. Computing the substitution σ is completely backtrack-free in this case. Moreover, no termination proofs are required at rewrite-time.

4.3 Completion Inferences and Strategies

In this abstract we will only briefly describe the basic inference rules and fairness requirements in completion of annotated, application-restricted equations. For details we refer to the full version [Gan88b] and to [Gan87], [Gan88] and [BG88].

Standard completion CC in the conditional case according to the concepts in [Gan87] and further refined in [BG88] consists of three inference rules for adding consequences, one rule for simplification and one rule for elimination of conditional equations. Conseqences are added by

- paramodulating an equation on the consequent of an equation (i.e. computation of contextual critical pairs)
- paramodulating an equation on some condition of an equation
- resolving some condition of an equation with $x \doteq x$.

Paramodulation is restricted in that terms are never replaced by bigger terms in the reduction ordering. Also, superposition is limited to the nonvariable part of a literal.

Equations $C \Rightarrow s \doteq t$ are simplified by rewriting with quasi-reductive equations, using the (skolemized) condition equations C as additional rewrite rules.

An equation $C \Rightarrow s \doteq t$ can be eliminated, if the current set E of equations admits a proof of $C \vdash s \doteq t$ which is simpler wrt. the proof ordering than the proof in which $C \Rightarrow s \doteq t$ is applied under the identity substitution to the hypotheses C. In practice, the different proofs of $C \vdash_E s \doteq t$ have to be enumerated to a certain depth and their complexities compared against the complexity of the proof which has lead to the creation of $C \Rightarrow s \doteq t$, cf. [Gan88].

The fairness requirements in CC-inference rule application depend on the annotations of the equations. The general case is described in [BG88] and [Gan88b]. As a particularly interesting subcase we mention the following result:

Theorem 4.6 *A CC-derivation $E_0, E_1, \ldots$ is* **fair**, *i.e. the final system $E_\infty = \bigcup_j \bigcap_{k \geq j} E_k$ is complete, if the following holds true:*

1. *E_∞ does not contain any unconditional equation annotated as nonoperational.*
2. *Any operational equation in E_∞ is quasi-reductive.*
3. *$\bigcup_k E_k$ contains all instances of each nonoperational equation $\eta \in E_\infty$ which can be obtained by paramodulating operational equations of E_∞ on one selected condition of η.*
4. *$\bigcup_k E_k$ contains all instances of each nonoperational equation $\eta \in E_\infty$ which can be obtained by resolving the same selected condition of η by $x \doteq x$.*
5. *$\bigcup_k E_k$ contains all critical pairs between operational equations in E_∞.*
6. *$\bigcup_k E_k$ contains all instances of each operational equation $\eta \in E_\infty$ which can be obtained by paramodulating operational equations of E_∞ on the right sides of the oriented condition equations of η.*

5 Order-Sorted Completion: The Many-Sorted Way

In this section we illustrate by means of examples that our techniques of completion for conditional equations can be successfully applied to order-sorted specifications. In the examples, equations will be operational, unless labelled by a "-". Moreover, we rearrange conditions of nonoperational equations such that the first condition is always the one which is selected for superposition. *Operational equations will all be reductive or quasi-reductive with the given orientation of literals and ordering of conditions.*

5.1 Smolka's Example

Our first example is due to Smolka and shows the incompleteness of order-sorted replacement of equals by equals, cf. [SNGM87] or [KKM88], in the case of non sort-decreasing rules.

Example 5.1

```
sort  s1 < s2
op  a:s1, b:s1, d:s2, f:s1 -> s1
axioms
    a = d
    b = d
```

In this example, $f(a) \doteq f(b)$ can be derived by order-sorted deduction, however it cannot be proven by replacement of equals by equals. The many-sorted equivalent $E^{\#}$ consists of the folllowing equations:

Example 5.2

```
1     i(a) = d
2     i(b) = d
3-    i(x) = i(y)   => x = y
```

where $i : s1 \rightarrow s2$ is the injection $i_{s1 \subset s2}$. Axiom 3 is the injectivity property of i. Orienting 1 and 2 from left to right creates the following final system of equations:

Example 5.3

```
1     i(a) = d
3-    i(x) = i(y)   => x = y
4-    i(x) = d      => x = a
5     b = a
```

Equation 4 is generated from superposing equation 1 on the condition of the nonoperational injectivity axiom 3 (cf. fairness constraint 3). We have here decided to classify 4 as nonoperational although it becomes a quasi-reductive equation when orienting its literals from right to left. After this, 4 generates equation 5 from superposition with equation 2. If $b > a$ in precedence, equation 5 is reductive, allowing to eliminate equation 2 by reduction. Any other superposition on the condition of 3 or 4 does only generate equations which can later be eliminated by the inference rules of simplification and elimination.

5.2 Squares of Integers

We return to the specification of integers as given in example 3.20. The result of translating into many-sorted and completing this system is the following:

Example 5.4

```
1     -(0) = int(0)
2     - (- (i:int))      = i
2a    - (- (X1:nzInt)) = X1
2b    - (- (X1:nat))     = int(X1:nat)
2c    - (- (X1:nzNat)) = nzInt(X1:nzNat)
3     (i:int)+int(0) = i
3a    (X2:nat)+0       = X2
3b    (X:nzNat)+0      = nat(X:nzNat)
4     int(0)+ (i:int) = i
4a    0+ (X1:nat)       = X1
5     (k:nat)+nat(s(m:nat))                      = nat(s((k:nat)+ (m:nat)))
5a    (X2:nzNat)+nat(s(m:nat))                   = nat(s((X2:nzNat)+ (m:nat)))
6     int(-s(k:nat))+int(nzInt(s(m:nat))) = - (k:nat)+int(m:nat)
7     (i:int)+ (- (j:int))           = - (- (i:int)+ (j:int))
7a    (i:int)+int(- (X1:nzInt)) = - (- (i:int)+int(X1:nzInt))
7b    (i:int)+int(- (X1:nzNat)) = - (- (i:int)+int(nzInt(X1:nzNat)))
7c    (i:int)+ (- (X1:nat))          = - (- (i:int)+int(X1:nat))
8     (i:int)*int(0)    = int(0)
8a    (X2:nat)*0         = 0
9     int(0)* (i:int) = int(0)
```

```
9a   O* (X1:nat)     = 0
10  (i:int)*int(nzInt(s(m:nat))) = (i:int)*int(m:nat)+ (i:int)
10a (X2:nat)*nat(s(m:nat))       = (X2:nat)* (m:nat)+ (X2:nat)
11  (i:int)* (- (j:int))        = -(i:int)* (j:int)
11a (i:int)*int(- (X1:nzInt)) = -(i:int)*int(X1:nzInt)
11b (i:int)* (- (X1:nat))     = -(i:int)*int(X1:nat)
11c (i:int)*int(- (X1:nzNat)) = -(i:int)*int(nzInt(X1:nzNat))
12  (- (i:int))* (j:int)      = -(i:int)* (j:int)
12a int(- (X1:nzInt))* (j:int) = -int(X1:nzInt)* (j:int)
12b (- (X1:nat))* (j:int)      = -int(X1:nat)* (j:int)
12c int(- (X1:nzNat))* (j:int) = -int(nzInt(X1:nzNat))* (j:int)
13  (i:int)* (i:int) =⃗ int(k:nat)             => square(i:int) = k
13a (i:int)* (i:int) =⃗ int(nzInt(X:nzNat)) => square(i:int) = nat(X:nzNat)
 -----
I1  -int(X1:nat)      = - (X1:nat)
I2  -int(X1:nzInt)    = int(- (X1:nzInt))
I3  -nat(X1:nzNat)    = int(- (X1:nzNat))
I5  -nzInt(X1:nzNat) = - (X1:nzNat)
I6  int(X2:nat)+int(X1:nat) = int((X2:nat)+ (X1:nat))
I7  nat(X2:nzNat)+ (X1:nat) = (X2:nzNat)+ (X1:nat)
I8  int(X2:nat)*int(X1:nat) = int((X2:nat)* (X1:nat))
I9  int(nat(X:nzNat)) = int(nzInt(X:nzNat))
I10 int(X2:nat)+int(nzInt(X:nzNat)) = int((X2:nat)+nat(X:nzNat))
I11 int(nzInt(X:nzNat))+int(X1:nat) = int((X:nzNat)+ (X1:nat))
I12 int(nzInt(X:nzNat))+int(nzInt(Y:nzNat)) = int((X:nzNat)+nat(Y:nzNat))
I13 int(X2:nat)*int(nzInt(X:nzNat)) = int((X2:nat)*nat(X:nzNat))
I14 int(nzInt(X:nzNat))*int(X1:nat) = int(nat(X:nzNat)* (X1:nat))
I15 int(nzInt(X:nzNat))*int(nzInt(Y:nzNat)) = int(nat(X:nzNat)*nat(Y:nzNat))
 -----
i1-    nat(X:nzNat) = nat(Y:nzNat) => X = Y
i2-    int(X:nat)    = int(Y:nat)    => X = Y
i3-    nzInt(X:nzNat) = nzInt(Y:nzNat) => X = Y
i4-    int(X:nzInt)    = int(Y:nzInt)    => X = Y
i5-    int(nzInt(X:nzNat))  = int(Y:nat) => nat(X:nzNat) = Y
i6-    int(nzInt(X1:nzNat)) = int(nzInt(X:nzNat))
          and int(nzInt(X:nzNat)) = (i1:int)* (i1:int) => X = X1
```

The initial set of many-sorted equations $E^{\#}$ consists of the equations 1–13, $I1$–$I9$, and $i1$–$i4$.[3] The remaining equations are generated during completion. The (INH)-equations have a number in $I1$–$I8$ or $I10$–$I15$, equation $I9$ is the only (CI)-axiom. The (nonoperational) injectivity axioms are the equations $i1$–$i4$. It is sufficient to start completion with just the (INH)-equations $I1$–$I8$ between any two neighboring operators (wrt. $<_t$), as the remaining ones are generated as critical pairs.

The final system also contains the lowest parses of many specializations (indicated by letters $a, b, c, \ldots$) of the initial order-sorted rules. This is in accordance with theorem 3.19. In a *reduced* final system like the one above, however, not all specializations need to be present. Equation 13*a* has been generated from superposing $I9$ on the right side of the condition of 13, cf. 4.6. In this example, completion has just verified the completeness of the initial system. No new order-sorted equation has been generated. The new equations on the many-sorted level serve to synchronize the application of order-sorted rules with the computation of lowest parses.

Acknowledgements. The author is grateful to H. Bertling for many discussions on the subject of this paper.

6 References

[Bac87] Bachmair, L.: Proof methods for equational theories. PhD-Thesis, U. of Illinois, Urbana Champaign, 1987.

[BDH86] Bachmair, L., Dershowitz, N. and Hsiang, J.: Proof orderings for equational proofs. Proc. LICS 86, 346-357.

[3] Injections $i_{s \subset s'}$ are ambiguously denoted by their codomain. s', e.g. *int* denotes both $i_{nzInt \subset int}$ and $i_{nat \subset int}$. The order-sorted rather than the many-sorted function symbols are used. $x : s$ denotes a variable x of sort s.

[BG88] Bertling, H. and Ganzinger, H.: Completion-time optimization of rewrite-time goal solving. Report M.1.3-R-12.0, PROSPECTRA-Project, FB Informatik, U. Dortmund, 1988.

[BG89] Bertling, H. and Ganzinger, H.: Completion of application-restricted conditional equations. Report, FB Informatik, U. Dortmund, 1989, to appear.

[BGS88] Bertling, H., Ganzinger, H. and Schäfers, R.: CEC: A system for conditional equational completion. User Manual Version 1.4, PROSPECTRA-Report M.1.3-R-7.0, U. Dortmund, 1988.

[DOS88] Dershowitz, N., Okada, M. and Sivakumar, G.: Confluence of conditional rewrite systems. Proc. 1st Int'l Workshop on Conditional Term Rewriting, Orsay, 1987, Springer LNCS 308, 1988, 31–44.

[Gan87] Ganzinger, H.: A Completion procedure for conditional equations. Report 234, U. Dortmund, 1987, also in: Proc. 1st Int'l Workshop on Conditional Term Rewriting, Orsay, 1987, Springer LNCS 308, 1988, 62–83 (revised version to appear in J. Symb. Computation).

[Gan88] Ganzinger, H.: Completion with History-Dependent Complexities for Generated Equations. In Sannella, Tarlecki (eds.): Recent Trends in Data Type Specifications. Springer LNCS 332, 1988, 73–91.

[Gan88b] Ganzinger, H.: Order-sorted completion: the many-sorted way (full version). Report 274, FB Informatik, Univ. Dortmund, 1988.

[GJM85] Goguen, J.A., Jouannaud, J.-P. and Meseguer, J.: Operational semantics for order-sorted algebra. Proc. 12th ICALP, 1985, 221–231.

[GKK87] Gnaedig, I., Kirchner, C. and Kirchner, H.: Equational completion in order-sorted algebra. Proc. CAAP '88, Springer LNCS 299, 1988, 165–184.

[GM87] Goguen, J.A. and Meseguer, J.: Order-sorted algebra I: partial and overloaded operators, errors and inheritance. Comp. Sci. Lab, SRI International, 1987.

[Gog78] Goguen, J.A.: Order-sorted algebra. Semantics and theory of computation. Report No. 14, UCLA Computer Science Department, 1978.

[Gogo86] Gogolla, M.: Partiell geordnete Sortenmengen und deren Anwendung zur Fehlerbehandlung in abstrakten Datentypen. PhD Thesis, FB Informatik, U. Braunschweig, West Germany, 1986.

[HR87] J. Hsiang, M. Rusinowitch: On word problems in equational theories, Int. Coll. on Automata Languages and Programming, Springer LNCS, 1987.

[JW86] Jouannaud, J.P. and Waldmann, B.: Reductive conditional term rewriting systems. Proc. 3rd TC2 Working Conference on the Formal Description of Prog. Concepts, Ebberup, Denmark, Aug. 1986, North-Holland.

[Kap84] Kaplan, St.: Fair conditional term rewrite systems: unification, termination and confluence. Report 194, U. de Paris-Sud, Centre d'Orsay, Nov. 1984.

[Kap87] Kaplan, St.: A compiler for conditional term rewriting. Proc. RTA 1987, LNCS 256, 1987, 25-41.

[KaR87] Kaplan, St. and Remy J.-L.: Completion algorithms for conditional rewriting systems. MCC Workshop on Resolution of Equations in Algebraic Structures, Austin, May 1987.

[KKM88] Kirchner, C, Kirchner, H., and Meseguer, J.: Operational semantics of OBJ3. Proc. ICALP 88, Springer LNCS, 1988.

[KR88] Kounalis, E. and Rusinowitch, M.: On word problems in Horn theories. Proc. 1st Int'l Workshop on Conditional Term Rewriting, Orsay, 1987, Springer LNCS 308, 1988, 144–160.

[Poi88] Poigné, A.: Partial algebras, subsorting, and dependent types. In Sannella, Tarlecki (eds.): Recent Trends in Data Type Specifications. Springer LNCS 332, 1988, 208–234.

[Rus87] Rusinowitch, M.: Theorem-proving with resolution and superposition: an extension of Knuth and Bendix procedure as a complete set of inference rules. Report 87-R-128, CRIN, Nancy, 1987.

[SNGM87] Smolka, G., Nutt, W., Goguen, J.A. and Meseguer, J.: Order-sorted equational computation. SEKI Report SR-87-14, U. Kaiserslautern, 1987.

[WB83] Winkler, F. and Buchberger, B.: A criterion for eliminating unnecessary reductions in the Knuth-Bendix algorithm. Coll. on Algebra, Combinatorics and Logic in Comp. Sci., Györ, 1983.

[ZR85] Zhang, H. and Remy, J.L.: Contextual rewriting. Conf. on Rewriting Techniques and Applications, Dijon 1985, Springer LNCS 202, 46–62.

Algebraization and Integrity Constraints for an Extended Entity-Relationship Approach

martin gogolla
Technische Universität Braunschweig
Informatik / Datenbanken
Postfach 3329
D-3300 Braunschweig
Federal Republic of Germany
e-mail: gogolla@infbs.uucp *or* gogolla@dbsinf6.bitnet

Abstract:

An extended entity-relationship model concentrating nearly all concepts of known "semantic" data models and especially allowing arbitrary user defined data types is introduced. The semantics of the model is described purely in algebraic terms mainly based on the notions of signature, algebra and extension. On this basis a calculus making intensive use of abstract data types is defined and employed for the formulation of typical integrity constaints like functional restrictions and key specifications.

Keywords:

Theory of data bases, data model, entity-relationship model, formal semantics, calculus, abstract data type, aggregate function, relational completeness, integrity constraint.

1. Introduction

Among the different steps for the design of a database the conceptual design plays a mayor role [TF 82, Ce 83]. Here all requirements of later database users are collected and described in a formal way. Many authors (among them [Ch 76]) agree that the Entity-Relationship model is the most adequate data model to be used in this phase of database design. Quite a number of ER languages and ER algebras [PS 85] (where the notion algebra is used analogously to relational algebra) have been proposed, but up to now nearly no work has been done in order to define an ER calculus (again, analogously to relational calculi). It is also rather an exception [Su 87, Bü 87] in the area of databases that languages have a formal semantics, more common is the definition by examples. Special interest in the formal description of database (and especially ER) languages has to be paid to aggregate functions like this has been done for the relational approach [Kl 82, ÖÖM 87].

On the other hand, some papers have tried to combine ideas developed in the field of algebraic specification (see for instance [EM 85]) with relational database design [EKW 78, DMW 82]. Also the modal systems of algebras as proposed in [GMS 83, KMS 85] supported the conceptual modelling by algebraic tools. Algebraic techniques were also successfully employed for the description of certain standard "universes" in connection with key specifications [EDG 86, SSE 87].

This paper tries to bridge the gap between the entity-relationship and the algebraic specification communities. We propose an extended ER model having an algebraic semantics and use (a polished version of) our extended ER calculus [HG 88] to formulate typical integrity constraints (to be used in the conceptual design). Let us finally remark that our calculus is not restricted to the ER model, it can be applied to other approaches [Sh 81, JS 82, SS 86] as well.

The paper is organized as follows : Chapter 2 gives an informal introduction into the data model and the calculus by means of an example. Chapter 3 formally defines the model and points out how to formulate it in algebraic terms. In chapter 4 the calculus is defined and it is applied in chapter 5 to formulate integrity constraints.

2. The Basic Idea

Before we explain the formal details of our approach we point out the basic ideas by means of an example. We consider a simple geo-scientific database where information about towns, countries and rivers has to be stored. First, we have entity types which are described by attributes returning values of given data types (possibly set-, bag- or list-valued), e.g., a TOWN has a name, a population and a geometry.

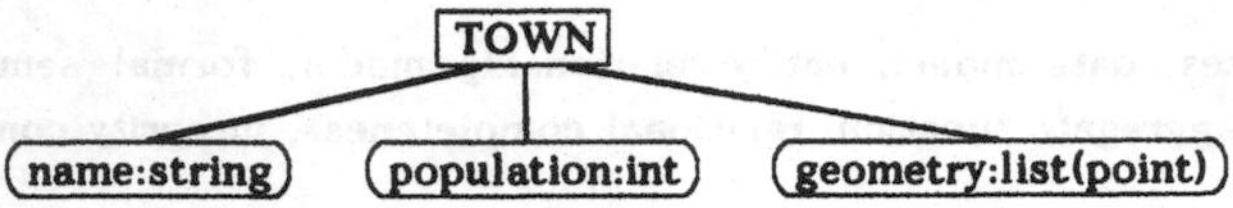

Secondly, relationships can exist between entity types, e.g., RIVERS flow through COUNTRIES, TOWNS lie at RIVERS and lie in COUNTRIES. These relationships can also have attributes.

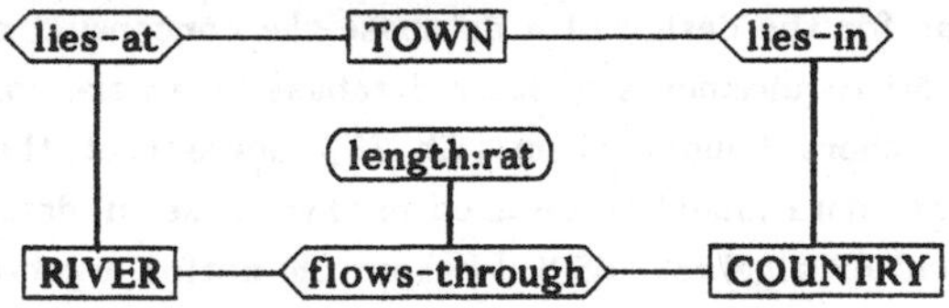

Thirdly, entity types can have other entity types as components, e.g., a TOWN has as component the set of its DISTRICTS and each district has as component the set of STREETS lying in the district.

Last, but not least, entity types can be constructed from other entity types, e.g., an entity of type WATERS is constructed from an entity of type RIVER, SEA or LAKE. RIVER, SEA and LAKE are the input types of the construction, WATERS the output type.

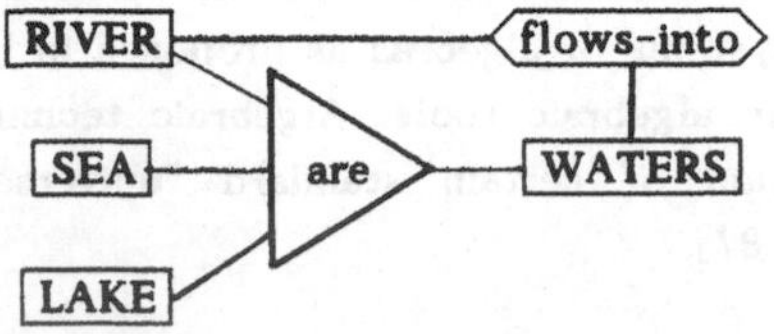

Due to space limitations only a short sketch of our model can be given. Full motivation of all concepts can be found in [HG 88]. For an entity-relationship schema (as introduced above) we shall define a signature (in the sense of abstract data type theory) and a database state for such a schema will then be an algebra for this signature (with carrier sets, functions and relations).

We will also define a calculus for such entity-relationship schemas which can be employed to express queries and integrity constraints. The calculus especially distinguishes between sets and bags (multisets). Therefore it is well suited to formulate aggregation properties not expressible for instance in the "classical" relational tuple or domain calculus [Ma 83]. For example the query "Give me for each country its name and the average of the population of its towns" is formulated as

-{ name(c) , AVG -{ population(t) | t : TOWN ∧ lies-in(t,c) }- | c : COUNTRY }-

Terms of the form -{ ... }- are bag-valued (retaining duplicates). Please notice, this is essential for the calculation of the average in the subquery, if two towns have the same population. It is also possible to restrict variables to the finite set of all stored values : Assume 'government' is an attribute for COUNTRY of sort string (for instance "socialistic", "democratic", etc.). Then the query "Give me every (stored) form of government and for every (stored) form of government the sum of country populations having this form" will be expressed as

-{ g , SUM -{ population(c) | c : COUNTRY ∧ government(c)=g }- |
g : BTS -{ government(c) | c : COUNTRY }- }-

The standard function BTS converts a Bag To a Set. The calculus is also employed to formulate integrity constraints. For example, relationships can be required to be functional :

(∀ fi , fi' : flows-into) fi.RIVER = fi'.RIVER ⇒ fi.WATERS = fi'.WATERS

This formula means that flows-into is functional from RIVER to WATERS, i.e., a river flows into (at most) one water. Another application is the specification of key attributes, key components and key relations :

(∀ r , r' : RIVER)
r ≠ r' ⇒
name(r) ≠ name(r') ∨ (∃ c : COUNTRY) flows-through(r,c) *xor* flows-through(r',c)

This formula, where *xor* stands for the exclusive or, expresses that a river is identified by its name and the countries through which it flows : {name,flows-through} is a key for RIVER. In other words, two different rivers must have different names or (if their names coincide) there must be a country such that one river flows through this country and the other one not: the rivers have to be observably inequivalent with respect to "name" or "flows-through". Let us finally mention the possibility of cardinality constraints :

(∀ d : DISTRICT) 50 ≤ CNT(streets(d)) ∧ CNT(streets(d)) ≤ 100

The above line (CNT stands for COUNT) says that districts have at least 50 and at most 100 streets.

3. Algebraization of the Extended Entity-Relationship Model

3.1 Axiomatic conventions: Let **|SET|** denote the class of sets, **|FISET|** the class of finite sets, **|FUN|** the class of total functions and **|REL|** the class of relations. There are the obvious inclusions |FISET| ⊆ |SET| and |FUN| ⊆ |REL| ⊆ |SET|. Assume sets S, S_1, ... , S_n ∈ |SET| are given. Then ***F*(S)** denotes the restriction of the powerset ***P*(S)** of S to finite sets, $\mathbf{S^*}$ the set of finite lists over S, $\mathbf{S^+}$ the set of finite non-empty lists over S, and $\mathbf{S_1}$ **x ... x** $\mathbf{S_n}$ the cartesian product of the sets $S_1,...,S_n$. The set of finite multisets or bags over S is given by ***B*(S)**. A bag can be considered as a (finite) set S together with a counting function *occur* : S → ℕ, giving for each element the number of occurrences in the bag.

Finite sets are written as $\{c_1,...,c_n\}$, lists as $<c_1,...,c_n>$, elements of the cartesian product as $(c_1,...,c_n)$, and bags as $\{\{c_1,...,c_n\}\}$. For a set $\{c_1,...,c_n\}$, i≠j implies $c_i \neq c_j$. But this is not necessarily true for bags: If we have a bag $\{\{c_1,...,c_n\}\}$ with occur(c)=k, this implies that there are k distinct indices $i_1,...,i_k$ ∈ 1..n with $c_{i_j} = c$ for j∈1..k. □

3.2 Definition: (data signature)

The **syntax of a data signature DS** is given by

- the sets DATA, OPNS, PRED ∈ |FISET|,
- a function *source*: OPNS → DATA*,
- a function *destination*: OPNS → DATA, and
- a function *arguments* : PRED → DATA⁺.

If σ ∈ OPNS, *source*(σ) = $<d_1,...,d_n>$, and *destination*(σ) = d, this is notated as $\sigma : d_1 \times ... \times d_n \rightarrow d$.

If π ∈ PRED with *arguments*(π) = $<d_1,...,d_n>$, this is notated as $\pi : d_1 \times ... \times d_n$.

The **semantics of a data signature DS** is given by

- a function μ[DATA] : DATA → |SET| and ⊥ ∈ μ[DATA](d) for every d ∈ DATA,
- a function μ[OPNS] : OPNS → |FUN| and $\sigma : d_1 \times ... \times d_n \rightarrow d$ implies
  $\mu[OPNS](\sigma) : \mu[DATA](d_1) \times ... \times \mu[DATA](d_n) \rightarrow \mu[DATA](d)$ for every σ∈OPNS, and
- a function μ[PRED] : PRED → |REL| and $\pi : d_1 \times ... \times d_n$ implies
  $\mu[PRED](\pi) \subseteq \mu[DATA](d_1) \times ... \times \mu[DATA](d_n)$ for every π∈PRED.

The set $\mathbf{OPNS^{2,1}}$ denotes all operators σ, σ ∈ OPNS, having source(σ) = <d,d> and destination(σ) = d for some d ∈ DATA. Additionally the corresponding function μ[OPNS](σ) has to be commutative and associative. □

3.3 Remarks: Throughout the paper all sets mentioned in definitions have to be disjoint, except when common elements are explicitly allowed. Thus, e.g., DATA, OPNS, and PRED are disjoint as well as (the interpretations of) all data sorts with the exception of ⊥, i.e., $\mu[DATA](d_1) \cap \mu[DATA](d_2) = \{\bot\}$, if $d_1 \neq d_2$. Furthermore, throughout the paper the greek letter μ stands for meaning (and hopefully not for mysterious).

We have required every data sort d to contain the value ⊥ ∈ μ[DATA](d), because it is useful to have an 'undefined' value as result for incorrect applications of operations. Thus, we obtain for incorrect applications of operations this special value, for example μ[OPNS](/)(c,0) = ⊥. In most cases it is useful to define the propagation of ⊥ in the following way: An operation σ ∈ OPNS with $\sigma : d_1 \times ... \times d_n \rightarrow d$ evaluates $\mu[OPNS](\sigma)(c_1,...,c_n)$ to ⊥, if there is a c_i with $c_i = \bot$. For predicates π ∈ PRED with $\pi : d_1 \times ... \times d_n$ $(c_1,...,c_n) \in \mu[PRED](\pi)$ does not hold, if there is a c_i with $c_i = \bot$.

For the rest of the paper we assume one fixed data signature and algebra including (among others) the sorts <u>int</u>, <u>rat</u> (for integer and rational numbers) and <u>string</u> together with adequate operations and predicates to be given.

3.4 Example: Our example presented in chapter 2 uses the data sorts int, rat, string and point. As will be seen later in the context of aggregations like SUM or AVG, the addition + on integers and on rationals has to be in the set $OPNS^{2,1}$. We also need the division / on rationals.

3.5 Fact: The syntax of the data signature directly corresponds to a signature DS = (DATA, OPNS, PRED) in the sense of abstract data type theory. The semantics of a data signature is equivalent to an algebra $\mu[DS]$ = ($\mu[DATA]$, $\mu[OPNS]$, $\mu[PRED]$) with carrier sets, functions and relations.

3.6 Definition: (sort expressions)
Let a data signature DS and a set $S \in |SET|$ with DATA $\subseteq$ S together with a (semantic) function $\mu[S] : S \to |SET|$ (such that $\mu[S](d) = \mu[DATA](d)$ for $d \in DATA$ and $\perp \in \mu[S](s)$ for every $s \in S$) be given. The **syntax of the sort expressions over S** is given by the set SORT-EXPR(S) determined by the following rules.

(i) If $s \in S$, then **s** $\in$ SORT-EXPR(S).
(ii) If $s \in$ SORT-EXPR(S), then **set(s)** $\in$ SORT-EXPR(S).
(iii) If $s \in$ SORT-EXPR(S), then **list(s)** $\in$ SORT-EXPR(S).
(iv) If $s \in$ SORT-EXPR(S), then **bag(s)** $\in$ SORT-EXPR(S).
(v) If $s_1,...,s_n \in$ SORT-EXPR(S), then **prod($s_1,...,s_n$)** $\in$ SORT-EXPR(S).

The **semantics of the sort expressions** is a function $\mu[SORT\text{-}EXPR(S)] : SORT\text{-}EXPR(S) \to |SET|$ determined by the following rules.

(i) $\mu[SORT\text{-}EXPR(S)](s) := \mu[S](s)$
(ii) $\mu[SORT\text{-}EXPR(S)](set(s)) := F(\mu[SORT\text{-}EXPR(S)](s)) \cup \{\perp\}$
(iii) $\mu[SORT\text{-}EXPR(S)](list(s)) := (\mu[SORT\text{-}EXPR(S)](s))^* \cup \{\perp\}$
(iv) $\mu[SORT\text{-}EXPR(S)](bag(s)) := B(\mu[SORT\text{-}EXPR(S)](s)) \cup \{\perp\}$
(v) $\mu[SORT\text{-}EXPR(S)](prod(s_1,...,s_n)) :=$
$$(\mu[SORT\text{-}EXPR(S)](s_1) \times ... \times \mu[SORT\text{-}EXPR(S)](s_n)) \cup \{\perp\}$$

$\hat{\mu}[SORT\text{-}EXPR(S)]$ is an abbreviation for $\bigcup_{s \in SORT\text{-}EXPR(S)} \mu[SORT\text{-}EXPR(S)](s)$, i.e., the set of all instances belonging to the sort expression over the set S. □

3.7 Example: In our example S is equal to the set {int, rat, point, string, TOWN, RIVER, COUNTRY, SEA, LAKE, WATERS, DISTRICT, STREET}. The diagrams use three multi-valued sort expression, namely list(point), set(DISTRICT) and set(STREET).

3.8 Fact: The set S together with the semantic function $\mu[S]$ induces a signature DS $\cup$ S = (S, OPNS, PRED) and an algebra $\mu[DS \cup S]$ = ($\mu[S]$, $\mu[OPNS]$, $\mu[PRED]$), which is a conservative and complete extension of $\mu[DS]$ (due to the fact that $\mu[DS](d) = \mu[S](d)$ for $d \in DATA$).

3.9 Remark: The same semantics of sort expressions can also be specified algebraically, if we view a sort expression as a new sort and introduce generating operations in the following way.

$EMPTY_{set(s)} : \to set(s)$
$ADD_{set(s)} : set(s) \times s \to set(s)$

$EMPTY_{bag(s)} : \rightarrow bag(s)$
$ADD_{bag(s)} : bag(s) \times s \rightarrow bag(s)$
$EMPTY_{list(s)} : \rightarrow list(s)$
$ADD_{list(s)} : list(s) \times s \rightarrow list(s)$
$MAKE_{prod(s_1,...,s_n)} : s_1 \times ... \times s_n \rightarrow prod(s_1,...,s_n)$

Additionally, error constants representing $\perp$ could be added. For sets, two equations must be valid, whereas the interpretation of bags is restricted by only one equation. Lists and products are generated "freely" (without restricting equations).

$ADD_{set(s)}(ADD_{set(s)}(S,X),Y) = ADD_{set(s)}(ADD_{set(s)}(S,Y),X)$
$ADD_{set(s)}(ADD_{set(s)}(S,X),X) = ADD_{set(s)}(S,X)$
$ADD_{bag(s)}(ADD_{bag(s)}(S,X),Y) = ADD_{bag(s)}(ADD_{bag(s)}(S,Y),X)$

Because the second set equation is not valid for bags, bags and sets fulfill the following rules (with respect to the {{...}}- and {...}-notation).

$\{\{c_1\}\} \cup \{\{c_2\}\} = \{\{c_2\}\} \cup \{\{c_1\}\} = \{\{c_1,c_2\}\} = \{\{c_2,c_1\}\}$ if $c_1 \neq c_2$
$\{\{c\}\} \cup \{\{c\}\} = \{\{c,c\}\}$
$\{c_1\} \cup \{c_2\} = \{c_2\} \cup \{c_1\} = \{c_1,c_2\} = \{c_2,c_1\}$ if $c_1 \neq c_2$
$\{c\} \cup \{c\} = \{c\}$

3.10 Definition: (operations and predicates induced by the sort expressions)
Let the sort expressions as defined above be given. The **syntax of the operations and predicates induced by the sort expressions** is given by the following sets and functions (s, s_1, ... , s_n refer to arbitrary sort expressions and σ to an element of $OPNS^{2,1}$ with $\sigma : d \times d \rightarrow d$) :

OPNS(S)/PRED(S)	source/arguments	destination	informal description
$CNT_{set(s)}$	: set(s)	$\rightarrow$ <u>int</u>	counts the elements
$IND_{set(s)}$	: set(s)	$\rightarrow$ set(<u>int</u>)	set of indices of a set
$APL_{\sigma.set(d)}$	: set(d)	$\rightarrow$ d	applies a binary operation to a set
$IN_{set(s)}$	: prod(set(s),s)		element-of relation
$CNT_{list(s)}$	: list(s)	$\rightarrow$ <u>int</u>	counts the elements
$IND_{list(s)}$	: list(s)	$\rightarrow$ set(<u>int</u>)	set of indices of a list
$LTB_{list(s)}$	: list(s)	$\rightarrow$ bag(s)	converts a list to a bag
$SEL_{list(s)}$	: prod(list(s),int)	$\rightarrow$ s	selects the i-th element
$POS_{list(s)}$	: prod(list(s),s)	$\rightarrow$ set(<u>int</u>)	set of indices of an element
$APL_{\sigma,list(d)}$	: list(d)	$\rightarrow$ d	applies a binary operation to a list
$IN_{list(s)}$	: prod(list(s),s)		element-of relation
$CNT_{bag(s)}$	: bag(s)	$\rightarrow$ <u>int</u>	counts the elements
$IND_{bag(s)}$	: bag(s)	$\rightarrow$ set(<u>int</u>)	set of indices of a bag
$BTS_{bag(s)}$	: bag(s)	$\rightarrow$ set(s)	converts a bag to a set
$OCC_{bag(s)}$	: prod(bag(s),s)	$\rightarrow$ <u>int</u>	counts the occurences
$APL_{\sigma,bag(d)}$	: bag(d)	$\rightarrow$ d	applies a binary operation to a bag
$IN_{bag(s)}$	: prod(bag(s),s)		element-of relation
$PRJ_{prod(s_1,...,s_n)}$	: prod(s_1,...,s_n)	$\rightarrow$ s_i ($i \in 1..n$)	projection

The **semantics of the operations induced by the sort expressions** is a function μ[OPNS(S)] : OPNS(S) $\rightarrow$ |FUN| and a function μ[PRED(S)] : PRED(S) $\rightarrow$ |REL| determined by the following lines. Domain and codomain of semantic functions and predicates are not given explicitly, but are determined by the following rules : If $\sigma : s_1 \to s_2$, then $\mu[OPNS(S)](\sigma) : \mu[SORT\text{-}EXPR(S)](s_1) \to \mu[SORT\text{-}EXPR(S)](s_2)$, and if $\pi : s$, then $\mu[PRED(S)](\pi) \subseteq \mu[SORT\text{-}EXPR(S)](s)$. Furthermore we abbreviate $\mu[OPNS(S)](\sigma)$ by $\mu(\sigma)$ and $\mu[PRED(S)](\pi)$ by $\mu(\pi)$. All these functions preserve the undefined value $\perp$ ($\mu(\sigma)(\perp) = \perp$ and $\mu(\sigma)(c_1,c_2) = \perp$, if $c_1 = \perp$ or $c_2 = \perp$) and the predicates do not hold for $\perp$ (not $\perp \in \mu(\pi)$).

$$\mu(CNT_{set(s)}) : \{c_1,...,c_n\} \mapsto n \qquad c_i \in \mu[SORT\text{-}EXPR(S)](s)$$

$$\mu(IND_{set(s)}) : \{c_1,...,c_n\} \mapsto \{1,...,n\}$$

$$\mu(APL_{\sigma,set(d)}) : \{c_1,...,c_n\} \mapsto \begin{cases} \perp & \text{if } n=0 \\ c_1 & \text{if } n=1 \\ \mu(\sigma)(c_1,\mu(APL_{\sigma,set(d)})(\{c_2,...,c_n\})) & \text{if } n\geq 2 \end{cases}$$

$$\mu(IN_{set(s)}) := \left\{ (\{c_1,...,c_n\} , c) \mid c \in \{c_1,...,c_n\} \right\}$$

$$\mu(LTB_{list(s)}) : \langle c_1,...,c_n \rangle \mapsto \{\{c_1,...,c_n\}\}$$

$$\mu(SEL_{list(s)}) : (\langle c_1,...,c_n \rangle , i) \mapsto \begin{cases} c_i \text{ if } 1 \leq i \leq n \\ \perp \text{ otherwise} \end{cases}$$

$$\mu(POS_{list(s)}) : (\langle c_1,...,c_n \rangle , c) \mapsto f_{list(s)}(\langle c_1,...,c_n \rangle , c , \{\})$$

The function $f_{list(s)}$: list(s) x s x set(<u>int</u>) $\to$ set(<u>int</u>) is not an operation induced by the sort expressions. It is only part of the definition of POS :

$$f_{list(s)} : (\langle c_1,...,c_n \rangle , c , r) \mapsto \begin{cases} r & \text{if } n=0 \\ f_{list(s)}(\langle c_1,...,c_{n-1} \rangle ,c,r) & \text{if } n\geq 1 \text{ and } c_n \neq c \\ f_{list(s)}(\langle c_1,...,c_{n-1} \rangle ,c,r \cup \{n\}) & \text{if } n\geq 1 \text{ and } c_n = c \end{cases}$$

$$\mu(BTS_{bag(s)}) : \{\{c_1,...,c_n\}\} \mapsto \begin{cases} \{\} & \text{if } n=0 \\ \{c_1\} \cup \mu(BTS_{list(s)})(\{\{c_2,...,c_n\}\}) & \text{if } n\geq 1 \end{cases}$$

$$\mu(OCC_{bag(s)}) : (\{\{c_1,...,c_n\}\} , c) \mapsto \begin{cases} 0 & \text{if } n=0 \\ \mu(OCC_{bag(s)})(\{\{c_2,...,c_n\}\},c) & \text{if } n\geq 1 \text{ and } c \neq c_1 \\ \mu(OCC_{bag(s)})(\{\{c_2,...,c_n\}\},c)+1 & \text{if } n\geq 1 \text{ and } c=c_1 \end{cases}$$

$$\mu(PRJ_{prod(s_1,...,s_n),i}) : (c_1,...,c_n) \mapsto c_i$$

$\mu[APL_{\sigma,list(s)}]$ and $\mu[APL_{\sigma,bag(s)}]$ are defined just as $\mu[APL_{\sigma,set(s)}]$ (analogously for the other functions not mentioned explicitly). □

3.11 Remarks: {...} $\cup$ {...} refers to the union of sets (this guarantees the elemination of duplicates, while {{...}} $\cup$ {{...}} is the union of bags respecting duplicates). For all above functions equational specifications can be given.

We use the following conventions and abbreviations for frequently used operations. If no ambiguities occur, we drop the list-, set-, bag- or prod-indices of the operation symbols and leave out parenthesis in accordance with the usual rules. **SUM** refers to APL_+ dependent on the context (the idea of applying a binary operator to a list of values has also been proposed in [Bi 87]). **MAX** means APL_{max} with max(x,y) = <u>*if*</u> x>y <u>*then*</u> x <u>*else*</u> y. **MIN** is defined analogously. $AVG_{set(d)}(x)$ refers to $APL_{+,set(d)}(x)$ / $CNT_{set(d)}(x)$, analogously for lists and bags. With this convention we have of course AVG(∅) = $\perp$. Instead of $PRJ_i(x)$ or $PRJ_i((x_1,...,x_n))$ we also use the more suggestive **x.i** or **$(x_1,...,x_n)$.i**, where i is a constant between 1 and n. LTS stands for BTS o LTB.

3.12 Example: In the first query of the example presented in chapter 2 AVG -{ ... }- stands for $APL_{+,bag(int)}$ -{ ... }- / $CNT_{bag(int)}$ -{ ... }-. In the second query SUM refers to $APL_{+,bag(int)}$ and BTS was short for $BTS_{bag(string)}$

3.13 Fact: The syntax of the sort expressions introduces a signature

DS ∪ SORT-EXPR(S) = (SORT-EXPR(S), OPNS ∪ OPNS(S), PRED ∪ PRED(S))

and the semantics corresponds to an algebra

μ[DS ∪ SORT-EXPR(S)] = (μ[SORT-EXPR(S)], μ[OPNS ∪ OPNS(S)], μ[PRED ∪ PRED(S)]).

Again, μ[DS ∪ SORT-EXPR(S)] is a conservative and complete extension of μ[DS ∪ S]. Of course, there is an infinite number of sorts and operations in this algebra.

3.14 Definition: (extended entity-relationship schema)

Let a data signature DS be given. The **syntax of an extended entity-relationship schema EER(DS) over DS** is given by

- the sets ENTITY, RELATION, ATTRIBUTE, COMPONENT, CONSTRUCTION ∈ |FISET| and
- the functions *participants, asource, adestination, csource, cdestination, input, output* such that *participants* : RELATION → $ENTITY^+$,
  asource: ATTRIBUTE → ENTITY ∪ RELATION,
  adestination : ATTRIBUTE → { d, set(d), list(d), bag(d) | d ∈ DATA},
  csource: COMPONENT → ENTITY,
  cdestination : COMPONENT → { e, set(e), list(e), bag(e) | e ∈ ENTITY },
  input: CONSTRUCTION → *F*(ENTITY) , and
  output : CONSTRUCTION → *F*(ENTITY).

If r ∈ RELATION with *participants*(r) = $<e_1,...,e_n>$, this is notated as $r(e_1,...,e_n)$. If a ∈ ATTRIBUTE with *asource*(a) = e or *asource*(a) = r and *adestination*(a) = d, this is notated as a : e → d or a : r → d, respectively. If c ∈ COMPONENT with *csource*(c) = e and *cdestination*(c) = e', this is notated as c : e → e'. If c ∈ CONSTRUCTION with *input*(c) = $\{i_1,...,i_n\}$ and *output*(c) = $\{o_1,...,o_m\}$, this is notated as $c(i_1,...,i_n ; o_1,...,o_m)$.

For two distinct constructions c_1,c_2 ∈ CONSTRUCTION the following conditions must hold:

(i) $output(c_1) \cap output(c_2) = \emptyset$

(ii) It is not allowed that $connection^+(e,e)$ holds for some e∈ENTITY, where $connection^+$ is the transitive closure of the relation *connection* defined by: if e_{in} ∈ *input*(c) and e_{out} ∈ *output*(c) for some c ∈ CONSTRUCTION, then $connection(e_{in},e_{out})$ holds.

The **semantics of an extended entity-relationship schema EER(DS)** is given by

- a function μ[ENTITY] : ENTITY → |FISET| such that ⊥ ∈ μ[ENTITY](e) for e ∈ ENTITY,
- a function μ[RELATION] : RELATION → |FISET| such that $r(e_1,...,e_n)$ implies μ[RELATION](r) ⊆ (μ[ENTITY](e_1)-{⊥} x ... x μ[ENTITY](e_n)-{⊥}) ∪ {⊥}, and ⊥∈μ[RELATION](r) for r∈RELATION,
- a function μ[ATTRIBUTE] : ATTRIBUTE → |FUN| such that
  a : e → d implies μ[ATTRIBUTE](a) : μ[ENTITY](e) → μ[SORT-EXPR(DATA)](d), and
  a : r → d implies μ[ATTRIBUTE](a) : μ[RELATION](r) → μ[SORT-EXPR(DATA)](d),
- a function μ[COMPONENT] : COMPONENT → |FUN| such that
  c : e → e' implies μ[COMPONENT](c) : μ[ENTITY](e) → μ[SORT-EXPR(ENTITY)](e'), and
- a function μ[CONSTRUCTION] : CONSTRUCTION → |FUN| such that $c(i_1,...,i_n ; o_1,...,o_m)$ implies $\mu[CONSTRUCTION](c) : \bigcup_{j=1}^{m} \mu[ENTITY](o_j) \rightarrow \bigcup_{k=1}^{n} \mu[ENTITY](i_k)$,
  where each μ[CONSTRUCTION](c) is injective.

Every function μ[ATTRIBUTE](a), μ[COMPONENT](c), and μ[CONSTRUCTION](c) has to preserve the undefined value, e.g., μ[ATTRIBUTE](a)(⊥) = ⊥ and so on. □

3.15 Remarks: The elements of ENTITY are called **entity** (or object) **types**, the elements of RELATION are the **relationship types**, and the elements of ATTRIBUTE are the **attributes names**, all known from the ER Model defined in [Ch 76]. μ[ENTITY](e) is the set of entities belonging to the entity type e, μ[RELATION](r) defines which entities are related by r, and μ[ATTRIBUTE](a) gives the attributes of entities or relationships (related entities).
Additionally, we have a set COMPONENT of **component names** to model complex entity types. μ[COMPONENT](c) gives the components of entities, i.e., an entity may have as part (component) of itself another entity, a set, list, or bag of entities.
To provide modelling primitives for specialization and generalization [SS 77] we introduce the set CONSTRUCTION. Its elements are called **type constructions**. A type construction c may be regarded as a rearrangement of entity types. Starting with non-constructed or already defined entity types in *input*(c) the new entity types in *output*(c) are constructed. Although all introduced entity types are disjoint by definition, the constructed entity types may be considered as a new classification of the entities in the input types. Formally, we express this fact by the function

$$\mu[\text{CONSTRUCTION}](c) : \bigcup_{j=1}^{m} \mu[\text{ENTITY}](o_j) \rightarrow \bigcup_{k=1}^{n} \mu[\text{ENTITY}](i_k)$$

yielding for an output entity the corresponding input entity it refers to. Since this function is injective, every output entity corresponds to exactly one input entity. But an input entity need not appear in any output type at all, because the function is not required to be surjective. This semantics of CONSTRUCTION is equivalent to demanding that every output entity "is" an input entity:

$$\bigcup_{j=1}^{m} \mu[\text{ENTITY}](o_j) \subseteq \bigcup_{k=1}^{n} \mu[\text{ENTITY}](i_k)$$

The restricting conditions for constructions guarantee that we have for a constructed entity e uniquely determined constructions $c_1,\ldots,c_n$, other constructed entities $e_1,\ldots,e_{n-1}$ and a non-constructed (basic) entity e_n, such that

$$e \xrightarrow{\mu[c_1]} e_1 \xrightarrow{\mu[c_2]} e_2 \xrightarrow{\mu[c_3]} e_3 \;\cdots\; e_{n-1} \xrightarrow{\mu[c_n]} e_n$$

In other words, for a constructed entity there is a uniquely determined construction rule.
As for DATA, we assume the element ⊥ to be element of every entity type in ENTITY. Thus, every attribute and component is per default optional. Indeed, this can be excluded by additional integrity constraints.

3.16 Example: In our example the following identities hold : ENTITY = {TOWN, RIVER, COUNTRY, SEA, LAKE, WATERS, DISTRICT, STREET}, RELATION = {lies-at, lies-in, flows-through, flows-into}, COMPONENT = {districts, streets} and CONSTRUCTION = {are}. The corresponding function participants, etc. can be derived from the diagrams.

3.17 Fact: The algebraization of an extended entity-relationship schema will be done in two steps. First we introduce the signature

EER-BASE = (SORT-EXPR(DATA) ∪ ENTITY ∪ RELATION,
OPNS ∪ OPNS(DATA) ∪ OPNS(EER-BASE),
PRED ∪ PRED(DATA)) with

OPNS(EER-BASE) = $\{ mk_r : e_1 \times ... \times e_n \to r \mid r(e_1,...,e_n) \in EER(DS) \} \cup$
$\{ a : e \to d \in EER(DS) , a : r \to d \in EER(DS) \} \cup$
$\{ c_{o_j;i_k} : o_j \to i_k \mid c(i_1,...,i_n;o_1,...,o_m) \in EER(DS), j \in 1..m, k \in 1..n \}$

Given the semantics of an EER schema, an algebra μ[EER-BASE] can be defined as follows. The carriers are given by μ[SORT-EXPR(DATA)] (which is based on μ[DATA]), μ[ENTITY] and μ[RELATION]. The functions μ[OPNS] and μ[OPNS(DATA)] and the predicates μ[PRED] and μ[PRED(DATA)] are fixed by the algebra μ[DS ∪ SORT-EXPR(DATA)]. It remains to define the functions μ[OPNS(EER-BASE)] :

- $\mu(mk_r) : (\underline{e}_1,...,\underline{e}_n) \mapsto$ *if* $(\underline{e}_1,...,\underline{e}_n) \in \mu(r)$ *then* $(\underline{e}_1,...,\underline{e}_n)$ *else* $\perp$
- μ[OPNS(EER-BASE)](a) is given by μ[ATTRIBUTE](a).
- $\mu[c_{o_j;i_k}] : \underline{o}_j \mapsto$ *if* μ[CONSTRUCTION](c)($\underline{o}_j$) ∈ μ[i_k] *then* μ[CONSTRUCTION](c)($\underline{o}_j$) *else* ⊥

In the second step we define the final signature by

EER = (SORT-EXPR(DATA ∪ ENTITY) ∪ RELATION,
OPNS ∪ OPNS(DATA ∪ ENTITY) ∪ OPNS(EER-BASE) ∪ OPNS(EER),
PRED ∪ PRED(DATA ∪ ENTITY)) with

OPNS(EER) = { c : e → e' ∈ EER(DS) | c ∈ COMPONENT }.

The corresponding algebra μ[EER] has carriers equal to μ[EER-BASE] except that the additional carriers are determined by μ[SORT-EXPR(DATA ∪ ENTITY)] (which is based on μ[DATA] of the DS level and μ[ENTITY] of the EER-BASE level). The additional functions in OPNS(EER) are determined by μ[COMPONENT](c).

Again, the considered algebras are strongly related : μ[EER] is a conservative and complete extension of μ[EER-BASE] which extends μ[DS ∪ SORT-EXPR(DATA)].

3.18 Example: In the signature EER for our example we will have (among others) the following sorts and functions.

name : TOWN → <u>string</u> ; population : TOWN → <u>int</u> ; geometry : TOWN → list(<u>point</u>)
$mk_{flows\text{-}through}$: RIVER x COUNTRY → flows-through
length : flows-through → <u>rat</u>
districts : TOWN → set(DISTRICT)
$are_{WATERS;RIVER}$: WATERS → RIVER ; $are_{WATERS;SEA}$: WATERS → SEA

4. The Entity-Relationship Calculus

4.1 Notations: Before discussing the EER calculus in more detail, we introduce some abbreviations. Let an EER schema EER(DS) be given. SORT refers to the union of all data, entity, and relationship sorts : SORT := DATA ∪ ENTITY ∪ RELATION. Furthermore, we extend the notions of operations (OPNS) and predicates (PRED) introduced for data signatures to sort expressions and EER schemas resulting in the following notations:

$OPNS_{EXPR}$:= { CNT, LTB, BTS, LTS, SEL, IND, POS, OCC, PRJ_i, AVG, SUM, MIN, MAX }
∪ { APL_σ | $\sigma \in OPNS^{2,1}$ }
$PRED_{EXPR}$:= { IN }
$OPNS_{EER}$:= ATTRIBUTE ∪ COMPONENT
$PRED_{EER}$:= RELATION

Please note that (for example) CNT in the above definition of $OPNS_{EXPR}$ is short for

$CNT_{list(s)}$, $CNT_{set(s)}$, $CNT_{bag(s)}$ with s∈SORT-EXPR(SORT). We can drop the indices, because no ambiguities shall occur. All operations and predicates are now concentrated to

OPNS := $OPNS_{DS}$ ∪ $OPNS_{EER}$ ∪ $OPNS_{EXPR}$

PRED := $PRED_{DS}$ ∪ $PRED_{EER}$ ∪ $PRED_{EXPR}$

μ[SORT], μ[$OPNS_{EXPR}$], μ[$PRED_{EXPR}$] μ[$OPNS_{EER}$], μ[$PRED_{EER}$], μ[OPNS], and μ[PRED], are determined by the corresponding μ's on the right hand side.

Let us remind you that the (interpretation of the) sets DATA, ENTITY, and RELATION are disjoint as well as all (interpretations of) set-/list-/bag-constructed sort expressions (except the special value ⊥). Due to the disjointness there is a unique function

sort : $\hat{\mu}$[SORT-EXPR(SORT)] - {⊥} → SORT-EXPR(SORT)

yielding for an instance of $\hat{\mu}$[SORT-EXPR(SORT)] the sort it belongs to; the value ⊥ belongs to every sort. However, we shall only build sort expressions over DATA ∪ ENTITY. Thus, we use the function *sort* only in the restricted form

sort : $\hat{\mu}$[SORTEXPRS] - {⊥} → SORTEXPRS

with SORTEXPRS := SORT-EXPR(DATA ∪ ENTITY) ∪ RELATION.

4.2 Definition: (variables and assignments)

Let a set VAR of *variables* and a function *type* : VAR → SORTEXPRS be given. The ***set of assignments*** ASSIGN is defined by

ASSIGN := { α∈|FUN| | α : VAR → $\hat{\mu}$[SORTEXPRS] and α(v)≠⊥ implies type(v)=sort(α(v)) }

The special assignment ε : VAR → {⊥} is called the ***empty assignment***. In the following v∈VAR_s stands for v∈VAR and type(v)=s. □

4.3 Remark: Our calculus has to leave the usual hierarchical structure of calculi (terms - atomic formulas - formulas), because we will allow arbitrary bag-valued terms of the form

$$\{\!\{ t_1 , \dots , t_n \mid \delta_1 \wedge \dots \wedge \delta_n \wedge \varphi \}\!\}$$

The terms t_1 , ... , t_n compute the target information, $\delta_1 \wedge \dots \wedge \delta_n$ are declarations of variables which can again use bag-valued terms of the above form to restrict the domain of a variable to a finite set (like this has been done in the second query in chapter 2).

4.4 Definition: (ranges)

The **syntax of ranges** is given by a set *RANGE* and functions *domain* : RANGE → SORTEXPRS and *free* : RANGE → *F*(VAR) (ρ∈$RANGE_s$ stands for ρ∈RANGE and domain(ρ)=s).

(i) If s∈ENTITY or s∈RELATION, then ***s∈RANGE$_s$*** and $free_{RANGE}$(s):=∅.

(ii) If t∈$TERM_{set(s)}$ with s∈SORTEXPRS, then ***t∈RANGE$_s$*** and $free_{RANGE}$(t):=$free_{TERM}$(t).

The **semantics of ranges** is a function μ[RANGE] : RANGE x ASSIGN → $\hat{\mu}$[SORTEXPRS].

(i) μ[RANGE](s,α):=μ[SORT](s).

(ii) μ[RANGE](t,α):=μ[TERM](t,α). □

4.5 Example: The queries in chapter 2 used the following ranges: TOWN, COUNTRY and BTS {{ government(c) | c : COUNTRY }}

4.6 Definition: (declarations)

The **syntax of declarations** is given by a set *DECL* and functions *free*, *decl* : DECL → *F*(VAR).

(i) If v∈VAR, $\rho_1,\dots,\rho_n$∈$RANGE_{type(v)}$, and not v∈free(ρ_1)∪...∪free(ρ_n), then ***v:ρ$_1$∪...∪ρ$_n$∈DECL***, free(v:ρ_1∪...∪ρ_n):=free(ρ_1)∪...∪free(ρ_n), and decl(v:ρ_1∪...∪ρ_n):={v}.

(ii) If $v \in VAR$, $\rho_1,\ldots,\rho_n \in RANGE_{type(v)}$, $\delta \in DECL$ with $free(\rho_1) \cup \ldots \cup free(\rho_n) \subseteq decl(\delta)$, and not $v \in free(\delta)$, then $v{:}\rho_1 \cup \ldots \cup \rho_n;\delta \in DECL$, $free(v{:}\rho_1 \cup \ldots \cup \rho_n;\delta) := free(\delta)$, and $decl(v{:}\rho_1 \cup \ldots \cup \rho_n;\delta) := \{v\} \cup decl(\delta)$.

The **semantics of declarations** is a relation $\mu[DECL] \subseteq DECL \times ASSIGN$.

(i) $(v{:}\rho_1 \cup \ldots \cup \rho_n, \alpha) \in \mu[DECL]$ ***iff*** $\alpha(v) \in \mu[RANGE](\rho_1,\alpha)$ or ... or $\alpha(v) \in \mu[RANGE](\rho_n,\alpha)$.

(ii) $(v{:}\rho_1 \cup \ldots \cup \rho_n;\delta,\alpha) \in \mu[DECL]$ ***iff*** ($\alpha(v) \in \mu[RANGE](\rho_1,\alpha)$ or ... or $\alpha(v) \in \mu[RANGE](\rho_n,\alpha)$) and $(\delta,\alpha) \in \mu[DECL]$. □

4.7 Remarks: At first, we do not only allow the form $(v{:}\rho)$ but also the more general (i) $(v{:}\rho_1 \cup \ldots \cup \rho_n)$. This form is necessary to express every term of the relational algebra [Ma 83], especially the union of sets, in our calculus. In declarations of the form (ii) $(v_1{:}\rho_1)\ ;\ldots;\ (v_n{:}\rho_n)$ the variables $v_1,\ldots,v_n$ (in front of the colons) are declared (each $(v_i{:}\rho_i)$ may be a union of form (i)). Each range ρ_i (i=1..n-1) may contain free variables, but only from the set of previously declared variables $\{v_{i+1},\ldots,v_n\}$, except ρ_n, which can have other free variables. Indeed, the later ones are the free variables of the declaration, which must be disjoint from the variables declared in the declaration.

Each variable v is bound to a finite set of values determined in the following way.

v_n can take each value from $\mu[RANGE](\rho_n,\alpha)$,

v_{n-1} can take each value from $\mu[RANGE](\rho_{n-1},\alpha)$ "possibly dependent on the value assigned to v_n by α", and so on.

4.8 Example: The queries of chapter 2 declared the following variables:

t : TOWN ; c : COUNTRY ; g : BTS -{ government(c) | c : COUNTRY }-

Declarations with free variable t of type TOWN are the following ones :

- "p : LTS(geometry(t))" declares the variable p (of type <u>point</u>).
- "s : streets(d) ; d : districts(t)" declares s (of type STREET) and d (of type DISTRICT).
- "n : BTS -{ name(r') | r' : RIVER ∧ r : RIVER ∧ lies-at(t,r) ∧ flows-into(r',WATERS(r)) }- ∪ BTS -{ name(r) | r : RIVER ∧ lies-at(t,r) }-" declares n (of type <u>string</u>).

4.9 Definition: (terms)

The **syntax of terms** is given by a set *TERM*, and functions *sort* : TERM → SORTEXPRS and *free* : TERM → *F*(VAR) ($t \in TERM_s$ stands for $t \in TERM$ and $sort(t)=s$).

(i) If $v \in VAR_s$, then $v \in TERM_s$ and $free(v) := \{v\}$.

(ii) If $v \in VAR_r$ and $r(e_1,\ldots,e_n) \in RELATION$, then $v.i \in TERM_{e_i}$ (i=1..n) and $free(v.i) := \{v\}$.

(iii) If $c \in CONSTRUCTION$, $s_{in} \in input(c)$, $s_{out} \in output(c)$, $t_{in} \in TERM_{s_{in}}$, and $t_{out} \in TERM_{s_{out}}$, then $s_{out}(t_{in}) \in TERM_{s_{out}}$ with $free(s_{out}(t_{in})) := free(t_{in})$ and $s_{in}(t_{out}) \in TERM_{s_{in}}$ with $free(s_{in}(t_{out})) := free(t_{out})$.

(iv) If $\omega{:}s_1 \times \ldots \times s_n \to s \in OPNS$ and $t_i \in TERM_{s_i}$ (i=1..n), then $\omega(t_1,\ldots,t_n) \in TERM_s$ with $free(\omega(t_1,\ldots,t_n)) := free(t_1) \cup \ldots \cup free(t_n)$.

(v) If $t_i \in TERM_{s_i}$ (i=1..n), $\delta_j \in DECL$ (j=1..k), $\varphi \in FORM$, and $decl(\delta_i) \cap decl(\delta_j) = decl(\delta_i) \cap free(\delta_j) = \emptyset$ ($i \neq j$), then -{ $t_1,\ldots,t_n$ / $\delta_1 \wedge \ldots \wedge \delta_k \wedge \varphi$ }- $\in TERM_{bag(prod(s_1,\ldots,s_n))}$ and free(-{ $t_1,\ldots,t_n$ | $\delta_1 \wedge \ldots \wedge \delta_k \wedge \varphi$ }-) := $(free(\delta_1) \cup \ldots \cup free(\delta_k) \cup free(t_1) \cup \ldots \cup free(t_n) \cup free(\varphi)) - (decl(\delta_1) \cup \ldots \cup decl(\delta_k))$.

The **semantics of terms** is a function $\mu[TERM] : TERM \times ASSIGN \to \hat{\mu}[SORTEXPRS]$.

(i) $\mu[TERM](v,\alpha) := \alpha(v)$.

(ii) $\mu[TERM](v.i,\alpha) :=$ <u>*if*</u> $\mu[TERM](v,\alpha) = (x_1,\ldots,x_n)$ <u>*then*</u> x_i <u>*else*</u> $\perp$.

(iii) μ[TERM]($s_{out}(t_{in}),\alpha$):=*if* there is e$\in\mu$[ENTITY](s_{out}) : μ[CONS.](c)(e)=μ[TERM](t_{in},α) *then* e *else* $\perp$.

μ[TERM]($s_{in}(t_{out}),\alpha$):=*if* μ[CONS.](c)(μ[TERM](t_{out},α)) $\in\mu$[ENTITY](s_{in}) *then* μ[CONS.](c)(μ[TERM](t_{out},α)) *else* $\perp$.

(v) μ[TERM]($\omega(t_1,\ldots,t_n),\alpha$):=$\mu$[OPNS]($\omega$)($\mu$[TERM]($t_1,\alpha$),...,$\mu$[TERM]($t_n,\alpha$)).

(vi) μ[TERM]($\{\!\!\{ t_1,\ldots,t_n\ \delta_1\wedge\ldots\wedge\delta_k\wedge\varphi \}\!\!\}$, α) := {{ (μ[TERM](t_1,α'),...,μ[TERM](t_n,α')) | there is $\alpha'\in$ASSIGN with $\alpha'(v)=\alpha(v)$ for $v\in$VAR-(decl(δ_1)$\cup\ldots\cup$decl(δ_k)) and (δ_1,α')$\in\mu$[DECL] and ... and (δ_k,α')$\in\mu$[DECL] and (φ,α')$\in\mu$[FORM] }}. □

4.10 Remark: If no ambiguity arises, we use entity types instead of integers in the case of terms according to point (ii), e.g., ft.RIVER instead of ft.1 (with type(ft)=flows-through).

4.11 Example: The queries in chapter 2 are term according to point (vi) of the above definition and have sort bag(prod(<u>string</u>,<u>rat</u>)) and bag(prod(<u>string</u>,<u>int</u>)), respectively.

- $\{\!\!\{$ population(c) | c : COUNTRY $\wedge$ government(c)=g $\}\!\!\}$ is a terms of sort bag(<u>int</u>) with free variable g. It was used in the second query in chapter 2.
- $\{\!\!\{$ r | r : RIVER $\wedge$ lies-in(r,c) $\}\!\!\}$ is a term of sort bag(RIVER) with free variable c.

4.12 Definition: (formulas)

The **syntax of formulas** is given by a set *FORM* and a function *free* : FORM $\to$ *F*(VAR).

(i) If π:s_1x...x$s_n\in$PRED and $t_i\in$TERM$_{s_i}$ (i=1..n), then ***$\pi(t_1,\ldots,t_n)\in$FORM*** and free($\pi(t_1,\ldots,t_n)$):=free(t_1)$\cup\ldots\cup$free(t_n).

(ii) If $t_1,t_2\in$TERM$_s$, then ***t_1=$t_2\in$FORM*** and free(t_1=t_2):=free(t_1)$\cup$free(t_2).

(iii) If t$\in$TERM, then ***UNDEF(t)$\in$FORM*** and free(UNDEF(t)):=free(t).

(iv) If $\varphi\in$FORM, then ***$\neg(\varphi)\in$FORM*** and free($\neg(\varphi)$):=free(φ).

(v) If $\varphi_1,\varphi_2\in$FORM, then ***$(\varphi_1\vee\varphi_2)\in$FORM*** and free(($\varphi_1\vee\varphi_2$)):=free($\varphi_1$)$\cup$free($\varphi_2$).

(vi) If $\varphi\in$FORM and $\delta\in$DECL, then ***$\exists\delta(\varphi)\in$FORM*** and free($\exists\delta(\varphi)$):=(free($\varphi$)-decl($\delta$))$\cup$free($\delta$).

The **semantics of formulas** is a relation μ[FORM] $\subseteq$ FORM x ASSIGN.

(i) ($\pi(t_1,\ldots,t_n),\alpha$)$\in\mu$[FORM] *iff* ($\mu$[TERM]($t_1,\alpha$),...,$\mu$[TERM]($t_n,\alpha$))$\in\mu$[PRED]($\pi$).

(ii) (t_1=t_2,α)$\in\mu$[FORM] *iff* μ[TERM](t_1,α)=μ[TERM](t_2,α).

(iii) (UNDEF(t),α)$\in\mu$[FORM] *iff* μ[TERM](t,α)=$\perp$.

(iv) ($\neg(\varphi),\alpha$)$\in\mu$[FORM] *iff* not (φ,α)$\in\mu$[FORM].

(v) (($\varphi_1\vee\varphi_2$),α)$\in\mu$[FORM] *iff* (φ_1,α)$\in\mu$[FORM] or (φ_2,α)$\in\mu$[FORM].

(vi) ($\exists\delta(\varphi),\alpha$)$\in\mu$[FORM] *iff* there is $\alpha'\in$ASSIGN with $\alpha'(v)=\alpha(v)$ for $v\in$VAR-decl(δ) and (φ,α')$\in\mu$[FORM] and (δ,α')$\in\mu$[DECL]. □

4.13 Remark: We also use the other logical connectives $\wedge$, $\Rightarrow$, *xor*, etc. and the quantifier $\forall$ with the usual semantics. They can be defined by means of the above definitions.

4.14 Definition: (queries)

The **syntax of queries** is the set QUERY:

- If t$\in$TERM with sort(t)$\in$SORT-EXPR(DATA) and free(t) = $\emptyset$, then t$\in$QUERY.

The **semantics of queries** is a function μ[QUERY] : QUERY $\to$ $\hat{\mu}$[SORT-EXPR(DATA)].

- μ[QUERY](t) := μ[TERM](t,ε). □

4.15 Facts:

- Every multi-valued term $t \in TERM$ is evaluated to a finite set, bag, or list for every assignment α.
- Since a query is a special case of a term, any query yields a finite result: The EER calculus is safe.
- The EER calculus is relationally complete (if every relation is modelled by an entity type).

The proofs can be found in [HG 88]. Thus our calculus preserves nice properties of the relational calculi [Ma 83], but on the other hand it is also more expressive. In the classical calculi it is not possible to compute for instance the cardinality (i.e., the number of acually stored tuples) of a relation $R\ (\ a_1 : D_1\ , \ldots\ , a_n : D_n\ )$. Of course, this presents no problem in our approach, because we can use data type operations : $CNT_{bag(R)}$ $\{\!\!\{\ r \mid r : R\ \}\!\!\}$. We here assume the relation R is modelled by an entity type R with attributes $a_1,\ldots,a_n$.

5. Integrity Constraints

5.1 Concept: We now use our calculus to formulate integrity constraints. An integrity constraint is a formula of the calculus without free variables. These formulas restrict the class of EER-algebras to algebras where the formulas hold.

5.2 Definition: (functional restriction)

The **syntax of a functional restriction** for a relation $r(e_1,\ldots,e_n)$ is a pair $(f=f_1,\ldots,f_m, g=g_1,\ldots,g_k)$ of subsets of $e_1,\ldots,e_n$. The **semantics of a functional restriction** is the following formula.

$$(\ \forall\ v\ ,\ w : r\ )\quad v.f_1 = w.f_1 \wedge \ldots \wedge v.f_m = w.f_m \quad\Rightarrow\quad v.g_1 = w.g_1 \wedge \ldots \wedge v.g_k = w.g_k \quad \square$$

5.3 Remark: Please notice, it is possible to formulate more than one functional restriction for a single relationship. For instance, if we have a relationship r(A,B,C) with three entity types, one can demand that r is a function from A to C as well as from C x B to A.

5.4 Example: In the example in chapter 2 the functional restriction ({RIVER},{WATERS}) on the relationship flows-into was given. It also makes sense to require the functional restriction ({TOWN},{COUNTRY}) for lies-in in order to express that a town lies in exactly one country.

5.5 Definition: (key specification)

The **syntax of a key specification** for an entity type e is given by a subset $a_1,\ldots,a_n$ of the attributes of e, a subset $c_1,\ldots,c_m$ of the components of e and a subset $r_1,\ldots,r_m$ of the relationships of e. The **semantics of a key specification** is given by the formula

$$(\ \forall\ v\ ,\ w : e\ )\ \ v \neq w\ \ \Rightarrow$$
$$[\ a_1(v) \neq a_1(w) \vee \ldots \vee a_n(v) \neq a_n(w) \vee$$
$$c_1(v) \neq c_1(w) \vee \ldots \vee c_m(v) \neq c_m(w) \vee$$
$$(\ \exists *_1\ )\ r_1(v,*_1)\ \mathit{xor}\ r_1(w,*_1) \vee \ldots \vee (\ \exists *_m\ )\ r_m(v,*_m)\ \mathit{xor}\ r_m(w,*_m)\ ] \qquad \square$$

5.6 Remarks: The $*$-notation has to be explained shortly : Suppose r(A,B,C) is given. Then the relationship r is part of a key for the entity type B, if the following formula holds.

$(\forall\, v , w : B)$ ...

$$(\exists\, x_A : A) (\exists\, x_C : C)\ r(x_A,v,x_C)\ \mathit{xor}\ r(x_A,w,x_C)$$

...

Please notice, the other direction of the implication in the key specification formula trivially holds. Thus, key specifications can be regarded as characterizations of equality.

Key relations also allow more subtle objects than key functions [EDG 86]. Consider the following simple example:

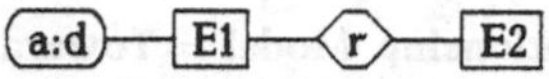

Suppose the data type d has only two values 0 and 1, the attribute a is the only key for E1 and the relationship r the only one for E2. Because the attribute a can only take two values there are (up to isomorphism) only two possible entities $\underline{el}_1$ and $\underline{el}_2$ of type E1, which can be observed to be inequivalent : a($\underline{el}_1$)=0 and a($\underline{el}_2$)=1. In the case of a key function r from E2 to E1 there are also only two observable inequivalent entities for E2. But, if r is an arbitrary relation, there are (up to isomorphism) four possible entities for E2 : $\underline{e2}_1$ related with no E1-entity, $\underline{e2}_2$ related with $\underline{el}_1$, $\underline{e2}_3$ related with $\underline{el}_2$ and $\underline{e2}_4$ related with $\underline{el}_1$ and $\underline{el}_2$. It is an open problem how to construct "universes" [EDG 86, SSE 87] for key relations. The above example suggests some kind of power set construction yielding a final algebra analogously to [EDG 86, SSE 87].

5.7 Example: The example in chapter 2 required that {name,flows-through} is a key for RIVER. For the entity type TOWN on can demand {name,lies-in} as well as {geometry} to be a key :

$(\forall\, t , t' : TOWN)$

$$t \neq t' \;\Rightarrow\; name(t) \neq name(t') \vee (\exists\, c : COUNTRY)\ lies\text{-}in(t,c)\ \mathit{xor}\ lies\text{-}in(t',c)$$

$$\wedge$$

$$t \neq t' \;\Rightarrow\; geometry(t) \neq geometry(t')$$

The last line says : If two towns have the same geometry (the same list of points representing the town's border), then the towns are identical. This example shows that more than one key for a single entity type can make sense.

5.8 Definition: (cardinality constraint)

The syntax of a cardinality constraint for a set-, list- or bag-valued attribute or component f of entity type e is given by a pair of integers (low,high). The semantics is given by

$(\forall\, v : e)\quad low \leq CNT(f(v)) \;\wedge\; CNT(f(v)) \leq high.$ □

5.9 Example: The example in chapter 2 restricted the number of streets for a district. These kind of constraints cannot be expressed in the "classical" relational calculi. Another such typical example again involves aggregate functions. Domain or tuple calculus cannot express a constraint like "The average age of the ministers of a country should not be greater than 65". If there is a component "ministers : COUNTRY → set(PERSON)" and an attribute "age : PERSON → int", this constraint reads in our calculus like

$\forall$ c:COUNTRY (AVG -[age(p) | p:PERSON $\wedge$ p IN ministers(c)]- $\leq$ 65).

Acknowledgement

Thanks go to Uwe Hohenstein for our collaboration on the model and the calculus. Special acknowledgement goes to Jugoslavia : For the kindness of the people of "Starigrad Paklenica" and for the uniqueness of the "Plitvička Jezera" region.

References

[Bi 87] Bird, R.S.: An Introduction to the Theory of Lists. Logic of Programming and Calculi of Discrete Design. Nato ASI Series, Vol. F36, Springer-Verlag Berlin Heidelberg 1987 (M. Broy, ed.) (pp. 5-42)

[Bü 87] Bültzingsloewen, G. v.: Translating and Optimizing SQL Queries Having Aggregates. 13th VLDB, Brighton 1987

[Ce 83] Ceri, S. (ed.): Methodology and Tools for Database Design. North-Holland, Amsterdam 1983

[Ch 76] Chen, P.P.: The Entity-Relationship Model - Towards a Unified View of Data. ACM Transactions on Database Systems, Vol. 1, No. 1, March 1976 (pp. 9-36)

[DMW 82] Dosch, W. / Mascari, G. / Wirsing, M. : On the Algebraic Specification of Databases. Proc. 8th VLDB, 1982

[EDG 86] Ehrich, H.-D. / Drosten, K. / Gogolla, M.: Towards an Algebraic Semantics for Database Specification. Proc. IFIP TC2 Working Conference on Knowledge & Data "DS-2", Albufeira (Portugal), November 1986

[EKW 78] Ehrig, H. / Kreowski, H.-J. / Weber, H. : Algebraic Specification Schemes for Data Base Systems. Proc. 4th VLDB, 1978

[EM 85] Ehrig, H. / Mahr, B.: Fundamentals of Algebraic Specification I. Springer-Verlag, Berlin 1985

[GMS 83] Golshani, F. / Maibaum, T. / Sadler, M. : A Modal System of Algebras for Database Specification and Query/Update Language Support. Proc. 9th VLDB, 1983

[HG 88] Hohenstein, U. / Gogolla, M. : Towards a Semantic View of an Extended Entity-Relationship Model. TU Braunschweig, Informatik-Bericht Nr. 88-02, 1988. Short Version: A Calculus for an Extended Entity-Relationship Model Incorporating Arbitrary Data Operations and Aggregate Functions. Proc. 7th Int. Conf. on Entity-Relationship Approach, North Holland, 1988 (to appear)

[JS 82] Jaeschke, G. / Schek, H.-J.: Remarks on the Algebra of Nonfirst Normal Form Relations. Proc. of the 1st ACM SIGACT-SIGMOD Symp. on Principles of Database Systems. Los Angeles (California), March 1982 (pp. 124-146)

[Kl 82] Klug, A.: Equivalence of Relational Algebra and Relational Calculus Query Languages Having Aggregate Functions. Journal of the ACM, Vol. 29, No. 3, July 1982 (pp. 699-717)

[KMS 85] Khosla, S. / Maibaum, T. / Sadler, M. : Database Specification. Proc. IFIP Working Conference on Database Semantics DS-1, 1985

[Ma 83] Maier, D: The Theory of Databases. Computer Sciene Press, Rockville MD 1983

[ÖÖM 87] Özsoyoglu, G. / Özsuyoglu, Z.M. / Matos, V.: Extending Relational Algebra and Relational Calculus with Set-Valued Attributes and Aggregate Functions. ACM Transactions on Database Systems, Vol. 12, No. 4, December 1987 (pp. 566-592)

[PS 85] Parent, C. / Spaccapietra, S.: An Algebra for a General Entity-Relationship Model. IEEE Transactions on Software Engineering, Vol. 11, No. 7, July 1985 (pp. 634-643)

[Sh 81] Shipman, D.W.: The Functional Data Model and the Data Language DAPLEX. ACM Transactions on Database Systems, Vol. 6, No. 1, 1981 (pp. 140-173)

[Su 87] Subieta, K. : Denotational Semantics of Query Languages. Information Systems, Vol. 12, No. 3, 1987 (pp. 69-82)

[SS 77] Smith, J.M. / Smith, D.C.P.: Database Abstractions: Aggregation and Generalization. ACM Transactions on Database Systems, Vol. 2, No. 2, 1977 (pp. 105-133)

[SS 86] Schek, H.-J. / Scholl, M.H.: The Relational Model with Relation-Valued Attributes. Information Systems, Vol. 11, No. 2, 1986 (pp. 137-147)

[SSE 87] Sernadas, A. / Sernadas, C. / Ehrich, H.-D.: Object-Oriented Specification of Databases: An Algebraic Approach. 13th VLDB, Brighton 1987

[TF 82] Teorey, F.J. / Fry, J.P.: Design of Database Structures.Prentice-Hall, Engelwood Cliffs (N.J.), 1982

Decidable Boundedness Problems for Hyperedge–Replacement Graph Grammars

Annegret Habel *, Hans-Jörg Kreowski *, Walter Vogler **

Abstract

Consider a class C of hyperedge–replacement graph grammars and a numeric function on graphs like the number of edges, the degree (i.e., the maximum of the degrees of all nodes of a graph), the number of simple paths, the size of a maximum set of independent nodes, etc. Each such function induces a Boundedness Problem for the class C: Given a grammar HRG in C, are the function values of all graphs in the language $L(HRG)$, generated by HRG, bounded by an integer or not? We show that the Boundedness Problem is decidable if the corresponding function is compatible with the derivation process of the grammars in C and if it is composed of maxima, sums, and products in a certain way. This decidability result applies particularly to the examples listed above.

1. Introduction

Context–free graph grammars (like edge– and hyperedge–replacement grammars as investigated, e.g., by Bauderon and Courcelle [BD 87] or in [HK 85+87b] or like boundary NLC grammars as introduced by Rozenberg and Welzl [RW 86a]) have been studied intensively for some time now because of – at least – two reasons:

(1) Although their generative power is intentionally restricted, they cover many graph languages interesting from the point of view of applications as well as of graph theory (for example, certain types of flow diagrams, PASCAL syntax diagrams, certain types of Petri nets, graph representations of functional expressions, series–parallel graphs, outerplanar graphs, k–trees, graphs with cyclic bandwidth $\leq k$).

(2) Of all classes of graph grammars discussed in the literature, they seem to render the most attractive theory with a variety of results on structure, decidability and complexity (see, e.g., Arnborg, Lagergren and Seese [ALS 88], Bauderon and Courcelle [BC 87, Co 87], Della Vigna and Ghezzi [DG 78], Lautemann [La 88], Lengauer and Wanke [LW 88], Rozenberg and Welzl [86a+b], Slisenko [Sl 82], and [HK 83 + 85 + 87b, HKV 87, Kr 79]).

In particular, Courcelle [Co 87], Arnborg et al. [ALS 88], Lengauer and Wanke [LW 88], and [HKV 87] present syntactic and semantic conditions such that, for a graph property P satisfying the conditions, the following hold for all context–free graph grammars of the types considered in the respective papers:

(1) It is decidable whether (or not) some graph with property P is generated.

(2) It is decidable whether (or not) all generated graphs have property P.

(3) It is decidable in linear time whether (or not) a generated graph represented by a derivation (or something equivalent) has property P.

The results apply to properties such as connectivity, planarity, k–colorability, existence of Hamiltonian and Eulerian paths and cycles.

*** Universität Bremen, Fachbereich Mathematik und Informatik, Postfach 330440, D-2800 Bremen 33**

**** Technische Universität München, Institut für Informatik, Postfach 202420, D-8000 München 2**

Based on the framework of hyperedge-replacement graph grammars, we continue this line of consideration in this paper. We are going to investigate the decidability of a different type of problems concerning functions on graphs and above all numeric quantities like the numbers of nodes, edges and paths, the node degree, maximum and minimum lengths of paths and cycles, etc. The kind of question we ask for a class of grammars may be called *Boundedness Problem*. It is as follows:

(4) Is it decidable whether (or not), concering a particular quantity, the values of all graphs generated by a grammar are bounded?

For example, we want to know whether the node degree or the number of paths grow beyond any bound within a graph language. In the main result, we show that such a Boundedness Problem is decidable for a class of hyperedge-replacement grammars if the corresponding quantity function is built up by maxima, sums and products and if the function is compatible with the derivation process of the given grammars. Examples of this kind are the bounded-node-degree problem, the bounded-maximum-path-length problem, the bounded-maximum-number-of-paths problem and others. It should be mentioned here that the only result of the same nature occurring in the literature is the decidability of the bounded-degree problem for NLC grammars (see [JRW 86]).

The paper is organized in the following way. Sections 2 and 3 comprise the preliminaries on (hyper)graphs and hyperedge-replacement grammars as needed. In Section 4, we discuss several examples of numeric functions which are compatible with the derivation process of our grammars in a certain way. In Section 5, we introduce the general notion of compatible functions, and we relate them with our earlier notion of compatible predicates [HKV 87]. Finally, we show in the main result in Section 6 that the Boundedness Problem corresponding to a numeric function is decidable if the function is pointwise defined as the maximum of sums and products and if it is compatible.

Except for the proof of the main theorem in Section 6, we omit proofs in this version. They can be found in the long version of the paper which will appear elsewhere. While the general results work for arbitrary classes of hyperedge-replacement grammars, we have to admit that most of our examples are formulated for the class of edge-replacement grammars. But we are confident that all of them can be adapted to more general classes of hyperedge-replacement grammars.

2. Preliminaries

This section provides the basic notions on graphs and hypergraphs as far as needed in the paper. The key construction is the replacement of some hyperedges of a hypergraph by hypergraphs yielding an expanded hypergraph. In our approach, a hyperedge is an atomic item with an ordered set of incoming tentacles and an ordered set of outgoing tentacles where each tentacle grips at a node through the source and target functions. Correspondingly, a hypergraph is equipped with two sequences of distinguished nodes so that it is enabled to replace a hyperedge.

2.1 Definition (hypergraphs)

1. Let C be an arbitrary, but fixed alphabet, called a set of ***labels*** (or ***colors***).

2. A ***hypergraph*** over C is a system (V, E, s, t, l) where V is a finite set of ***nodes*** (or ***vertices***), E is a finite set of ***hyperedges***, $s : E \to V^*$ and $t : E \to V^*$ [1] are two mappings assigning a sequence of ***sources*** $s(e)$ and a sequence of ***targets*** $t(e)$ to each $e \in E$, and $l : E \to C$ is a mapping ***labeling*** each hyperedge.

3. A hyperedge $e \in E$ of a hypergraph (V, E, s, t, l) is called an ***(m,n)-edge*** for some $m, n \in \mathbb{N}$ if $|s(e)| = m$ and $|t(e)| = n$. [2] The pair (m, n) is the ***type*** of e, denoted by $type(e)$. e is said to be ***well-formed*** if its sources and targets are pairwise distinct.

---

[1] For a set A, A^* denotes the set of all words over A, including the empty word λ.

[2] For a word $w \in A^*$, $|w|$ denotes its length.

4. A ***multi-pointed hypergraph*** over C is a system $H = (V, E, s, t, l, begin, end)$ where the first five components define a hypergraph over C and $begin, end \in V^*$. Components of H are denoted by $V_H, E_H, s_H, t_H, l_H, begin_H, end_H$, respectively. The set of all multi-pointed hypergraphs over C is denoted by $\mathcal{H}_C$.
5. $H \in \mathcal{H}_C$ is said to be an ***(m, n)-hypergraph*** for some $m, n \in \mathbb{N}$ if $|begin_H| = m$ and $|end_H| = n$. The pair (m, n) is the ***type*** of H, denoted by $type(H)$. H is said to be ***well-formed*** if all hyperedges are well-formed and the begin-nodes and end-nodes of H are pairwise distinct.
6. Let $H \in \mathcal{H}_C$, $begin_H = begin_1 \ldots begin_m$ and $end_H = end_1 \ldots end_n$ with $begin_i, end_j \in V_H$ for $i = 1, \ldots, m$ and $j = 1, \ldots, n$. Then $EXT_H = \{begin_i | i = 1, \ldots, m\} \cup \{end_j | j = 1, \ldots, n\}$ denotes the set of ***external nodes*** of H. Moreover, $INT_H = V_H - EXT_H$ denotes the set of ***internal*** nodes of H.

Remarks: 1. There is a 1-1-correspondence between hypergraphs and (0,0)-hypergraphs so that hypergraphs may be seen as special cases of multi-pointed hypergraphs.
2. An (m, n)-hypergraph over C with (1,1)-edges only is said to be an ***(m, n)-graph***. The set of all (1,1)-graphs over C is denoted by $\mathcal{G}_C$.

2.2 Definition (special hypergraphs)

1. A multi-pointed hypergraph H is said to be a ***singleton*** if $|E_H| = 1$ and $|V_H - EXT_H| = 0$. $e(H)$ refers to the only hyperedge of H and $l(H)$ refers to its label.
2. A singleton H is said to be a ***handle*** if $s_H(e) = begin_H$ and $t_H(e) = end_H$. If $l_H(e) = A$ and $type(e) = (m, n)$ for some $m, n \in \mathbb{N}$, then H is called an ***(m, n)-handle*** induced by A.

Remark: Given $H \in \mathcal{H}_C$, each hyperedge $e \in E_H$ induces a handle $e^\bullet$ by restricting the mappings s_H, t_H, and l_H to the set $\{e\}$, restricting the set of nodes to those ones occurring in $s_H(e)$ and $t_H(e)$, and choosing $begin_{e^\bullet} = s_H(e)$ and $end_{e^\bullet} = t_H(e)$.

2.3 Definition (subhypergraphs and isomorphic hypergraphs)

1. Let $H, H' \in \mathcal{H}_C$. Then H is called a ***(weak) subhypergraph*** of H', denoted by $H \subseteq H'$, if $V_H \subseteq V_{H'}$, $E_H \subseteq E_{H'}$, and $s_H(e) = s_{H'}(e)$, $t_H(e) = t_{H'}(e)$, $l_H(e) = l_{H'}(e)$ for all $e \in E_H$.
[Note that nothing is assumed on the relation of the distinguished nodes.]
2. Let $H, H' \in \mathcal{H}_C$ and $i_V : V_H \to V_{H'}$, $i_E : E_H \to E_{H'}$ be bijective mappings. Then $i = (i_V, i_E) : H \to H'$ is called an ***isomorphism*** from H to H' if $i_V^*(s_H(e)) = s_{H'}(i_E(e))$, $i_V^*(t_H(e)) = t_{H'}(i_E(e))$, $l_H(e)) = l_{H'}(i_E(e))$ for all $e \in E_H$ as well as $i_V^*(begin_H) = begin_{H'}$, $i_V^*(end_H) = end_{H'}$ [3]. H and H' are said to be ***isomorphic***, denoted by $H \cong H'$, if there is an isomorphism from H to H'.

Now we are ready to introduce how hypergraphs may substitute hyperedges. An (m, n)-edge can be replaced by an (m, n)-hypergraph in two steps:
(1) Remove the hyperedge,
(2) add the hypergraph except the external nodes and hand over each tentacle of a hyperedge (of the replacing hypergraph) which grips to an external node to the corresponding source or target node of the replaced hyperedge.
Moreover, an arbitrary number of hyperedges can be replaced simultaneously in this way.

2.4 Definition (hyperedge replacement)

Let $H \in \mathcal{H}_C$ be a multi-pointed hypergraph, $B \subseteq E_H$, and $repl : B \to \mathcal{H}_C$ a mapping with $type(repl(b)) = type(b)$ for all $b \in B$. Then the ***replacement*** of B in H through ***repl*** yields the multi-pointed hypergraph X given by

---

[3] For a mapping $f : A \to B$, the free symbolwise extension $f^* : A^* \to B^*$ is defined by $f^*(a_1 \ldots a_k) = f(a_1) \ldots f(a_k)$ for all $k \in \mathbb{N}$ and $a_i \in A$ $(i = 1, \ldots, k)$.

- $V_X = V_H + \sum_{b\in B}(V_{repl(b)} - EXT_{repl(b)})$, [4]
- $E_X = (E_H - B) + \sum_{b\in B}(E_{repl(b)})$,
- each hyperedge of $E_H - B$ keeps its sources and targets,
- each hyperedge of $E_{repl(b)}$ (for all $b \in B$) keeps its internal sources and targets and the external ones are handed over to the corresponding sources and targets of b, i.e., $s_X(e) = h^*(s_{repl(b)}(e))$ and $t_X(e) = h^*(t_{repl(b)}(e))$ for all $b \in B$ and $e \in E_{repl(b)}$ where $h : V_{repl(b)} \to V_X$ is defined by $h(v) = v$ for $v \in V_{repl(b)} - EXT_{repl(b)}$, $h(b_i) = s_i$ $(i = 1,\ldots,m)$ for $begin_{repl(b)} = b_1\ldots b_m$ and $s_H(b) = s_1\ldots s_m$, $h(e_j) = t_j$ $(j = 1,\ldots,n)$ for $end_{repl(b)} = e_1\ldots e_n$ and $t_H(b) = t_1\ldots t_n$.
- each hyperedge keeps its label,
- $begin_X = begin_H$ and $end_X = end_H$.

The resulting multi-pointed hypergraph X is denoted by $REPLACE(H, repl)$.

Remark: The construction above is meaningful and determines (up to isomorphism) a unique hypergraph X if h is a mapping. This is automatically fulfilled whenever the *begin*-nodes and *end*-nodes of each replacing hypergraph are pairwise distinct. If one wants to avoid such a restriction, one has to require that the following application condition is satisfied for each $b \in B$: If $begin_{repl(b)} = x_1\ldots x_m$ and $end_{repl(b)} = x_{m+1}\ldots x_{m+n}$ as well as $s_H(b) = y_1\ldots y_m$ and $t_H(b) = y_{m+1}\ldots y_{m+n}$, then, for $i,j = 1,\ldots,m+n$, $x_i = x_j$ implies $y_i = y_j$.

3. Hyperedge–Replacement Grammars and Languages

In this section we give a short summary of the basic notions on hyperedge–replacement grammars generalizing edge–replacement grammars as investigated e.g. in [HK 83+85] and context–free string grammars. Details and examples can be found in [HK 87a+b].

Based on hyperedge replacement, one can derive multi–pointed hypergraphs from multi–pointed hypergraphs by applying productions of a simple form.

3.1 Definition (productions and derivations)

1. Let $N \subseteq C$. A *production* (over N) is an ordered pair $p = (A, R)$ with $A \in N$ and $R \in \mathcal{H}_C$. A is called *left–hand side* of p and is denoted by $lhs(p)$, R is called *right–hand side* and is denoted by $rhs(p)$. The *type* of p, denoted by $type(p)$, is given by the type of R.
2. Let $H \in \mathcal{H}_C$, $B \subseteq E_H$, and P be a set of productions. A mapping $prod : B \to P$ is called a *production base* in H if $l_H(b) = lhs(prod(b))$ and $type(b) = type(rhs(prod(b)))$ for all $b \in B$.
3. Let $H, H' \in \mathcal{H}_C$ and $prod : B \to P$ be a production base in H. Then H *directly derives* H' *through prod* if H' is isomorphic to $REPLACE(H, repl)$ where $repl : B \to \mathcal{H}_C$ is given by $repl(b) = rhs(prod(b))$ for all $b \in B$. We write $H \underset{P}{\Longrightarrow} H'$ or $H \Longrightarrow H'$ in this case.
4. A sequence of direct derivations $H_0 \Longrightarrow H_1 \Longrightarrow \ldots \Longrightarrow H_k$ is called a *derivation* from H_0 to H_k (of length k). Additionally, in the case $H \cong H'$, we speak of a *derivation* from H to H' of length 0. A derivation from H to H' is shortly denoted by $H \overset{*}{\underset{P}{\Longrightarrow}} H'$ or $H \overset{*}{\Longrightarrow} H'$. If the length of the derivation should be stressed, we write $H \overset{k}{\underset{P}{\Longrightarrow}} H'$ or $H \overset{k}{\Longrightarrow} H'$.
5. A direct derivation through $prod : \emptyset \to P$ is called a *dummy*. [5] A derivation is said to be *valid* if at least one of its steps is not a dummy.

---

[4] The sum symbols $+$ and $\sum$ denote the disjoint union of sets; the symbol $-$ denotes the set–theoretic difference.

[5] A production base $prod : B \to P$ in H may be *empty*, i.e., $B = \emptyset$. In this case $H \Longrightarrow H'$ through $prod$ implies $H \cong H'$, and there is always a trivial direct derivation $H \Longrightarrow H$ through $prod$.

Remarks: 1. The application of a production $p = (A,R)$ of type (m,n) to a multi-pointed hypergraph H requires the following two steps only:
(1) Choose a hyperedge e of type (m,n) with label A.
(2) Replace the hyperedge e in H by R.
2. Some significant properties of direct derivations are: On the one hand, the definition of a direct derivation includes the case that no hyperedge is replaced. This dummy step derives a hypergraph isomorphic to the initial one. On the other hand, it includes the case that all hyperedges are replaced in one step. Moreover, whenever some hyperedges can be replaced in parallel, they can be replaced one after the other leading to the same derived hypergraph.

Using the introduced concepts of productions and derivations hyperedge–replacement grammars and languages can be introduced in a straightforward way.

3.2 Definition (hyperedge–replacement grammars and languages)

1. A ***hyperedge–replacement grammar*** is a system $HRG = (N,T,P,Z)$ where $N \subseteq C$ is a set of ***nonterminals***, $T \subseteq C$ is a set of ***terminals***, P is a finite set of ***productions*** over N, and $Z \in \mathcal{H}_C$ is the ***axiom***. The class of all hyperedge–replacement grammars is denoted by $\mathcal{HRG}$.
2. HRG is said to be ***typed*** if there is a mapping $ltype : N \cup T \to \mathbb{N} \times \mathbb{N}$ such that, for each production $(A,R) \in P$, $ltype(A) = type(R)$ and $ltype(l_R(e)) = type(e)$ for all $e \in E_R$ and $ltype(l_Z(e)) = type(e)$ for all $e \in E_Z$. HRG is said to be ***well-formed*** if the right–hand sides of the productions are well-formed and all hyperedges in Z are well-formed.
3. The ***hypergraph language*** $L(HRG)$ ***generated by*** HRG consists of all hypergraphs which can be derived from Z applying productions of P and which are terminally labeled:

$$L(HRG) = \{H \in \mathcal{H}_T \mid Z \overset{*}{\underset{P}{\Longrightarrow}} H\}.$$

Remarks: 1. Even if one wants to generate graph languages rather than hypergraph languages, one may use nonterminal hyperedges because the generative power of hyperedge-replacement grammars increases with the maximum number of tentacles of a hyperedge involved in the replacement (see [HK 87b]).
2. Without effecting the generative power, we will assume in the following that N and T are finite, $N \cap T = \emptyset$, and Z is a singleton with $l(Z) \in N$. Furthermore, we will assume that the hyperedge–replacement grammars considered in this paper are typed and well–formed.

The results presented in the following sections are mainly based on some fundamental aspects of hyperedge–replacement derivations. Roughly speaking, hyperedge–replacement derivations cannot interfere with each other as long as they handle different hyperedges. On the one hand, a collection of derivations of the form $e^\bullet \overset{*}{\Longrightarrow} H(e)$ for $e \in E_R$ can be simultaneously embedded into R leading to a single derivation $R \overset{*}{\Longrightarrow} H$. On the other hand, restricting a derivation $R \overset{*}{\Longrightarrow} H$ to the handle $e^\bullet$ induced by the hyperedge $e \in E_R$ one obtains a so–called "restricted" derivation $e^\bullet \overset{*}{\Longrightarrow} H(e)$ where $H(e) \subseteq H$. Finally, restricting a derivation to the handles induced by the hyperedges, and subsequently embedding them again returns the original derivation. In other words, hyperedge–replacement derivations can be distributed to the handles of the hyperedges without losing information. We state and use this result in the following recursive version concerning terminal hypergraphs which are derivable from handles.

3.3 Theorem

Let $HRG = (N,T,P,Z)$ be a typed and well–formed hyperedge–replacement grammar, $A \in N \cup T$, and $H \in \mathcal{H}_T$. Then there is a derivation $A^\bullet \Longrightarrow R \overset{k}{\Longrightarrow} H$ for some $k \geq 0$ [6] if and only if $A^\bullet \Longrightarrow R$ and, for each $e \in E_R$, there is a derivation $l_R(e)^\bullet \overset{k}{\Longrightarrow} H(e)$ with $H(e) \subseteq H$ such that $H \cong REPLACE(R, repl)$ with $repl(e) = H(e)$ for $e \in E_R$.

[6] For a symbol $A \in N \cup T$ with $ltype(A) = (m,n)$, $A^\bullet$ denotes an (m,n)–handle induced by A. [Note that (m,n)–handles induced by a symbol A are isomorphic].

Remarks: 1. The derivation $l_R(e)^\bullet \overset{k}{\Longrightarrow} H(e)$ may be valid or not. In the first case, it has the same form as the original derivation, but it is shorter as the original one. In the latter case, $H(e)$ is isomorphic to $e^\bullet$ (resp. $l_R(e)^\bullet$) and hence a terminal handle.

2. Given a derivation $R \overset{k}{\Longrightarrow} H$, the derivation $l_R(e)^\bullet \overset{k}{\Longrightarrow} H(e)$ for each $e \in E_R$ is called the *fibre* of e and — the other way round — the given derivation is the *joint embedding* of its fibres.

4. Some Graph–Theoretic Functions Compatible With Derivations

A hyperedge–replacement grammar as a generating device specifies a (hyper)graph language. Unfortunately, in a finite amount of time, the generating process only produces a finite section of the language explicitly (and even this may consume much time). Hence one may wonder what the hyperedge–replacement grammar can tell us about the generated language. As a matter of fact, by Theorem 3.3, we have the following nice situation. Given a hyperedge–replacement grammar and an arbitrary terminal (hyper)graph H with derivation $A^\bullet \Longrightarrow R \overset{*}{\Longrightarrow} H$, we get a decomposition of H into "smaller" components which are derivable from the handles of the hyperedges in R. If one is interested in values of graph–theoretic functions of derived (hyper)graphs, one may ask how a certain value of a derived (hyper)graph depends on values of the components. A function is said to be "compatible" with the derivation process of hyperedge–replacement grammars if it can be computed for each derived (hyper)graph H by computing the values (or related values) for the components and composing the values to the value of H.

In this section, we pick up several graph–theoretic functions and show that they are "compatible" with the replacement process of hyperedges. A formal definition of compatibility is given in the next section. We discuss the number of nodes and hyperedges, the number of paths and cycles, the length of a shortest path, the length of a longest simple path,and the minimum and maximum degree.

Let $HRG = (N, T, P, Z)$ be a typed and well–formed hyperedge–replacement grammar, $H \in \mathcal{H}_T$, $A^\bullet \Longrightarrow R \overset{*}{\Longrightarrow} H$ a derivation of H in HRG, and, for $e \in E_R$, $l_R(e)^\bullet \overset{*}{\Longrightarrow} H(e)$ be the fibre of $R \overset{*}{\Longrightarrow} H$ induced by e. Then the number of nodes in H can be computed from the number of nodes in R and the number of internal nodes in the $H(e)$'s. Similarly, the number of internal nodes can be computed. Even simpler, the number of hyperedges in H can be determined by the number of hyperedges in the $H(e)$'s.

4.1 Theorem (Number of Nodes and Hyperedges)

For a hypergraph $H \in \mathcal{H}_C$, let $|V_H|$ denote the number of nodes, $|INT_H|$ the number of internal nodes, and $|E_H|$ the number of hyperedges in H. Then

$$|V_H| = |V_R| + \sum_{e \in E_R} |INT_{H(e)}|,$$

$$|INT_H| = |INT_R| + \sum_{e \in E_R} |INT_{H(e)}|,$$

$$|E_H| = \sum_{e \in E_R} |E_{H(e)}|.$$

Remarks: 1. Similarly, the composed function size given by $\mathit{size}(H) = |V_H| + |E_H|$ can be handled. It makes use of the auxiliary function $\mathit{intsize}$ given by $\mathit{intsize}(H) = |INT_H| + |E_H|$.

2. The density function $dens$ given by $dens(H) = \frac{|E_H|}{|V_H|}$ if $|V_H| > 0$ (and $dens(H) = \diamond$ [7] otherwise) can also be expressed in such a way:

$$dens(H) = \frac{\sum_{e \in E_R} |E_{H(e)}|}{|V_R| + \sum_{e \in E_R} |INT_{H(e)}|}$$

The expression for computing $dens(H)$ makes use of the possibility to compute the number of internal nodes as well as the number of hyperedges of the $H(e)$'s. It does not make use of the density of some of the $H(e)$'s. □

For simplifying the technicalities, we restrict our following consideration to the class $\mathcal{ERG}$ of edge-replacement grammars in the sense of [HK 83+85]. To be more explicit, a typed and well-formed hyperedge-replacement grammar $HRG = (N, T, P, Z)$ is in $\mathcal{ERG}$ if and only if the right-hand sides of the productions as well as the axiom are (1,1)-graphs. Note that, in this case, each $G \in L(HRG)$ is a (1,1)-graph, i.e., a graph with two distinguished nodes $begin_G$ and end_G.

Let G be a graph. A ***path*** joining v_0 and v_n is a sequence $p = v_0, e_1, v_1, e_2, \ldots, e_n, v_n$ of alternating nodes and edges such that for $1 \leq i \leq n$, v_{i-1} and v_i are the nodes incident with e_i. If $v_0 = v_n$ then p is said to be a ***cycle***. If in a path each node appears once, then the sequence is called a ***simple*** path. If each node appears once except that $v_0 = v_n$ and $n \geq 3$ then p is a ***simple cycle***. The ***length*** of a path or a cycle p, denoted $length(p)$, is the number of edges it contains. "e on p" denotes the fact that e occurs in p.

4.2 Theorem (Number of Simple Paths, Minimum and Maximum Simple-Path Length)

For a (1,1)-graph G, let $PATH_G$ denote the set of simple paths joining $begin_G$ and end_G and $numpath(G)$ the number of these paths in G. Moreover, let $minpath(G)$ and $maxpath(G)$ denote the minimum resp. maximum simple-path length, if any (and $minpath(G) = maxpath(G) = \diamond$ otherwise [8]). Then

$$numpath(G) = \sum_{p \in PATH_R} \prod_{e \text{ on } p} numpath(G(e)),$$

$$minpath(G) = \min_{p \in PATH_R} \sum_{e \text{ on } p} minpath(G(e)),$$

$$maxpath(G) = \max_{p \in PATH_R} \sum_{e \text{ on } p} maxpath(G(e)).$$

The number of simple cycles, the minimum cycle length, and the maximum simple-cycle length of a graph can be determined using the computation of the number of simple paths, the minimum path length, and the maximum simple-path length, respectively.

---

[7] $dens$, $minpath$, and $maxpath$ are defined to be functions with values in $\mathbb{N} \cup \{\diamond\}$, the set of all non-negative integers plus a special symbol $\diamond$. We use this special symbol $\diamond$, if the considered function has no sensible integer value. We calculate with $\diamond$ as follows: $\forall i \in I\ \forall n_i \in \mathbb{N} \cup \{\diamond\}$,

- $\sum_{i \in I} n_i = \diamond$ and $\prod_{i \in I} n_i = \diamond$ if and only if $n_j = \diamond$ for some $j \in I$,
- $\min_{i \in I} n_i = \min_{i \in I'} n_i$ and $\max_{i \in I} n_i = \max_{i \in I'} n_i$ for $I' = \{i \in I | n_i \neq \diamond\}$, and $\min_{i \in I} n_i = \diamond$ and $\max_{i \in I} n_i = \diamond$ for $I = \emptyset$.

4.3 Theorem (Number of Simple Cycles, Minimum and Maximum Simple–Cycle Length)

For a (1,1)–graph G, let $CYCLE_G$ denote the set of simple cycles and $numcycle(G)$ the number of these cycles in G. Moreover, let $mincycle(G)$ and $maxcycle(G)$ denote the minimum resp. maximum simple–cycle length, if any; otherwise, let $mincycle(G) = \diamond = maxcycle(G)$. Then

$$numcycle(G) = \sum_{c \in CYCLE_R} \prod_{e \text{ on } c} numpath(G(e)) + \sum_{e \in E_R} numcycle(G(e)),$$

$$mincycle(G) = \min \left\{ \min_{c \in CYCLE_R} \sum_{e \text{ on } c} minpath(G(e)) \; , \; \min_{e \in E_R} mincycle(G(e)) \right\},$$

$$maxcycle(G) = \max \left\{ \max_{c \in CYCLE_R} \sum_{e \text{ on } c} maxpath(G(e)) \; , \; \max_{e \in E_R} maxcycle(G(e)) \right\}.$$

4.4 Theorem (Minimum and Maximum Degree)

For a graph G, let $mindegree(G)$ and $maxdegree(G)$ denote the minimum resp. maximum degree among the nodes of G. Moreover, let $minintdegree(G)$ denote the minimum degree among the internal nodes of G and $bdegree(G)$ and $edegree(G)$ the degree of $begin_G$ resp. end_G. Then

$$mindegree(G) = \min \left\{ \min_{v \in V_R} D_G(v) \; , \; \min_{e \in E_R} minintdegree(G(e)) \right\}$$

$$minintdegree(G) = \min \left\{ \min_{v \in INT_R} D_G(v) \; , \; \min_{e \in E_R} minintdegree(G(e)) \right\}$$

$$maxdegree(G) = \max \left\{ \max_{v \in V_R} D_G(v) \; , \; \max_{e \in E_R} maxdegree(G(e)) \right\}$$

$$bdegree(G) = D_G(begin_G) \text{ and } edegree(G) = D_G(end_G)$$

where, for $v \in V_R$, $D_G(v) = \sum_{e \in s_R^{-1}(v)} bdegree(G(e)) + \sum_{e \in t_R^{-1}(v)} edegree(G(e)).$

5. Compatible Functions

In this section we introduce the notion of compatible functions in such a way that all functions considered in the previous section are special cases. Roughly speaking, a function f_0 on hypergraphs is said to be compatible with the derivation process of hyperedge–replacement grammars if, for each hypergraph H and each derivation of it, the value of H, $f_0(H)$, can be computed from the values of some specific subhypergraphs $H(e)$ determined by the fibres of the derivation. As the examples will show, this view is oversimplified for most applications. To compute the value of H, it might be necessary to compute the values of some other related functions for the $H(e)$'s. Therefore, we use families of functions indexed by some finite set I and we need a mapping *assign* which determines the values for the $H(e)$'s with respect to the different value functions.

The notion of compatible functions generalizes obviously our earlier notion of compatible predicates (see [HKV 87]). More interesting, a certain type of compatible functions that are composed of minima, maxima, sums, and products induce compatible predicates of the form: the function value of a graph exceeds a given fixed integer, or the function value does not exceed a fixed integer. Consequently, we get the decidability of the problems (1), (2), and (3) in the introduction for these predicates as a corollary.

5.1 Definition (compatible functions)

1. Let C be a class of hyperedge-replacement grammars, I a finite index set, VAL a set of values, $f : \mathcal{H}_C \times I \to VAL$ a function [8], and f' a function defined on triples $(R, assign, i)$ with $R \in \mathcal{H}_C$, $assign : E_R \times I \to VAL$, and $i \in I$. Then f is called *(C, f')-compatible* if, for all $HRG = (N, T, P, Z) \in C$ and all derivations of the form $A^{\bullet} \Longrightarrow R \stackrel{*}{\Longrightarrow} H$ with $A \in N$ and $H \in \mathcal{H}_T$, and for all $i \in I$,

$$f(H, i) = f'(R, assign, i)$$

where $assign : E_R \times I \to VAL$ is given by $assign(e, j) = f(H(e), j)$ for all $e \in E_R$ and all $j \in I$.
2. A function $f_0 : \mathcal{H}_C \to VAL$ is called *C-compatible* if functions f and f' and an index i_0 exist such that $f_0 = f(-, i_0)$ and f is (C, f')-compatible. [9]

Remark: Intuitively, a function is compatible if it can be computed for a large hypergraph derived by a fibre by computing some values for the smaller components of the corresponding shorter fibres. Such a function must be closed under isomorphisms because the derivability of hypergraphs is independent of the representation of nodes and hyperedges.

5.2 Examples

By Theorem 4.1, the following functions on hypergraphs are $\mathcal{HRG}$-compatible: the number of nodes, the number of hyperedges, and the density of a hypergraph. By Theorems 4.2-4.4, the following functions on graphs are $\mathcal{ERG}$-compatible: the number of simple paths connecting the external nodes, the minimum-path length (of paths connecting the external nodes), the maximum-simple-path length (of paths connecting the external nodes), the number of simple cycles, the minimum-cycle length, the maximum-simple-cycle length, the minimum degree, and the maximum degree of a graph.

We recall now the notion of compatible predicates and relate it with compatible functions.

5.3 Definition (compatible predicates)

1. Let C be a class of hyperedge-replacement grammars, I a finite index set, $PROP$ a decidable predicate[10] defined on pairs (H, i) with $H \in \mathcal{H}_C$ and $i \in I$, and $PROP'$ a decidable predicate on triples $(R, assign, i)$ with $R \in \mathcal{H}_C$, a mapping $assign : E_R \to I$, and $i \in I$. Then $PROP$ is called *$(C, PROP')$-compatible* if, for all $HRG = (N, T, P, Z) \in C$ and all derivations $A^{\bullet} \Longrightarrow R \stackrel{*}{\Longrightarrow} H$ with $A \in N$ and $H \in \mathcal{H}_T$, and for all $i \in I$, $PROP(H, i)$ holds if and only if there is a mapping $assign : E_R \to I$ such that $PROP'(R, assign, i)$ holds and $PROP(H(e), assign(e))$ holds for all $e \in E_R$.
2. A predicate $PROP_0$ on $\mathcal{H}_C$ is called *C-compatible* if predicates $PROP$ and $PROP'$ and an index i_0 exist such that $PROP_0 = PROP(-, i_0)$ [11] and $PROP$ is $(C, PROP')$-compatible.

Remarks: 1. Intuitively, a property is compatible if it can be tested for a large hypergraph with a long fibre by checking the smaller components of the corresponding shorter fibres.
2. Examples of compatible properties are: connectivity, planarity, existence of Hamiltonian and Eulerian paths and cycles, k-colorability for each $k \geq 0$ (see [HKV 87] and [Ha 88]).

---

[8] We assume that all considered functions are *closed under isomorphisms*, i.e., for a function f, if $H \cong H'$ for some $H, H' \in \mathcal{H}_C$, then $f(H, i) = f(H', i)$ (resp. $f(H, assign, i) = f(H', assign, i)$) for all $i \in I$.

[9] For $i \in I$, $f(-, i)$ denotes the unary function defined by $f(-, i)(H) = f(H, i)$ for all $H \in \mathcal{H}_C$.

[10] We assume that all considered predicates are *closed under isomorphisms*, i.e., if a predicate Φ holds for $H \in \mathcal{H}_C$ and $H \cong H'$, then Φ holds for H', too.

[11] For $i \in I$, $PROP(-, i)$ denotes the unary predicate defined by $PROP(-, i)(H) = PROP(H, i)$ for all $H \in \mathcal{H}_C$.

3. In [HKV 87] it is shown, that, for all C–compatible properties $PROP_0$, it is decidable whether, given any hyperedge–replacement grammar $HRG \in C$, $PROP_0$ holds for some $H \in L(HRG)$ and $PROP_0$ holds for all $H \in L(HRG)$.

5.4 Theorem

Let $PROP_0$ be a C–compatible predicate. Then the function $f_0 : \mathcal{H}_C \to \{0,1\}$ given by

$$f_0(H) = \begin{cases} 1 & \text{if } PROP_0(H) \text{ holds} \\ 0 & \text{otherwise} \end{cases}$$

is C–compatible.

Certain C–compatible functions with values in $\mathbb{N}^\diamond = \mathbb{N} \cup \{\diamond\}$ induce specific C–compatible predicates.

5.5 Definition

1. A function $f : \mathcal{H}_C \times I \to \mathbb{N}^\diamond$ is said to be $(C, \min, \max, +, \cdot)$*–compatible* if there exists an f' such that for each right–hand side R of some production in C and each $i \in I$, $f'(R, -, i)$ corresponds to an expression formed with variables $assign(e, j)$ ($e \in E_R$, $j \in I$) and constants from $\mathbb{N}$ by addition, multiplication, minimum, and maximum, and f is (C, f')–compatible. The function is $(C, \max, +, \cdot)$*–compatible* if the operation min does not occur.
2. A function $f_0 : \mathcal{H}_C \to \mathbb{N}^\diamond$ is $(C, \min, \max, +, \cdot)$*–compatible* (resp. $(C, \max, +, \cdot)$*–compatible*) if a function f and an index i_0 exist such that $f_0 = f(-, i_0)$ and f is $(C, \min, \max, +, \cdot)$–compatible (resp. $(C, \max, +, \cdot)$–compatible).

5.6 Theorem

Let $f_0 : \mathcal{H}_C \to \mathbb{N}^\diamond$ be a $(C, \min, \max, +, \cdot)$–compatible function for some class C of hyperedge–replacement grammars. Moreover, let $n \in \mathbb{N}^\diamond$. Then the predicates given by "$f_0(H) \leq n$" and "$f_0(H) > n$" are C–compatible. [12]

5.7 Corollary

Let f_0 be a $(C, \min, \max, +, \cdot)$–compatible function for some class C of hyperedge–replacement grammars. Moreover, let $n \in \mathbb{N}^\diamond$. Then, for all $HRG \in C$ the following statements hold.

(1) It is decidable whether (or not) there is some $H \in L(HRG)$ with $f_0(H) \leq n$.
(2) It is decidable whether (or not), for all $H \in L(HRG)$, $f_0(H) \leq n$.
(3) It is decidable in linear time whether (or not) a generated hypergraph $H \in L(HRG)$ represented by a derivation (resp. a derivation tree) has a value $f_0(H) \leq n$.

Proof: Corollary 5.7 follows immediately from the C–compatibility of the predicate "$f_0(-) \leq n$" (see Theorem 5.6) and the theorems for C–compatible predicates given in [HKV 87]). □

6. A Metatheorem for Boundedness Problems

Given a graph-theoretic function f_0 and a class C of hyperedge–replacement grammars, we are going to study the following type of questions for all $HRG \in C$: "Is it decidable whether (or not) the values of all hypergraphs generated by HRG are bounded?" The question turns out to be decidable provided that f_0 is $(C, \max, +, \cdot)$–compatible. We call this result "metatheorem" because of its generic character: Whenever one can prove the $(C, \max, +, \cdot)$–compatibility of a function (and

[12] We assume that, for all $n \in \mathbb{N}^\diamond$, $\diamond \leq n$.

we have given various examples in section 4), one gets a particular decision result for this function as corollary of the metatheorem.

6.1 Theorem

Let f_0 be a $(\mathcal{C}, \max, +, \cdot)$–compatible function for some class $\mathcal{C}$ of hyperedge–replacement grammars. Then, for all $HRG \in \mathcal{C}$, it is decidable whether or not there is a natural number $n \in \mathbb{N}$ such that $f_0(H) \leq n$ for all $H \in L(HRG)$.

Proof: Let f_0 be a $(\mathcal{C}, \max, +, \cdot)$–compatible function. Let f and f' be the corresponding functions over the index set I so that f is $(\mathcal{C}, f')$–compatible and $f_0 = f(-, i_0)$ for some $i_0 \in I$.
Let $HRG = (N, T, P, Z)$ be a typed and well-formed hyperedge–replacement grammar in $\mathcal{C}$. By Definition 5.1, we may assume that, for each $A \in N$, the grammar $HRG(A) = (N, T, P, A^\bullet)$ is in $\mathcal{C}$, too. ($\mathcal{C}$–compatibility is concerned with productions of a grammar, not with the axiom.)
The proof is based on the following idea. We construct a directed graph D containing all relevant information on derivations in HRG and look for certain cyclic structures in D. This enables us to decide whether or not the values may grow beyond any bound.
Let $J = \{\diamond, 0, 1, big\}$ and $[-] : \mathbb{N}^\diamond \to J$ be the mapping given by $[m] = big$ if $m \geq 2$ and $[m] = m$ otherwise. By (a generalized version of) Corollary 5.7, we can effectively determine the set $EXIST = \{(A, p : I \to J) | \exists H \in \mathcal{H}_T : A^\bullet \overset{*}{\Longrightarrow} H \ \wedge \forall j \in I : [f(H, j)] = p(j)\}$. $H \in \mathcal{H}_T$ is said to be an *(A,p)–hypergraph* if H can be derived from the handle induced by A and, for all $j \in I$, $[f(H, j)] = p(j)$. We define a directed graph D with two types of edges, called ***greaterequal–edges*** and ***greater–edges*** as follows. Let $V = \{(A, p, i) | (A, p) \in EXIST \wedge p(i) = big\}$ be the node set of D. The edge set of D is determined as follows: Let $(A, R) \in P$ be a production of HRG, $q : E_R \times I \to J$ a function such that, for all $e \in E_R$, $(l_R(e), q(e, -)) \in EXIST$, $i \in I$ an index, and $[f'(R, q, i)] = big$. Moreover, let $p : I \to J$ be the function given by $p(j) = [f'(R, q, j)]$ for $j \in I$.
By assumption, the function f is $(\mathcal{C}, \max, +, \cdot)$–compatible. Since multiplication distributes over addition and maximum and addition distributes over maximum, we may assume that $f'(R, -, i)$ is a maximum of sums, each formed from products of constants and variables $assign(e, j)$ ($e \in E_R$, $j \in I$). Substitute $assign(e, j)$ by $q(e, j)$, if $q(e, j) \in \{\diamond, 0, 1\}$, and simplify, i.e., delete all sums that evaluate to $\diamond$, all products that evaluate to 0 and all factors that evaluate to 1.

- If some sum simply is $assign(e, j)$, then we add an edge from (A, p, i) to $(l_R(e), q(e, -), j)$ in D, a so called ***greaterequal–edge***, denoted by $(A, p, i) \Rightarrow (l_R(e), q(e, -), j)$.
- If some sum contains $assign(e, j)$, but also a non–trivial factor or some other product, then we add a so called ***greater–edge*** from (A, p, i) to $(l_R(e), q(e, -), j)$ in D, denoted by $(A, p, i) \to (l_R(e), q(e, -), j)$.

In the following, we will show that the graph D contains all information to decide whether or not some function values grow beyond any bound. It turns out that the greater–edges of D play an important role. Remember that for each (B, p', j) in D, there is at least one derivation $B^\bullet \overset{*}{\Longrightarrow} G$ in HRG with $[f(G, -)] = p'$ and $f(G, j) \geq 2$. We will show that, whenever we have a derivation $B^\bullet \overset{*}{\Longrightarrow} G$ in HRG with $[f(G, -)] = p'$ and $f(G, j) \geq 2$ and there is a greater–edge $(A, p, i) \to (B, p', j)$ in D, then there exists a derivation $A^\bullet \overset{*}{\Longrightarrow} H$ in HRG with $[f(H, -)] = p$ and $f(H, i) > f(G, j)$.

Claim 1: Let $(A, p, i), (B, p', j) \in V$ and G be a (B, p')–hypergraph.
(1) If $(A, p, i) \to (B, p', j)$, then there is an (A, p)–hypergraph H with $f(H, i) > f(G, j)$.
(2) If $(A, p, i) \Rightarrow (B, p', j)$, then there is an (A, p)–hypergraph H with $f(H, i) \geq f(G, j)$.

Proof of Claim 1: Let $(A, p, i) \to (B, p', j)$ be a greater–edge in D and G be an arbitrary (B, p')–hypergraph. By construction of D, there is some production $(A, R) \in P$ and some $q : E_R \times I \to J$ such that, for all $e \in E_R$, $(l_R(e), q(e, -) \in EXIST$. Moreover, there is some $e' \in E_R$ with $l_R(e') = B$ and $q(e', -) = p'$. By definition of $EXIST$, for each $e \in E_R$, there exists a derivation $l_R(e)^\bullet \overset{*}{\Longrightarrow} H(e)$ such that $[f(H(e), -)] = q(e, -)$. Since $l_R(e') = B$ and G

is an (B,p')–hypergraph, there exists a derivation $l_R(e')^\bullet \stackrel{*}{\Longrightarrow} G$ such that $[f(G,-)] = p'$. Joint Embedding of the derivations $l_R(e)^\bullet \stackrel{*}{\Longrightarrow} H(e)$ for $e \in E_R - \{e'\}$ and the derivation $l_R(e')^\bullet \stackrel{*}{\Longrightarrow} G$ – instead of $l_R(e')^\bullet \stackrel{*}{\Longrightarrow} H(e')$ – into R yields a derivation $R \stackrel{*}{\Longrightarrow} H$. Combining it with the direct derivation $A^\bullet \Longrightarrow R$, we get a derivation $A^\bullet \stackrel{*}{\Longrightarrow} H$. By the $(\mathcal{C},\max,+,\cdot)$–compatibility of f, H is an (A,p)–hypergraph: For all $j \in I$, we have $[f(H,j)] = [f'(R,assign',j)] = [f'(R,[assign'],j)] = [f'(R,[assign],j)] = [f'(R,q,j)] = p(j)$ where $assign'(e,-) = assign(e,-) = f(H(e),-)$ for $e \in E_R - \{e'\}$, $assign'(e',-) = f(G,-)$, and $assign(e',-) = f(H(e'),-)$. Moreover, by the special choice of the edges of D, $f(H,i) = f'(R,assign',i) > assign'(e',j) = f(G,j)$. [Observe that in the sum leading to the creation of $(A,p,i) \rightarrow (B,p',j)$ all remaining variables are substituted by at least 2.] Analogously, if $(A,p,i) \Rightarrow (B,p',j)$ is a greaterequal–edge in D, we get $f(H,i) \geq f(G,j)$.
□

In the following, we will look for special structures in D, called lasso structures. A subgraph L of D is called a ***lasso structure*** if it contains for each node a unique outgoing edge and each cycle contains a greater–edge. A node (A,p,i) of D is said to be ***unbounded***, if, for all $n \in \mathbb{N}$, there is an (A,p)–hypergraph H with $f(H,i) > n$; otherwise it is said to be ***bounded***.

Claim 2: Let L be a lasso structure in D. Then every (A,p,i) in L is unbounded.

Proof of Claim 2: Assume to the contrary and let k be minimal such that, for some (A,p,i) in L, for every (A,p)–hypergraph H we have $f(H,i) \leq k$. By the above claim we have for the unique successor (B,p',j) of (A,p,i) in L and every (B,p')–hypergraph H that $f(H,j) \leq k$. By choice of k, there must exist a (B,p')–hypergraph H with $f(H,j) = k$ and we have $(A,p,i) \Rightarrow (B,p',j)$. Repeating this consideration we eventually get a lasso [13] in L whose cycle has greaterequal–edges only, a contradiction. □

Claim 3: There exists a lasso structure L in D containing all unbounded (A,p,i).

Proof of Claim 3: Let k be the maximal $f(H,i)$, where H is an (A,p)–hypergraph such that (A,p,i) is bounded, but at least 2. Moreover, let $\Phi(k)$ be determined as follows: Note that each $f'(R,-,i)$ with R a right–hand side of some production of HRG, $i \in I$, can be expressed as a maximum of sums of products of variables and constants. Evaluate each of all these sums by replacing each variable by $k \in \mathbb{N}^\diamond$, and let $\Phi(k)$ be the maximum of these values plus 1. In the following, we define a subgraph L of D iteratively using sets OK and NOK, such that the following properties hold after each step:

(1) $OK \cup NOK = \{(A,p,i) \in V | (A,p,i) \text{ is unbounded}\}$;
(2) $OK \cap NOK = \emptyset$;
(3) $OK \subseteq V_L \subseteq OK \cup NOK$;
(4) each node in OK has a unique outgoing edge in L;
(5) each cycle of L contains a greater–edge;
(6) each maximal path [14] of L ends with a greater–edge.

Initially, let $OK = \emptyset$, $NOK = \{(A,p,i) \in V | (A,p,i) \text{ is unbounded}\}$, and L be the empty graph. For the iteration step, choose a derivation $A^\bullet \stackrel{*}{\Longrightarrow} H$ of minimal length such that H is an (A,p)–hypergraph with $f(H,i) \geq \Phi(k)$ and $(A,p,i) \in NOK$. Let (A,R) be the first production of this derivation. We have hypergraphs $H(e)$, $e \in E_R$, and some $q : E_R \times I \rightarrow J$ such that $H(e)$ is an $(l_R(e),q(e,-))$–hypergraph for $e \in E_R$. $f'(R,-,i)$ in its simplified normal form is a maximum of sums, and, by definition of k and Φ, the maximum is attained for a sum containing a variable $assign(e,j)$ such that $f(H(e),j) > k$. Put $(B,p',j) = (l_R(e),q(e,-),j)$, $OK = OK \cup \{(A,p,i)\}$, $NOK = NOK - \{(A,p,i)\}$, add to L the corresponding edge from (A,p,i) to (B,p',j) and – if necessary – (A,p,i) and/or (B,p',j). The first four conditions on L given above hold true (we have $f(H(e),j) > k$, therefore, $(B,p',j) \in OK \cup NOK$). If the new edge is a greater–edge, then

---

[13] If we add to a path $v_1 \ldots v_n$, which has distinct nodes by definition, an edge $v_n v_i$, $i \in \{1,\ldots,n-1\}$, then the resulting graph is called a ***lasso***.

[14] We call a path *maximal*, if its last node has outdegree 0.

each new cycle contains it, each new non–trivial maximal path ends with it $((B,p',j) \notin OK)$ or ends with a non–trivial maximal path that already existed $((B,p',j) \in OK)$. If the new edge is a greaterequal–edge, we must have $f(H(e),j) \geq \Phi(k)$, thus $(B,p',j) \in OK$, since we have chosen a shortest derivation. Hence any new non-trivial maximal path ends with an old one starting at (B,p',j). If there are new cycles, then we already had $(A,p,i) \in V_L$ and any edge leading to (A,p,i) is a greater–edge.
Since the set $\{(A,p,i) \in V | (A,p,i) \text{ is unbounded}\}$ is finite, the construction is finished after a finite number of steps. After these steps, $OK = \{(A,p,i) \in V | (A,p,i) \text{ is unbounded}\}$ and $NOK = \emptyset$. Moreover, by (3), (4), and (5), $V_L = OK$, each node of L has a unique outgoing edge in L, and each cycle of L contains a greater–edge. Consequently, the constructed L is a lasso structure and, since $V_L = \{(A,p,i) \in V | (A,p,i) \text{ is unbounded}\}$, L contains all unbounded (A,p,i). □
Now we may proceed as follows: (1) Construct the graph D for HRG. (2) Check for each subgraph of D whether it is a lasso structure. (3) Check for each lasso structure L whether it contains $(l(Z),p,i_0)$ for some $p : I \to J$.
If there is a lasso structure L in D containing $(l(Z),p,i_0)$ (for some p), then, by Claim 2, $(l(Z),p,i_0)$ is unbounded, meaning that, for all $n \in \mathbb{N}$, there is an $(l(Z),p)$–hypergraph H with $f(H,i_0) > n$. Hence, for all $n \in \mathbb{N}$, there is a hypergraph $H \in L(HRG)$ with $f_0(H) > n$.
Conversely, if, for all $n \in \mathbb{N}$, there is a hypergraph $H \in L(HRG)$ with $f_0(H) > n$, then, for all $n \in \mathbb{N}$, there is a p and an $(l(Z),p)$–hypergraph H with $f(H,i_0) > n$. Since the number of p's is finite, we can find some p such that, for all $n \in \mathbb{N}$, there is an $(l(Z),p)$–hypergraph H with $f(H,i_0) > n$. Therefore, $(l(Z),p,i_0)$ is unbounded and, by Claim 3, there exists a lasso structure containing $(l(Z),p,i_0)$. This completes the proof of the theorem. □

Combining the compatibility results of Section 4 and Theorem 6.1, one obtains a list of decidability results concerning boundedness problems.

6.2 Corollary

For each edge–replacement grammar $ERG \in \mathcal{ERG}$ and each function in the following list, it is decidable whether (or not) the function values of the graphs in $L(ERG)$ grow beyond any bound: the number of nodes, the number of edges, the number of simple paths connecting the external nodes, the number of simple cycles, the maximum–simple–path length of paths connecting the external nodes, the maximum–simple–cycle length, and the maximum degree of a graph.

Proof: The statements follow directly from the theorems 4.1-4.4 and 6.1. □

Remarks: 1. Remember that the functions "number of nodes" and "number of hyperedges" are compatible for arbitrary hyperedge–replacement grammars $HRG \in \mathcal{HRG}$.
2. Although we avoided the troublesome technicalities in this paper, we are convinced that the other considerations of this section work for more general types of hyperedge–replacement grammars, too. For example, all the statements should hold even if the class $\mathcal{ERG}$ is replaced by the class of all hyperedge–replacement grammars which generate ordinary graph languages and use hyperedges with a bounded number of tentacles as nonterminals. We even think that the considered functions are compatible for arbitrary hyperedge–replacement grammars if their definition is properly adapted to hypergraphs.

Finally, let us mention that some problems — like the connectivity problem, the maximum–clique–size problem, and the chromatic–number problem — are trivial in the following sense: for all hyperedge–replacement grammars HRG, there is a bound (depending only on HRG) such that the function values of all graphs do not exceed the bound. This knowledge can be used to show that other boundedness problems — as the minimum–clique–covering problem and the maximum–indepentent–set problem — are decidable.

The *clique partition number* of a graph G, $C(G)$, is the smallest number of cliques that form a partition of the node set V_G. A set of nodes in a graph G is *independent* if no two of them are adjacent. The largest number of nodes in such a set is called the *independence number* of G and is denoted by $I(G)$.

6.3 Theorem

For each hyperedge–replacement grammar $HRG \in \mathcal{HRG}$ generating a set of graphs, it is decidable whether (or not) the clique partition number and the independence number of graphs in $L(HRG)$ grows beyond any bound.

Proof: Since for each hyperedge–replacement grammar HRG, the maximum clique size is bounded on $L(HRG)$, say by $c(HRG) \geq 1$, and, for each $G \in L(HRG)$,

$$\frac{|V_G|}{c(HRG)} \leq C(G) \leq |V_G|,$$

the clique partition number is bounded on $L(HRG)$ if and only if the number of nodes is bounded on $L(HRG)$. Since for each hyperedge–replacement grammar HRG the chromatic number is bounded on $L(HRG)$, say by $k(HRG) \geq 1$, for each $H \in L(HRG)$, the maximum number of equally colored nodes in a $k(HRG)$–coloring of G, $MAX(G)$, is a lower bound of $I(G)$. On the other side, $|V_G| \leq k(HRG) \cdot MAX(G)$. Thus,

$$\frac{|V_G|}{k(HRG)} \leq I(G) \leq |V_G|.$$

Therefore, the independence number is bounded on $L(HRG)$ if and only if the number of nodes is bounded on $L(HRG)$. □

7. Discussion

Each class $\mathcal{C}$ of graph grammars and each function f on graphs with integer values establish a Boundedness Problem:

> Is it decidable, for all graph languages $L(GG)$ generated by GG in $\mathcal{C}$, whether or not there is a bound n such that $f(G) \leq n$ for all $G \in L(GG)$?

In this paper, we have been able to show that the Boundedness Problem is solvable for classes of hyperedge–replacement grammars and functions that are compatible with the derivation process and where the values of derivable graphs are composed of maxima, sums, and products of component values. Although this result applies to a variety of examples it seems to be strangely restricted. Further research should clarify the situation:

(1) We would expect that the metatheorem holds under more general or modified assumptions. Especially, we would like to know how functions given by minima or differences or divisions work.
(2) We suspect that certain combinations of arithmetic operations are not allowed. For instance, maxima and minima seem to antagonize each other — at least sometimes.
(3) Compatible functions are defined for arbitrary domains. But we have got significant results only for boolean and integer values. What about other domains? How can be arbitrary compatibility be exploited? How do other meaningful interpretations look like?

References

[ALS 88] S. Arnborg, J. Lagergren, D. Seese: Problems Easy for Tree–Decomposable Graphs, Proc. ICALP'88, Lect. Not. Comp. Sci. 317, 38-51, 1988

[BC 87] M. Bauderon, B. Courcelle: Graph Expressions and Graph Rewriting, Math. Systems Theory 20, 83-127, 1987

[Co 87] B.Courcelle: On Context–Free Sets of Graphs and Their Monadic Second–Order Theory, Lect. Not. Comp. Sci. 291, 133-146, 1987

[DG 78] P. Della Vigna, C. Ghezzi: Context–Free Graph Grammars, Inf. Contr. 37, 207-233, 1978

[Ha 88] A. Habel: Graph–Theoretic Properties Compatible with Graph Derivations, to appear in: Proc. Graph–Theoretic Concepts in Computer Science 1988 (WG'88), Lect. Not. Comp. Sci., 1988

[HK 83] A. Habel, H.-J. Kreowski: On Context–Free Graph Languages Generated by Edge Replacement, Lect. Not. Comp. Sci. 153, 143-158, 1983

[HK 85] A. Habel, H.-J. Kreowski: Characteristics of Graph Languages Generated by Edge Replacement, University of Bremen, Comp. Sci. Report No. 3/85, also in: Theor. Comp. Sci. 51, 81-115, 1987

[HK 87a] A. Habel, H.-J. Kreowski: May We Introduce to You: Hyperedge Replacement, Lect. Not. Comp. Sci. 291, 15-26, 1987

[HK 87b] A. Habel, H.-J. Kreowski: Some Structural Aspects of Hypergraph Languages Generated by Hyperedge Replacement, Proc. STACS'87, Lect. Not. Comp. Sci. 247, 207-219, 1987

[HKV 87] A. Habel, H.-J Kreowski, W. Vogler: Metatheorems for Decision Problems on Hyperedge Replacement Graph Languages, to appear in Acta Informatica, short version with the title "Compatible Graph Properties are Decidable for Hyperedge Replacement Graph Languages" in: Bull. EATCS 33, 55-62, 1987

[JRW 86] D. Janssens, G. Rozenberg, E. Welzl: The Bounded Degree Problem for NLC Grammars Is Decidable, Journ. Comp. Syst. Sci. 33, 415-422, 1986

[Kr 79] H.-J. Kreowski: A Pumping Lemma for Context–Free Graph Languages, Lect. Not. Comp. Sci. 73, 270-283, 1979

[La 88] C. Lautemann: Decomposition Trees: Structured Graph Representation and Efficient Algorithms, Proc. CAAP'88, Lect. Not. Comp. Sci. 299, 28-39, 1988

[LW 88] T. Lengauer, E. Wanke: Efficient Analysis of Graph Properties on Context–Free Graph Languages, Proc. ICALP'88, Lect. Not. Comp. Sci. 317, 379-393, 1988

[RW 86a] G. Rozenberg, E. Welzl: Boundary NLC Graph Grammars — Basic Definitions, Normal Forms, and Complexity, Inf. Contr. 69, 136-167, 1986

[RW 86b] G. Rozenberg, E. Welzl: Graph Theoretic Closure Properties of the Family of Boundary NLC Graph Languages, Acta Informatica 23, 289-309, 1986

[Sl 82] A.O. Slisenko: Context–Free Graph Grammars as a Tool for Describing Polynomial–Time Subclasses of Hard Problems, Inf. Proc. Lett. 14, 52-56, 1982

Implementation of Parameterized Observational Specifications

Rolf Hennicker
Fakultät für Mathematik und Informatik
Universität Passau
Postfach 2540
D-8390 Passau

Abstract
An observational approach to the modular construction of algebraic implementations is presented. Based on the theory of parameterized observational specifications an implementation relation is defined which formalizes the intuitive idea that an implementation is correct if it produces correct observable output. It is shown that observational implementations compose vertically and (under appropriate conditions) horizontally. To be useful in practice proof theoretic criteria for parameterized observational implementations are given which are based on the notion of observable parameter context.

1. Introduction

An implementation concept for parameterized algebraic specifications is presented which is based on an observational approach to software development. The basic assumption is that from the software user's point of view a software product is a correct implementation if it satisfies the desired input/output behaviour, independently of the internal properties of a program which may not satisfy a given specification. For example the familiar array-pointer realization of stacks does not satisfy the stack equation pop(push(x, s)) = s or the usual implementation of sets by lists does not satisfy the characteristic set equations. Nevertheless, these implementations are considered to be correct since they produce correct observable output.
In the framework of algebraic specifications this means informally that a concrete specification SP1 is a correct implementation of an abstract specification SP if it preserves the observable properties of SP. To formalize this principle one needs a formal notion of observability which allows to abstract from the internal (non observable) details of data structures. In the literature several behavioural approaches were studied which principally agree in their view of behavioural equivalence of algebras (either with respect to a set of visible (or primitive) sorts (cf. [Giarratana et al. 76], [Reichel 81], [Goguen, Meseguer 82], [Broy et al. 84], [Schoett 87], [Nivela, Orejas 87]) or wrt. a set of observable terms (cf. [Wirsing 86], [Sannella, Tarlecki 87]) or wrt. a set of observable formulas (cf. [Sannella, Tarlecki 85], [Pepper 83])). Our approach is based on observational specifications (cf. [Hennicker, Wirsing 85]) which allow a direct axiomatization of the observable properties of data structures by means of an observability predicate. Having specified a behaviour one can abstract from the model class (of an observational specification) by constructing its behaviour class which is the closure of the model class under observational equivalence. Based on this abstraction principle a simple formalization of the notion of observational implementation is possible:

An observational specification SP1 is an observational implementation of SP if the behaviour class of SP1 (after appropriate restriction) is a subclass of the behaviour class of SP.

Obviously, this implementation relation is transitive, i.e. observational implementations compose *vertically* (cf. [Goguen, Burstall 80]) which is a basic requirement for the construction of correct programs by stepwise refinement.
In practice often large specifications have to be implemented. In this case it should be possible to decompose the (abstract) specification and to implement parts of it independently from each other (e.g. by different programmers). In order to get a correct implementation of the whole system it is necessary that the single "implementation pieces" compose horizontally (cf. [Goguen, Burstall 80]). To support the modular construction of observational implementations we introduce parameterized observational specifications and their implementation.
As usual parameterized observational specifications have a formal parameter specification and a body which in our framework are both observational specifications. Semantically, a parameterized specification is viewed as a function which can be applied to an argument specification and produces a specification as result. The definitions of admissible actual parameter and instantiation are defined similarly to well-known constructions (cf. e.g. [Ganzinger 83], [Ehrig, Mahr 85]). Since parameterized observational specifications

are treated as (partial) functions the observational implementation relation can be extended pointwise to the parameterized case:

A parameterized observational specification P1 is called (parameterized) observational implementation of P (with common formal parameter) if for all admissible parameters SP the application P1(SP) is an observational implementation of P(SP).

This pointwise implementation definition is different from the (proof theoretic) approach of [Ganzinger 83] and the concept of [Ehrig, Kreowski 82] which is based on the initial algebra approach and the free functor semantics. Further implementation concepts for parameterized specifications using a pointwise definition like ours (but based on different implementation notions for (non parameterized) specifications) are provided in [Sannella, Wirsing 82], [Wirsing 86] and [Sannella, Tarlecki 87].

The simplicity of our basic implementation concept results from the fact that observational specifications describe the observable behaviour of data structures and hence determine (on an abstract level) already all possible implementations. On the contrary, implementation concepts adopting the "forget-restrict-identify" method (cf. e.g. [Ehrig et al. 82], [Sannella, Wirsing 82], [Broy et al. 86]) require to connect the models of the implementation with models of the abstract specification e.g. by means of a congruence relation or an abstraction homomorphism.
Related to our notion of implementation are the implementation concepts of [Goguen, Meseguer 82] and [Sannella, Tarlecki 85]. While in [Goguen, Meseguer 82] abstract specification and implementation are persistent extensions of the same specification with all sorts visible, [Sannella, Tarlecki 85] adopt a simple notion of implementation which is based on abstraction with respect to a set of observable formulas. More generally, [Sannella, Tarlecki 87] propose implementations with respect to an arbitrary abstractor. As a main difference to our approach abstractors are determined semantically by an equivalence relation on classes of algebras whereas observational implementations are based on an axiomatization of the observable behaviour and hence are appropriate for the development of proof theoretic implementation criteria.
Based on the notion of observable context a proof theoretic criterium for observational implementations is presented which roughly says that a specification SP1 is an observational implementation of SP if all applications of observable contexts to the axioms of SP yield a sentence which is deducible from the axioms of SP1 (e.g. an observational implementation of the usual specification of stacks (with observable top elements) not necessarily has to satisfy the stack equation pop(push(x, s)) = s but has to satisfy all applications of observable contexts to this equation as e.g. the equation top(pop(push(x, s))) = top(s)). It is shown that this condition can be extended to a context criterium for (parameterized) observational implementations by using observable parameter contexts. In particular the context criteria can be applied to specifications with fixed sets of observable (or visible) sorts (as e.g. in [Goguen, Meseguer 82]).
As already mentioned it is an important issue whether implementations compose horizontally. In our framework this means: If P1 and P are parameterized observational specifications such that P1 implements P and if SP1 and SP are actual parameters such that SP1 implements SP, is the application P1(SP1) an observational implementation of P(SP)?
It is shown that observational implementations compose horizontally if one of the parameterized specifications involved is *monotonic*. For the *monotonicity* of parameterized specifications a sufficient criterium is provided (using the notions of *parameter completeness* and *parameter tolerancy*). The stepwise and modular construction of observational implementations is demonstrated by an example.
The paper is organized as follows: In section 2 the basic notions of observational specifications (cf. [Hennicker, Wirsing 85]) are summarized which are necessary for the following sections. In section 3 the concept of observational implementations is defined and a proof theoretic context criterium is provided. Section 4 introduces parameterized observational specifications and their instantiation and in section 5 the observational implementation relation is extended to the parameterized case. In section 6 the horizontal composition of observational implementations is discussed and illustrated by an example.

2. Basic notions

We assume the reader to be familiar with the basic notions of algebraic specifications (cf. e.g. [Ehrig, Mahr 85]), that are the notions of ***signature*** $\Sigma = (S, F)$, ***signature morphism*** σ, ***total*** Σ***-algebra*** $A = ((A_s)_{s\in S}, (f^A)_{f\in F})$, where A_s denotes the carrier sets of A and f^A the total operations of A, ***term algebra*** $W_\Sigma(X)$ over a S-sorted family $X = (X_s)_{s\in S}$ of sets of identifiers, ***ground term algebra*** W_Σ, ***term*** $t \in W_\Sigma(X)$, ***ground term*** $t \in W_\Sigma$, ***substitution*** $\sigma: X \to W_\Sigma(X)$, ***instantiation*** $\sigma(t) = t[\sigma(x_1)/x_1,\ldots, \sigma(x_n)/x_n]$ (i.e. replacement of the identifiers $x_1,\dots, x_n \in X$ occurring in t by the terms $\sigma(x_1),\dots, \sigma(x_n)$); *valuation* $\alpha: X \to A$, *interpretation* of a term t wrt. α, and *finitely generated* (or *term generated*) Σ-algebra.

Moreover, a total Σ-algebra B is called Σ-*subalgebra* of A if $B_s \subseteq A_s$ for all $s \in S$ and $f^A|_B = f^B$ for all function symbols $f \in F$, where $f^A|_B$ denotes the restriction of f^A to the elements of B. For every Σ-algebra A there exists a smallest finitely generated Σ-subalgebra.

A signature $\Sigma' = (S', F')$ is called *subsignature* of Σ if $S' \subseteq S$ and $F' \subseteq F$. The *restriction* of a total Σ-algebra A to Σ' is the Σ'-algebra $A|_{\Sigma'} = ((A_s)_{s\in S'}, (f^A)_{f\in F'})$.

In the following we give a short overview of the theory of observational specifications introduced in [Hennicker, Wirsing 85]. Observational specifications provide an axiomatic description of the observable behaviour of data structures. For that purpose conditional equational specifications are extended by an *observability predicate* "Obs" for specifying the observable objects of data structures.

2.1 Definition An *observational specification* SP is a pair SP = (Σ, E) consisting of a signature Σ and a set E of Horn formulas of the form (*)

(*) $\phi_1 \wedge \dots \wedge \phi_n \Rightarrow \phi_{n+1}$,

where $\phi_1,\dots, \phi_{n+1}$ are atomic formulas. An atomic formula is either an equation $t_i = r_i$ or an *observation* of the form $Obs(t_i)$ (with terms $t_i, r_i \in W_\Sigma(X)$). The Horn formulas of E are called *axioms* of SP. ◊

The semantics of an observational specification SP = (Σ, E) is defined to be the class af all observational Σ-algebras satisfying the axioms of E:

2.2 Definition Let $\Sigma = (S, F)$ be a signature. An *observational Σ-algebra* is a pair (A, Obs^A) consisting of a total Σ-algebra A and a family $Obs^A = (Obs^A{}_s)_{s\in S}$ of subsets $Obs^A{}_s \subseteq A_s$.
Obs^A is called *observable part* of A. The elements of Obs^A are called *observable objects*.

An observational Σ-algebra (A, Obs^A) is called *finitely generated* (or *term generated*) if the total Σ-algebra A is finitely generated. An observational Σ-algebra (B, Obs^B) is called *observational Σ-subalgebra* of (A, Obs^A) if the total Σ-algebra B is a Σ-subalgebra of A and if for all sorts $s\in S$: $Obs^B{}_s = Obs^A{}_s \cap B_s$.

Let $\Sigma' = (S', F')$ be a subsignature of Σ (i.e. $S'\subseteq S$, $F' \subseteq F$). The *restriction* of (A, Obs^A) to Σ' is the observational Σ'-algebra $(A|_{\Sigma'}, Obs^A|_{\Sigma'})$ where $A|_{\Sigma'}$ is the restriction of A to Σ' (see above) and $Obs^A|_{\Sigma'} = (Obs^A{}_s)_{s\in S'}$. The finitely generated Σ'-subalgebra of $(A|_{\Sigma'}, Obs^A|_{\Sigma'})$ is denoted by $\langle(A, Obs^A)\rangle_{\Sigma'}$.
The *satisfaction relation* is the classical one of first order predicate calculus, whereby all axioms of a specification are assumed to be universally quantified. In particular, an observation Obs(t) ($t \in W_\Sigma(X)$) is valid in an observational Σ-algebra (A, Obs^A) (written $(A, Obs^A) \models Obs(t)$) iff for all valuations $\alpha: X \to A$ the interpretation of t wrt. α is an observable object of (A, Obs^A). ◊

2.3 Definition A finitely generated observational Σ-algebra (A, Obs^A) is called *model* of an observational specification SP if (A, Obs^A) satisfies all axioms of SP. The *model class* of SP is denoted by Mod(SP). ◊

2.4 Definition Let (A, Obs^A) and (B, Obs^B) be finitely generated observational Σ-algebras.
(A, Obs^A) and (B, Obs^B) are called *observationally equivalent* (written $(A, Obs^A) \sim_{obs} (B, Obs^B)$) iff for all ground terms $t, r \in W_\Sigma$:

$((A, Obs^A) \models t = r$ and $(A, Obs^A) \models Obs(t))$ iff $((B, Obs^B) \models t = r$ and $(B, Obs^B) \models Obs(t))$ ◊

Observational equivalence defines an equivalence relation on the class of finitely generated observational Σ-algebras.

2.5 Fact
Model classes of observational specifications are in general <u>not</u> closed under observational equivalence. (For example lists are observational equivalent to sets if (only) the results of the *iselement operation* "$\in$" are specified as observable. But lists are not a model of a specification of sets with the usual set equations as axioms.)

2.6 Definition Let SP = (Σ, E) be an observational specification. The closure of the model class of SP under behavioural equivalence is called *behaviour class* of SP and denoted by Beh(SP), i.e.

$$\text{Beh(SP)} = \{(B, Obs^B) \mid (B, Obs^B) \text{ is a finitely generated observational } \Sigma\text{-algebra and there exists a model } (A, Obs^A) \in \text{Mod(SP) such that: } (A, Obs^A) \sim_{obs} (B, Obs^B)\}.$$

Algebras (B, Obs^B) $\in$ Beh(SP) are called *behaviours* of SP. ◊

The construction of the behaviour class of an observational specification provides a uniform abstraction principle for observational specifications which is the basis for the definition of observational implementations.

3. Observational implementations

An important application domain for algebraic specifications is in the formal development of programs by stepwise refinement, a programming discipline which has been proposed already in the beginning of the seventies by Wirth and Dijkstra. Starting from an abstract problem specification one proceeds by constructing step by step more concrete specifications, whereby each step refines the step before by making some design decisions (e.g. choice of data representations, choice of algorithms, etc.) and by elaborating a more detailed description of the problem. For achieving formally the correctness of the final product of a development process a formal notion of implementation is necessary.
From the observational point of view one obtains intuitively the following simple principle of correct implementation:

A specification SP1 is an observational implementation of a specification SP if SP1 preserves the "observable behaviour" of SP, i.e. all algebras satisfying the observable properties of SP1 satisfy the observable properties specified by SP as well.

This informal notion of implementation can be simply formalized by means of the abstraction principle for observational specifications. For technical simplicity we assume that the signature $\Sigma 1$ of the concrete specification SP1 comprises the signature Σ of the abstract specification SP. (i.e. the construction of the implementing specification SP1 by an appropriate enrichment of a given specification, say SP1', is assumed to be already done). Now the implementation principle from above can be formalized as follows:

An observational specification SP1 is an observational implementation of SP if for any behaviour (B, Obs^B) $\in$ Beh(SP1) some appropriately defined restriction of (B, Obs^B) to an observational Σ-algebra belongs to the behaviour class of SP, i.e.

$$\text{Beh(SP1)}|_{\text{"appropriate restriction"}} \subseteq \text{Beh(SP)}.$$

In the simplest case this restriction is just $\langle(B, Obs^B)\rangle_\Sigma$, that is forgetting the sorts and operation symbols of $\Sigma 1$ not belonging to Σ and then constructing the finitely generated Σ-subalgebra (cf. definition 2.2). These steps are usually called *forget* and *restrict*.
In order to get enough generality we allow a further restriction of the observable part Obs^B to a subset $B_0 \subseteq Obs^B$. Informally, the restriction of the observable part means that the implementing specification may specify more objects observable than required by the abstract specification. This coincides with our intuition that implementations fix more and more details whereby the degree of possible abstraction may be limited.
In summary, we get the following definition of observational implementation:

3.1 Definition Let $SP1 = (\Sigma1, E1)$ and $SP = (\Sigma, E)$ be observational specifications with $\Sigma \subseteq \Sigma1$. Let S1 be the set of sorts of $\Sigma1$. SP1 is called *observational implementation* of SP (written SP1 <~~~ SP) if for all behaviours $(B, Obs^B) \in Beh(SP1)$ there exists a family $B_o = ((B_o)_s)_{s \in S1}$ of subsets $(B_o)_s \subseteq Obs^B{}_s$ such that:

$$\langle(B, B_o)\rangle_\Sigma \in Beh(SP).$$

($\langle(B, B_o)\rangle_\Sigma$ denotes the finitely generated Σ-subalgebra of the restriction of (B, B_o) to Σ (cf. definition 2.2)) ◊

If one wishes to rule out trivial implementations one could simply restrict the class of admissible models (and behaviours) to those algebras satisfying true ≠ false and require consistency of the implementing specification. Under this requirements the criteria for observational implementations (given below) remain valid if the abstract specification SP and the concrete specification SP1 are assumed to contain the basic type BOOL with observable truth values.

The next lemma is an immediate consequence of definition 3.1, definition 2.4 and the transitivity of observational equivalence:

3.2 Lemma SP1 is an observational implementation of SP iff for all models $(B, Obs^B) \in Mod(SP1)$ there exists a model $(A, Obs^A) \in Mod(SP)$ and a family of subsets $B_o \subseteq Obs^B$ such that for all ground terms $t, r \in W_\Sigma$:

$$((B, B_o) \models t = r \text{ and } (B, B_o) \models Obs(t)) \text{ iff } ((A, Obs^A) \models t = r \text{ and } (A, Obs^A) \models Obs(t)). \quad ◊$$

To be useful in software development by stepwise refinement it is necessary that the composition of consecutive implementation steps yields a single correct implementation step (i.e. composes *vertically* in the sense of [Goguen, Burstall 80]). It is a direct consequence of the definition that observational implementations satisfy this basic requirement:

3.3 Fact The observational implementation relation is transitive.

The notion of observational implementation differs essentially from concepts adopting the "forget-restrict-identify" approach (cf. e.g. [Ehrig et al. 82], [Sannella, Wirsing 82], [Broy et al. 86]). Although the forget-restrict steps correspond to the restriction of behaviours to the signature of the abstract specification the main step in those concepts is the identification of concrete objects which represent the same abstract objects (e.g. by an abstraction function or a congruence relation). Since observational specifications provide a more abstract view of (the semantics of) algebraic specifications this identification can be simply omitted.
Related to our notion of implementation are the implementation concepts of [Goguen, Meseguer 82] and [Sannella, Tarlecki 85]. While in [Goguen, Meseguer 82] abstract specification and implementation are persistent extensions of the same specification with all sorts visible, [Sannella, Tarlecki 85] adopt a simple notion of implementation which is based on abstraction with respect to a set of observable formulas. Compared with [Sannella, Tarlecki 85] (and more generally with [Sannella, Tarlecki 87]) observational implementations correspond to the implementation of an abstract specification after having applied an appropriate abstractor. In our framework no manipulation of the original abstract specification is necessary since observational specifications give already an axiomatization of a behaviour. This axiomatic basis leads to proof theoretic criteria for implementation relations.

An important issue for the application of formal implementation notions in practice is the question for appropriate proof methods (for implementation relations). Since it is highly desirable that correctness proofs are supported by machine we are particularly interested in proof theoretic conditions for implementation relations.
Observational specifications give an axiomatization of the observable behaviour of data structures and hence are appropriate for the development of proof theoretic criteria for implementations. As proof system for observational specifications we generalize the proof system of [Selman 72] to conditional formulas of the form (*) (see definition 2.1) and then extend it by the rule

$$\frac{\phi_1 \wedge \ldots \wedge \phi_n \Rightarrow t = r, \quad \phi_1 \wedge \ldots \wedge \phi_n \Rightarrow Obs(t)}{\phi_1 \wedge \ldots \wedge \phi_n \Rightarrow Obs(r)}$$

which asserts the compatibility of the observability predicate with equality. If a Horn formula $\phi_1 \wedge \ldots \wedge \phi_n \Rightarrow \phi_{n+1}$ is deducible from a set of axioms E we write $E \vdash \phi_1 \wedge \ldots \wedge \phi_n \Rightarrow \phi_{n+1}$. As a simple generalization of theorem 2 in [Selman 72] it can be shown that the proof system "$\vdash$" is sound and complete (wrt. the model class of an observational specification).
For the development of a proof theoretic criterium for observational implementation relations we need the following definitions:

3.4 Definition Let $SP = (\Sigma, E)$ be an observational specification, let S be the set of sorts of Σ, and let $Z = \{z_s \mid s \in S\}$ be an S-sorted set of identifiers.

1.) A term $c \in W_\Sigma(Z)$ is called *context* over Σ, if c contains exactly one identifier $z_s \in Z$. To indicate the identifier occurring in c we often write $c[z_s]$ instead of c.
The application of a context $c[z_s]$ to a term $t \in W_\Sigma$ of sort s is defined by the substitution of z_s by t. Instead of $c[t/z_s]$ we write briefly $c[t]$.

2.) A sort $s \in S$ is called *observable sort* of SP if there exists an axiom $\phi_1 \wedge \ldots \wedge \phi_n \Rightarrow Obs(t)$ of SP such that t is of sort s.

3.) A context $c \in W_\Sigma(Z)$ is called *observable context* of SP if the sort of c is an observable sort of SP.
◊

Using the notion of observable context one can show that an observational specification SP1 implements an observational specification SP (with observable premises in the axioms) if SP1 preserves observability of objects and if SP1 satisfies all Horn formulas of the form $\phi_1 \wedge \ldots \wedge \phi_n \Rightarrow c[t] = c[r]$ where $\phi_1 \wedge \ldots \wedge \phi_n \Rightarrow t = r$ is (a ground instance of) an axiom of SP and c is an observable context of SP. In particular, SP1 not necessarily satisfies all equational axioms $t = r$ of SP but all applications of observable contexts to (ground instances of) $t = r$. For example a usual stack specification where the top elements of stacks are specified as observable may be implemented by a specification SP1 which does not satisfy the equation $pop(push(x, s)) = s$ but satisfies all applications of observable contexts to this equation as e.g. the equation $top(pop(push(x, s))) = top(s)$.
Formally, we obtain the following criterium for observational implementations:

3.5 Proposition Let $SP1 = (\Sigma1, E1)$ and $SP = (\Sigma, E)$ be observational specifications with $\Sigma \subseteq \Sigma1$ and let all premises of the axioms of SP be of observable sort (i.e. if $p = q$ is a premises of an axiom of SP then p and q are of observable sort of SP).

SP1 is an observational implementation of SP if the following conditions are satisfied:

a) If $(\phi_1 \wedge \ldots \wedge \phi_n \Rightarrow Obs(t)) \in E$ $(n \geq 0)$ then $E1 \vdash \sigma^*(\phi_1 \wedge \ldots \wedge \phi_n \Rightarrow Obs(t))$ for all (ground) substitutions $\sigma: X \to W_\Sigma$.

b) For all observable contexts $c[z_s]$ of SP holds:
If $(\phi_1 \wedge \ldots \wedge \phi_n \Rightarrow t = r) \in E$ $(n \geq 0)$ and if t is of sort s
then $E1 \vdash \sigma^*(\phi_1 \wedge \ldots \wedge \phi_n \Rightarrow c[t] = c[r])$ for all (ground) substitutions $\sigma: X \to W_\Sigma$.

(σ^* denotes the instantiation of formulas wrt. σ. The proof of proposition 3.5 is given in [Hennicker 88], pp. 143.)
◊

Proposition 3.5 gives a practically applicable criterium for implementation relations. For the verification of condition b) in examples the proof technique of *context induction* is appropriate (cf. [Hennicker 88]).

4. Parameterized observational specifications

Parameterized specifications provide a flexible tool for achieving generality and reusability in the process of software specification and development. Similar to a function declaration in some ordinary programming language parameterized specifications have formal parameters and a body which defines the effect of the application to an actual parameter. Actual parameters are just specifications and the result of an application yields again a specification (the instantiation of the body specification). There are several parameterization concepts for algebraic specifications in the literature to some of which it will be referred later in comparison with our approach.
In the following parameterized specifications are studied from the observational point of view. In particular the notion of observational implementation is extended to parameterized specifications.

4.1 Definition A *parameterized observational specification* P is a pair P = (PA, B) consisting of

— an observational specification PA = (Σ_{PA}, E_{PA}) (called *formal parameter*) and

— an observational specification B = (Σ_B, E_B) (called *body*)

such that $\Sigma_{PA} \subseteq \Sigma_B$ and $E_{PA} \subseteq E_B$. ◊

4.2 Example The parameterized observational specification SET describes properties of finite sets which are parameterized with respect to their elements. The formal parameter specification ELEM requires observability of the elements (of sets) and of the boolean values. In particular the results of the *iselem* operation which tests the membership of an element in a set are observable. ELEM contains a subspecification BOOL with the usual laws of the Boolean algebra as axioms (x or x = x, x or y = y or x, etc.), an equality test *eq* for elements which is specified by the axioms of an equivalence relation, and a constant *const*.

Formal parameter:

spec ELEM = **enrich** BOOL **by**
  sorts: elem
  functs: const: → elem
    eq: elem x elem → bool
  axioms:
    $Obs(x_{bool})$, $Obs(x_{elem})$,
    eq(x, x) = true,
    eq(x, y) = eq(y, x),
    eq(x, y) = true ∧ eq(y, z) = true ⇒ eq(x, z) = true

(x_{bool} and x_{elem} are identifiers of sort *bool*, resp. *elem*)

Body:

spec SET (**parameter** ELEM) = **enrich** ELEM **by**
  sorts: set
  functs: empty: → set
    add: elem x set → set
    iselem: elem x set → bool
  axioms:
    iselem(x, empty) = false,
    iselem(x, add(y, s)) = eq(x, y) or iselem(x, s),
    add(x, add(x, s)) = add(x, s),
    add(x, add(y, s)) = add(y, add(x, s)).

◊

Semantically, a parameterized observational specification is considered as a (partial) function which takes an observational specification as argument and yields an observational specification as result. This view coincides with [Sannella, Tarlecki 87] and is basically related to the theory procedures in CLEAR (cf. [Burstall, Goguen 80]) or the λ-calculus like approach of ASL (cf. [Wirsing 86]).
The application of a parameterized observational specification is defined for all actual parameters satisfying the requirements of the formal parameter. Thereby actual parameters are connected to a formal parameter via a signature morphism (from the formal parameter signature to the actual parameter signature) and an actual parameter is called *admissible* if all its models satisfy the (renamed) axioms of the formal parameter specification. Since models are assumed to be term generated this means that all instantiations of the axioms of the formal parameter with ground terms over the signature of the actual parameter are deducible from the axioms of the actual parameter:

4.3 Definition Let P = (PA, B) be a parameterized observational specification with formal parameter PA = (Σ_{PA}, E_{PA}) and body B = (Σ_B, E_B). Let SP = (Σ, E) be an observational specification such that $(\Sigma_B \setminus \Sigma_{PA}) \cap \Sigma = \emptyset$ (i.e. no name clashes).
SP is called *admissible actual parameter* of P with respect to a signature morphism $\rho: \Sigma_{PA} \to \Sigma$ if for all axioms $(\phi_1 \wedge \ldots \wedge \phi_n \Rightarrow \phi_{n+1}) \in E_{PA}$ holds:

$$E \vdash \sigma^*(\rho^*(\phi_1 \wedge \ldots \wedge \phi_n \Rightarrow \phi_{n+1})) \text{ for all (ground) substitutions } \sigma: X \to W_\Sigma .$$

(ρ^* denotes the extension of ρ to formulas (over Σ_{PA}) and σ^* denotes the instantiation of formulas wrt. σ.) ◊

As an equivalent definition one could require that all (renamed) axioms of PA are in the inductive theory of SP, i.e. are deducible (from the axioms of SP) by infinite induction. This definition of admissible parameter is more liberal than other notions (cf. e.g. [Ehrig, Mahr 85], [Goguen, Meseguer 82]) which require that all axioms of the formal parameter (not only ground instantiations) are provable from the actual ones. Equivalent to our definition is the notion of [Ganzinger 83] (if it is restricted to non parameterized actual parameters). For the CLEAR procedures the more liberal notion of actual parameter can be achieved by using "fitting" morphisms between data theories.
The application of a parameterized observational specification to an actual parameter (also called *parameter passing*) is defined as follows:

4.4 Definition Let P = (PA, B) be a parameterized observational specification with formal parameter PA = (Σ_{PA}, E_{PA}) and body B = (Σ_B, E_B). Let SP = (Σ, E) be an admissible parameter of SP wrt. a signature morphism $\rho: \Sigma_{PA} \to \Sigma$.
The *application* of P to SP wrt. ρ is the observational specification

$$P_\rho(SP) = B[SP/PA]_\rho$$

which is defined by replacing the formal parameter PA by the actual parameter SP while all sorts and function symbols of Σ_{PA} occurring in B \ PA are renamed wrt. to ρ. (To simplify the notation we often omit the index ρ and write simply P(SP) instead of $P_\rho(SP)$.) ◊

This syntactic definition of instantiation corresponds to the instantiation concept in [Ehrig, Mahr 85] which is equivalent to a pushout construction (as e.g. in [Goguen, Meseguer 82]).
The application P(SP) of a parameterized observational specification P is called *parameter protecting* if the properties of the actual parameter SP are preserved, i.e. no additional elements of parameter sort are generated by the body ("no junk") and the application does neither introduce new identities for objects of parameter sort ("no confusion") nor additional observable objects (of parameter sort). A necessary and sufficient condition for a parameterized observational specification P to be parameter protecting for all actual parameters is the *parameter completeness* and *parameter tolerancy* of P:

4.5 Definition Let P = (PA, B) be a parameterized observational specification with formal parameter PA = (Σ_{PA}, E_{PA}) and body B = (Σ_B, E_B). Let X_{PA} be a countably infinite set of identifiers of parameter sorts (i.e. sorts of PA).

1.) P is called *parameter complete* if for all terms $t \in W_{\Sigma_B}(X_{PA})$ of parameter sort s there exists a term $p \in W_{\Sigma_{PA}}(X_{PA})$ of sort s such that: $E_B \setminus E_{PA} \vdash^{B} t = p$.

2.) P is called *parameter tolerant* if for all observational Σ_{PA}-algebras (A, Obs^A) satisfying the axioms E_{PA} there exists an observational Σ_B-algebra (B, Obs^B) such that (A, Obs^A) is an observational Σ_{PA}-subalgebra of the restriction of (B, Obs^B) to Σ_{PA}. ◊

Parameter completeness together with parameter tolerancy corresponds to the various notions of *persistency* which can be found in the literature. It guarantees that for any actual parameter SP = (Σ, E) of P and for any model (A, Obs^A) of SP there exists an extension to a model $(B, Obs^B) \in Mod(P(SP))$ such that $(B, Obs^B)|_\Sigma = (A, Obs^A)$, i.e. SP is protected by the application.

4.6 Remark Parameter completeness is a necessary condition for achieving that for all actual parameters SP the application P(SP) is sufficiently complete (wrt. ground terms of parameter sort) over SP. Consider for example the specification PA' which is obtained from the formal parameter PA by adding a constant for each parameter sort and by replacing the axioms of PA by all their ground instantiations. Then PA' is an admissible parameter of P but P(PA') is not sufficiently complete over PA' if P is not parameter complete.
The condition of parameter completeness could be weakend to the condition "$E_B \vdash t = p$" if either a less liberal notion of admissible parameter is adopted (see above) or if the instantiation of an actual parameter would be defined by adding the (renamed) axioms of the formal parameter to the actual application (as in [Ganzinger 83]).

5. Implementation of parameterized observational specifications

In this section the notion of observational implementation is extended to parameterized observational specifications. If P1 and P are parameterized observational specifications with common formal parameter then P1 is called *(parameterized) observational implementation* of P if for all admissible parameters SP the application P1(SP) is an observational implementation of P(SP). (For a discussion of different formal parameters see [Sannella, Wirsing 82].) This pointwise definition corresponds exactly to our view of parameterized specifications as functions. Analogous definitions based on different implementation notions (forget-restrict-identify, model class inclusion, abstractor and constructor implementations) are given in [Sannella, Wirsing 82], [Wirsing 86], resp. [Sannella, Tarlecki 87]. A purely proof theoretic implementation concept for parameterized specifications is provided in [Ganzinger 83]. As a major step towards the applicability of algebraic specifications to the modular construction of software systems in [Ehrig, Kreowski 82] it is shown that parameterized implementations based on the initial algebra approach and the free functor semantics are compatible with parameter passing.

5.1 Definition Let P = (PA, B) and P1 = (PA, B1) be parameterized observational specifications. P1 is called *(parameterized) observational implementation* of P (written P1 <~~~ P) if for all admissible parameters SP (wrt. a signature morphism ρ) holds:

$P1_\rho(SP)$ is an observational implementation of $P_\rho(SP)$. ◊

Obviously, the transitivity of the implementation relation for the non parameterized case extends to the parameterized case:

5.2 Fact
The observational implementation relation for parameterized observational specifications is transitive.

As in the non parameterized case for practical applications one is particularly interested in proof theoretic criteria for (parameterized) implementation relations. Proposition 3.5 implies that a parameterized specification P1 is an observational implementation of P if the conditions a) and b) of the proposition can be proved for all applications P1(SP) and P(SP) to actual parameters SP. Hence for the verification all actual parameters have to be considered.
To get rid of reasoning over all actual parameters we are interested in context conditions which only depend on the properties of P and P1 and can be proved independently from actual applications. The basic idea is that an instantiation may impose the observability of *all* parameter sorts and hence we consider (instead of observable contexts (cf. definition 3.4)) all contexts which are either of observable sort *or* of parameter sort, Moreover, since instantiations may introduce arbitrary many ground terms of parameter sort we have to consider contexts which contain arbitrary identifiers of parameter sort.

5.3 Definition Let P = (PA, B) be a parameterized observational specification with formal parameter $PA = (\Sigma_{PA}, E_{PA})$ and body $B = (\Sigma_B, E_B)$ and let S_B be the set of sorts of Σ_B.
Moreover, let $Z = \{z_s \mid s \in S_B\}$ be an S_B-sorted set of identifiers and let X_{PA} be a set of identifiers of parameter sort (as in defintion 4.5) such that Z and X_{PA} are disjoint.
A term $c \in W_{\Sigma_B}(Z \cup X_{PA})$ is called *parameter context* if c contains exactly one identifier $z_s \in Z$ (and arbitrary identifiers of X_{PA}). If the sort of c is either a parameter sort or an observable sort (cf. definition 3.4) of the body B then c is called *observable parameter context* of P. (Context application is defined analogously to definition 3.4 and the notation $c[z_s]$ is used to indicate the identifier z_s of c.) ◊

Using this definition the conditions of proposition 3.5 can be extended to a criterium for parameterized implementation relations. As a prerequisite we require that the implementing parameterized specification is parameter complete. (This could be omitted if we had chosen the alternative definition of parameter passing where the (renamed) axioms of the formal parameter are added to the axioms of the actual application.)

5.4 Proposition Let P1 = (PA, B1) and P = (PA, B) be parameterized observational specifications with bodies B1 = (Σ_{B1}, E_{B1}) and B = (Σ_B, E_B) such that $\Sigma_B \subseteq \Sigma_{B1}$ and let all premises of the axioms of B be of observable sort. Moreover, let P1 be parameter complete.

P1 is a (parameterized) observational implementation of P if the following conditions are satisfied:

a) If $(\phi_1 \wedge \ldots \wedge \phi_n \Rightarrow Obs(t)) \in E_B$ $(n \geq 0)$ then $E_{B1} \vdash \sigma^*(\phi_1 \wedge \ldots \wedge \phi_n \Rightarrow Obs(t))$ for all substitutions $\sigma: X \to W_{\Sigma_B}(X_{PA})$.

b) For all observable parameter contexts $c[z_s]$ of P holds:
If $(\phi_1 \wedge \ldots \wedge \phi_n \Rightarrow t = r) \in E_B$ $(n \geq 0)$ and if t is of sort s
then $E_{B1} \vdash \sigma^*(\phi_1 \wedge \ldots \wedge \phi_n \Rightarrow c[t] = c[r])$ for all substitutions $\sigma: X \to W_{\Sigma_B}(X_{PA})$.

(As before σ^* denotes the instantiation of formulas wrt. σ. The proof of proposition 5.4 is given in [Hennicker 88], pp. 117.) ◊

5.5 Example In this example (parameterized) sets are implemented by (parameterized) lists which in turn are implemented by (parameterized) arrays with pointers. Thus, by the vertical composition property (cf. fact 5.2) this yields an observational implementation of sets by arrays with pointers.
In the first step the parameterized specification SET of example 4.2 is implemented by the following parameterized specification LIST which specifies lists in a usual way. Since all elements are specified as observable by the formal parameter ELEM in particular the first elements of lists can be observed by the operation *first*.

spec LIST (**parameter** ELEM) = **enrich** ELEM **by**

sorts:	list	**axioms:**	first(empty) = const,
functs:	empty: → list		rest(empty) = empty,
	add: elem x list → list		first(add(x, s)) = x,
	first: list → elem		rest(add(x, s)) = s,
	rest: list → list		iselem(x, empty) = false,
	iselem: elem x list → bool		iselem(x, add(y, s)) = eq(x, y) or iselem(x, s).

Fact LIST is a (parameterized) observational implementation of SET.

Informally, this fact is clear since lists and sets have the same behaviour wrt. the observable results of the SET operation *iselem*. For a formal proof proposition 5.4 can be applied. The validity of condition b) can be shown by induction on the structure of the parameter contexts. (A detailed proof using the principle of *context induction* is given in [Hennicker 88], pp. 155.)
Note that the sort *set* has to be identified here with the sort *list*. An explicit renaming could simply be treated as in [Sannella, Tarlecki 87].

In the next step lists are implemented by arrays with pointers. The following specification ARRAY_POINTER specifies lists as pairs consisting of an array a (which is parameterized with respect to its entries) and a natural number p (called pointer). The empty list is implemented by the empty array *vac* together with the pointer 0. Adding an element x to a list is implemented by putting x into the (p+1)-th component of the array (if the pointer has value p) and by incrementing the pointer. The *rest* operation is simply implemented by decrementing the pointer (without deleting the last entry) and the first element of a list is obtained by accessing the p-th component of the array (if the pointer has value p).
For simplicity only those array operations are specified here which are necessary for the implementation (*vac* for the empty array, *put* for putting a new element on the array and *get* for selecting an element of the array).

spec ARRAY_POINTER (**parameter** ELEM) =
enrich ELEM, NAT **by**
sorts: array, list
functs: vac: → array
put: array x nat x elem → array
get: nat x array → elem
pair: array x nat → list
empty: → list
add: elem x list → list
first: list → elem
rest: list → list
iselem: elem x list → bool

axioms:
get(p, vac) = const,
get(p, put(a, p, x) = x,
eq_nat(p, q) = false ⇒ get(p, put(a, q, x)) = get(p,a),
empty = pair(vac, 0),
add(x, pair(a, p)) = pair(put(a, p+1, x), p+1),
first(pair(a, p)) = get(p, a),
rest(pair(a, p)) = pair(a, p–1),
iselem(x, pair(a, 0)) = false,
iselem(x, pair(a, p+1)) =
= eq(x, get(p+1, a)) or iselem(x, pair(a, p)).

Note that ARRAY_POINTER does not satisfy the LIST-axiom rest(add(x, s)) = s since the *rest* operation only decrements the pointer and does not delete the element x in the array. But the observable behaviour of LIST is preserved by ARRAY_POINTER since ARRAY_POINTER satisfies all equations between terms (containing only identifiers of parameter sort) of the observable sorts *elem* or *bool* which are satisfied by LIST. In particular ARRAY_POINTER satisfies the equations first(rest(add(x, t))) = first(t) and first(add(x, t)) = x with an identifier x of sort *elem* and a term t with identifiers of parameter sort.

Fact ARRAY_POINTER is a (parameterized) observational implementation of LIST.
(A formal proof by context induction is given in [Hennicker 88], pp. 138.)

By vertical composition of the two implementation steps one obtains

Fact ARRAY_POINTER is a (parameterized) observational implementation of SET. ◊

6. Horizontal composition of observational implementations

It is one of the main issues for the use of formal implementation notions in practice whether they support the modular construction of implementations. More precisely this means that local implementations of parts of a (structured) specification should compose to an implementation of the whole specification. This property is called *horizontal composition* (cf. [Goguen, Burstall 80]). In the framework of parameterized observational specifications the situation can be described as follows:
Given two parameterized observational specifications P1 and P such that P1 <~~~ P and two admissible parameters SP1 and SP such that SP1 <~~~ SP, the question is whether the application P1(SP1) is an observational implementation of P(SP). To study this issue we distinguish two cases for the construction of the implementation P1(SP1) which can be illustrated by the following diagram:

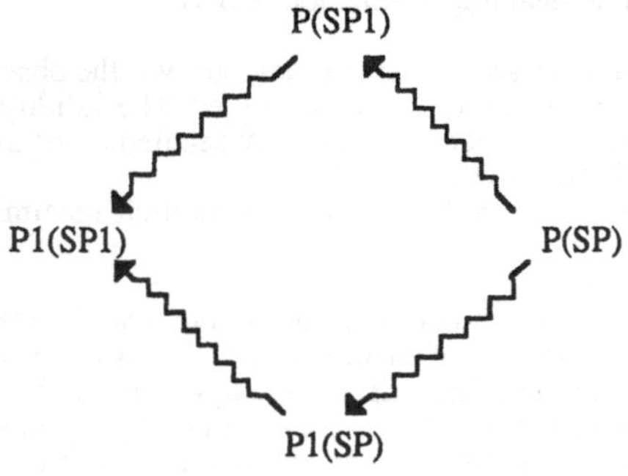

Corresponding to the upper part of the diagram at first the actual parameter SP is replaced locally by its implementation SP1. The resulting specification P(SP1) is an observational implementation of P(SP) if P preserves implementation relations for arguments, i.e. if P is *monotonic*. In a next step P(SP1) is implemented by P1(SP1) which is obviously a correct implementation since it is assumed that P1 implements P.

In the second case corresponding to the lower part of the diagram first P(SP) is implemented by P1(SP) and then P1(SP) is implemented locally by replacing SP by SP1. In this case the resulting specification P1(SP1) is an observational implementation of P1(SP) (and hence of P(SP)) if P1 is *monotonic*.
This discussion shows that the crucial point for achieving horizontal composition is the monotonicity of one of the parameterized specifications involved.

6.1 Definition Let P = (PA, B) be a parameterized observational specification. P is called *monotonic* if for all admissible parameters SP1 and SP which are related by the implementation relation SP1 <~~~ SP holds: P(SP1) <~~~ P(SP).
(More precisely it is assumed that the signature morphisms connecting the formal parameter PA with the actual parameters SP and SP1 coincide.) ◊

We now obtain the following two conditions for the horizontal composition of observational implementations (corresponding to the upper resp. lower part of the above implementation diagram).

6.2 Proposition Let P1 = (PA, B1) and P = (PA, B) be parameterized observational specifications such that P1 <~~~ P. Moreover, let SP and SP1 be admissible parameters such that SP1 <~~~ SP.

1.) If P is monotonic then P1(SP1) <~~~ P(SP).

2.) If P1 is monotonic then P1(SP1) <~~~ P(SP). ◊

Monotonicity is sufficient for the horizontal composition of observational implementations. Compared with the horizontal composition of abstractor implementations in [Sannella, Tarlecki 87] the requirement of monotonicity corresponds to the preservation of abstraction equivalences by the parameterized specifications. Based on a module oriented approach horizontal compositions are studied in [Schoett 87]. The basic requirement there is the stability of the "cells" involved in the program development process. Since our approach admits proof theoretic properties (like parameter completeness) we can give the following sufficient condition for monotonicity:

6.3 Proposition Let P be a parameter complete and parameter tolerant parameterized observational specification. Moreover, let all observable sorts of P be parameter sorts. Then P is monotonic. ◊

(Proposition 6.3 can be proved by a simple generalization of the proof in [Hennicker 88], pp. 153.)

6.4 Example In the following the implementation of sets by arrays with pointers is horizontally composed with an implementation of the integers by sequences of bits. The composition yields an observational implementation of sets of integers by pairs consisting of an array over bit sequences and a pointer.
We start with the specification INT of the intergers. INT specifies the boolean values and all integers (which are constructed by *zero*, *succ*, and *pred*) as observable. For the definition of the equality test *eq_int* the auxiliary function *non_negative* is used. Note that *eq_int* is completely specified for all integers since the equality test can always be reduced to the equality test for non negative integers.

```
spec INT = enrich BOOL by
   sorts:  int
   functs: zero: → int
           succ, pred: int → int
           eq_int: int x int → bool
           non_negative: int → bool
   axioms:
           Obs(true), Obs(false),
           Obs(i) ⇒ Obs(succ(i)),
           Obs(i) ⇒ Obs(pred(i)),
           pred(succ(i)) = i,
           succ(pred(i) = i,
           eq_int(zero, zero) = true,
           eq_int(i, j) = eq_int(j, i),
           eq_int(succ(i), succ(j)) = eq_int(i, j),
           non_negative(i) = true ⇒ eq_int(zero, succ(i)) = false,
           non_negative(zero) = true,
           non_negative(pred(zero)) = false,
           non_negative(i) = true ⇒ non_negative(succ(i)) = true,
           non_negative(i) = false ⇒ non_negative(pred(i)) = false.
```

Next we give an implementation of integers by pairs $\langle x, s\rangle$ consisting of a sign $x \in \{O, L\}$ and a bit sequence over $\{O, L\}$ (constructed by the constant ε and the operation &). The sign x represents the sign of an integer and the bit sequence s is the usual binary representation of the absolute value. Sequences with leading O and the pairs $\langle O, \varepsilon\rangle$, $\langle L, \varepsilon\rangle$ (representing *zero*) are identified.
For the implementation of *succ* and *pred* the auxiliary function *compl* is used which complements the sign of a bit sequence. That way the computation of the successor (predecessor) of a negative integer can be reduced to the predecessor (successor) of its complement. The equality test *eq_int* is implemented by means of the equality test *eq_seqbit* for bit sequences (taking into account that sequences with leading O are identified). We hope that the basic idea of the implementation is sufficiently illustrated and drop a detailed discussion of the axioms.

```
spec SEQBIT = enrich BOOL by
    sorts:  bit, seqbit, int
    functs: O, L: → bit
            eq_bit: bit x bit → bool
            ε: → seqbit
            . & .: seqbit x bit → seqbit
            eq_seqbit: seqbit x seqbit → bool
            ⟨.,.⟩: bit x seqbit → int
            zero: → int
            succ, pred, compl: int → int
            eq_int: int x int → bool
            non_negative: int → bool
  axioms:
            Obs(true), Obs(false),
            Obs(i) ⇒ Obs(succ(i)),
            Obs(i) ⇒ Obs(pred(i)),
            eq_bit(x, x) = true,
            eq_bit(O, L) = false,
            eq_bit(L, O) = false,
            ε&O = ε,
            eq_seqbit(ε, ε) = true,
            eq_seqbit(ε, s&L) = false,
            eq_seqbit(ε, s&O) = eq_seqbit(ε, s),
            eq_seqbit(s&x, s´&y) =
                = eq_bit(x, y) and eq_seqbit(s, s´),
            eq_seqbit(s, s´) = eq_seqbit(s´, s),
            ⟨O, ε⟩ = ⟨L, ε⟩,
            compl⟨L, s⟩ = ⟨O, s⟩,
            compl⟨O, s⟩ = ⟨L, s⟩,
            zero = ⟨L, ε⟩,
            succ⟨L, ε⟩ = ⟨L, ε&L⟩,
            succ⟨L, s&O⟩ = ⟨L, s&L⟩,
            succ⟨L, s⟩ = ⟨L, s´⟩ ⇒ succ⟨L, s&L⟩ = ⟨L, s´&O⟩,
            succ⟨O, s&x⟩ = compl(pred⟨L, s&x⟩),
            pred⟨L, ε⟩ = ⟨O, ε&L⟩,
            pred⟨L, s&L⟩ = ⟨L, s&O⟩,
            pred⟨L, s⟩ = ⟨L, s´⟩ ⇒ pred⟨L, s&O⟩ = ⟨L, s´&L⟩,
            pred⟨O, s&x⟩ = compl(succ⟨L, s&x⟩),
            eq_int(⟨x, s⟩, ⟨x, s´⟩) = eq_seqbit(s, s´),
            eq_int(⟨O, s⟩, ⟨L, s´⟩) = eq-seqbit(s, ε) and eq-seqbit(s´, ε),
            eq_int(i, j) = eq_int(j, i),
            non_negative⟨L, s⟩ = true,
            non_negative⟨O, s⟩ = eq_seqbit(s, ε).
```

It can be shown that SEQBIT <~~~ INT and that INT and SEQBIT are admissible parameters of the parameterized specifications SET, LIST, and ARRAY_POINTER (wrt. the signature morphism ρ with $\rho(elem) = int$, $\rho(const) = zero$, $\rho(eq) = eq\_int$). (For the proof one uses the fact that all ground terms over SEQBIT of sort *int* can be reduced to a normal form $\langle L, \varepsilon\rangle$ or $\langle L, \varepsilon\&L\&a_1\&\ldots\&a_n\rangle$ or $\langle O, \varepsilon\&L\&a_1\&\ldots\&a_n\rangle$ with $a_i \in \{O, L\}$.)

Moreover, one can show that the parameterized specifications SET, LIST, and ARRAY_POINTER are parameter complete and parameter tolerant. Since by example 5.5 ARRAY_POINTER <~~~ SET proposition 6.2 and 6.3 can be applied and one obtains by horizontal composition: ARRAY_POINTER(SEQBIT) <~~~ SET(INT).

All implementation relations which can be constructed for the specifications in our example by horizontal and vertical composition are illustrated by the following diagram:

```
ARRAY_POINTER(INT)      <~~~  LIST(INT)     <~~~  SET(INT).
        ⇣                        ⇣                   ⇣
ARRAY_POINTER(SEQBIT)   <~~~  LIST(SEQBIT)  <~~~  SET(SEQBIT)
```

◊

7. Discussion

In the examples given above the formal parameter ELEM specifies not only the boolean values as observable but also all objects of sort *elem*. Hence it would be meaningless e.g. to construct the specification SET(SET(INT)) (with an appropriately defined equality operation for sets) since sets are not specified as observable and hence SET(INT) is not an admissible parameter of SET. As a solution there are (at least) two possibilities:
First, one could simply omit the axiom Obs(x_{elem}) in the formal parameter ELEM. Then, of course, all implementation relations from above remain valid. Compared with our example there is only one difference concerning the parameterized specifications LIST and ARRAY_POINTER since now the elements of lists can not more be observed via the operation *first* (but still are implicitly observable by the operations *eq* and *iselem*). This means that we have adopted a more abstract view which allows to abstract from the non observable properties of the elements. That way it would cause no problems to construct e.g. the following implementations: ARRAY_POINTER(ARRAY_POINTER(INT)) <~~~ LIST(ARRAY_POINTER(INT)) <~~~ LIST(LIST(INT)) <~~~ SET(LIST(INT)) <~~~ SET(SET(INT)).
The next possibility is to omit again the axiom Obs(x_{elem}) in the formal parameter ELEM but to introduce the axiom Obs(first(s)) in the body of the parameterized specifications LIST and ARRAY_POINTER (yielding parameterized specifications, say LIST' and ARRAY_POINTER'). Then still the implementation relations ARRAY_POINTER' <~~~ LIST' <~~~ SET hold but LIST' and ARRAY_POINTER' are not more parameter tolerant since they introduce more objects as observable than specified by the formal parameter. In fact, LIST' and ARRAY_POINTER' are not monotonic. (This is not surprising since an observational implementation SP1 of SP (where SP and SP1 are admissible parameters of LIST') may preserve only identities between terms of the observable sort *bool* whereas e.g. the implementation relation LIST'(SP1) <~~~ LIST'(SP) only holds if LIST'(SP1) preserves additionally all identities between terms of sort *elem*.) However, for the parameterized specifications SET and LIST' (resp. SET and ARRAY_POINTER') horizontal composition works since proposition 6.2 says that it is enough if one of the parameterized specifications involved is monotonic. Since SET is monotonic regardeless whether the axiom Obs(x_{elem}) belongs to the formal parameter or not one obtains for all admissible parameters SP and SP1 such that SP1 <~~~ SP, the implementation relation LIST'(SP1) <~~~ SET(SP) (resp. ARRAY_POINTER'(SP1) <~~~ SET(SP)).

A further issue to be discussed is more generally concerned with our notion of admissible parameter and its consequences with respect to the horizontal composition property. Consider e.g. a parameterized specification P, an admissible parameter SP of P, and an observational implementation SP1 of SP. If one wishes to implement P(SP) by P(SP1) this may be not feasible in our framework if SP1 is not an admissible parameter of P. To prevent this situation the notion of admissible parameter could be appropriately generalized such that observational implementations automatically preserve admissibility of parameters. From the observational point of view we suggest that an actual parameter should be considered as "observationally admissible" if it respects the observable behaviour specified by the formal parameter. Formally, this notion could be defined similarly to the observational implementation relation by requiring behaviour class inclusion wrt. the behaviour classes of the actual and the formal parameter. In contrast to observational implementations in this case the definition of behaviour classes has to be extended to non term generated algebras.

8. Concluding remarks

The present study shows that observational specifications provide a semantically well founded and flexible tool for the formal development of software. Based on the theory of observational specifications the observational implementation relation formalizes the intuitively clear idea that an implementation is correct if it produces correct observable output. It has been shown that observational implementations compose vertically and (under certain conditions) horizontally and hence are useful for the modular and stepwise construction of implementations. An advantage of the observational specification technique is their axiomatic description of behaviours which gives rise to the development of proof theoretic criteria for implementation relations and horizontal composability. Since it is highly desirable to support implementation proofs by machine a further step should be the development of algorithms for the verification of implementation relations. Based on the principle of context induction a first attempt into this direction has been undertaken in [Hennicker 88].

Acknowledgements
I would like to thank Martin Wirsing who supported this work by many ideas and valuable inspirations and Manfred Broy for many fruitful discussions. I gratefully acknowledge a number of useful comments made by the referees.

References

[Broy et al. 84]
M. Broy, C. Pair und M. Wirsing: A systematic study of models of abstract data types. *Theoretical Computer Science* **33**, 139-174 (1984).

[Broy et al. 86]
M. Broy, B. Möller, P. Pepper und M. Wirsing: Algebraic implementations preserve program correctness. *Science of Computer Programming* **7**, 1, 35-54 (1986).

[Burstall, Goguen 80]
R.M. Burstall, J.A. Goguen: The semantics of Clear, a specification language. *Proc. of Advanced Course on Abstract Software Specifications*, Kopenhagen. Springer Lecture Notes in Computer Science **86**, 292-332 (1980).

[Ehrig, Kreowski 82]
H. Ehrig, H.J. Kreowski: Parameter passing commutes with implementation of parameterized data types. In: M. Nielsen, E.M. Schmidt (eds.): *Proc.* ICALP 82, *9th Coll. on Automata, Languages and Programming,* Aarhus, July 1982. Springer Lecture Notes in Computer Science **140**, 197-211 (1982).

[Ehrig, Mahr 85]
H. Ehrig, B. Mahr: Fundamentals of algebraic specification 1. EATCS Monographs on Theor. Comp. Science, Vol. 6, Springer Verlag (1985).

[Ehrig et al. 82]
H. Ehrig, H.J. Kreowski, B. Mahr und P. Padawitz: Algebraic implementation of abstract data types. *Theoretical Computer Science* **20**, 209-263 (1982).

[Futatsugi et al. 85]
K. Futatsugi, J.A. Goguen, J.P. Jouannaud und J. Meseguer: Principles of OBJ2. *Proc. 12th* ACM *Symposium on Principles of Programming Languages*, New Orleans, 52-66 (1985).

[Ganzinger 83]
H. Ganzinger: Parameterized specifications: parameter passing and implementation with respect to observability. ACM Trans. on Prog. Lang. and Systems **5**, 3, 318-354 (1983).

[Geser, Hussmann 86]
A. Geser, H. Hussmann: Experiences with the RAP system – a specification interpreter combining term rewriting and resolution. In: B. Robinet, R. Wilhelm (eds.): *Proc.* ESOP 86, *Europ. Symp. on Programming*, Saarbrücken, March 1986. Springer Lecture Notes in Computer Science **213**, 339-350

[Giarratana et al. 76]
V. Giarratana, F. Gimona und U. Montanari: Observability concepts in abstract data type specification. In: A. Mazurkiewicz (ed.): *Proc.* MFCS 76, *5th Internat. Symp. on Mathematical Foundations of Comp. Science*, Gdansk, Sept. 1976. Springer Lecture Notes in Computer Science **45**, 576-587 (1976).

[Goguen, Burstall 80]
J.A. Goguen, R.M. Burstall: CAT, a system for the structured elaboration of correct programs from structured specifications. Technical report CSL-118, Computer Science Laboratory, SRI International

[Goguen, Meseguer 82]
J.A. Goguen, J. Meseguer: Universal realization, persistent interconnection and implementation of abstract modules. In: M. Nielsen, E.M. Schmidt (eds.): *Proc.* ICALP 82, *9th Coll. on Automata, Languages and Programming,* Aarhus, July 1982. Springer Lecture Notes in Computer Science **140**, 265-281 (1982).

[Hennicker 88]
R. Hennicker: Beobachtungsorientierte Spezifikationen. Dissertation, Fakultät für Mathematik und Informatik, Universität Passau (1988).

[Hennicker, Wirsing 85]
R. Hennicker, M. Wirsing: Observational specification: a Birkhoff-theorem. In: H.J. Kreowski (ed.): Recent Trends in Data Type Specification. *3rd Workshop on Theory and Appl. of Abstract Data Types*, Selected Papers. Informatik Fachberichte **116**, 119-135, Springer Verlag (1985).

[Nivela, Orejas 87]
Mª P. Nivela, F. Orejas: Initial behaviour semantics for algebraic specifications. *Proc. 5th Workshop on Algebraic Specifications of Abstract Data Types*, Gullane, September 1987, Springer Lecture Notes in Computer Science **332**, 184-207 (1988).

[Pepper 83]
P. Pepper: On the correctness of type transformations. Talk at *2nd Workshop on Theory and Appl. of Abstract Data Types*, Passau, May 1984.

[Reichel 81]
H. Reichel: Behavioural equivalence -- a unifying concept for initial and final specification methods. In: M. Arato, L. Varga (eds.): Math. Models in Comp. Systems, *Proc. 3rd Hungarian Computer Science Conf.*, Budapest, January 1981, 27-39 (1981).

[Sannella, Tarlecki 85]
D.T. Sannella, A. Tarlecki: On observational equivalence and algebraic specification. In: H. Ehrig, C. Floyd, M. Nivat, J. Thatcher (eds.): *Proc.* TAPSOFT 85, *Joint Conf. on Theory and Practice of Software Development*, Berlin, March 1985. Springer Lecture Notes in Computer Science **185**, 308-322 (1985).

[Sannella, Tarlecki 87]
D.T. Sannella, A. Tarlecki: Toward formal development of programs from algebraic specifications: implementations revisited. *Proc.* TAPSOFT 87, *Joint Conf. on Theory and Practice of Software Development*, Pisa, March 1986. Springer Lecture Notes in Computer Science **249**, 96-110 (1987).

[Sannella, Wirsing 82]
D.T. Sannella, M. Wirsing: Implementation of parameterized specifications. In: M. Nielsen, E.M. Schmidt (eds.): *Proc.* ICALP 82, *9th Coll. on Automata, Languages and Programming*, Aarhus, July 1982. Springer Lecture Notes in Computer Science **140**, 473-488 (1982).

[Schoett 87]
O. Schoett: Data abstraction and the correctness of modular programming. Ph.D. thesis, CST–42–87, Department of Comp. Science, University of Edinburgh (1987).

[Selman 72]
A. Selman: Completeness of calculii for axiomatically defined classes of algebras. *Algebra universalis* **2**,

[Wirsing 86]
M. Wirsing: Structured algebraic specifications: a kernel language. *Theoretical computer science* **42**, 123-249 (1986).

Priority Controlled Incremental Attribute Evaluation in Attributed Graph Grammars*

Simon M. Kaplan and Steven K. Goering
Department of Computer Science
University of Illinois

Abstract: We show how incremental attribute change propagation can be efficiently done on Graph Grammars. The order of evaluation is controlled by priority numbers assigned to each attribute vertex. The priority numbers are assigned in linear time as directed by static grammar analysis. We also develop a powerful set-theoretic based embedding description format, and introduce the idea of `roles` as a mechanism to aid in graph grammar engineering. Roles are also a powerful aid to static analysis of grammar properties. A key to our approach is the uniformity gained by collapsing the substrate (structure) graph and the attribute dependency graph into a single directed graph structure. Related groups of attributes are simultaneously replaced by *cluster rewrites.*

1 Introduction

State-of-the-art program development environments use a structured representation for programs rather than a text stream. Such a representation allows the tool to provide incremental feedback to the user concerning the syntactic correctness of his code, and also to check that semantic constraints on the program (for example type-correctness) are met. Usually this representation is a tree [15] [2] [13]; a drawback of such a representation is that the tools do not seem to scale up to supporting more than programming-in-the-small. To correct this deficiency, graphs have been proposed as an alternative structure [9] [12] [1].

We are developing a family of tools which use *attributed graph grammars* to define the legal structures that can be built using the tools and to specify the constraints on the structures. We refer to "structures" rather than "programs", because our graph-grammar based tools support phases of the software development cycle other than programming-in-the-small [8]. An example we will use throughout this paper is that of *module interconnection structures.* A graph grammar is analogous to a string grammar, except that the grammar generates graphs rather than parse trees. Attribution is a well-know mechanism for describing semantic constraints on sentences in the language of a grammar [11], but research on the efficient evaluation of attributes has largely been confined to the case of string grammars [16] [10].

The purpose of this paper is to define an attributed graph grammar model and show how to perform efficient incremental attribute evaluation on the graphs constructed from a grammar that meets certain simple restrictions. Testing if a grammar meets these restrictions can be performed in polynomial time.

We are not the first to suggest the use of graph grammars in software development environments; The IPSEN project [12] [3] has investigated generating tools from graph grammars; in this project semantic constraints are specified using action routines [13]. A major problem with action routines is that they are

---

***Supported in part by the National Science Foundation under grant CCR-8809479 and by the AT&T Illinois Software Engineering Project.** **For more information contact Simon M. Kaplan, Department of Computer Science, University of Illinois, Urbana, Illinois 61801, USA. *email:* `kaplan@a.cs.uiuc.edu` *or* `uunet!uiucdcs!kaplan`.**

not declarative, making them much harder to write and forcing the programmer to define explicitly actions to be taken in the face of editor actions such as insertions, deletions, etc. A declarative notation such as an attribute grammar does not have such problems [14].

Gottler [4] [5] has investigated the use of attribution of graph grammars to help in layout for diagram editing, but does not address the question of attribute evaluation.

Others have investigated environments in which the internal representations are attributed graphs, but do not use grammars to impose structure on the graphs [1]. Their work introduces the idea of using *priority numbering* to schedule attribute evaluations, but because of the lack of structure in the graphs, checking of semantic constraints after an edit cannot even be guaranteed to complete in polynomial time. Further, because of the lack of structure in their graphs, it is not clear how useful their approach will be; there is no way to describe abstractions on structures, or generate tools automatically from formal specifications.

While our approach is greatly influenced by this work, especially the idea of priority numbering, our use of grammars to impose structure on the graph allows a great improvement on these results; the cost of renumbering the graph after an edit is restricted to the size of the subgraph inserted during the rewrite, and the worst-case cost of the attribute evaluation is $O(|\mathit{Influenced}| \cdot \log(|\mathit{Influenced}|))$, where *Influenced* is the set of attributes which receive a new value as a result of the attribute evaluations, or are the immediate descendents of such attributes.

This paper makes several contributions to the theory and practice of building environments for software development. It proposes a powerful model of attributed graph grammars, and a new methodology of defining embeddings in rewrites. It introduces the idea of *roles* as a mechanism which aids in the engineering of the graph grammar and permits efficient attribute evaluation. It gives algorithms for the efficient evaluation of constraints on the structures provided the grammar meets certain properties, and describes tests to determine if the grammar does indeed have these properties.

The remainder of this paper is structured as follows. We begin in section 2 with an informal example of the use of attributed graph grammars to describe the module interconnection structure in a large program. Section 3 introduces our graph grammar model. Section 4 describes how to incrementally reevaluate attributes after an edit. Details of the static grammar analysis algorithms are deferred to an appendix.

2 An Introductory Example

We illustrate our ideas by considering a *module interconnection structure*. Each module has a name, *imports* sets of variable names from other modules, and in turn *exports* a set of variable names. For simplicity we ignore the bodies of modules, and assume that both imports and exports are sets of tuples (the name of the module and the variable). It is trivial to rewrite the syntax of "real" languages to this form. The example will check two semantic constraints: all module names must be unique, and any variable imported from a module must be explicitly exported by that module.

We want to build a structure that will represent the interconnections among modules, and modify it each time a module's interface is changed or a new module is added to the program. After modifying the structure, we will need to check that the semantic constraints are still enforced. Figure 1 shows two sample interconnection structures; part (a) shows a single-module structure, and part (b) the structure after addition of a module.

We represent a module by a *cluster* of six vertices. One vertex is labeled by the symbol M and acts as a *center* for graph rewriting. Each of the other five vertices hold attribution information, including the name of the attribute, the *attribution rule* (i.e. function used to compute its value), the last value computed,

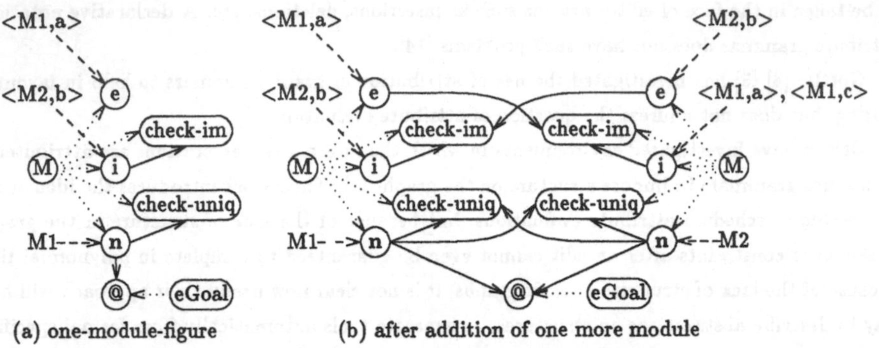

Figure 1: Example Graphs

and some housekeeping information. Three of the attributes represent the module name (`n`), its import list (`i`) and its export list (`e`). In each case the attribution rule to compute the attribute value is the constant function returning the appropriate information, shown in the diagrams as a literal connected to the appropriate vertex by a dashed line. The final two attribute vertices (labeled `check-im` and `check-uniq`) represent the computation necessary to check the semantic constraints. The operations are not shown to save space; they simply verify the conditions stated above. The center of the cluster is connected to all the attribute vertices by *cluster edges*, shown as dotted lines.

Edges are placed to the `check-im` attribute of each module to reflect the import/export dependencies among modules and other edges are placed to connect all the name attributes to the `check-uniq` attribute of each module in order to check that module names are unique.

Figure 1(a) shows a simple structure with one module cluster. The module has name `M1`, imports `b` from `M2` and export `a`. The `n` attribute vertex is connected to the `@` attribute vertex of a `eGoal` cluster. This is used in expanding the graph and will be described later. Note that the `check-im` attribute vertex is connected to no other modules; thus, the importation of `b` from module `M2` will be in error. Because there are no other modules, the uniqueness constraint is satisfied.

Figure 1(b) shows the graph after addition of a second module, `M2`, which exports `b` and imports `a` and `c` from `M1`. Edges are placed from `M2`'s `n` vertex to the `check-uniq` vertex of every other module cluster, and from every other module's `n` vertex to `M2`'s `check-uniq` vertex. Edges are also placed from the `e` vertex of any module exporting variables which are imported by `M2` to `M2`'s `check-im` vertex, and from the `e` vertex of `M2` to the `check-im` vertex in the cluster of any other module exporting from it.

In this instance, the constraint on importation is satisfied for `M1`, but is not satisfied for `M2`, as `M1` does not export `c`. Further modifications to the structure would be needed to satisfy this constraint.

The rewrite of the graph just described is controlled by a `graph grammar`. The grammar for the module interconnection structure is given in figure 2. Each production consists of three parts: A name (with parameters), a goal and a bodygraph. The parameters are constant functions and are bound to the attribution rules of some of the attribute vertices; in our example, the module name, import and export lists are given values in this way. The goal is a symbol, and the bodygraph is analogous to the body of a production in string grammars. Each attribute vertex in the bodygraph is labeled by several things, including an attribution rule (either a constant or some other function), a name, and a *role*, R_i. Rewriting replaces a cluster by a bodygraph, as follows:

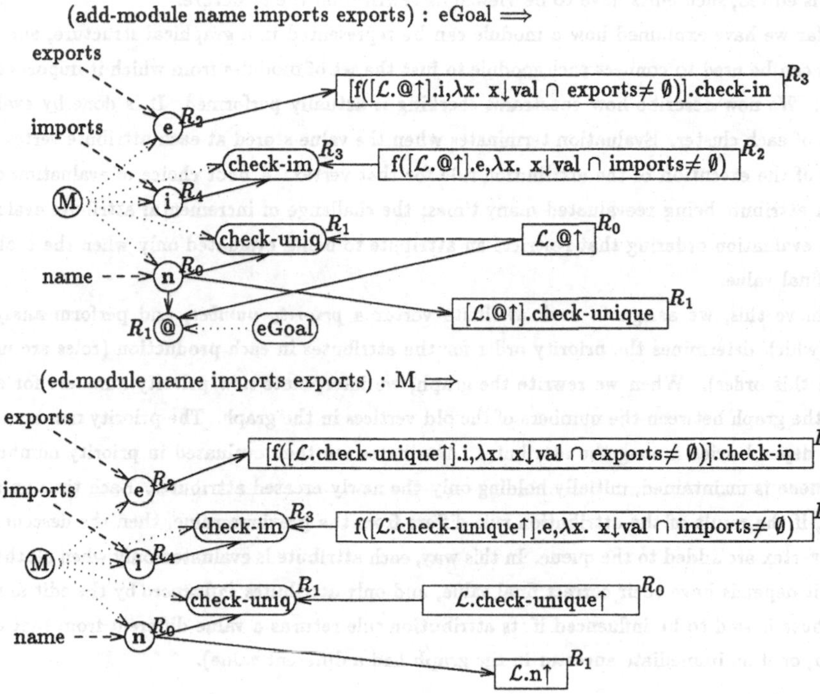

Figure 2: Module Interconnection Grammar

- The symbol labeling the center of the cluster is identified. This must the be same as the goal symbol of the production to be used.
- The vertices in the cluster are identified by following the cluster (dotted) edges.
- The *neighborhood* (all clusters adjacent to the cluster about to be rewritten) is identified. Two clusters are adjacent if there is an edge from some vertex in one to some vertex in the other.
- The cluster is removed from the graph, along with all incident edges.
- The bodygraph of the production is instantiated into the graph.
- The vertices in the instantiation of the bodygraph are *embedded* into the graph by placing edges from (to) vertices in the bodygraph to (from) vertices in the graph. The bodygraph can only be connected to vertices in the neighborhood identified above.

Embedding is controlled by *embedding expressions*, shown in rectangular boxes in the figure. An embedding expression, when evaluated, returns a set of vertices in the neighborhood to which edges may be placed. Embedding expression and attribute vertices are labeled by *roles*. Roles are used to engineer the grammar, and play an important part in static grammar analysis. To see how the embedding expressions are used, consider the expression "f([$\mathcal{L}$.@↑].e, λx.x↓val ∩ imports ≠ ∅)". It evaluates to the set of attribute vertices which export variables that are imported by the new module being added. This is explained in detail below.

Figure 1(b) is derived from figure 1(a) by applying the production `add-module`. The `eGoal` cluster is rewritten to a new module (`M2`), along with a new `eGoal` cluster, which can later be further rewritten to add more modules. Editing a module (for example, to add `c` to the export list of `M1` would involve rewriting the cluster for M1 by the ed-module production. Because the placement of edges in the graph can change if a module is edited, such edits have to be viewed as rewrites on the structure.

Thus far we have explained how a module can be represented in a graphical structure, and how graph rewritings can be used to connect each module to just the set of modules from which it imports or to which it exports. We now describe how constraint checking is actually performed. It is done by evaluating the attributes of each cluster. Evaluation terminates when the value stored at each attribute vertex is equal to the result of the execution of the attribution rule for that vertex. A poor choice of evaluation order could lead to an attribute being reevaluated many times; the challenge of incremental attribute evaluation is to choose an evaluation ordering that restricts an attribute to being evaluated only when the evaluation will yield the final value.

To achieve this, we assign to each attribute vertex a *priority number*, and perform analysis on the grammar which determines the priority order for the attributes in each production (roles are used to help determine this order). When we rewrite the graph, we interpolate new priority numbers for each vertex added to the graph between the numbers of the old vertices in the graph. The priority numbers capture all transitive dependencies among the attributes. Attributes are then evaluated in priority number order. A priority queue is maintained, initially holding only the newly created attributes. Each time an attribute is evaluated, if the result of the attribution rule differs from the previous value, then the descendants of the attribute vertex are added to the queue. In this way, each attribute is evaluated only when all the attributes on which it depends have their correct final value, and only attributes *influenced* by the edit are evaluated. (An attribute is said to be influenced if its attribution rule returns a value different from that of the prior evaluation, or if an immediate ancestor in the graph had a different value).

3 Attributed Graph Grammar Definition

In the notation below, we use x_Y to represent the field x of tuple Y. Where the tuple is obvious from context the subscript is omitted to reduce clutter.

Definition 1 *A* World *is a tuple* $W = < sym, attr, role, val, fns, A >$ *where:*

- *sym is a finite set of symbols;*
- *attr is a finite set of attribute names;*
- *role is a finite set of roles;*
- *val is an arbitrary domain;*
- *fns is a finite set of functions. Each element of* fns *takes zero or more values from val and produces one or more values of val, i.e. is a multivalued function of multiple arguments; and*
- $A : sym \rightarrow \wp(attr)$ *is a function that associates a set of attribute names with each symbol, where* $\wp$ *is the powerset operator.*

Sym*, *attr*, and *role are name sets that will be used to label grammar productions and graph vertices. To simplify description we assume these to be pairwise disjoint. Val is the domain of legal values that attributes may have. We do not preclude the use of attributes of different types, *val* is the sum domain of all relevant types. *Fns* is the set of attribution rules to do the computations. Finally, A declares attribute names for each symbol.

$\langle e4 \rangle ::=$	$\mathcal{L}$	base set
	$\mid \langle e4 \rangle . \langle \text{attrname} \rangle$	narrow
	$\mid \langle e4 \rangle \uparrow$	follow
	$\mid$ '[' $\langle e4 \rangle$ ']'	widen
	$\mid \mathtt{f}(\langle e4 \rangle, \langle \text{pred} \rangle)$	filter
	$\mid \langle e4 \rangle \cup \langle e4 \rangle$	union
	$\mid \langle e4 \rangle \cap \langle e4 \rangle$	intersection
	$\mid \langle e4 \rangle - \langle e4 \rangle$	set difference

<pred> is a predicate of one vertex parameter

operator	type	description
$\mathcal{L}$	1 → set:symb	singleton set containing symbol vertex of the goal cluster
.	set:symb × attrname → set:attr	follow C-edges to attr vertices of given name
↑	set:symb → set:attr set:attr → set:attr	follow all D-edges incident on any element of the set (either to source or target)
[ ]	set:attr → set:symb	return all clusters to which one or more attrs being considered belongs
f	set:a × predicate → set:a	return all elements that satisfy predicate
∪, ∩, −	set:a × set:a → set:a	standard set operations

Notes:

- 'set:symb' means "set of symbol vertices", 'set:attr' means "set of attribute vertices", and 'set:a' means "set of vertices of one of the two kinds".
- Predicates for filters take a single vertex as a parameter. They are allowed reference the vertex label (and thus symbol name, attribute name, role, value, etc.) only. Syntax to reference a field is '↓$\langle$ *fieldname* $\rangle$'.

Figure 3: Syntax and semantics for Edge-End Embedding Expressions

of the production. The rewrite finishes by embedding the daughter graph into the *host graph* (the original with the cluster removed). Embedding is restricted to the distance one neighborhood of the goal cluster (1-NCE) [6]. Daughter graph vertices may only be connected to other daughter graph vertices and to vertices that were originally in the neighborhood of the goal cluster (i.e. adjacent to the goal cluster). We provide a language, called *Edge-End Embedding Expressions* (E^4), to refer to particular subsets of the vertices in the neighborhood of the goal cluster.

Definition 3 Edge-End Embedding Expressions *have syntax and semantics as shown in figure 3.*

As examples we explain some of the expressions used in the rectangles of the Module Interconnection grammar of the previous section.

The expression $\mathcal{L}.n \uparrow$ is evaluated by taking the the n attribute of the goal cluster, i.e. $\mathcal{L}.n$, and following all D-edges incident on it out of the cluster. This expression is used to embed the n attribute of the replacement cluster (for the module editing production). We can determine by inspection that all D-edges are out-directed from it, and that is exactly how the embedding expression is being used to replace D-edges, so this will do an "identity embedding", that is the replacement cluster will be connected identically as the old cluster was connected at this attribute vertex.

The rolename acts as an additional filter to control embedding. In the example grammar, rolenames nearly duplicate attribute names (although several different attributes, @ and check-uniq, take role R_1 indicating that they accept 0 or more module names). In the example productions the rolenames provide

3.1 Attributed Graphs

Definition 2 *A* graph *with respect to world W is a tuple* $G = < Sv, Av, sl, al, E >$ *where:*

- *Sv is a finite set of symbol vertices;*
- *Av is a finite set of attribute vertices;*
- $sl : Sv \rightarrow sym$ *is a labeling function assigning a symbol to every symbol vertex;*
- $al : Av \rightarrow attr_W \times role_W \times Q \times fns_W \times val_W$ *is a labeling function assigning a 5-tuple to every attribute vertex. For every vertex* $v \in Av$*, the label* $al(v) = < a, r, p, o, v >$ *gives its name, role, priority (a rational number[1]), attribution rule, and value respectively;*
- *E is a finite set of (directed) edges, where each edge is specified as a source vertex and a target vertex. An edge is called a* dependency *(D-edge) if its source is an attribute vertex and a* cluster edge *(C-edge) if its source is a symbol vertex.*

G is well-formed *if:*

- *The target of every edge is an attribute vertex;*
- *For each symbol vertex v with label sl(v), let B be the set of all attribute vertices adjacent to v through C-edges. Then no two elements of B have the same name and the set of all the names of elements of B equals* $A_W(sl(v))$*;*
- *Every attribute vertex is the target of either zero or one C-edge;*

For convenience the term graph *will imply well-formedness below unless otherwise stated.*

Informally, a graph is a set of *clusters* (attribute vertices tied to a central symbol vertex with out-directed C-edges, i.e. dotted arrows) and some attribute vertices that are not part of any cluster; plus D-edges between the attribute vertices (drawn by continuous arrows). Example clusters are centered around the symbols `M` (with attributes `e`, `i`, `n`, `check-in`, and `check-uniq`) and `eGoal` (with attribute `@`).

If a symbol s has a set of attribute names $A(s)$ associated to it, by the second requirement every vertex instance of s has attribute vertices of those same names associated with it *via* C-edges. The vertex s and its attribute vertices (and the C-edges connecting them) are a cluster. By the third requirement, clusters are disjoint, but do not necessarily cover the graph.

We allow attribute vertices to be part of no cluster as a convenient shorthand for building an extra (dummy) symbol vertex and making them all a cluster around this. The purpose that symbol vertices serve is as rewrite points, so any vertex analogous to a terminal symbol can be omitted to simplify the grammar. Any attribute vertices that need to remain are then not part of a cluster.

We define adjacency with respect to clusters as the set of all clusters (and attributes not part of any cluster) where some vertex in them is adjacent to some vertex in the original cluster.

As part of each attribute vertex label there is a priority, an operator, and a value. These are concerned with attribute evaluation so we defer full discussion of them until section 4. The rest of this section describes how graphs are rewritten by grammar productions.

3.2 Graph Rewriting

Productions are used to control *cluster rewriting.* Starting from a symbol vertex, all C-edges are followed to identify (in linear time) the cluster. This is replaced by the *daughter graph* instantiated from the *bodygraph*

[1] We make the assumption that operations on rational numbers can always be done in unit time. [1] justifies this assumption and discusses its implications in depth.

visual clue as to meaning (or would if space permitted mnemonic names for them) but do not affect the embedding expression evaluation.

The expression $[\mathcal{L}.@ \uparrow]$.check-uniq is evaluated to yield the attributes named check-uniq in clusters adjacent through D-edges from the @ attribute vertex of the goal cluster. Starting from $\mathcal{L}$ (the goal symbol vertex) we "narrow" consideration to the cluster's @ vertex, then follow all D-edges, then "widen" consideration to the cluster centers at which we have arrived. Finally we narrow back to the check-uniq attribute vertices of all of these clusters.

Finally we explain the expression

$$f([\mathcal{L}.@ \uparrow].e, \lambda x.x \downarrow val \cap imports \neq \emptyset)$$

This must refer to all export lists of modules that we import from. $\mathcal{L}.@ \uparrow$ gets to the name attribute of all modules (except the one currently being created). We widen to get to the modules' centers and narrow to focus on their export lists. Then we filter out those that are not exporting anything that we want to import, by comparing their attribute values to parameters given the production. The attribute vertices that remain are precisely those export lists to which we must embed.

Embedding of the daughter graph is specified by including special vertices called *sockets* in the bodygraph (denoted by retangles in the Module Interconnection example productions). Sockets are not instantiated into vertices to be added as part of the daughter graph into the host graph. Rather they are labeled by Edge-End Embedding Expressions and represent each vertex in the goal cluster's neighborhood to which their label evaluates. An edge incident on a socket is instantiated into one edge for each vertex that the socket represents. These are the edges that embed the daughter graph into the host graph.

Definition 4 Bodygraph *extends the definition of graph by having a third set of vertices (sockets) and a labeling function to assign an Edge-End Embedding Expression and a rolename to each socket.*

A bodygraph is well-formed *if:*

- *The graph resulting from removing all the sockets and their incident edges is well-formed;*
- *All E^4s (i.e. socket labels) evaluate to attribute vertex sets (the typing system makes this easy to check statically);*
- *Sockets are only adjacent to attribute vertices. An implication is that there are no edges between sockets, which is necessary for being 1-NCE.*

We extend the previous definition of D-edges in bodygraphs: a D-edge is allowed to have source or target be a socket. By the third requirement, and since C-edges originate from symbol vertices, we do not have to extend that definition.

Technically, bodygraphs must also have priorities and values for each attribute vertex, but these are calculated and used dynamically, so we don't have to specify them here.

Definition 5 *A* production *is a pair $< L, R >$ where L is a symbol and R is a well-formed bodygraph.*

In the examples of section 2, productions also have parameters that allow them to customize the constant values of some of the new attributes created. Using such parameters is a shorthand for an infinite set of productions, one for each possible combination of values. This is just a way to implement the "intrinsic" attribute values [14] of classical attribute grammar theory, and is ignored in the sequel.

Definition 6 Rewrite *of a graph G by a production p starting from a symbol vertex v to produce a graph H (denoted $G \rightarrow_v^p H$) is accomplished in the following steps:*

- *Verify that the label of v equals L_p (the rewrite cannot be done otherwise);*
- *Identify the cluster C around v by following all C-edges;*
- *Evaluate each E^4 that appears as a label of a socket in R_p. The result of each is some subset of attribute vertices and we further restrict these to the neighborhood of C;*
- *Produce an intermediate graph $I = (G - C) + (R_p - S)$ (where S is the sockets of R_p). I is the host graph plus an instantiation of the bodygraph;*
- *For each edge from a vertex w in R_p to a socket s in R_p, placed edges from the vertex in I corresponding to w to each vertex in I corresponding to vertices in G that were identified by evaluating s's label;*
- *Repeat this for edges that go the other way between sockets and other vertices in the bodygraph;*
- *Call the result (of adding the embedding to I) H.*

Theorem 7 *Within a given world, rewrites preserve graph well-formedness.*

Proof is tedious but straightforward. Since a production's bodygraph is well-formed it only allows placement of new vertices and edges into the host graph so that the result satisfies the well-formedness criteria.

Embedding during rewrites is restricted to the 1-neighborhood. When symbols act as abstractions for complex computations to be expanded later, it may be useful to have placeholding adjacencies so that needed attributes are in their neighborhood to be embedding to when expansion occurs. This is the purpose of the `e` attribute of the module interconnection example. It has edges from each name so that when `eGoal` is rewritten to create a new module, the new module can get correct import and export adjacencies.

3.3 Obligations and Roles

Consider an edge in a graph between two attributes. The attribute vertex at the source of the edge is said to have an *obligation* to the attribute at the target of the edge, *viz.* to produce information which the target can consume. Each attribute vertex therefore has an *obligation set*, its immediate successors in the graph, and a *resource* set, its immediate predecessors in the graph. When the graph is rewritten, in order for the rewrite to be meaningful it is necessary that the obligation/resource relations from the cluster being rewritten to its neighborhood are satisfied after the embedding of the new daughter-graph. In other words, if v is a vertex in a cluster about to be rewritten, w is a vertex in the neighborhood of the cluster, and v has an obligation to w, then after the rewrite a new vertex z must have been introduced into the graph such that z now has an obligation to w.

In practice, obligation/resource relations may be very complex. For example, in our module interconnection example, an `e` vertex has an obligation to an indeterminate number of `check-im` vertices, which depends on what variables the other modules actually import. Other operations may require a specific number of inputs. We therefore characterize operations as requiring an exact number of inputs, 0 or more inputs (as is the case in our example), or 1 or more inputs. A purpose of the static analysis algorithms in appendix A is to check that rewriting maintains obligation/resource relations.

To be able to perform such checks, we introduce the idea of *roles*. Roles are values from a preselected set of rolenames, and are used to annotate all attribute and embedding expression vertices in the grammar according to their *purpose*. In our example, roles are $R_0, R_1, \ldots$ (Space on the diagrams did not permit more mnemonic role names). An n attribute vertex has role R_0, and any embedding expression for such a vertex must also have role R_0. We can think of R_0 as "the obligation to supply a name". `check-uniq` vertices have role R_1; this role may be called "the possibility to check names". We want all vertices with role R_0 to be connected to vertices with role R_1. Roles help in grammar engineering by forcing the grammar-writer to focus clearly on such issues.

When performing embeddings, roles must match, *i.e.* if an embedding expression labeling a socket with role r evaluates to a vertex v in the graph, then v must have role r. This is the key to our grammar analysis; we know when analyzing the grammar that only vertices with appropriate roles can be possible connection points, and then check that all possible connection points satisfy the obligations. Further analysis to check that attribution rules get the correct numbers of inputs can then be checked in a secondary phase.

3.4 Attributed Graph Grammars

Definition 8 *A* grammar *is a triple* $R =< W, P, Z >$ *where:*

- *W is a world;*
- *P is a finite set of productions with respect to W; and*
- *$Z \in sym_W$ such that $A_W(Z) = \emptyset$ (i.e. Z has no attribute names).*

A single vertex labeled by Z_R is a well-formed graph. We call this the *axiom* (graph) of R. A *sentence* with respect to a grammar R is a graph that can be derived through a finite number of rewrites starting from the axiom and using productions of R.

In the next section we define some properties to guarantee attribute evaluability and facilitate efficient incremental change propagation. We also give some conditions and results through which static grammar analysis may check for satisfaction of these properties.

4 Grammar and Graph Properties

Recall from definition 2 that every attribute vertex has a priority, an attribution rule, and a value.

Definition 9 *A graph is called* prioritized *if every topological sort of its attribute vertices produces a sequence of their priorities which is non-decreasing.*

Definition 10 *An attribute vertex is called* evaluable *if D-edges exist to provide argument values to each parameter of its attribution rule and if those arguments are of appropriate types.*

A graph is evaluable *if every attribute vertex in it is evaluable.*

Definition 11 *An attribute vertex is called* consistent *if it is evaluable and its value is equal the the result of evaluating its attribution rule.*

A graph is consistent *if every attribute vertex in it is consistent.*

We now give conditions sufficient to establish that a grammar produces only prioritized and evaluable graphs and show how attribute vertex evaluation can be done efficiently on a prioritized evaluable graph in order to make it consistent.

4.1 Prioritizing Graphs

Method 12 *The static analysis algorithm in appendix A will either fail (indicating that there is a potential cyclic dependency among the attributes of a bodygraph), or return a partial ordering on the attribute and socket vertices for each bodygraph, such that the ordering captures all possible transitive dependencies among attribute vertices instantiated by the production. We call this ordering a* relative priority numbering.

The orderings returned by the algorithm are a conservative approximation of the transitive dependencies that may arise in practice. The algorithm is a generalization of Kastens analysis algorithm for attributed string grammars [10].

Method 13 (Priority number interpolation) *When a production p is used to rewrite a cluster C, examine the priority numbers of the attribute vertices in the neighborhood of C that are represented by sockets of R_p. For every socket s determine the maximum and minimum of all priorities of vertices represented by s. In practice we examine the bodygraph to decide for each socket whether we need the maximum or minimum (or both) and only find what is needed. This saves some work, but does not change the time complexity. Call these rational numbers* $\max(s)$ *and* $\min(s)$.

If there are sockets s and t where the $priority_s < priority_t$ *and* $\max(s) \not< \min(t)$ *then the algorithm fails.*

For every attribute vertex v instantiated by the rewrite, assign a priority number n such that, if there is a transitive dependence (as determined by method 12 from socket s then $n > \max(s)$ *and if there is a transitive dependence to socket t then* $n < \min(t)$ *(Because we use rational numbers this is always possible; see [1]). n must also be consistent with the assignment of priorities to other instantiated attribute vertices according to the relative priorities computed for the bodygraph. This condition can be met by simply interpolating numbers to vertices in the order identified by method 12.*

This method describes how to dynamically assign priority numbers consistent with the relative priority numbers of the bodygraph and the dependencies generated by the embedding.

Lemma 14 *Given that a grammar R completes static analysis (method 12) priority number interpolation will not fail when rewriting a sentence by a production of R.*

The static analysis algorithm predicts all possible cases where a rewrite may introduce a transitive dependency between two already existing attribute vertices. When static analysis succeeds the relative priority numbering fixed for each production's bodygraph encodes not only those dependencies that are used for the given rewrite, but those dependencies that may be created in any possible recursive rewriting of clusters introduced. Therefore, even if a production is introducing new transitive dependencies, this eventuality was prepared for earlier so that interpolation is possible (i.e. the actual priority numbers on both sides of the new transitive dependence have the correct sense of inequality).

Since the introduction of new transitive dependencies is the only situation that can make interpolation fail and static analysis (if it succeeds) covers this case, we guarantee that interpolation is always possible.

Lemma 15 *Method 13 preserves the prioritized property in graphs.*

According to the method new priorities are interpolated between attribute vertex priorities in the neighborhood of the rewrite. Also assignment follows the dependencies internal to the daughter graph that is instantiated. If the original graph was prioritized and the method succeeds, then the result will be prioritized.

Theorem 16 *Given a grammar that passes static analysis and applying method 13 uniformly (i.e. for every rewrite starting from the axiom) every graph produced is prioritized.*

Proof is by induction using lemmas 14 and 15 and noting that the axiom graph is prioritized trivially.

As a result of this theorem we assume in the sequel that all graphs are prioritized. The cost of keeping graphs prioritized is of the same time complexity as the cost of generating the graphs.

Theorem 17 *Method 13 is optimal.*

The cost of priority number interpolation is linear in the number of attribute vertices added and the size of the goal cluster's neighborhood. Rewriting requires this much work anyway. As E^4s are evaluated to consider subsets of the neighborhood, min and max information is gathered with a constant factor additional time. Similarly, the relative priority numbers of the bodygraph give us an efficient precomputed order in which to instantiate attribute vertices and give them priority numbers.

4.2 Attribute Evaluation

We use the standard definition of attribute evaluation, as found, for example, in [16]. After some change to a graph, optimal incremental change propagation is performed guided by static analysis. This will work whenever all attribute vertices *evaluable*, as defined above.

Requirement 18 *When an attribute vertex is instantiated from the bodygraph of some production by a rewrite, it must immediately be evaluable.*

This constrains static analysis to check that each attribution rule gets arguments either locally from the production's bodygraph or remotely from a socket. If from a socket, static analysis must guarantee that the socket label will always evaluate to a set able to satisfy the attribution rule.

Requirement 19 *Every rewrite must fulfill the obligations of the goal cluster, i.e. every D-edge target outside the cluster (where the edge originated from some attribute vertex of the cluster) is a part of the set that some E^4 evaluates to. The socket labeled by that E^4 must have a D-edge in-incident on it.*

Again we use static analysis to check whether a grammar will always satisfy this run-time requirement.

Theorem 20 *Any graph resulting from a grammar satisfying the previous two requirements is evaluable.*

Proof is by induction on rewrites. After each rewrite requirement 18 guarantees that newly created attributes are evaluable and requirement 19 guarantees that attributes in the neighborhood remain evaluable.

Method 21 (Attribute change propagation) *Given a prioritized evaluable graph, incrementally reestablishing consistency from an arbitrary initial working-set p of attributes*[2] *that are known to need evaluation proceeds by:*

- *Place elements of p into a priority queue q according to their priority numbers;*
- *Repeat the following until q is empty—*
- *Remove the front of q and evaluate it, putting the result as the value of the attribute vertex;*
- *If the value of the vertex changed, add all its graph successors to q (inserted according to their priority numbers).*

After a rewrite the attribute change propagation algorithm is run. Forming the initial priority queue is linear time since the attribute vertices' priorities are established from relative priority numbers so their sorted order is precomputed by static analysis. A variant of the model is to allow multiple rewrites between attribute reevaluation, for efficiency. In this case forming the queue is by merging sequences rather than sorting all the attribute vertices.

The cost of method 21 is computed as follows. If *Influenced* is the set of attributes reevaluated by method 21, then the number of evaluations is $|\mathit{Influenced}|$. Each evaluation is followed by the cost of an

[2]For this paper, p is just the attribute vertices instantiated from the bodygraph of the production used in the rewrite.

update to the priority queue. Clearly, the queue can be no larger than the size of the *Influenced* set, so the worst-case complexity of the algorithm is $O(|\mathit{Influenced}| \cdot \log |\mathit{Influenced}|$. In practice the size of the priority queue will usually be significantly smaller than $|\mathit{Influenced}|$, especially for sparse graphs, so the algorithm will approach linear-time behavior. Note that because of the flexibility of the embeddings, we cannot in general bound the out-degree of any attribute vertex, so, unlike the standard evaluation algorithm on trees [14] the cost of maintaining the worklist of attributes needing reevaluation is not bound by a constant.

References

[1] B. Alpern, A. Carle, B. Rosen, P. Sweeney, and K. Zadeck. *Incremental Evaluation of Attributed Graphs.* Technical Report CS-87-29, Department of Computer Science, Brown University, December 1987.

[2] Veronique Donzeau-Gouge, Gilles Kahn, Bernard Lang, and Bertrand Melese. Documents structure and modularity in mentor. In *Proceedings of the ACM SIGPLAN/SIGSOFT Symposium on Practical Software Development Environments*, pages 141–148, Pittsburgh, Pa, May 1984.

[3] Gregor Engels and Wilhelm Schafer. Graph grammar engineering: a method used for the development of an integrated programming environment. In Hartmut Ehrig, Christiane Floyd, Maurice Nivat, and James Thatcher, editors, *Proceedings of the International Joint Conference on Theory and Practice of Software Development (TAPSOFT), LNCS 186*, pages 179–193, Springer-Verlag, 1985.

[4] Herbert Gottler. Attributed graph grammars for graphics. In Hartmut Ehrig, Manfred Nagl, and Grzegorz Rozenberg, editors, *Proceedings of the second International Workshop on Graph Grammars and their Application to Computer Science, LNCS 153*, pages 130–142, Springer-Verlag, 1982.

[5] Herbert Gottler. Graph grammars and diagram editing. In G. Rozenberg H. Ehrig, M. Nagl and A. Rosenfeld, editors, *Proceedings of the Third International Workshop on Graph Grammars and their Application to Computer Science, LNCS 291*, pages 216–231, Springer-Verlag, Heidelberg, 1987.

[6] D. Janssens and G. Rozenberg. Graph grammars with neighbourhood-controlled embeddings. *Theoretical Computer Science*, 21:55–74, 1982.

[7] Simon M. Kaplan. *Incremental Attribute Evaluation on Graphs (Revised Version).* Technical Report UIUC-DCS-86-1309, University of Illinois at Urbana-Champaign, December 1986.

[8] Simon M. Kaplan and Roy H. Campbell. Designing and prototyping in grads. In *Proceedings of the Second IEE/BCS Conference on Software Engineering, Conference Publication 290*, pages 55–59, IEE, Liverpool, July 1988.

[9] Simon M. Kaplan, Steven K. Goering, and Roy H. Campbell. Supporting the software development process with attributed nlc graph grammars. In G. Rozenberg H. Ehrig, M. Nagl and A. Rosenfeld, editors, *Proceedings of the Third International Workshop on Graph Grammars and their Application to Computer Science, LNCS 291*, pages 309–325, Springer-Verlag, Heidelberg, 1987.

[10] Uwe Kastens. Ordered attribute grammars. *Acta Informatica*, 13(3):229–256, 1980.

[11] Donald E. Knuth. Semantics of context-free languages. *Mathematical Systems Theory*, 2(2):127–145, June 1968.

[12] Manfred Nagl. A software development environment based on graph technology. In G. Rozenberg H. Ehrig, M. Nagl and A. Rosenfeld, editors, *Proceedings of the Third International Workshop on Graph Grammars and their Application to Computer Science, LNCS 291*, pages 458–478, Springer-Verlag, Heidelberg, 1987.

[13] David Notkin. The gandalf project. *Journal of Systems and Software*, 5(2):91–106, May 1985.

[14] Thomas Reps. *Generating Language-Based Environments.* MIT Press, 1984.

[15] Thomas Reps and Tim Teitelbaum. The synthesizer generator. In *Proceedings of the SIGSOFT/SIGPLAN Software Engineering Symposium on Practical Software Development Environments*, Pittsburgh, PA, April 1984.

[16] William M. Waite and Gerhard Goos. *Compiler Construction.* Springer-Verlag, New York, 1984.

A Static Grammar Analysis

There are two reasons for analyzing a grammar:

- To determine an approximation of transitive dependencies that may occur in sentences derived by the grammar (that includes at least those dependencies that will arise in practice). We do this in support of method 12 to assign relative priority numbers; and
- To verify that rewrites preserve evaluability (i.e. to show that requirements 18 and 19 hold).

These two goals are unified through the concept of role. By giving each vertex and each socket a role the analysis algorithm has more information[3], so that it can get a tighter approximation of transitive dependencies. This reduces (and in practice eliminates) the problem of inducing cycles of dependencies where no such cycles exist in practice. Also, specifying roles for attribute and socket vertices gives a handle on obligations and resources for each attribute.

Roles are not attribute names. Attribute names are assigned on the basis of type information. Roles are given to declare intended use for the attribute values.

A.1 Inducing Transitive Dependency Information

Definition 22 *A socket* matches *an attribute vertex if and only if their roles are equal.*

This is a conservative approximation of sockets to attribute vertices because at run time the socket role must equal the attribute vertex's role and its E^4 must evaluate to include that attribute vertex for it to represent the vertex.

It is possible to generalize roles to *role sets*, so that an attribute can have multiple roles, which is often useful in practice. This generalization is straightforward and is not considered in this paper.

Method 23 *Let $B_1, B_2, \ldots, B_n$ be the production bodygraphs of a grammar, R. Let $I_1, I_2, \ldots, I_n$ be* induced dependency bodygraphs *with values as assigned during the algorithm. Let D be a set of triples where the first element is a symbol and the other two elements are socket vertices of a bodygraph that is the right-hand-side*

[3] In an earlier work [7], embedding expressions were simply (nonterminal) symbols and there were no roles; attempting to do static analysis in this model often induced so many false cycles that the class of grammars which passed the analysis was too restricted. The approach presented in this paper eliminates this problem, by using roles, powerful embedding expressions, and embedding to attributes directly rather than to their associated terminal or nonterminal symbol.

of a production with the symbol as goal. D will be built by the algorithm as dependencies across sockets within each bodygraph are discovered.

Initialize each I_i to `transitive-closure(` B_i `)`.
For each edge $(s_1, s_2) \in I_i$ between sockets, add the triple
  $< L_{p_i}, s_1, s_2 >$ *to D, where L_{p_i} is the goal of production i.*
While D continues growing do
  For each I_i do
    For each symbol vertex $v \in I_i$, where $sl(v) =$ 'X' do
      For each triple $< X, a, b > \in D$ do
        If there exists vertices y, z in the cluster neighborhood of v
          and y matches a and z matches b and $(y, z) \notin I_i$
          then add (y, z) to I_i.
    If any new edges were added to I_i do
      Let $I_i =$`transitive-closure(`I_i`)`
      If I_i is cyclic then the algorithm fails
      Add newly induced socket dependencies $< L_{p_i}, s_1, s_2 >$ to D.

The result is a set of graphs $I_1, I_2, \ldots, I_n$ with the same vertices as the bodygraphs and a superset of the transitive closures of the edges.

This algorithm is a generalization of the first phase of Kastens algorithm for analyzing attributed string grammars to test if they are *ordered* [10]. As with Kastens algorithm, experience thus far suggests that "real" grammars which are actually acyclic pass the test.

Given an acyclic graph I_i for production p_i, we assign relative priority numbers to attribute vertices and sockets of B_i. Priority interpolation (method 13) requires that numbers assigned to sockets be equal unless there is a dependence induced between them. Since I_i is transitively closed, it is trivial to assign numbers 0, 1, *etc.* so that both this, and monotonicity of numbering across dependencies are preserved.

A.2 Deriving Obligation Information

Static analysis can determine if a vertex meets its obligations, and whether the resources it needs are supplied to it, in the face of all possible rewrites on the graph. Passing this analysis guarantees that requirements 18 and 19 are met for the grammar.

Checking these requirements involves an inductive algorithm. First the conditions are shown to hold for rewrite of the initial graph (the base case), and then for any other rewrite. In both cases, the methodology is the same: For each attribute vertex v with role r, consider all the sockets adjacent to it. Suppose that s is the role of the socket. Identify all attribute vertices with role s. Now check that the rewrite of the cluster associated with each vertex will place an edge to vertices with role r. Finally, check that all rewrites of v will introduce a vertex with role r connected via a socket to vertices with role s.

Note that since roles are designed to correspond to the *purpose* of an attribute, there should be a close correlation between the vertices identified in the algorithm as potential connection points, and vertices to which connections are actually made. Roles therefore act to reduce the number of possible connection points. In practice we find this correlation to be a 1-to-1 relation; if this is not true, then the grammar is probably improperly engineered.

SOME APPLICATIONS AND TECHNIQUES FOR GENERATING FUNCTIONS

P.Massazza - N.Sabadini
Dip. Scienze dell'Informazione - Università di Milano

Abstract

In this paper we apply generating functions techniques to the problem of deciding whether two probabilistic finite state asynchronous automata define the same events. We prove that the problem can be solved by an efficient parallel algorithm, in particular showing that it is in the class DET. Furthermore, we develop some methods for studying properties of generating functions, in particular from the point of view of the algebricity.

1. Introduction

Generating functions have been applied successfully in combinatorics, probability theory and formal languages theory. In combinatorial theory, they represent one of the basic techniques for dealing with problems of enumeration [St][DRS][Go]. In formal language theory, Schuetzenberger developped a theory of generating functions in a finite number of non commuting variables to solve problems in context free languages (CFL) [CS].

In particular, in some cases generating functions techniques give an analytic method for approaching the problem of determining whether a language L belongs to a given class C. Informally, the idea is to associate to every language in C a generating function and to state properties that these functions must satisfy. If the generating function of L does not satisfy these properties then L does not belong to C. For example, it is well known from a classical result due to [CS] that every unambiguous CFL has an algebraic generating function. So, it is possible to prove that some CFL's are inherently ambiguous, by showing that their generating functions are not algebraic: this method has been successfully used in [Fl] to solve some open problems on the ambiguity of CFL's. Some conditions for the applicability of this method to classes of trace languages have been recently given in [BS2].

Furthermore, generating functions techniques can be used to solve decision problems on languages. For example, in [BPS] it is proved that the equivalence problem for some classes of unambiguous regular trace languages is decidable, by showing that two languages coincide iff a suitable rational generating function is identically zero.

Hence, it is possible to individuate two research topics:

1) to find new and significant applications of generating functions techniques to decision problems,

2) to develop suitable methods for studying properties of classes of generating functions (for example, a natural question is whether the generating functions of CFL's lie in a particular class of trascendental functions [Fl] or whether the algebricity of the generating function for subclasses of CFL's is decidable).

In this paper we consider both directions: as a main application in the first one, we extend the notion of finite state asynchronous automaton (introduced in [Zi] in order to characterize the class of recognizable trace languages) to the probabilistic case and we study equivalence problems on the events defined by these automata. In particular, we introduce a notion of distance between events, showing that this distance function is computable. Furthermore we prove that the problem of deciding equality between events is in the class DET.

In the second direction, we study the algebricity problem for generating functions. Although the general problem of determining whether the generating function of a CFL is algebraic is undecidable [BS1], nevertheless for a suitable subclass of languages it is possible to develop proper techniques. In particular, we study the problem of deciding whether the linear space of the generating functions, whose coefficients are the solutions of linear homogeneous recurrence equations (having polynomial coefficients), only contains algebraic functions. We develop an algorithm for this problem, based on Klein's method for solving differential equations of the second order having polynomial coefficients.

2. Preliminary definitions on generating functions

In this section we recall some basic definitions and results about generating functions of languages. For all the definitions on formal languages we refer to a classical book as [Sa].

Given a finite alphabet Σ, we associate with a language $L \subseteq \Sigma^*$ its *enumeration* or *counting sequence* $\{a_n\}$, defined by $a_n = \#\{x \in L \mid |x| = n\}$. This sequence univocally defines $F_L(z) = \sum_{n=0}^{\infty} a_n z^n$, called the *generating function* of the language L. This function is an analytic function in a neighbourhood of the origin and its convergence radius is at least $1/\#\Sigma$. In this paper we are interested in *algebraic* analytic functions; we recall that a complex function f(z) is algebraic if there exists a finite sequence of polynomials $q_0(x),\dots, q_d(x) \in \mathbf{C}[x]$ (where $\mathbf{C}$ is the complex field) such that for every $z \in \mathbf{C}$ $\sum_{j=0}^{d} q_j(z)f(z)^j = 0$. The degree of the algebraic function is the least integer d such that the above relation holds. Useful necessary conditions for a function to be algebraic are given in [Fl]; in particular we recall the following result of Comtet [Com]:

Fact 2.1- Let $f(z) = \sum_{n=0}^{\infty} c_n z^n$ be an algebraic function of degree d. Then there exist an integer n° and a finite sequence of polynomials $p_0(x),\dots, p_q(x) \in \mathbf{Z}[x]$ such that:

1) $p_q(x) \neq 0$
2) the degree of $p_j(x) < d$ for every j
3) for every $n \geq n^\circ$ $p_0(n)c_n + p_1(n)c_{n-1} + \dots + p_q(n)c_{n-q} = 0$.

3. Preliminary definitions on trace languages

In this section we recall some basic definitions about traces and trace languages, introduced by Mazurkiewicz [Ma] in order to formally describe the behaviour of concurrent systems, and extensively studied by many authors [see AR for a review]. Given a finite alphabet Σ and a concurrency relation on Σ (i.e. a symmetric and irreflexive relation $C \subseteq \Sigma \times \Sigma$) we consider the *concurrent alphabet* $<\Sigma, C>$ and the *free*

partially commutative monoid $F(\Sigma, C)$ generated by $<\Sigma, C>$, defined as the initial object in the category of monoids generated by Σ and satisfying, besides the usual monoid axioms, the set of commutativity laws $\{ab = ba \mid aCb\}$. From standard results $F(\Sigma, C)$ is the quotient structure $F(\Sigma, C) = \Sigma^*/=_C$, where $=_C$ is the least congruence on Σ^* extending C. A *trace* t is an element of $F(\Sigma, C)$ which can be interpreted as an equivalence set of words, a *trace language* is a subset of $F(\Sigma, C)$; given a string $x \in \Sigma^*$ we will denote by $[x]_C$ the equivalence class of x (i.e. the trace generated by x) and by $|x|$ its length. So, given a language $L \subseteq \Sigma^*$, the trace language $T = [L]_C$ generated by L is defined as $T = \{t = [x]_C \mid x \in L\}$.
The notion of *regular trace language* (introduced by Mazurkiewicz [Ma]), is defined as follows: T is regular iff $T = [L]_C$ and L is a regular language on Σ. We denote by $R(\Sigma, C)$ the class of regular trace languages. An interesting subclass of the class of regular trace languages is the class $R_1(\Sigma, C)$ of *deterministic* or *unambiguous* languages [BMS], defined as the class of trace languages T for which there exists a regular language $L \subseteq \Sigma^*$ such that it holds the following:
a) $T = [L]_C$ b) $\forall x \in \Sigma^* \ (\#\{[x]_C \cap L\} \leq 1)$.
Given a trace language $T \subseteq F(\Sigma, C)$, we define the *generating function* $f_T(z)$ of T as :

$$f_T(z) = \sum_{k=0}^{\infty} \#\{t \mid t \in T, |t| = k\} z^k.$$

An analysis of some properties of generating functions of regular trace languages has been carried out in [BS2].
Finally, we present two simple examples of solution of decision problems over languages by means of generating functions techniques. Let us consider the following problems:

Problem 1:

Instance: A context free grammar G on Σ

Question: to decide whether $L_G = \Sigma^*$, where L_G is the language generated by G

Problem 2:

Instance: A regular grammar R on Σ for a concurrent alphabet $<\Sigma, C>$

Question: to decide whether $T_R = F(\Sigma, C)$, where $T_R = [L_R]_C$ and L_R is the language generated by R

It is well known that these problems are undecidable [Sa][AH]. Let $F_{L_G}(z)$ and $F_{T_R}(z)$ be the generating functions of L_G and T_R; it is immediate to prove that:

1) $L_G = \Sigma^*$ iff $F_{L_G}(z) = \frac{1}{1 - \#\Sigma z}$

2) $T_R = F(\Sigma, C)$ iff $F_{T_R}(z) = \frac{1}{\sum_k (-)^k C_k z^k}$, where C_k is the number of k-cliques in the concurrency relation [BBMS].

We recall that if G is unambiguous then $F_{L_G}(z)$ is algebraic and if L_R generates T_R deterministically then $F_{T_R}(z)$ is rational; in these cases the problems consist of verifying the equality of algebraic (rational) functions. Hence, we can conclude that the previous problems restricted to unambiguous CFL's and deterministic regular trace languages become decidable.

4. Probabilistic Asynchronous Automata

In this section we introduce the notion of probabilistic finite state asynchronous automaton (PFSAA), which can be considered as an interesting type of stochastic Petri Nets [Mo] We study the equivalence problem for this model using suitable generating functions techniques. In particular we prove that the "distance" between behaviours of PFSAA's with rational probabilities is computable and that the "equality" of behaviours is decidable by means of an efficient parallel algorithm.

Def.4.1- A *probabilistic finite state asynchronous automaton* (PFSAA) with n processes is a tuple $A = <P_1, P_2, ..., P_n, \Delta, F>$ where:

- $\forall i\ 1 \le i \le n\ P_i = <\Sigma_i, S_i, s_{i0}>$ is the i-th process, Σ_i being a finite non empty alphabet, S_i denoting a finite set of (local) states s.t. $S_i \cap S_j = \emptyset$ for $j \neq i$, $s_{i0} \in S_i$ standing for the initial state of P_i,
- $F \subseteq \prod_{i=1,n} S_i$ is the set of final states,
- $\Delta = \{\delta_\sigma \mid \sigma \in \bigcup_{i=1,n}\Sigma_i\}$ is a set of stochastic transition matrices (we recall that a stochastic matrix is a square matrix with nonnegative real entries and with row sums equal to 1).

Let us consider the following notations:

- $\Sigma = \bigcup_{i=1,n}\Sigma_i$,
- Proc = {1,..., n},
- $Q = \prod_{i=1,n} S_i$ (the set of global states of A),
- $q_o = (s_{1o}, ..., s_{no})$ (the initial state of A).

$P(\sigma) = \{i \in \text{Proc} \mid \sigma \in \Sigma_i\}$ is the set of processes which can "execute" an action $\sigma \in \Sigma$. For every action $\sigma \in \Sigma$ the set Δ contains exactly one stochastic matrix δ_σ whose indices are the elements of $\prod_{i\in P(\sigma)} S_i$; furthermore, $\delta_\sigma((s_{j_1}, ..., s_{j_k}), (s'_{j_1}, ..., s'_{j_k}))$ gives the probability that the automaton, while executing the action σ, modifies the local states of the processors which can execute the action from $(s_{j_1}, ..., s_{j_k})$ to $(s'_{j_1}, ..., s'_{j_k})$.

For each symbol σ in Σ, we extend δ_σ to the matrix δ'_σ, which has as indices the elements of $\prod_{i\in \text{Proc}} S_i$. In order to do this, we introduce the following notation: let $v = (s_1, ..., s_n)$ be an element of $\prod_{i\in \text{Proc}} S_i$, we denote by $\text{Pro}_\sigma(s_1, ..., s_n)$ the element of $\prod_{i\in P(\sigma)} S_i$ obtained from v by erasing the states of the processors which cannot execute the action σ. Now the extended matrix δ'_σ can be defined as follows:

$$\delta'_\sigma((s_1, ..., s_n), (s_1', ..., s_n')) = \underline{\text{if}}\ \forall i \notin P(\sigma)\ s_i = s_i'\ \underline{\text{then}}\ \delta_\sigma(\text{Pro}_\sigma(s_1, ..., s_n), \text{Pro}_\sigma(s_1', ..., s_n'))\ \underline{\text{else}}\ 0.$$

It is clear that if actions α and β operate on disjoint sets of processes then they may be executed independently, so the *concurrency (independency) relation* of A is defined as $C_A = \{(\alpha,\beta) \in \Sigma x \Sigma \mid P(\alpha) \cap P(\beta) = \emptyset\}$ and A defines a concurrent alphabet $<\Sigma, C_A>$ in a natural way. It is immediate to observe the following fact:

Fact 4.1- Given two symbols $\alpha, \beta \in \Sigma$, $\alpha C_A \beta$ implies that $\delta'_\alpha \circ \delta'_\beta = \delta'_\beta \circ \delta'_\alpha$, where $\circ$ denotes the usual product of matrices.

As a consequence, if the words $x_1...x_n$, $x'_1...x'_n$ are in the same trace (interpreted as equivalence class of words), then $\delta'_{x_1} \circ ... \circ \delta'_{x_n} = \delta'_{x'_1} \circ ... \circ \delta'_{x'_n}$.
Furthermore, we observe that the concurrency relation defined by A can be interpreted as a graph and that to every *clique* $C = \{\sigma_1,..., \sigma_s\}$ of C_A we can associate the matrix $\delta'(C) = \delta'_{\sigma_1} \circ ... \circ \delta'_{\sigma_n}$.
Given a PFSAA A, let π and μ denote the stochastic vectors which have as indices the elements of $\prod_{i \in Proc} S_i$, defined as:
$\pi (s_1,..., s_n) =$ if $(s_1,..., s_n) = (s_{1\,0},..., s_{n\,0})$ then 1 else 0
$\mu (s_1,..., s_n) =$ if $(s_1,..., s_n) \in F$ then 1 else 0
(π and μ are the characteristic vectors of the initial state and of final states of the automaton).
Now we can define the *behaviour* of the PFSAA A with input $t = [x_1...x_n]_{C_A} \in F(\Sigma, C_A)$, as the probability that A, starting from the initial state, reaches a final one; this probability is obtained as $\pi \circ \delta'_{x_1} \circ ... \circ \delta'_{x_n} \circ \mu_T$, where μ_T is the transposed vector of μ. More formally we have:

Def.4.2- Given a PFSAA A, the *event* realized by A is the function $S_A : F(\Sigma, C_A) \to [0,1]$ where $S_A(t) = \pi \circ \delta'_{x_1} \circ ... \circ \delta'_{x_n} \circ \mu_T$ and $[x_1...x_n]_{C_A} = t$.

Now, given two PFSAA A_1 and A_2 on the same concurrent alphabet $<\Sigma, C>$, we are interested in the Equivalence Problem stated as follows:

Equivalence Problem for PFSAA (EPFSAA)
Given a concurrent alphabet $<\Sigma, C>$:
Instance: two PFSAA A_1, A_2 on $<\Sigma, C>$
Question: $S_{A_1} = S_{A_2}$?

A more general question consists of defining a suitable notion of distance between the events realized by two PFSAA's, and of giving a procedure to compute it.

Def.4.3 - $d(S_{A_1},S_{A_2}) = \mathrm{Sup}_n \dfrac{\sum_{|t|=n}\left(S_{A_1}(t) - S_{A_2}(t)\right)^2}{F_n}$, where F_n is the number of traces in $F(\Sigma, C)$.

The corresponding problem is:

Distance between events

Instance: two PFSAA A_1, A_2 on $<\Sigma, C>$

Question: to compute $d(S_{A_1}, S_{A_2})$.

In order to solve this problem, we consider the following generating function in the complex variable z:

$$f_{A_1,A_2}(z) = \sum_{k=0}^{\infty}\sum_{|t|=k}\left(S_{A_1}(t) - S_{A_2}(t)\right)^2 z^k.$$

As a first result, by using formal power series in partially commutative variables [La], we show that $f_{A_1,A_2}(z)$ is a rational function.

We recall some basic definitions on formal power series in partially commutative variables: let $<A, +, ., 1>$ be a ring with unity (not necessarily commutative) and $F(\Sigma, C)$ the free partially commutative monoid, then a *formal power series in partially commutative variables* is a function $\psi : F(\Sigma, C) \rightarrow A$, represented as the formal sum $\sum_{t\in F(\Sigma,C)} \psi(t)t$. The set of formal power series on $F(\Sigma, C)$ form a ring $A[[F(\Sigma, C)]]$, by considering the usual operations of sum (+) and Cauchy product (.):

$(\phi + \psi)(t) = \phi(t) + \psi(t)$

$(\phi . \psi)(t) = \Sigma_{xy=t}\, \phi(x)\psi(y)$

with the identity **1** defined as follows:

$\mathbf{1}(t) =$ if $t = [\varepsilon]_C$ then 1 else 0 (being ε the empty word of Σ^*).

Given the concurrent alphabet $<\Sigma, C>$, let $\mathcal{C}$ be the set of cliques of C (we recall that every clique in $\mathcal{C}$ individuates a monomial in which every two symbols commute). We will use the following result (it can be immediately obtained by applying the techniques used to prove the Moebius inversion formula):

Lemma 4.1- Let $\varphi : F(\Sigma, C) \rightarrow <A, ., 1>$ be a monoid morphism, then it holds:

$$\left(\sum_t \varphi(t)t\right)\left(\sum_{c\in \mathcal{C}} (-)^{|c|}\varphi(c)c\right) = 1.$$

In particular, by considering the ring **Z** of integers and the morphism $\varphi(t) = 1$, we obtain the usual Moebius inversion formula:

$$\left(\sum_t t\right)\left(\sum_{c\in \mathcal{C}} (-)^{|c|} c\right) = 1.$$

Now, we can state our main result:

Theor.4.1- Given two PFSAA A_1, A_2 on a concurrent alphabet $<\Sigma, C>$, the associated generating function $f_{A_1,A_2}(z)$ is a rational function in z.

Proof (outline): Let M_N be the monoid of NxN square matrices with real components with the usual product operation, and $\varphi : F(\Sigma, C) \rightarrow M_N$ be a monoid morphism. We observe that the formal series $\Sigma_t\varphi(t).t$ can be interpreted as a formal series on the ring $<M_N, +, .>$ or as a matrix with formal series as components. Given two 1xN vectors π, μ with real components, it holds:

$$\sum_t(\pi\varphi(t)\mu_T)t = \pi\left(\sum_t\varphi(t)t\right)\mu_T$$

By lemma 4.1 and by considering the morphism from the ring of formal power series on $<\Sigma, C>$ and the ring of series in the variable z induced by the substitutions $\sigma \rightarrow z$ ($\sigma \in \Sigma$), we obtain:

$$\left(\sum_0^{\infty}\left(\sum_{|t|=n}\varphi(t)\right)z^n\right)\left(\sum_{c\in C}(-)^{|c|}\varphi(c)z^{|c|}\right) = 1,$$

and finally

$$\sum_n\left(\sum_{|t|=n}\pi\varphi(t)\mu_T\right)z^n = \pi\left(\sum_{c\in C}(-)^{|c|}\varphi(c)z^{|c|}\right)^{-1}\mu_T. \qquad (*)$$

Given two PFSAA A_1,A_2 on $<\Sigma, C>$, we denote by π_i, μ_i, $\delta'_i(\sigma)$ (for $\sigma \in \Sigma$), respectively, the characteristic vector of initial states, of final states (with dimensions $1xn_i$) and the stochastic (n_ixn_i) matrices associated to A_i (i = 1, 2), where n_i is the number of global states of the automaton.

Let be:

$$\pi = (\pi_1, \pi_2) \otimes (\pi_1, \pi_2) \qquad \mu = (\mu_1, \mu_2) \otimes (\mu_1, -\mu_2)$$

$$\forall\sigma \in \Sigma \quad \varphi(\sigma) = (\delta'_1(\sigma) \oplus \delta'_2(\sigma)) \otimes (\delta'_1(\sigma) \oplus \delta'_2(\sigma))$$

where $\oplus$ and $\otimes$ are the usual direct sum and Kronecker's product, and (π_1, π_2) is the $(1x(n_1+n_2))$ vector obtained by joining the vectors π_1, π_2.

Let us consider the monoid morphism $\varphi : F(\Sigma, C) \rightarrow M_N$ defined as before, where $N = (n_1 + n_2)^2$, and a trace $t = [x_1...x_p]_C$; by using the usual properties of $\oplus$ and $\otimes$ we obtain:

$$\pi\varphi(t)\mu_T = \pi\varphi(x_1)...\varphi(x_p)\mu_T = (\pi_1\delta'_1(x_1)...\delta'_1(x_p)\mu_{1T} - \pi_2\delta'_2(x_1)....\delta'_2(x_p)\mu_{2T})^2$$
$$= (S_{A_1}(t) - S_{A_2}(t))^2$$

Hence $\sum_0^{\infty}\left(\sum_{|t|=n}\pi\varphi(t)\mu_T\right)z^n = f_{A_1,A_2}(z)$ and from relation (*) the thesis follows

($\sum_{c\in C}(-)^{|c|}\varphi(c)z^{|c|}$ is a matrix whose components are polynomials in the variable z).

Now, given two PFSAA's A_1, A_2 on $F(\Sigma, C)$, let be $G_n = \sum_{|t|=n} \left(S_{A_1}(t) - S_{A_2}(t)\right)^2$; we recall that F_n is the number of traces of length n. If the associated stochastic matrices have rational number as entries, since $\Sigma_{0,\infty}\, G_n.z^n$ and $\Sigma_{0,\infty}\, F_n.z^n$ are rational functions with rational coefficients, then we can conclude that there exist polynomials with algebraic coefficients $p_1,..,p_m$, $q_1,..,q_p$ and algebraic numbers $a_1,..,a_m$, $b_1,..,b_p$ such that $G_n = \sum_{k=1}^{m} p_k(n)a_k^n$, $F_n = \sum_{j=1}^{p} q_j(n)b_j^n$.

Hence, there is an integer w such that $\mathrm{Lim}_{k\to\infty} \frac{G_{kw+s}}{F_{kw+s}} = r_s$ (s = 1,..., w), where r_s is an algebraic number. This is the main step which allows to conclude:

Fact.4.2 - The distance function d for PFSAA's, whose associated stochastic matrices have rational number as entries, is computable.

The previous result implies that the problem EPFSAA for PFSAA's with rational entries is decidable.
More generally, by observing that $f_{A_1,A_2}(z) = 0$ iff $S_{A_1} = S_{A_2}$, we can reduce the problem of deciding the equivalence of PFSAA to the problem of determining whether the rational function $f_{A_1,A_2}(z) = \pi\left(\sum_{c\in C} (-)^{|c|}\varphi(c)z^{|c|}\right)^{-1} \mu_T$ is identically zero.
Let F be a subfield of **R** whose elements admit a representation such that, firstly, the operations of sum and product are total recursive, secondly the equality is decidable (for example, we could consider the field of rational (or algebraic) numbers). The following fact holds:

Fact 4.3 - The Equivalence problem for PFSAA defined by stochastic matrices, with probability values in F, is decidable.

We are now interested in a classification of EPFSAA from the complexity point of view: we consider the case of automata such that the associated stochastic matrices have rational numbers as entries. We will prove that in this case EPFSAA is "efficiently parallelizable"; in order to do that we recall the definitions of the hierarchy $\{NC^k\}$ and of the class DET [Co]:

Def.4.4 - For every integer k, NC^k is the set of functions computable by a family of uniform boolean circuits of depth $O(\log^k n)$ and size $n^{O(1)}$. The class NC is defined as: $NC = \cup_k NC^k$.

Uniform boolean circuits are considered as a standard model of parallel computations and are widely studied in literature [Co]; size and depth of a circuit are the number of nodes and the length of the longest path from input to output nodes respectively. It is well known that NC is contained in the class P of problems solvable in polynomial sequential time; although it seems reasonable that the inclusion is proper, to prove this is an open problem. A central notion for investigating the structure of the hierarchy $\{NC^k\}$ and the complexity of problems that lie between NC^1 and NC^2 is the NC^1-reducibility [Co]. In particular we are interested in the following class:

Def.4.5- DET is the class of the functions belonging to NC^1 which are reducible to computing the determinant of a nxn matrix of n-bits integers.

Fact 4.4- $NC^1 \subseteq DET \subseteq NC^2$.

In [Co] it is proved that the problem of computing the inverse of a nxn matrix A with n-bits integer entries is in DET; furthermore, the problem remains in DET even if the matrix entries are polynomials with a fixed number of variables (or rational functions) over the rationals. Using this result, it is easily proved the following:

Fact 4.5- EPFSAA is in the class DET if the stochastic matrices associated to the automata have rational number as entries.

5. Generating functions and algebricity

In this section we study the problem of determining whether the function $F_L(x) = \Sigma a_n x^n$ is algebraic or not, $\{a_n\}$ being the counting sequence for a language L; in particular we consider counting sequences satisfying linear recurrence equations with polynomial coefficients:

$$\sum_{k=0}^{d} a_{n-k} P_k(n) = 0$$

As a matter of fact the interest in this type of conditon is motivated by the following observations:

1) the coefficients of every algebraic function verify an equation of this type [Com],
2) there exists a sufficiently large class of CFL's which have generating functions whose coefficients are given through such an equation [Fl].

Examples

5.1a) The language $L_1 = \{x \in \{a,b\}^* \mid n_a(x) = n_b(x)\}$, where $n_\sigma(x)$ is the number of occurrences of the symbol σ in the string x, has generating function

$$F_{L_1}(x) = \sum_{n=0}^{\infty} \binom{2n}{n} x^{2n} \quad ;$$

taking $\phi(x) = \sum_{n=0}^{\infty} \binom{2n}{n} x^n$ $(F_{L_1}(x) = \phi(x^2))$ then $na_n = 2(2n-1)a_{n-1}$

5.2a) the language $L_2 = \{x \in \{a,b,c\}^* \mid n_a(x) = n_b(x) = n_c(x)\}$ has generating function

$$F_{L_2}(x) = \sum_{n=0}^{\infty} \binom{3n}{n}\binom{2n}{n} x^{3n} = \sum_{n=0}^{\infty} \frac{3n!}{(n!)^3} x^{3n} \;;$$

taking $\varphi(x) = \sum_{n=0}^{\infty} \frac{3n!}{(n!)^3} x^n$ ($F_{L_2}(x) = \varphi(x^3)$) then $n^2 a_n = 3(3n-1)(3n-2)a_{n-1}$

The problem of deciding algebricity for a single function seems to be very difficult: in the following we study a simpler problem which gives sometime an answer to the previous one.

Problem (space-algebricity):

Instance: a linear homogeneous recurrence equation with polynomial coefficients

$$\sum_{k=0}^{d} a_{n-k} P_k(n) = 0$$

Question: does the linear space of the generating functions $F(x) = \sum a_n x^n$,where $\{a_n\}$ satisfy the recurrence equation, contains only algebraic functions?

The following result allows us to transform the problem in that of studying equations on generating functions:

Fact 5.1- Given a linear homogeneous recurrence equation with polynomial coefficients, let $\{a_n\}$ be a solution (regardless of the initial conditions). Then $F(x)= \sum a_n x^n$ satisfies a linear homogeneous differential equation with polynomial coefficients, and viceversa.

The proof of this fact is constructive and it is based on two simple correspondences: the first one is that which exists between the displacement operator (on the space of the numerical successions) and the multiplication by a monomial x^k (on the space of the generating functions), the second one being that between the multiplication by n and the operator xD (D stands for the derivation operator).

Examples

5.1b) Taking $\phi(x) = \sum_{n=0}^{\infty} a_n x^n$ then $x\phi'(x) = \sum_{n=0}^{\infty} n a_n x^n$,

$$x(x\phi(x))' = \sum_{n=1}^{\infty} n a_{n-1} x^n \quad , \quad x\phi(x) = \sum_{n=1}^{\infty} a_{n-1} x^n .$$

The differential equation for example 5.1a is $(4x - 1)\phi(x)' + 2x\phi(x) = 0$

5.2b) Taking $\varphi(x) = \sum_{n=0}^{\infty} a_n x^n$ then $x(x\varphi'(x))' = \sum_{n=0}^{\infty} n^2 a_n x^n$,

$$x(x(x\varphi(x))')' = \sum_{n=1}^{\infty} n^2 a_{n-1} x^n .$$

Recalling example 5.2a we obtain $(27x^2 - x)\varphi''(x) + (54x - 1)\varphi'(x) + 6\varphi(x) = 0$.

It can be noted that the above theorem provides us with an easy proof of Comtet's theorem since all the algebraic functions satisfy differential equations with polynomial coefficients.

We recall here Klein's method for solving differential equations of the second order having polynomial coefficients; this is the starting point for an algorithm for testing algebricity.

The basic idea is that, given a differential equation of order m on the complex field with a fundamental system of integrals $w_1,\dots,w_m$, the effect caused by a description of a closed path enclosing one or more singularities is to replace the system with one of the type

$$w_1' = a_{11}w_1 + \dots + a_{1m}w_m$$
$$\vdots$$
$$w_m' = a_{m1}w_1 + \dots + a_{mm}w_m$$

that is $w' = S(w)$ where S is a linear substitution.

The set of linear substitutions (obtained by considering infinitely many contours passing through a fixed point) forms a group with respect to the composition. Moreover it is possible to prove the following

Fact 5.2- The group of linear substitutions associated with a linear differential equation (with polynomial coefficients) is finite <u>iff</u> the space of solutions only contains algebraic functions.

Klein gives a method for determining linear differential equations of the second order with algebraic integrals, by associating them with the finite groups of linear substitutions of two variables:

$$w_1' = \alpha w_1 + \beta w_2,\ w_2' = \gamma w_1 + \delta w_m$$
$$(w_1', w_2') = S(w_1, w_2).$$

In such a case he studies the finite groups of homographic substitutions by taking

$$s = \frac{w_1}{w_2},\ S = \frac{w_1'}{w_2'} = \frac{\alpha s + \beta}{\gamma s + \delta}$$

and he succeeds in computing the orders of the groups, identifying only five cases. His results can be summarized in the following theorem [Fo]:

Theor. 5.1- (Klein) Given a linear homogeneous differential equation of the second order with polynomial coefficients

$$A(z)\frac{d^2w}{dz^2} + B(z)\frac{dw}{dz} + C(z)w = 0,$$

his integrals are algebraic functions if and only if,taking $p(z) = \frac{B(z)}{A(z)}$, $q(z) = \frac{C(z)}{A(z)}$, one has:

1) $p(z) = \frac{1}{u}\frac{du}{dz}$ where u is an algebraic function of z;

2) the invariant $I(z) = q(z) - \frac{1}{4}p^2(z) - \frac{1}{2}\frac{dp}{dz}$ is either of the form

$$(*) \quad \frac{1}{4}\left[\frac{1-\frac{1}{\nu_2^2}}{Z^2} + \frac{1-\frac{1}{\nu_1^2}}{(Z-1)^2} + \frac{\frac{1}{\nu_1^2}+\frac{1}{\nu_2^2}-\frac{1}{\nu_3^2}-1}{Z(Z-1)}\right]\left(\frac{dZ}{dz}\right)^2 + \frac{1}{2}\{Z,z\}$$

or of the form

$$(**) \quad \frac{1}{4}\,\frac{1-\frac{1}{N^2}}{Z^2}\left(\frac{dZ}{dz}\right)^2 + \frac{1}{2}\{Z,z\}$$

where Z is a rational function of z, N is an integer, and $\{Z,z\} = \frac{Z'''}{Z'} - \frac{3}{2}\left(\frac{Z''}{Z'}\right)^2$ is the Schwarzian derivative. ν_1,ν_2,ν_3 are integers which are related to the orders of the groups of substitutions, their values being determined for each case.

case	ν_1	ν_2	ν_3
1	N	N	-
2	2	2	N
3	2	3	3
4	2	3	4
5	2	3	5

(*) for the cases 2,3,4,5
(**) for the case 1

6. An algorithm for the algebricity

We outline here an algorithm which is based on the results of the theorem 5.1. The main goal is to test the existence of a rational function Z(z) satisfying equations (*) or (**): this can be done by explicitly constructing Z(z).

Let Z(z) and Z(z) - 1 be in the form

$$Z(z) = A\,\frac{\prod_{i=1}^{m}(z-a_i)^{\alpha_i}}{\prod_{j=1}^{n}(z-c_j)^{\gamma_j}}, \quad Z(z) - 1 = B\,\frac{\prod_{i=1}^{m}(z-b_i)^{\beta_i}}{\prod_{j=1}^{n}(z-c_j)^{\gamma_j}} \quad \text{A,B constants}$$

then by an analysis of the poles of the second order in the expression for I(z) (the invariant cannot have poles of order greater than 2) we obtain

$$(*)\ I(z) = \sum_{i=1}^{m} \frac{\frac{1}{4}\left(1 - \frac{\alpha_i^2}{\nu_2^2}\right)}{(z - a_i)^2} + \sum_{i=1}^{m} \frac{\frac{1}{4}\left(1 - \frac{\beta_i^2}{\nu_1^2}\right)}{(z - b_i)^2} + \sum_{j=1}^{n} \frac{\frac{1}{4}\left(1 - \frac{\gamma_j^2}{\nu_3^2}\right)}{(z - c_j)^2} - \sum_{k=1}^{p} \frac{\frac{1}{2}\tau_k + \frac{1}{4}\tau_k^2}{(z - t_k)^2} + \ldots$$

$$(**)\ I(z) = \sum_{i=1}^{m} \frac{\frac{1}{4}\left(1 - \frac{\alpha_i^2}{N^2}\right)}{(z - a_i)^2} + \sum_{j=1}^{n} \frac{\frac{1}{4}\left(1 - \frac{\gamma_j^2}{N^2}\right)}{(z - c_j)^2} - \sum_{k=1}^{p} \frac{\frac{1}{2}\tau_k + \frac{1}{4}\tau_k^2}{(z - t_k)^2} + \ldots$$

where t_k is a zero of $Z'(z)$ of multiplicity τ_k which is not a repeated zero of $Z(z)$.

The algorithm works in the following steps:

1) it computes the coefficients of the poles of the second order in $I(z)$,

2) it guesses a correspondence between the values found (which must be rational numbers) and those in the symbolic expression at the right side of (*) or (**). If such a correspondence does not exist it halts with a negative answer (This means that there is no rational function $Z(z)$ satisfying the conditions of the theorem 5.1, hence the space of solutions contains at least one function which is not algebraic),

3) for each existing correspondence (it might not be unique - so it has to try for all the cases) - i.e. for each set of acceptable values for $\alpha_i,\beta_i,\gamma_j,\tau_k$ - it constructs an algebraic equation in the variables z,a_i,b_i,g_j,t_k by manipulating the equations (*) or (**),

4) it evaluates each algebraic equation at p+1 points where p is the degree of the equation in the variable z , obtaining a set of systems of algebraic equations in the variables a_i,b_i,g_j,t_k,

5) it tests whether at least one system admits a solution: if this is the case then the space of the solutions only contains algebraic integrals otherwise not.

<u>Example</u>

6.1a) For the equation $(27x^2 - x)\varphi''(x) + (54x - 1)\varphi'(x) + 6\varphi(x) = 0$ we have

$$p(x) = \frac{54x - 1}{27x^2 - x}, \quad q(x) = \frac{6}{27x^2 - x}.$$

The first condition is easily satisfied (take $u = 27x^2 - x$), then we compute the invariant $I(x)$.

$$I(x) = \frac{648x^2 - 24x + 1}{x^2(27x - 1)^2}$$

In this case the coefficients of the poles of the second order turn out to be 1, and this suffices to state that the space of the solutions doesn't contain only algebraic integrals, since for each case there is no way for an equation in $\alpha_i,\beta_i,\gamma_j$ to hold.

A preliminary version of this algorithm has been implemented in the muMATH system. At the moment it handles differential equations having polynomial coefficients of limited degree and it does not deal with the last step. The first limitation is of technical reason

(steps 1 through 4 have a complexity which is a polynomial in the maximum degree of the coefficients), while the second one is related to the inherent difficulty of the problem (from the complexity point of view): a possible solution could be that of using techniques based on Groebner bases which seem to work well in this case.
It should be noted that whenever the answer of the algorithm is negative, it might be the case that one of the integrals still be algebraic, so that we cannot decide for the ambiguity of the given language.

7. Generating functions and contour integration

In this section we give a technique which is sometimes useful to establish properties of generating functions. This method works well when there is a suitable relation between the coefficients of the generating function we want to study and those of another generating function being known.
In this setting we study generating functions in two variables, which are a natural extension of the univariate case, and we consider them as analytic functions. The basic result is in the following

Fact 7.1- Let $F(x,y) = \sum_{i,j} F_{ij} x^i y^j$ be an analytic function in the neighbourhood of the origin. Consider any function of the form

$$\phi(x) = \sum_i F_{hi\,i} x^{ki} \quad \text{(h, k integers)},$$

then we have

$$\phi(x) = \frac{1}{2\pi i}\oint_c F(x^\alpha s^\beta, x^\gamma s^\delta)\frac{ds}{s}$$

where α, β, γ, δ are integers such that $h = \frac{-\delta}{\beta}$, $k = \alpha h + \gamma$, and c is a circle centered at the origin.

Example

7.1a) It is quite easy to show that the language $L_3 = \{x \in \{a,b,c\}^* \mid n_a(x) = n_b(x)\}$ has the generating function (in two variables)

$$F_{L_3}(x,y) = \sum_{i,j} F_{ij} x^i y^j = \sum_{i,j} F_{2i\,j} x^{2i} y^j = \frac{1}{\sqrt{(1-y)^2 - 4x^2}},$$

where $F_{2i\,j} = \#\{x \in \{a,b,c\}^* \mid n_a(x) = n_b(x) = i,\ n_c(x) = j\}$.

It is then immediate to see that $F_{L_2}(x) = \sum_{i=0}^{\infty} F_{2i\,i} x^{3i}$, therefore by fact 7.1 we have

$$F_{L_2}(x) = \frac{1}{2\pi i}\oint_c F_{L_3}(xs, xs^{-2})\frac{ds}{s} = \frac{1}{2\pi i}\oint_c \frac{s}{\sqrt{(s^2 - x)^2 - 4x^2 s^6}}\,ds$$

and finally by taking $s = \sqrt{z}$ we obtain

$$F_{L_2}(x) = \frac{1}{4\pi i}\oint_c \frac{1}{\sqrt{(z-x)^2 - 4x^2z^3}}\,dz$$

that is an elliptic integral of the first type. We can conclude that the generating function of L_2 is not algebraic since it has transcendental values at algebraic points (see for example [Sc]).

Acknowledgements

We warmly thank Alberto Bertoni for pointing us this topic and for stimulating discussions.

This work has been partially supported by MPI (Ministero della Pubblica Istruzione) funds.

References

[AR] IJ.J.Aalbersberg, G.Rozenberg, *Theory of traces*, tec. Rep. 86-16, Inst. of Appl.Math. and Comp. Sci. University of Leiden, 1986

[AH] IJ.J.Aalbersberg, H.J.Hogeboom, *Decision problems for regular trace languages*, Proc.14th ICALP, Lect.Not.Comp.Sci. 267, 251-259, Springer, 1987

[BBMS] A.Bertoni, M.Brambilla, G.Mauri, N.Sabadini, *An application of the theory of free partially commutative monoids: asymptotic densities of trace languages*, Lect. Not. Comp. Sci. 118, 205-215, Springer, 1981

[BMS1] A.Bertoni, G.Mauri, N.Sabadini, *Unambiguous regular trace languages*, Algebra Combinatorics and Logica in Computer Science, Colloquia Mathematica Societatis J. Bolyai, vol.42, 113-123, North Holland, Amsterdam 1985

[BPS] D.Bruschi, G.Pighizzini, N.Sabadini, *On the existence of the minimum asynchronous automaton and on decision problems for unambiguous regular trace languages*, Proc. STACS 88, Lect.Notes Comp.Sci. 294, pp.334-346, Springer, 1988

[BS1] A.Bertoni,N.Sabadini, *Algebricity of the generating function for context free languages*, Internal Rep. Dip.Scienze Informazione, Univ.Milano, 1985

[BS2] A.Bertoni,N.Sabadini, *Generating functions of trace languages*, Bulletin of EATCS 35, 1988

[Com] L.Comtet, *Calcul pratique des coefficient de Taylor d'une fonction algebrique*, Einsegnement Math.10, 267-270, 1964

[Co] S.A.Cook, *A taxonomy of problems with fast parallel algorithms*, Information and control 64, 2-22, 1985

[CS] N.Chomsky, M.Schuetzenberger, *The algebraic theory of context free languages*, Comp. Prog. and Formal Systems, North Holland,118-161, 1963

[DRS] P.Doubilet,G.C.Rota,R.Stanley, *On the foundations of combinatorial theory : the idea of generating functions*, VI° Berkeley Symp.Math.Stat.Prob.2, 267-318

[Fo] A.R.Forsyth, *Theory of differential equations*, Dover Publications, New York, 1959

[Fl] P.Flajolet, *Analytic models and ambiguity of context-free languages*, TCS.49, 283-309, 1987

[Go] J.Goldman, *Formal languages and enumeration*, Journal of Comb.Theory, series A 24, 318-338, 1978

[La] G.Lallement, *Semigroups and combinatorial applications*, J.Wiley and sons, New York, 1979

[Ma] A.Mazurkiewicz, *Concurrent program schemes and their interpretations*, DAIMI Rep.PB-78, Aarhus Univ., 1977

[Mo] H.K.Molloy, *On the integration of delay and throughput measures in distributed processing modes*, Ph.D. Thesis, Univ. of California, Los Angeles, 1981

[Pa] A.Paz, *Introduction to probabilistic automata*, Academic Press, New York London, 1971

[Sa] A.Salomaa, *Formal languages*, Academic Press, New York London, 1973

[Sc] T.Schneider, *Introduction aux nombres transcendants*, Gauthier-Villars, Paris, 1959

[St] P.Stanley, *Generating functions*, in Studies in Combinatorics, vol.17, ed. G.C.Rota, 100-141,1978

[Zi] W.Zielonka, *Notes on asynchronous automata*, RAIRO Inf.Théor.vol. 21 n.2, 99-135, 1987

Semi-constructive formal systems and axiomatization of abstract data types

Pierangelo Miglioli, Ugo Moscato, Mario Ornaghi
Department of Information Science - University of Milan

1. Introduction

In the area of abstract data types (ADT) the "isoinitial model" approach [2,3,4] based on (classical) model theory, has been proposed with two aims: to provide a simple treatment of the recursiveness problem and to allow ADT specifications less restrictive than the "algebraic" ones [6,10,12].

As for the latter aspect, the algebraic attitude is mainly interested in setting up "small" theories (i.e., with a small deductive power as compared, e.g., with the one of full first order arithmetic) in order to axiomatize simple ADT to be furtherly extended in a (possibly long) sequence of "small" refinement steps. This point of view, which has given rise to important developments, is surely adequate to (stepwise) program synthesis; however, it is not oriented to the so called "constructive" attitude [1,5,8,9,11,16,17,18,19,22] which looks at "proofs" (of a constructive formal system) as "programs".

On the contrary, the isoinitial approach is quite in line with the latter attitude. Accordingly, in [4] the problem has been considered of characterizing constructive formal systems providing, together with a powerful ADT specification-method, a set of "abstract" algorithms on the ADT, i.e., the set of (constructive) proofs definable in them. Of course, this requires "reasonably powerful" formal systems (i.e., formal systems with a deductive power comparable, e.g., with the one of intuitionistic first order arithmetic).

In this frame, the possibility of looking for "large" *constructive* and *classically sound* formal systems S=T+L has been considered by the authors, where:

- the notation "T+L" means that the system S consists of a mathematical part T (a first order theory in the sense of [7] which, interpreted according to classical semantics, has an isoinitial model [3,4]) and a superintuitionistic logic L (i.e., INT⊆L⊆CL, INT and CL being intuitionistic and classical logic respectively);
- the sense according to which S is *constructive* is that S satisfies the disjunction property DP and the explicit definability property EDP [20,21];

- the sense according to which *classical soundness* is assumed is that T+L is consistent iff T+CL is.

As it is known, there are constructive first order systems which are consistent but not classically consistent [20]. Perhaps, as far as only operational (procedural) interpretations are considered (where a logical formula is read as something as a λ-expression, see, e.g., [8,13,16]) the classical soundness is not important. But we are interested, as said before, in connecting the area of constructivism with the one of ADT: and the latter has been developed (by means of algebraic or model theoretic tools) in a classical context.

Also, we believe that the classical reading of formulas has a simplicity which hardly can be found in other kinds of interpretations; this simplicity (which has been taken as paradigmatic in fields of computer science such as program correctness and artificial intelligence) makes classical semantics the most natural "denotational semantics" (to be preserved by an operationally correct formal system).

According to the above, our "large" systems correspond to the attempt of setting up as great as possible recursively axiomatizable constructive systems S=T+L contained in the classical T-system T+CL, even if the following limitations cannot be avoided:
- the greatest constructive subsystem (for a given T) of T+CL doesn't exist [14,21] (there is a set of maximal, "constructively incompatible" systems);
- the maximal constructive subsystems of T+CL are not, in general, recursively axiomatizable (even if T+CL is).

In this line, some results oriented to a classification (concerning the mutual "constructive compatibility" of various powerful logical and mathematical principles) have been obtained by the authors. Also, classically consistent systems corresponding to a weaker notion (we have called "semi-constructiveness") have been found, where S=T+L is *semi-constructive* iff it satisfies the following *weak disjunction property* WDP and *weak explicit definability property* WEDP:

(WDP) $S \vdash A \vee B$ and $A \vee B$ is closed => $T+CL \vdash A$ or $T+CL \vdash B$

(WEDP) $S \vdash \exists x C(x)$ and $\exists x C(x)$ is closed => $T+CL \vdash C(t)$ for some closed t.

We call *sub-constructive* any system S=T+L contained in some (fully) constructive system S'=T+L' (with the **same** mathematical part T): it turns out that any sub-constructive system is semi-constructive (but the converse does not hold, as we will see).

We look at sub-constructiveness and, more generally, at semi-constructiveness as one of the two extremes within which a constructive point of view may range, the other being strong constructiveness, where (roughly speaking): by a *strongly constructive system* we will mean any system S such that any proof of $A \vee B$ in S ($A \vee B$ closed) contains sufficient information to build up a proof of A in S or a proof of B in S, and the like for $\exists x A(x)$.

Strong constructiveness is appropriately treated in a proof theoretical attitude, while sub-constructiveness and semi-constructiveness generally involve simpler model theoretic aspects. Form the point of view of program specification and construction the former is needed in contexts such as program synthesis, where proofs are taken as programs and must give rise to effective computations; on the other hand, the latter can be used if one is not interested in computational devices, but only in foreseeing (as a first approximation) that some functions or relations (definable in the frame of first order theories) are (in principle) computable, or in guaranteeing that some expansions of given "intended models" satisfy some general requirements (having a character more semantical than syntactical). In this sense, a notion such as constructiveness (as defined by DP and EDP), turns out to be, according to the cases, undercharacterized or overcharacterized, while sub-constructiveness and semi-constructiveness seem to be adequate to ADT specification and extension.

Thus, one of the aims of the present paper is to show how semi-constructiveness can be used in a stepwise method of ADT-definition. In particular, we will provide a rather general result (see THEOR.4 below) allowing to pass from a theory T with an isoinitial model M to a stronger T' with an isoinitial model M' expanding M. Results of this kind have been expounded in [4] only for definitory extensions of T; here we will provide a criterion allowing to obtain from T a T' which may be *non conservative* over T.

Also, we will discuss how these results can be used in the definition of classes of ADT, i.e. of families of ADT together with instantiation and extension methods (as typical in an "object oriented" attitude).

The more powerful the semi-constructive systems one uses are, the more powerful our extension method becomes. Thus, the development of axiomatic methods to set up great semi-constructive systems not only is interesting from a purely logical point of view but also might give rise to applications in the area of ADT. In this line, in the last part of the paper we will present five classes of (mutually "incompatible") or sub-constructive systems coming from our classification; this is only a first classification and we hope to find examples where the listed principles are useful.

2. The logic IKA and theories completely formalizing an ADT

In this section we briefly introduce two basic notions involved in the paper; a more detailed treatment can be found in [4] (where IKA is called CON) and in [3].

As said above, we are interested in semi-constructive and classically sound systems S=T+L. The classical soundness is automatically guaranteed by using superintuitionistic logics L containing the logic IKA so characterized: IKA is obtained by adding to INT (intuitionistic predicative logic with identity) the following axioms:

(K) $\forall x \sim\sim A(x) \rightarrow \sim\sim \forall x A(x)$ (Kuroda principle)
(A) $\sim\sim A \rightarrow A$ for any *atomic* A

Of course, the restriction to the atomic formulas in axiom (A) prevents the validity (in IKA) of the non constructive principle
$\sim\sim H \rightarrow H$ for any H.

We remark that the addition of (K) alone to INT is sufficient to obtain classically consistent systems.

We say that a model M of a theory T is *isoinitial in T* [2,3,4] iff, for every model M' of T, there is a unique isomorphic embedding [7] from M to M'. For instance, the standard structure of natural numbers is an isoinitial model of Peano Arithmetic and is the "intended" model of Peano axioms. We give the following definition:

DEF.1 A theory T completely formalizes an ADT iff T has a reachable isoinitial model M.

If an isoinitial model I of T exists, then: (1) I is recursive [2,3,4]; (2) any other isoinitial model of T is isomorphic to I; (3) T can be extended with the addition of a recursive diagram into a theory T' completely formalizing an expansion (with new constants) of I. By (2) we can choose any isoinitial M in order to represent the ADT formalized by T; we will say also that T formalizes the ADT I. We say that T is *atomically complete* iff $T+CL \vdash A$ or $T+CL \vdash \sim A$, for every closed atomic formula A.

The following theorem, whose proof is implicitly contained in [3,4], gives an useful criterion to study isoinitiality.

THEOR.1. T completely formalizes an ADT iff T has a reachable model and T is atomically complete.

In the above, we don't impose any restriction on the form of the axioms and we consider also many sorted first order languages. THEOR.1 allows to prove:

PROP.1. For every set C of "constructors" (i.e. constant and function symbols), the term-algebra generated by C is an isoinitial model of the theory T(C) = identity + injectivity axioms + induction principles.

The injectivity axioms state that different closed terms represent different objects. Identity and injectivity axioms are sufficient to obtain the isoinitiality result, but it is useful to introduce also various induction principles in order to use logical semi-constructive systems in a relevant way; in particular, we consider the usual structural induction and the descending chain principle based on the well founded order relation related to the structural complexity of the closed terms [4,16].

3. Towards an ADT-definition methodology within semi-constructive systems

Our approach to ADT is based on the previous notion of a theory completely formalizing an ADT and on THEOR.1. In this frame, we propose a method where the first order axioms characterizing an ADT are built up in one or more steps, in such a way that each step satisfies the requirements of THEOR.1. In particular, it may happen that condition of atomic completeness of THEOR.1 can be proved by showing that a suitable formula of the kind $\forall x(H(x) \vee \sim H(x))$ can be proved in a semi-constructive system T'+L (for a suitable T'⊃T and a suitable logic L). Also, a stepwise construction can be given for *classes* of ADT, as we will see later.

In the first step of the construction of an ADT one essentially uses THEOR.1. In particular, PROP.1 proposes a general way to start, but, of course, other theories T satisfying THEOR.1 can be taken. The subsequent steps are *extension steps,* where the extension of an ADT by new relations, functions and sorts is characterized by the following definition:

DEF.2 Let T be a theory completely formalizing an ADT M, let LT be the language of T and let LT' be a language extending LT by new sort, relation and function symbols; we say that a theory T'⊇T is an ADT-extension of T into LT' iff T' completely formalizes an ADT M' which is an *expansion* [7] of M into LT' (i.e. M' interprets the symbols of LT *exactly* as M and the old carriers are unchanged).

The requirements of DEF.2 are analogous to the safety conditions required to preserve "sufficient completeness" in the algebraic attitude [12].

Using THEOR.1, we can easily prove the following result.

THEOR.2. Let T be a theory in a language LT; let S be a sort symbol not in LT and $c_0,..,c_n,f_0,..,f_k$ be constant and function symbols of sort S not in

LT (the arities of the new function-symbols may contain also sorts of LT); let LT' be the extension of LT by S,c_0,..,c_n,f_0,..,f_k,= (= the identity on S) and T'=(T + identity and injectivity ax. + induction principles for the new symbols); then T' is an ADT-extension of T into LT'.

This theorem allows the enrichment by new sorts and also the definability of a class of parametric ADT (indeed, the starting theory T may be any theory completely formalizing an ADT).
To add new functions and relations, one can use explicit definitions [7]; but it may happen that an extension by explicit definition is *not* an ADT-extension. We give the following theorem (stated in [17] for particular systems based on the logic IKA and here extended to *any* semi-constructive system):

THEOR.3 Let T be a theory completely formalizing an ADT, let LT be the language of T and let L be an intermediate logic. If the system S=T+L is semi-constructive, then: (I) let us consider the definition axiom $\forall x(r(x) \leftrightarrow H(x))$ (r a new relation symbol, H any formula); if $S \vdash \forall x(H(x) \vee \sim H(x))$, then the theory $T' = T \cup \{\forall x(r(x) \leftrightarrow H(x))\}$ is an ADT-extension of T into $LT \cup \{r\}$; (II) let us consider the definition axiom $\forall x F(x, f(x))$ (f a new function symbol, F any formula); if $S \vdash \forall x \exists! z F(x,z)$, then the theory $T' = T \cup \{\forall x F(x, f(x))\}$ is an ADT-extension of T into $LT \cup \{f\}$.

If the starting theory T is too poor, generally one cannot characterize interesting functions by explicit definitions. To obtain a more general result, involving also *non conservative* extensions, the following definitions are in order:

A is *existentially compound* (e.c.) iff one of the following clauses applies:

1) A=B and B is quantifier free;
2) $A = B \wedge C$ or $A = B \vee C$, and B, C are e.c.;
3) $A = \exists x B$ and B is e.c.;
4) $A = B \rightarrow C$, C is e.c. and $B = \forall \mathbf{x} D$, where D is quantifier free and "$\forall \mathbf{x}$" indicates a possibly empty sequence of universal quantifications.

A *good* formula is any formula of the kind $\forall \mathbf{x}(A(\mathbf{x}) \rightarrow B(\mathbf{x}))$ where $A(\mathbf{x})$ is e.c. and $B(\mathbf{x})$ is quantifier free.

Now we can prove:

THEOR.4 Let T, LT and L be defined as in THEOR.3. We have: (I) let r be a new relation symbol, let $LT' = LT \cup \{r\}$, let $ax_r \subseteq LT'$ be a set of *good* formulas and let $T' = T \cup ax_r$; if T is consistent, S=T'+L is semi-constructive and $S \vdash \forall x(r(x) \vee \sim r(x))$, then T' is an ADT-extension of T into LT'; (II) let F be a new relation symbol, let f be a new function symbol, let $LT' = LT \cup \{F\}$ and $LT'' = LT \cup \{F,f\}$; let $T' = T \cup ax_F$, where ax_F is a set of *good* formulas of LT'; if T' is consistent, $S = T' + L$ is semi-constructive and $S \vdash \forall x \exists ! z F(x,z)$, then $T'' = T' \cup \{\forall x F(x,f(x))\}$ is an ADT-extension of T into LT''.

We remark that the use of powerful semi-constructive systems according to THEOR.4 allows to capture, *within the formal systems themselves,* aspects which in the above quoted algebraic attitude can be expressed only at a metatheoretical level.

We also remark that Theor.'s 3 and 4 *still hold* if one takes as ADT the *initial* models [3,10] instead of the isoinitial ones (on the other hand, as it happens for isoinitiality, arbitrary explicit definitions don't preserve initiality).

An example of ADT defined by extensions is:

(E1) Peano Arithmetic can be obtained starting with the set of constructors C={0,s} [so the theory T(C) contains the injectivity axioms $\forall x(\sim s(x)=0)$, $\forall x \forall y(s(x)=s(y) \rightarrow x=y)$ and the usual induction schema] and *extending* it by + and * using THEOR.4 and the logic IKA.

An extensive analysis of the applicability of semi-constructive systems stronger that IKA in ADT-extensions requires a great amount of "experimental" work; we are at the beginning in this direction.

However, we have sketched some schemas which seem to be promising. The following example shows a general way of applying the principle:

(WGRZ) $\sim\sim\forall x B(x) \wedge \forall x(A \vee B(x)) \rightarrow A \vee \forall x B(x)$

which is contained in all the sub-constructive systems we will present in the next section.

(E2) Let us assume that $\forall x \exists ! y B(x,y)$ can be classically proved from a T completely formalizing an ADT (as said above, the introduction in T of a corresponding explicit definition introducing a new function-symbol may not preserve isoinitiality or initiality).

There are cases where one can prove (e.g. by suitable inductive principles) that $T+L \vdash \forall x(A \vee \exists y B(x,y))$, for suitable formulas A and suitable logics L which contain (WGRZ) and are such that T+L is sub-constructive. In these cases, an application of (WGRZ) provides $T+L \vdash A \vee \forall x \exists y B(x,y)$.

Now, let $\sim B$ be consistent with T: then, by the subconstructiveness of T+L, one deduces that there is a logic L' (possibly non effective) such that $L \subseteq L'$, T+L' is constructive and $T+L' \vdash \forall x \exists ! y B(x,y)$.

Even if in general one cannot use T+L' to compute the defined function, one can be sure that the addition of the considered explicit definition to T preserves isoinitiality (preserves initiality).

Now, we come to the notion of a *class of ADT*. Firts of all, we remark that, if we have a theory T satisfying the conditions of THEOR.1 and we add to T the axioms (*) $\forall x(r(x) \vee \sim r(x))$ for any relation symbol r, we obtain a T'⊃T which, of course, completely formalizes an ADT (indeed, (*) is classically valid; thus, neither the class of the models of T nor the set of the classical theorems of T are affected). But formulas such as (*) become relevant axioms whenever one considers semi-constructive systems T+L instead of T+CL, as we are going to explain.

A theory T *formalizes a L-class* (or, simply, T *is a L-class*) iff the formulas (*) can be proved in T+L. A L-class T is *not* required to formalize an ADT and T+L *doesn't need* to be sub- or semi-constructive.

An *instance* of a L-class T is any $I = T_I \cup T$ such that I completely formalizes an ADT; we call T_I an "instantiation" for T. We don't require any constructiveness property on I; according to the cases, it may be semi-constructive, sub-contructive, strongly constructive or nonconstructive at all.

Let T be a L-class: we say that T'⊃T is a *L-class-extension* of T iff T' is an L-class (expanding T by new sorts, functions or relations) and, for every instantiation T_I for T and T', $T' \cup T_I$ is an ADT-extension of $T \cup T_I$. I.e., one has the following commutative diagram:

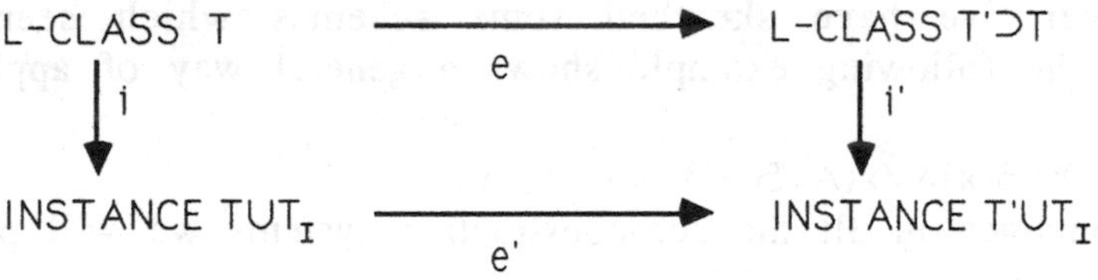

where e is a class-extension, i and i' are provided by the **same** instantiation T_I and e' is the ADT-extension corresponding to e, T_I.

Now, one can see that the formulas (*) are useful to prove that a T'⊃T is a *class-extension* of T; such a proof can be based on the above theorems and on the provability of the suitable formulas in T'+L.

The following example briefly shows an IKA-class and an IKA-class extension.

(E3). IKA-CLASS T_{seq} (finite sequences of generic elements).

CONSTRUCTORS: nil : seq; a : seq, elem -> seq;

RELATIONS: = : seq, seq; = : elem, elem; ≤ : elem, elem;

IDENTITY AXIOMS: the usual ones;

INJECTIVITY AXIOMS: (i1) $\forall x \forall y \sim nil=a(x,y)$;

(i2) $\forall x_1 \forall x_2 \forall y_1 \forall y_2 (a(x_1,y_1)=a(x_2,y_2) \rightarrow x_1=x_2 \wedge y_1=y_2)$.

IKA-CLASS AXIOMS: (c1) $\forall x \forall y(x=y \vee \sim x=y)$; (c2) $\forall x \forall y(x \leq y \vee \sim x \leq y)$.

STRUCTURAL INDUCTION RULE:

$$\dfrac{H(nil) \qquad \begin{array}{c}\cancel{H(j)}\\ \Pi \\ H(a(j,x))\end{array}}{H(t)}$$

The formula $\forall x_1 \forall x_2(x_1=x_2 \vee \sim x_1=x_2)$ can be proved by induction starting from (c1) and the other axioms.
For every T_{elem} with (at least) a sort elem, if T_{elem} completely formalizes an ADT (and if the obvious requirement to merge the languages of T_{elem} and T_{seq} in a sound way is satisfied) then $T_{elem} \cup T_{seq}$ completely formalizes an ADT (i.e., $T_{elem} \cup T_{seq}$ is an instance of T_{seq}).
An IKA-class-extension of T_{seq} can be obtained , e.g., by adding suitable axioms (ORD) defining an ordering $\leq$ on sequences induced by the ordering $\leq$ on elements; to be sure that $T'_{seq}=T \cup (ORD)$ is powerful enough to provide an IKA-class-extension, one has to prove in T'_{seq}+IKA the formula $\forall x_1 \forall x_2(x_1 \leq x_2 \vee \sim x_1 \leq x_2)$.

4. Some semi-constructiveness results

Here we explain some results which relate the form of the axioms of T to the logic L in order that S=T+L be semi-constructive. To do so, first of all we introduce the following superintuitionistic logical principles:
(MA) $\forall x(A(x) \vee \sim A(x)) \wedge \sim\sim\exists x A(x) \rightarrow \exists x A(x)$;
(KP$\vee$) $(A \rightarrow B \vee C) \rightarrow (A \rightarrow B) \vee (A \rightarrow C)$;
(KP$\exists$) $(A \rightarrow \exists x B(x)) \rightarrow \exists x(A \rightarrow B(x))$, with x not free in A;
(GRZ) $\forall x(A \vee B(x)) \rightarrow A \vee \forall x B(x)$, with x not free in A.
(WGRZ) $\sim\sim\forall x B(x) \wedge \forall x(A \vee B(x)) \rightarrow A \vee \forall x B(x)$, with x not free in A.

These principles, except (WGRZ), are well known: (MA) is Markov Principle [20,21]; (KP$\vee$) has been introduced by Kreisel and Putnam in order to provide a propositional logic stronger than the intuitionistic one and satisfying the disjunction property [15]; (KP$\exists$) is a predicative variant of (KP$\vee$) which is also known as (IP) [20,21]; (GRZ) is Grzegorczyk principle (see Smorinski's essay in [20]); finally, (WGRZ) is a weak variant of (GRZ) which turns out to have a much wider applicability than the former (differently from (GRZ), it can be combined with theories containing induction principles, as we will see).

Some of the above principles are separately meaningful for program synthesis (e.g., (MA) allows to deal with the minimalization operator), but not all are "constructively compatible": more precisely, (MA)+(KP$\exists$)+Intuit. arithmetic = Class. arithmetic [21]; also, one can show that (MA)+(KP$\vee$)+Intuit. arithmetic is not semi-constructive; finally, (GRZ) + Intuit. arithmetic=Class. arithmetic [20]. We also remark that

(GRZ), with the addition of the intuitionistic principles, allows to derive (MA).

Now, we are going to present five classes of semi-constructive systems which are mutually "constructively incompatible"; each class will contain one (or more) of the principles (MA), (KP∨), (KP∃) and (GRZ) together with (WGRZ) and appropriate new logical principles. We start with the following definitions:

a *∀-formula* is of the kind $\forall \mathbf{x} H$, a *∀∃-formula* is of the kind $\forall \mathbf{x} \exists \mathbf{y} H$, with H quantifier free and $\forall \mathbf{x}$, $\exists \mathbf{y}$ possibly empty sequences of quantifications;

a *∀∃~-formula* is inductively so defined:

every ∀∃-formula is a ∀∃~-formula; every formula such as $\sim A$ is a ∀∃~-formula; if A, B are ∀∃~-formulas and C is any formula, then $A \wedge B$, $C \rightarrow A$, $\forall x A$ are ∀∃~-formulas;

a *∀~-formula* is defined likewise taking in the basic clause the ∀-formulas instead of the ∀∃-formulas.

We remark that the set of the ∀∃~-formulas contains both the set of the ∀~-formulas and the set of the ∀∃-formulas; also, the set of the ∀~-formulas doesn't contain the set of the ∀∃-formulas, and conversely; finally, the set of the ∀~-formulas contains the set of the Harrop formulas (which are defined, e.g., in [20]).

To define the first class of formal systems, we introduce the two following logical principles:

(P1) $\exists x A(x) \vee \forall x(A(x) \wedge \sim\sim B(x) \rightarrow B(x))$

(P2) $A(t) \vee \exists x(A(x) \wedge \sim\sim B(x) \rightarrow B(x))$

Now the *1-systems* T+L are so defined:

T completely formalizes an ADT;

T (possibly) contains induction principles (also in the form of descending chain principles) and all its other axioms are ∀∃~-formulas;

L=IKA+(MA)+(WGRZ)+(P1)+(P2).

We can prove:

THEOR.5 The 1-systems are sub-constructive.

Remark 1. As far as only sub-constructiveness and, more generally, semi-constructiveness is involved, results such as THEOR.5 and the following THEOR.'s 6 and 7 can be proved in a reasonably simple way (much more complex proofs are required to establish the full constructiveness of some "large subsystems" of the ones presented here, discussed, e.g., in [17]). The proofs of our theorems can be seen simply as soundness proofs and involve the following aspects, where S=T+L is the system in hand and S_{CL}=T+CL is the corresponding classical system:

- one has to define a set F(T) of S_{CL}-provable formulas such that F(T) is closed under the IKA-inference rules and satisfies the disjunction property and the explicit definability property;
- one has to prove that all the S-provable formulas are contained in F(T).

We call F(T) a frame for the system S. The important aspect to be pointed out is that the frame is fully constructive but is not, in general, recursively enumerable; on the other hand, the recursively enumerable set of the S-provable formulas turns out to be sub-constructive but cannot be guaranteed to be (fully) constructive. As an example, we present the frame F1(T) to be used for the 1-systems considered in THEOR.5 (the frames for the systems considered in the subsequent theorems are omitted in this paper):
- if H is a closed formula, then $H \in F1(T)$ iff $T+CL \vdash H$ and one of the following clauses applies:
1) $H= \sim B$ or H is atomic;
2) $H=B \wedge C$ and $B \in F1(T)$ and $C \in F1(T)$;
3) $H=B \wedge C$ and $B \in F1(T)$ or $C \in F1(T)$;
4) $H=B \rightarrow C$ and $B \notin F1(T)$ or $C \in F1(T)$;
5) $H= \exists x B(x)$ and there is a closed term t such that $B(t) \in F1(T)$;
6) $H= \forall x B(x)$ and for every closed term t $B(t) \in F1(T)$;
- if $H=A(\mathbf{x})$ is an open formula, then $H \in F1(T)$ iff $A(\mathbf{t}) \in F1(T)$ for every $A(\mathbf{t})$ obtained from $A(\mathbf{x})$ by substituting the free variables **x** with closed terms.

One easily sees that F1(T) is closed under the inference rules and satisfies DP and EDP (one can also show that F1(T) is maximal in the following sense: if F(T) is any set of S_{CL}-provable formulas such that F(T) is closed under the IKA-inference rules, satisfies DP and EDP and contains F1(T), then F(T)=F1(T)). Then, the proof of THEOR.5 amounts to show that if S=T+L is a 1-system then all the S-provable formulas are contained in F1(T).

Remark 2. THEOR.5 can be used to extend an ADT according to THEOR.3: for, THEOR.3 requires the semi-constructiveness of the system T+L, where T is the starting theory.

On the other hand, THEOR.5 cannot be used, as such, to extend an ADT according to THEOR.4: for, one needs the semi-constructiveness of the system T'+L, where T' is the extended theory , just to be sure that T' completely formalizes an ADT; but the definition of a 1-system requires that the involved theory (here T') completely formalizes an ADT. The problem is related to the following aspects:

- the possibility of using $\forall\exists\sim$-axioms requires (in order that T'+L be semi-constructive) the atomic completeness of T' (this is an assumption weaker than the hypothesis that T' completely formalizes an ADT, but even the atomic completeness of T' is provided, in THEOR.4, by the semi-constructiveness of T'+L);
- the possibility of using (semi-constructively) (MA) (together with $\forall\exists\sim$ -axioms and induction principles) requires that T' has a reachable model (i.e., since T' must be atomically complete by the above point, that T' completely formalizes an ADT).

These difficulties can be partially overcome as follows.
Let LT' be the language of a theory T' and let LT be a language contained in LT'; let M be any structure for the language LT' (not necessarily a model of T): we say that M is an LT-model of T' if all formulas of T belonging to LT hold in M; we say that T' is LT-atomically complete iff $T' \vdash A$ or $T' \vdash \sim A$ for every closed atomic formula $A \in LT$; we say that a LT-model M of T' is LT-reachable if every element of the carrier of M is denoted by some closed term of LT; finally, we say that T' completely formalizes a LT-ADT iff T' is LT-atomically complete and T' has a reachable LT-model M.
Now, the LT-LT'-1-systems T'+L' are so defined:
- $LT \subseteq LT'$, where LT' is the language of T, and T' completely formalizes a LT-ADT;
- T' (possibly) contains induction principles (in the full language LT') and all other axioms of T' are $\forall\exists\sim$-formulas of LT or Harrop-formulas (in the full language LT');
- $L' = IKA + (MA)_{LT} + (P1) + (P2)$, where $(MA)_{LT}$ is the set of all the instances of (MA) in the language LT.
Then we can prove:
(a) The LT-LT'-1-systems are sub-constructive.
If one is not interested in using $(MA)_{LT}$, then one can consider the weak LT-LT'-1-systems: they are defined as the LT-LT'-1-systems, with the only difference that $(MA)_{LT}$ is not included in L' and T' must be LT-atomically complete but may not completely formalize an ADT. One can prove:
(b) The weak LT-LT'-1-systems are sub-constructive.

To introduce other kinds of semi-constructive systems, we need the following definitions:
a formula A is *stable* iff: A is atomic or negated, or A is of the form $B \wedge C$ or $B \vee C$ or $B \rightarrow C$ or $\forall x B$ with B, C stable;

A is a *formula in* U, $U \to V$ iff A is any formula constructed starting from U, U→V and using only the propositional connectives;
a formula A is *negatively saturated* iff every (possible) quantifier occurrence in it is in the scope of ~.

Also, we introduce the following families of principles:

(Q∨) $(\forall x(\sim\sim H(x) \wedge \sim\sim K(x) \to \forall y \sim U(y)) \to (\forall x(H(x) \to K(x)) \to C \vee D)) \to$
$(\forall x(\sim\sim H(x) \wedge \sim\sim K(x) \to \forall y \sim U(y)) \to$
$(\forall x(H(x) \to K(x)) \to C) \vee (\forall x(H(x) \to K(x)) \to D)),$

where C is any formula in U, U->V (D is arbitrary).

(Q∃) $(\forall x(\sim\sim H(x) \wedge \sim\sim K(x) \to \forall y \sim U(y)) \to (\forall x(H(x) \to K(x)) \to \exists w C(w))) \to$
$(\forall x(\sim\sim H(x) \wedge \sim\sim K(x) \to \forall y \sim U(y)) \to \exists(\forall x(H(x) \to K(x)) \to C(w)))$, where C is any formula in U, U -> V.

If in (Q∨) and (Q∃) we take a stable K(x), we obtain (SQ∨) and (SQ∃); if in (Q∨) we take a negatively saturated C, we obtain (NSQ∨).

We remark that (SQ∨) and (SQ∃) allow to deduce (KP∨) and (KP∃).

Now, the *2-systems* T+L are so defined:
T is atomically complete;
T (possibly) contains induction principles (including descending chain principles) and all its other axioms are ∀∃-formulas or ∀~-formulas;

L = IKA + **(WGRZ)** + (SQ∨) + (SQ∃).

The *3-systems* T+L are so defined:
T is atomically complete;
T (possibly) contains structural induction (but no descending chain principle) and all its other axioms are ∀∃-formulas or ∀~-formulas;
L = IKA + **(WGRZ)** + (KP∨) + (NSQ∨).

Remark 3. (NSQ) doesn't allow to deduce (KP∨).
We can prove:

THEOR.6 The 2-systems and the 3-systems are sub-constructive.

Remark 4. To apply this result to THEOR.4, we define the LT-LT'-2-systems T'+L' as follows:
for the LT-LT'-2-systems, L'=IKA+(SQ∨)+(SQ∃), while T' must be LT-atomically complete, may contain induction principles in the full language LT' (including descending chain principles) and all its other axioms are ∀∃-formulas of LT, or ∀~-formulas of LT, or Harrop-formulas (in the full language LT');
for the LT-LT'-3-systems, L'=IKA+(KP∨)+(NSQ∨), while T' satisfies the same conditions as in the LT-LT'-2-systems, with the only difference

that descending chain principles are not allowed (structural induction, on the other hand, is allowed).

We can prove:

(c) The LT-LT'-2-systems and the LT-LT'-3-systems are sub-constructive.

The addition of (KP∃) or of (MA) to the 3-systems does not preserve semi-constructiveness; thus, the 1-systems, the 2-systems and the 3-systems are mutually "incompatible".
Logics L more powerful than the ones considered above can be (semi-constructively) used together with sufficiently weak theories T. More precisely:
- we say that a theory T is a Harrop-theory iff all axioms of T are Harrop-formulas (no general induction principles are allowed!);
- we say that a system T+L is a ***4-system*** if T is a Harrop-theory and L=IKA+(GRZ)+(Q∨)+(Q∃) (remark that T is not necessarily atomically complete and that (Q∨) and (Q∃) are taken without restrictions);
- we say that a system T+L is a *5-system* if T is a Harrop-theory and L=IKA+(GRZ)+(P1)+(P2) (again, T may not be atomically complete).

We can prove:

THEOR.7 The 4-systems and the 5-systems are sub-constructive.

If L is the logic IKA+(GRZ)+(Q∨)+(Q∃)+(P1)+(P2) and T is the empty theory (hence, T is a Harrop-theory), then T+L collapses, i.e., T+L is not semi-constructive. Thus, the 4-systems and the 5-systems are constructively "incompatible".

Remark 5. The systems considered in the above THEOR.'s 5, 6 and 7 can be extended into (fully) constructive (possibly non effective, i.e., non recursively enumerable) systems: examples of the latter are the corresponding frames (in the sense of Remark 1). On the other hand, one can define recursively axiomatizable semi-constructive systems S=T+L (effective or not) which are not sub-constructive. A similar situation is found if only DP, e.g., is taken into account. For instance, let L=IKA+(KP∨)+(KP∃)+A∨(A→B∨∼B) and let T be any theory satisfying the conditions of THEOR.5: then, T+L satisfies WDP; on the other hand, there is a T (in this class of theories) such that T+L cannot be extended into a system T+L' satisfying DP.

This work has been supported by grants of the Ministero della Pubblica istruzione (40% and 60%).

REFERENCES.

[1] Bates J., Constable R. - Proofs as programs - ACM Transaction on Programming Languages and Systems, vol. 7, n.1, 1985.

[2] Bertoni A., Mauri G., Miglioli P., Wirsing M. - On different approaches to abstract data types and the existence of recursive models - EATCS bulletin vol. 9, oct. 1979.

[3] Bertoni A., Mauri G., Miglioli P. - On the power of model theory to specify abstract data types and to capture their recursiveness - Fundamenta Informaticae IV.2, 1983, pp. 127-170.

[4] Bertoni A., Mauri G., Miglioli P., Ornaghi M. - Abstract data types and their extension within a constructive logic Semantics of data types (Valbonne, 1984), Lecture Notes in Computer Science, vol. 173, Springer-Verlag, Berlin, 1984, pp. 177-195.

[5] Bresciani P., Miglioli P., Moscato U., Ornaghi M. - PAP: Proofs as Programs - (abstract), JSL, Vol. 51, n.3, 1986, pp. 852-853.

[6] Broy M., Wirsing M. - On the algebraic extension of abstract data types - in: Diaz J., Ramos I. (ed.) - Formalization of programming concepts - Lecture Notes in Comp. Sci. vol. 107, Springer-Verlag, Berlin, 1981.

[7] Chang C.C., Keisler H.J. - Model theory - North-Holland, 1973.

[8] Girard J. - The system F of variable types 15 years later - Report of CNRS, Paris, 1985.

[9] Goad C. - Computational uses of the manipulation of formal proofs - Rep. STAN-CS-80-819, Stanford University, 1980.

[10] Goguen J.A., Thatcher J.W., Wagner E.G. - An initial algebra approach to the specification, correctness and implementation of abstract data types - IBM Res. Rep. RC6487, Yorktown Heights, 1976.

[11] Goto S. - Program synthesis through Gödel's interpretation - Mathematical studies of information processing, (proceedings, Kyoto, 1978), Lecture Notes in Computer Science, vol.75, Springer-Verlag, Berlin, 1979, pp. 302-325.

[12] Guttag J., Horning J. - The algebraic specification of abstract data types - Acta Informatica 10, 27-52, 1978.

[13] Howard W.A. - The formulae-as-types notion of construction - in To Curry H.B.: essays on combinatory logic, lambda calculus and formalism, Academic Press, London, 1980.

[14] Kirk R.E. - A result on propositional logics having the disjunction property - Notre Dame Journal of Formal Logic, 23,1, 71-74, 1982.

[15] Kreisel G., Putnam H. - Eine unableitbarkeitsbeismethode für den intuitionistischen Aussagenkalkul - Archiv für Mathematische Logik und Grundlagenforschung, 3, 74-78, 1957.

[16] Martin-Löf P. - Constructive Mathematics and Computer Programming - Logic, Methodology and Philosophy of Science VI, L. Cohen, J. Los, H. Pfeiffer, K. Podewski (ed.), North-Holland, Amsterdam, 1982, pp.153-175.

[17] Miglioli P., Moscato U., Ornaghi M. - Constructive theories with abstract data types for program synthesis - Proceedings of the symposium Mathematical Logic and its Applications, Plenum Press, New York, 1988, pp.293-302.

[18] Miglioli P., Ornaghi M. - A logically justified model of computation I,II - Fundamenta Informaticae, IV.1,2, 1981.

[19] Nordstrom B., Smith J.M. - Propositions, Types and Specifications of Programs in Martin-Löf's Type Theory - BIT, Vol. 24, n.3, 1984, pp.288-301.

[20] Troelstra A.S. - Metamathematical investigation of Intuitionistic Arithmetic and Analysis - Lecture Notes in Mathematics, vol.344, Springer-Verlag, Berlin, 1973.

[21] Troelstra A.S. - Aspects of constructive mathematics - in: Barwise J. (ed.) - Handbook of Mathematical Logic, North Holland, Amsterdam 1977.

[22] Miglioli P., Moscato U., Ornaghi M. - PAP: a logic programming system based on a constructive logic - LNCS, n.306, Springer Verlag, 1988, pp.143-156.

Inductive Proofs
by Resolution and Paramodulation

Peter Padawitz
Fakultät für Mathematik und Informatik
Universität Passau
Postfach 2540
D-8390 Passau

Constructor-based sets of Horn clauses constitute a class of formulas for presenting verification problems occurring in data type specification as well as functional and logic programming. Inductive proofs of such clause sets can be carried out in a strict top-down manner by inductive expansion: the set is transformed via (linear) resolution and paramodulation into a case distinction, which covers all ground substitutions. Being a backward method, inductive expansion reduces the search space of corresponding forward proofs. The method does not put confluence or termination restrictions on the theorems to be proved such as procedures based on inductive completion do. Moreover, inductive expansion does not prescribe a strategy for controlling search trees so that the user may select "promising" paths according to specific applications.

1. Introduction

The mathematical models used in data type specification and program verification are *term-generated.* Each carrier element of the model is obtained by evaluating a *ground,* i.e. variable-free, functional expression. Hence a valid statement takes the form of an *inductive theorem,* which means that all ground instances are derivable. The proof is carried out by induction on the structure of ground terms (cf. [Bur69]) or, more generally, by induction with respect to a *Noetherian* relation on ground terms.

As one knows from inductive proofs in general, it might be difficult, not only to find a suitable Noetherian relation, but also to state an appropriate induction hypothesis, which often turns out to be a *generalization* of the theorem to be proved. While classical theorem proving provides explicit (more or less heuristic) induction rules to solve these problems, *inductive completion* (or *inductionless induction*) tries to get rid of induction steps by switching to *consistency* (or *conservative extension*) proofs (cf. [HH82], [JK86], [KM87], [Pad88a]).

Inductive completion puts strong restrictions not only on the underlying specification, but also on the theorems to be proved. Its requirement that axioms *and* theorems induce a *Church-Rosser* set of rewrite rules entails a number of syntactical restrictions, which might not be welcome, although some of these restrictions can be lowered if one uses weaker Church-Rosser criteria (cf. [HR87], [Pad88a]). In this paper, we describe an alternative method for proving inductive theorems based on traditional approaches like [BM79], [Hut86] and [GG88].

We start out from (Horn) clauses, written as $p \Leftarrow \gamma$, where p is an atom(ic formula) and γ is a finite set of atoms, called a *goal,* which consists of the premises under which p is required to hold. The existence of premises compels us to choose between two definitions of an inductive theorem:

Let AX be a set of axioms and $\vdash$ be a complete inference relation for valid clauses (cf. Section 2). For each clause $p \Leftarrow \gamma$ and each ground substitution f, let $p[f] \Leftarrow \gamma[f]$ denote the *instance of* $p \Leftarrow \gamma$ *by f,* i.e. the clause constructed from $p \Leftarrow \gamma$ by instantiating all variables according to f. By the first definition, $p \Leftarrow \gamma$ is an inductive theorem if for all ground substitutions f,

$$AX \cup \gamma[f] \vdash p[f]. \tag{1}$$

Alternatively, one may define: $p \Leftarrow \gamma$ is an inductive theorem if for all ground substitutions f,

$$AX \vdash \gamma[f] \text{ implies } AX \vdash p[f]. \tag{2}$$

(1) is equivalent to the validity of $p \Leftarrow \gamma$ in *all* term-generated models of AX (cf. [Pad88a], Cor. 4.3.3), while (2) characterizes the validity of $p \Leftarrow \gamma$ in the subclass of all *initial* models. (2) is weaker than (1): By (1), we may use $\gamma[f]$ in a proof of $p[f]$. By (2), we may also use formulas which occur in *every* derivation of $\gamma[f]$. When analyzing data type specifications one observes that crucial consequences of their axioms are valid in the sense of (2), but not in the sense of (1) (cf., e.g., [Pad88b] or [Pad88a], Ex. 4.3.4). In data base applications, the essence of (2) is known as the *closed world assumption* (cf. [Rei78]) that certain implications $p \Leftarrow q$ are in fact

equivalences. If, for deriving an instance of p *it is necessary* to derive the corresponding instance of q, then q⇐p holds true as well, but, in general, only in the sense of (2). It is often the case that, for proving an implication inductively, one needs the inverse of an axiom as a lemma (cf. Ex. 4.8).

We present the proof method of *inductive expansion* in three steps. First, the (meta-)implication involved in (2) is eliminated. Therefore, a set *IN* of variables, called *input variables*, is separated from all other variables, which are called *output variables*. *Input terms* contain only input variables, *output terms* contain only output variables. Some (weak) conditions are put on the theorems to be proved such that (2) becomes a consequence of the following non-implicational property: For all ground substitutions f,

$$AX \vdash ((p)\cup\gamma)[f|IN+g] \quad \text{for } \textit{some } g. \tag{3}$$

(Here f+g stands for the parallel composition of f and g, which maps the domains of f and g to the images of f and g, respectively.) Moreover, we distinguish clause *sets* M such that M consists of inductive theorems if and only if for all ground substitutions f,

$$AX \vdash ((p)\cup\gamma)[f|IN+g] \quad \text{for } \textit{some } p\Leftarrow\gamma \in M \text{ and some } g. \tag{4}$$

This characterization is valid if M is *constructor-based*, i.e.,

- for all p⇐γ ∈ M, γ is a set of equations with input terms on the left-hand side and *constructors* on the right-hand side,
- the set of premises over all clauses of M constitutes a *complete* and *minimal* case distinction,

where constructors are output terms such that

- each two constructor instances c[f] and d[g] are *decomposable*, i.e. c[f] and d[g] are equivalent (w.r.t. the underlying axioms) only if c = d and f and g are equivalent.

In a second step, we aim at reducing the infinite number of *forward* proofs involved in (4) to a finite number of *backward* proofs (here called *expansions*) of the form

$$\begin{gathered} \langle (p_1)\cup\gamma_1, id\rangle \vdash_{EX} \langle \delta_1, g_1\rangle, \\ \ldots \\ \langle (p_n)\cup\gamma_n, id\rangle \vdash_{EX} \langle \delta_n, g_n\rangle \end{gathered} \tag{5}$$

such that $\{p_1\Leftarrow\gamma_1,\ldots,p_n\Leftarrow\gamma_n\}$ covers the set of theorems to be proved and each ground substitution f is *subsumed* by some $\langle\delta_i,g_i\rangle$, i.e. $AX \vdash \delta_i[h]$ and $g_i[h] = f$ for some h. *id* denotes the identity substitution and $\vdash_{EX}$ stands for the inference relation generated by (linear) *resolution* [Rob65] and *paramodulation* [RW69].

The actual power of the approach is accomplished in a third step when *inductive* resolution and paramodulation rules are added to $\vdash_{EX}$. Applications of these rules simulate induction steps by resolving or paramodulating upon induction hypotheses. Furthermore, an inductive rule produces an atom of the form *fz >> z'* where f is the substitution obtained so far, z is the sequence of input variables, z' is a copy of z and >> is a Noetherian relation, which justifies the induction step. Regarding *fz >> z'* as a subgoal amounts to proving M *and* the soundness of the induction step simultaneously. This is what we call *inductive expansion*: resolving and paramodulating upon axioms, lemmas *and* induction hypotheses. The main result of the paper (Theorem 4.7) characterizes constructor-based inductive theorems as being provable by inductive expansion.

Section 2 presents basic notions concerning the syntax and semantics of Horn clause specifications with equality. Section 3 gives a precise definition of constructor sets and constructor-based clause sets along with their characterization as inductive theorems (Theorem 3.4). Section 4 starts from resolution and paramodulation as the basis of *backward* proofs and leads to the main result, given by Theorem 4.7 (see above).

At certain points, an inductive expansion relies on complete case distinctions (called *case matrices*). *The* requirement for completeness will be reduced to the question whether certain terms are *base-representable*. Theorem 5.3 tells us how this property can be proved by inductive expansion as well.

2. Preliminaries

Given a set S, an *S-sorted set* A is a family of sets, i.e. $A = \{A_s \mid s \in S\}$. For all $w = (s_1,...,s_n) \in S^*$, A_w denotes the cartesian product $A_{s1}\times...\times A_{sn}$. A *signature* SIG = (S,OP,PR) consists of a set S of *sorts* and two S^+-sorted sets OP and PR the elements of which are called *function* (or *operation*) *symbols* and *predicate symbols*, respectively. S-sorted function symbols are called *constants*.

We assume that for all $s \in S$, PR_{ss} implicitly contains a predicate symbol $\equiv_s$, called the *equality predicate* for s. We also fix an S-sorted set X of *variables* such that for all $s \in S$, X_s is countably infinite.

Example 2.1 The signature of our running example throughout the paper provides constructor functions for Boolean values, natural numbers, sequences and bags (multisets) together with operations that will be axiomatized later.

SORT		
sorts	bool, nat, seq, bag	
	symbol	*type*
opns	true	bool
	false	bool
	0	nat
	_+1	nat → nat
	ε	seq
	_&_	nat,seq → seq
	∅	bag
	add(_,_)	nat,bag → bag
	le(_,_)	nat,nat → bool
	seqToBag(_)	seq → bag
	insert(_,_)	nat,seq → seq
	sort(_)	seq → seq
preds	_ ≤ _	nat,nat
	_ > _	nat,nat
	sorted(_)	seq
	_ >> _	seq,seq

We hope that the notation is self-explanatory. ∎

Given a signature SIG = (S,OP,PR), a *SIG-structure* A consists of an S-sorted set, also denoted by A, a function $F^A : A_w \to A_s$ for each function symbol $F \in OP_{ws}$, $w \in S^*$, $s \in S$, and a relation $P^A \subseteq A_w$ for each predicate symbol $P \in PR_w$, $w \in S^+$. *T(SIG)* denotes the S^+-sorted set of terms (and term tuples) over SIG.

Given a term t, *root(t)*, *var(t)* and *single(t)* denote the leftmost symbol of t, the set of all variables of t, and the set of variables that occur exactly once in t, respectively. t is *ground* if var(t) is empty. *GT(SIG)* denotes the set of ground terms over SIG. We assume that SIG is *inhabited*, i.e. for each sort s there is a ground term t of sort s.

When speaking about terms in general, we use the prefix notation: *F* is placed in front of its argument list *t* to give the term *Ft*. In examples, however, the layout of terms is adapted to the underlying signature where infix, postfix or mixfix notations may occur as well.

Let A and B be S-sorted sets. An *S-sorted function* $f : A \to B$ is a family of functions, i.e. $f = \{f_s:A_s \to B_s \mid s \in S\}$. The set of S-sorted functions from A to B is denoted by B^A. The functions of $T(SIG)^X$ are called *substitutions*. Given a substitution f, *dom(f)*, the *domain of f*, is the set of all $x \in X$ such that $fx \neq x$. If dom(f) is empty, f is called the *identity substitution* and is denoted by *id*. If dom(f) is finite, say $dom(f) = \{x_1,...,x_n\}$, and if $fx_1 = t_1,..., fx_1 = t_n$, we also write $\{t_1/x_1,...,t_n/x_n\}$ instead of f. Given $V \subseteq X$, *f/V*, the restriction of f to V, is defined by $(f|V)(x) = fx$ for all $x \in V$ and by $(f|V)(x) = x$ for all $x \in X\text{-}V$. f is *ground* if the range of f consists of ground terms.

The *instance* of a term *t by f*, denoted by *t[f]*, is the term obtained from t by replacing all variables of t by their values under f. Conversely, one says that t *subsumes* t[f] or that t is a *prefix* of t[f]. f *unifies* t and t' if t[f] = t'[f]. The *sequential composition* of two substitutions f and g, denoted by f[g], is defined by (f[g])(x) = (fx)[g] for all x ∊ X. Accordingly, f[g] is an *instance* of f, and f *subsumes* f[g]. The *parallel composition* of f and g, denoted by f+g, is defined only if f and g have distinct domains. Then (f+g)(x) = fx if x ∈ dom(f), and (f+g)(x) = gx otherwise.

Given $w \in S^+$, $P \in PR_w$ and $u \in T(SIG)_w$, the expression Pu is called an *atom*. If P is an equality predicate and thus w = (s,s) for some s ∈ S and u = (t,t') for some t,t' ∊ $T(SIG)_w$, then Pu is called an *equation*, written as t≡t'. The notions *var*, *instance* and *unifier* extend from terms to atoms as if predicate symbols were function symbols.

Finite sets of atoms are called *goals*. A *clause* p⇐γ consists of an atom p, the *conclusion* of p⇐γ, and a goal $\gamma = \{p_1,...,p_n\}$, the *premise* of p⇐γ. If p is an equation, then p⇐γ is a *conditional equation*. If γ is empty, then p⇐γ is *unconditional* and we identify p⇐γ with the atom p. Note that unconditional clauses and goals are the same.

A *specification* is a pair (SIG,AX), consisting of a signature SIG and a set AX of clauses, comprising the axioms of the specification.

Example 2.1 (continued) The axioms of SORT, specifying *sort* as "insertion sort", are given by:

```
vars    x,y : nat;  s : seq;  b : bag
axms    seqToBag(ε) ≡ ∅                                    (BA1)
        seqToBag(x&s) ≡ add(x,seqToBag(s))                 (BA2)
        add(x,add(y,b)) ≡ add(y,add(x,b))                  (BA3)
        sort(ε) ≡ ε                                        (IS1)
        sort(x&s) ≡ insert(x,sort(s))                      (IS2)
        insert(x,ε) ≡ x&ε                                  (IN1)
        insert(x,y&s) ≡ x&y&s   ⇐   x ≤ y                  (IN2)
        insert(x,y&s) ≡ y&insert(x,s)   ⇐   x > y          (IN3)
        le(0,x) ≡ true
        le(x+1,0) ≡ false
        le(x+1,y+1) ≡ le(x,y)
        x ≤ y   ⇐   le(x,y) ≡ true                         (LE1)
        x > y   ⇐   le(x,y) ≡ false                        (LE2)
        sorted(ε)                                          (SO1)
        sorted(x&ε)                                        (SO2)
        sorted(x&y&s)   ⇐   x ≤ y, sorted(y&s)             (SO3)
        x&s >> s                                           (GR)  ▮
```

A clause $Pu \Leftarrow P_1u_1,...,P_nu_n$ is *valid* in a SIG-structure A if for all $b \in A^X$, $(\forall\ 1 \le i \le n : b^*u_i \in P_i^A)$ implies $b^*u \in P^A$, where b^* is the unique (SIG-) homomorphic extension of b to T(SIG). Given a clause set AX, A is a *SIG-model of AX* if each p⇐γ ∈ AX is valid in A and if for all s ∈ S, $\equiv_s^A$ is the identity on A_s.

Let us fix a specification (SIG,AX). The *cut calculus with equality* consists of the congruence axioms for all equality symbols (w.r.t. SIG) and two inference rules:

(SUB)	For all substitutions f, p⇐γ ⊢ p[f]⇐γ[f].
(CUT)	{p⇐γ∪{q}, q⇐δ} ⊢ p⇐γ∪δ.

$\vdash_C$ denotes the corresponding inference relation. The class of all SIG-models of AX satifies a clause p⇐γ if and only if p can be derived from AX∪γ via the cut calculus with equality such that the variables of γ need not be instantiated (cf. [Pad88a], Cor. 4.2.4).

Two terms t and t' are called *AX-equivalent* if AX $\vdash_C$ t≡t'. Two substitutions f and g are AX-equivalent if for all x ∈ X, fx and gx are AX-equivalent.

Definition A clause p⇐γ is called an *inductive AX-theorem* if for all ground substitutions f,

$$AX \vdash_C \gamma[f] \text{ implies } AX \vdash_C p[f].$$

The set of inductive AX-theorems is denoted by *ITh(AX)*. A set M of clauses is an inductive AX-theorem if all clauses of M are inductive AX-theorems. ∎

The model-theoretic counterpart of inductive theorems are *initial structures*:

A SIG-structure A is *initial w.r.t. AX* if A satisfies AX and each model B of AX admits a unique (SIG-) homomorphism from A to B. *Ini(AX)* denotes the (isomorphism) class of initial structures w.r.t. AX.

Theorem 2.2 (cf. [Pad88a], Thm. 4.4.3) *p⇐γ ∈ ITh(AX) iff Ini(AX) satisfies p⇐γ.* ∎

Corollary 2.3 *ITh(ITh(AX)) = ITh(AX).* ∎

3. Constructor-based Clause Sets

General Assumption (part 1) Let *IN* be a fixed finite set of variables, called *input variables*. The elements of the complement *OUT* = X-IN are called *output variables*. *Input terms* are terms containing only input variables, *output terms* are terms containing only output variables. ∎

Definition A set T of output terms is *ground complete for a term t* if for all ground substitutions f there is t' ∈ T such that t[f] is AX-equivalent to some instance of t'. T is *ground complete* if T is ground complete for all ground terms. T is a *set of constructors* if for all c,d ∈ T and ground substitutions f,g such that c[f] and d[g] are AX-equivalent, c equals d and flvar(c) is AX-equivalent to glvar(c). ∎

In many applications, the constructor property can be checked easily by referring to a given initial structure A w.r.t. AX: Ground terms are AX-equivalent iff they denote the same element of A in terms of which the property is obvious. A "syntactical" constructor criterion is given in Section 6.

In order to define constructor-based clause sets we need a schema for presenting case distinctions.

Definition A finite set CM of finite sets of equations is a *(constructor-based) case matrix with input* $IN_0 \subseteq X$ if either CM = {∅} or CM = {{t≡c_i} ∪ γ | 1≤i≤n, γ ∈ CM_i} for a term t, a ground complete set {c_1,...,c_n} of output terms (constructors) for t, and (constructor-based) case matrices CM_1,...,CM_n with input IN_0 ∪ var(c_1), ..., IN_0 ∪ var(c_n), respectively, such that var(t) ⊆ IN_0 and for all 1≤i≤n, var(c_i) ∩ IN_0 = ∅. ∎

For instance, a constructor-based case matrix using SORT (cf. Ex. 2.1) is given by

```
{{s ≡ ε},
 {s ≡ x&ε},
 {s ≡ x&y&s'}, {le(x,y) ≡ true},
 {s ≡ x&y&s'}, {le(x,y) ≡ false}}
```

The case matrix condition is purely syntactic except for the ground completeness of {c_1,...,c_n}. How to prove this property is the topic of Section 5.

A case matrix covers the set of ground substitutions:

Proposition 3.1 *Let CM be a case matrix with input IN_0. Then for all ground substitutions f there are γ ∈ CM and a substitution g such that* $AX \vdash_C \gamma[f|IN_0+g]$. ∎

Sometimes the case matrix condition is too restrictive (cf. the first covering derived in Ex. 4.8). In fact, it suffices to get a case matrix as an instantiation of a set of sets of equations:

Definition A set EM of sets of equations is *extendable to a case matrix* if there are output variables x_1,...,x_n and output terms c_1,...,c_n such that CM = {γ[c_1/x_1,...,c_n/x_n] | γ ∈ EM, 1≤i≤n} is a case matrix. ∎

Prop. 3.1 immediately implies

Proposition 3.2 *Let EM be extendable to a case matrix with input IN_0 = IN (see above). Then for all ground substitutions f there are γ ∈ CM and a substitution g such that* $AX \vdash_C \gamma[f|IN+g]$. ∎

For *un*conditional theorems, (2) and (3) (cf. Sect. 1) coincide. (Take IN = X.) If premises are involved, the equivalence of (2) and (3) is guaranteed only for constructor-based clause sets.

Definition A set M of clauses is *constructor-based* if there is a constructor-based case matrix CM with input IN such that

(a) $\gamma \in$ CM iff there is p with $p \Leftarrow \gamma \in$ M,
(b) for all $p \Leftarrow \gamma \in$ M, $var(p) \subseteq IN \cup var(\gamma)$,
(c) for all $p \Leftarrow \gamma$, $q \Leftarrow \delta \in$ M with $p \Leftarrow \gamma \neq q \Leftarrow \delta$, γ is not a subset of δ. ∎

Conditions (a)-(c) are purely syntactic. (c) forbids different clauses with subsuming premises. The "if" part of (a) can always be ensured by adding to M a clause $x \equiv x \Leftarrow \gamma$ for each "missing case" $\gamma \in$ CM, provided that CM exists, i.e., predicates are available for specifying a complete case distinction. If the set of premises does not cover all ground substitutions, one may decompose it into a case matrix and a common condition ϑ and treat ϑ by *premise elimination* (cf. Sect. 4).

Constructor-based clause sets turned out to comprise a language used very frequently for writing functional and logic programs. From sorting algorithms via tree and graph manipulating functions up to interpreters, this language is powerful enough for bringing them into a concise *and* executable form (cf. [Pad87,88b,c,e]). The translation of a suitable sublanguage into PASCAL is described in [GHM88].

The following lemma is crucial for characterizing constructor-based inductive theorems (cf. Thm. 3.4). It says that each ground substitution satisfies at most one clause of a constructor-based clause set.

Lemma 3.3 *Let M be a constructor-based clause set, $p \Leftarrow \gamma, q \Leftarrow \delta \in M$ and f,g be ground substitutions such that $AX \vdash_C (\{p\} \cup \gamma)[f|IN+g] \cup \delta[f]$. Then $\gamma = \delta$ and $AX \vdash_C q[f]$.* ∎

Theorem 3.4 *A constructor-based clause set M is an inductive AX-theorem iff for all ground substitutions f there are $p \Leftarrow \gamma \in M$ and a substitution g such that $AX \vdash_C (\{p\} \cup \gamma)[f|IN+g]$.* ∎

Theorem 3.4 provides the basis for inductive proofs of constructor-based clause sets. In the next section, we turn from forward proofs using the cut calculus to backward proofs based on resolution and paramodulation.

4. Inductive Expansion

Derivations via the cut calculus proceed bottom-up from axioms to the theorems to be proved. In contrast, *resolution* and *paramodulation* work top-down from a goal by applying axioms backwards until the empty goal is achieved, indicating that the initial goal is *solvable*. A solution is built up stepwisely in the course of the proof. We call such a derivation an *expansion* in order to stress the "procedural interpretation" of Horn clauses underlying this kind of proof.

For guaranteeing the completeness of paramodulation it is well-known that in some (rare) cases *functionally-reflexive* axioms of the form $Fx \equiv Fx$ must be applied. In [Pad88a], Chapter 5, we have shown that these additional axioms need only occur as superterms of instances of other axioms. Hence, instead of adding all functionally-reflexive axioms to AX, we replace AX by the set of *prefixed axioms of AX*.

Definition *Pre(AX)*, the set of *prefixed axioms of AX*, is the smallest set of clauses, which contains all conditional equations of AX and satisfies the following closure property:

- If $u \equiv u' \Leftarrow \vartheta \in Pre(AX)$ and t is a term of the form $F(x_1,...,x_n)$ such that $sort(x_i) = sort(u)$ for some $1 \le i \le n$, then $t[u/x_i] \equiv t[u'/x_i] \Leftarrow \vartheta \in Pre(AX)$. ∎

Definition The *expansion calculus* consists of three rules (given below) for transforming pairs consisting of a goal and a substitution. We assume that the variables of a goal subjected to a derivation step belong to a set *GV* of variables, which do not occur in axioms. If the step brings axiom variables into the goal, they must be renamed as variables of GV before the derivation continues.

Resolution Rule Let γ be a goal, p be an atom, $q \Leftarrow \vartheta \in AX$, f be a substitution and g be a unifier of p and q. Then $$\langle \gamma \cup \{p\}, f \rangle \vdash \langle (\gamma \cup \vartheta)[g], f[g]|GV \rangle.$$

Paramodulation Rule Let δ be a goal, $x \in$ single(δ), t be a term, $u \equiv u' \Leftarrow \vartheta$ (or $u' \equiv u \Leftarrow \vartheta$) $\in$ Pre(AX), f be a substitution and g be a unifier of t and u. Then

$$<\delta[t/x], f> \;\vdash\; <(\delta[u'/x] \cup \vartheta)[g], f[g]|GV>.$$

Unification Rule Let γ be a goal, f be a substitution and g be a unifier of terms t and t'. Then

$$<\gamma \cup \{t \equiv t'\}, f> \;\vdash\; <\gamma[g], f[g]>.$$

An *expansion* is a sequence $<\gamma_1,f_1>,...,<\gamma_n,f_n>$ of goal-substitution pairs such that for all $1 \le i < n$, $<\gamma_{i+1},f_{i+1}>$ is obtained from $<\gamma_i,f_i>$ by applying a rule of the expansion calculus.

$\vdash_{EX}$ denotes the corresponding inference relation. ∎

Theorem 4.1 ([Pad88a], Thm. 5.3.5) *Let γ be a goal and f be a substitution such that var(γ) $\cup$ dom(f) $\subseteq$ GV. Then $AX \vdash_C \gamma[f]$ if and only if $<\gamma, id> \vdash_{EX} <\emptyset, f>$.* ∎

$\vdash_{EX}$ uses only (prefixed) *axioms* to resolve or paramodulate upon. Cor. 2.3 allows us to apply *lemmas* as well, i.e., Pre(AX) can be extended to the set ITh(AX) of all inductive AX-theorems. For ground terms, Thm. 4.1 remains valid:

Corollary 4.2 *Let γ be a goal and f be a ground substitution such that var(γ) $\cup$ dom(f) $\subseteq$ GV. Then $AX \vdash_C \gamma[f]$ iff $<\gamma, id> \vdash_{EX} <\emptyset, f>$.* ∎

Suppose we have a set of expansions

$$<\gamma,id> \vdash_{EX} <\emptyset,g_1>,$$
$$<\gamma,id> \vdash_{EX} <\emptyset,g_2>,$$
$$...$$

such that each ground substitution f is subsumed by some g_i. Then, by Cor. 4.2, γ is an inductive theorem. Instead of expanding $<\gamma,id>$ into the empty goal one may stop in a situation like

$$<\gamma,id> \vdash_{EX} <\delta_1,g_1>,$$
$$<\gamma,id> \vdash_{EX} <\delta_2,g_2>,$$
$$...$$

where $<\delta_1,g_1>,<\delta_2,g_2>,...$ represents a *ground complete* case distinction.

Definition A set GS of goal-substitution pairs is *ground complete* if for all ground substitutions f there are $<\delta,g> \in$ GS and a substitution h such that $AX \cup EAX \vdash \delta[h]$ and $g[h]|IN$ is AX-equivalent to $f|IN$. ∎

The combination of Cor. 4.2 with the characterization of constructor-based clause sets (Thm. 3.4) leads to

Corollary 4.3 *Let M be a constructor-based clause set such that for all $p \Leftarrow \gamma \in M$ there is a ground substitution f with $AX \vdash_C \gamma[f]$. M is an inductive AX-theorem iff there is a countable set of expansions*

$$<\{p_1\} \cup \gamma_1, id> \vdash_{EX} <\delta_1,g_1>,$$
$$<\{p_2\} \cup \gamma_2, id> \vdash_{EX} <\delta_2,g_2>,$$
$$...$$

such that $M = \{p_1 \Leftarrow \gamma_1, p_2 \Leftarrow \gamma_2,...\}$ and $\{<\delta_1,g_1>,<\delta_2,g_2>,...\}$ is ground complete. ∎

For checking the ground completeness of $\{<\delta_1,g_1>,<\delta_2,g_2>,...\}$ one may, again, refer to case matrices:

Proposition 4.4 A finite set $\{<\delta_1,g_1>,...,<\delta_k,g_k>\}$ of goal-substitution pairs is ground complete if the set

$$\delta_1 \cup \{x \equiv g_1x \mid x \in IN\}$$
$$...$$
$$\delta_k \cup \{x \equiv g_kx \mid x \in IN\}$$

is extendable to a case matrix. ∎

With Prop. 4.4, the ground completeness of a set of goal-substitution pairs is reduced to the ground completeness of term sets (cf. Sect. 5).

So far, the proof procedure involved in Cor. 4.3 does not employ induction steps. Consequently, infinitely many expansions will often be needed in order to obtain a ground complete set of goal-substitution pairs. As in corresponding forward proofs, only the explicit use of induction hypotheses may reduce the search space to a finite proof tree. But how do induction hypotheses enter the expansion calculus?

In principle, the idea is as classical as the step from bottom-up derivations to top-down expansions. We find it, for instance, in Manna and Waldinger's deductive tableaus used for program synthesis (cf. [MW80], [MW87]), especially in the "formation of recursive calls". It amounts to including Noetherian relations into the specification, which allow us to distinguish certain instances of a clause as induction hypotheses.

Definition A binary relation R on a set A is *Noetherian* or *well-founded* if there are no infinite sequences $a_1,a_2,a_3,\ldots$ of elements of A such that for all $i\geq 1$, $\langle a_i,a_{i+1}\rangle \in R$. ∎

Here we are interested in relations on GT(SIG) which arise from a binary predicate >>, being part of the specification (SIG,AX).

Definition Let $s \in S$ and $>> \in PR_{ss}$ (cf. Sect. 2). Then

$$R(>>) = \{(t,t') \in GT(SIG)^2 \mid AX \vdash_C t >> t'\}. ∎$$

The Noetherian property of R(>>) can be reduced to *one* of its interpretations:

Proposition 4.5 *R(>>) is Noetherian iff there is a SIG-model A of AX such that $>>^A$, the interpretation of >> on A, is Noetherian.* ∎

General Assumption (part 2; cf. Sect. 2) We order a subset of IN, say $\{z_1,\ldots,z_n\}$, into a sequence, say $z = (z_1,\ldots,z_n)$, and assume a predicate symbol $>> \in PR_{ss}$ such that R(>>) is Noetherian.

For avoiding name clashes we also use a primed copy of $\{z_1,\ldots,z_n\}$. So let $z' = (z_1',\ldots,z_n')$, and for all clause sets M, let M' be M with all variables replaced by their primed counterparts. ∎

R(>>) is compatible with AX-equivalence: If $AX \vdash_C \{t>>t', t\equiv u, t'\equiv u'\}$, then by congruence axioms for $\equiv$, $AX \vdash_C u>>u'$. In particular, $AX \vdash_C \{t>>t', t\equiv t'\}$ implies $AX \vdash_C t'>>t'$, which means that R(>>) can only be well-founded if it is *disjoint* from AX-equivalence. Therefore, R(>>) cannot agree with a reduction ordering needed for inductive completion (cf. Sects. 1 and 7): a reduction ordering *contains* an "oriented" version of AX-equivalence. This does not contradict the fact that the definition of a reduction ordering may use (parts of) the "semantic" relation R(>>) (cf. the semantic path orderings in [Der87a]).

Semantic relations, which are compatible with AX-equivalence, on the one hand and reduction orderings on the other hand are employed for different purposes. The former are a means for ensuring that inductive proofs of semantic properties are sound. The latter guarantee a purely syntactic condition: the well-foundedness of rewrite sequences.

Now think of a *forward* proof of $q \Leftarrow \vartheta$ by using induction hypotheses. Usually, one reduces the set of all ground substitutions to a finite *covering*, say $\{f_1,\ldots,f_n\}$, presupposes the validity of all premise instances $\vartheta[f_i]$ and infers the corresponding conclusion instances $q[f_i]$. In the course of deriving $q[f_i]$ from $\vartheta[f_i]$, an induction step replaces a ground instance of ϑ, say $\vartheta[g]$, by $q[g]$, provided that gz is "less than" f_iz (see the General Assumption). In other words, the clause

$$q \Leftarrow \vartheta \cup \{f_iz>>z\} \qquad (*)$$

is regarded as an additional axiom: $q[g]$ is the result of cutting (*) with $\vartheta[g]$ and $f_iz>>gz$. Indeed, (*) represents an induction hypothesis.

The forward proof will succeed only if a suitable covering $\{f_1,\ldots,f_n\}$ has been guessed and if no *generalization* is needed, i.e., if $q \Leftarrow \vartheta$ is strong enough for generating induction hypotheseses. The backward proof, on the other hand, which proceeds by resolution and paramodulation on axioms, lemmas and induction hypotheses leads more or less automatically both to a covering and to necessary generalizations.

Definition Let M be a clause set. The *inductive expansion calculus (for M)* consists of the expansion calculus and two additional rules:

Inductive Resolution Rule Let γ be a goal, p be an atom, $q \Leftarrow \vartheta \in M'$, f be a substitution and g be a unifier of p and q. Then

$$\langle \gamma \cup \{p\}, f \rangle \quad \vdash \quad \langle (\gamma \cup \vartheta \cup \{fz >> z'\})[g], f[g] \rangle.$$

Inductive Paramodulation Rule Let δ be a goal, $x \in single(\delta)$, t be a term, $u \equiv u' \Leftarrow \vartheta$ (or $u' \equiv u \Leftarrow \vartheta$) $\in M'$, f be a substitution and g be a unifier of t and u. Then

$$\langle \delta[t/x], f \rangle \quad \vdash \quad \langle (\delta[u'/x] \cup \vartheta \cup \{fz >> z'\})[g], f[g] \rangle.$$

An application of the Inductive Resolution or Paramodulation Rule is called an *M-induction step*. An M-induction step is *closed* if the output variables of the hypothesis resolved or paramodulated upon are regarded as constants (and thus prevented from subsequent instantiations).

An *inductive M-expansion* is a sequence $<\gamma_1,f_1>,...,<\gamma_n,f_n>$ of goal-substitution pairs such that for all $1 \le i < n$, $<\gamma_{i+1},f_{i+1}>$ is obtained from $<\gamma_i,f_i>$ by applying a rule of the inductive expansion calculus. If all M-induction steps in the sequence are closed, the expansion is called a *closed inductive M-expansion*.

$\vdash_{EX(M)}$ denotes the corresponding inference relation. ∎

Sometimes several clauses can only be proved by simultaneous induction. Therefore let us generalize clauses to formulas $\psi \Leftarrow \gamma$ where ψ *and* γ are goals. $\psi \Leftarrow \gamma$ stands for the union of all $p \Leftarrow \gamma$ over all $p \in \psi$. As before, we use Greek letters for goals and small Latin letters for atoms.

The question remains whether *inductive* M-expansions are sound. As the reader might expect, this can be proved by Noetherian induction with respect to R(>>).

Lemma 4.6 *Let M be a constructor-based clause set. If for all ground substitutions f there are $\psi \Leftarrow \gamma \in M$, a substitution g and an inductive expansion $\langle \psi \cup \gamma, id \rangle \vdash_{EX(M)} \langle \emptyset, g \rangle$ such that g/IN and f/IN are AX-equivalent, then M is an inductive AX-theorem.* ∎

Of course, Lemma 4.6 does not *characterize* the set of those constructor-based clause sets which are inductive theorems. The inductive rules involved in $\vdash_{EX(M)}$ depend on the predicate >>. Instead, the important fact we conclude from Lemma 4.6 is the possibility of carrying out induction steps in backward proofs as well as in forward proofs, with the aim of achieving a *finite* proof. Moreover, backward induction improves over forward induction because it leads to linear proofs without any second-order arguments.

Yet we must cope with the restriction to constructor-based clause sets M, in particular with the requirement that the set of premises of M be a case matrix. In turn, this implies that the predicates used in the case matrix must be specified completely, the positive as well as the negative cases.

In fact, the restriction can be weakened. We can also handle conditional clause sets of the form $M \Leftarrow \vartheta$ where M is a constructor-based clause set, ϑ is an input goal, i.e., $var(\vartheta) \subseteq IN$, and $M \Leftarrow \vartheta$ stands for the set of all clauses $\psi \Leftarrow \gamma \cup \vartheta$ with $\psi \Leftarrow \gamma \in M$. The proof of $M \Leftarrow \vartheta$ proceeds as an inductive M-expansion, with possible applications of the following inference rule:

Premise Elimination Rule Let γ be a goal and f be a substitution. Then for all $\delta \subseteq \vartheta$,

$$\langle \gamma \cup \delta[f], f \rangle \quad \vdash \quad \langle \gamma, f \rangle.$$

As an immediate consequence of Lemma 4.6, Cor. 4.3 holds true for $\vdash_{EX(M)}$ as well as for $\vdash_{EX}$. Moreover, Prop. 4.4 provides a criterion for checking the ground completeness of the final set $\{<\delta_1,g_1>,<\delta_2,g_2>,...\}$ of goal-substitution pairs. In summary, this yields

Theorem 4.7 *Let M be a constructor-based clause set. M (or $M \Leftarrow \vartheta$; see above) is an inductive AX-theorem if there is a finite set of expansions*

$$\langle \psi_1 \cup \gamma_1, id \rangle \vdash_{EX(M)} \langle \delta_1, g_1 \rangle,$$

$$\ldots$$

$$\langle \psi_n \cup \gamma_n, id \rangle \vdash_{EX(M)} \langle \delta_n, g_n \rangle$$

such that $M = \{\psi_1 \Leftarrow \gamma_1, ..., \psi_n \Leftarrow \gamma_n\}$ and

$$\delta_1 \cup \{x \equiv g_1 x \mid x \in IN\},$$
$$\ldots$$
$$\delta_n \cup \{x \equiv g_n x \mid x \in IN\}$$

is extendable to a case matrix, called the derived covering. ∎

Example 4.8 Two equations capture the correctness of insertion sort as specified in Section 2: T1 says that *sort* returns a sorted sequence. T2 ensures that the sorted sequence is a permutation of the original one.

sorted(sort(s)) (T1)
seqToBag(sort(s)) ≡ seqToBag(s) (T2)

With IN = {s}, {{T1,T2}} is a constructor-based clause set. One obtains three inductive {{T1,T2}}-expansions using two lemmata, namely:

sorted(insert(x,s)) ⇐ sorted(s) (L1)
seqToBag(insert(x,s)) ≡ seqToBag(x&s) (L2)

	goal	*substitution*	*axioms and lemmas applied*
1	sorted(sort(s)) (T1) seqToBag(sort(s)) ≡ seqToBag(s) (T2)		
1.1	sorted(sort(ε)) seqToBag(sort(ε)) ≡ seqToBag(ε)	ε/s	
	sorted(ε) seqToBag(ε) ≡ seqToBag(ε)		IS1
	∅		SO1, unification
1.2	sorted(sort(x&s')) seqToBag(sort(x&s')) ≡ seqToBag(x&s')	x&s'/s	
	sorted(insert(x,sort(s'))) seqToBag(insert(x,sort(s'))) ≡ seqToBag(x&s')		IS2 IS2
	sorted(sort(s')) seqToBag(x&sort(s')) ≡ seqToBag(x&s')		L1 L2
	x&s' >> s' seqToBag(x&sort(s')) ≡ seqToBag(x&s')		T1 as induction hypothesis
	add(x,seqToBag(sort(s'))) ≡ add(x,seqToBag(s'))		GR, BA2
	x&s' >> s' add(x,seqToBag(s')) ≡ add(x,seqToBag(s'))		T2 as induction hypothesis
	∅		GR, unification

The first induction step in expansion 1.2 applies the Inductive *Resolution* Rule, while the second one is an application of the Inductive *Paramodulation* Rule: T2 is applied from left to right to the subterm seqToBag(sort(s')). The covering derived by expansions 1.1 and 1.2 is the set {{s≡ε}, {s≡x&s'}}, which is a case matrix because {ε,x&s'} is ground complete for s (cf. Ex. 5.4).

With IN = {x,s}, the following inductive {{sorted(insert(x,s))}}-expansions yield a proof of L1.

2	sorted(insert(x,s))		
2.1	sorted(insert(x,ε))	ε/s	
	sorted(x&ε)		IN1
	∅		SO2
2.2	sorted(insert(x,y&s'))	y&s'/s	
2.2.1	sorted(x&y&s') x ≤ y		IN2
	sorted(y&s') x ≤ y		SO3
	x ≤ y		premise elimination
	le(x,y) ≡ true		LE1
2.2.2	sorted(y&insert(x,s')) x > y		IN3
2.2.2.1	sorted(y&insert(x,ε)) x > y	ε/s'	
	sorted(y&x&ε) x > y		IN1
	sorted(x&ε) y ≤ x, x > y		SO2
	y ≤ x, x > y		SO1
	x > y		y ≤ x ⇐ x > y
	le(x,y) ≡ false		LE2
2.2.2.2	sorted(y&insert(x,z&s")) x > y	z&s"/s'	
2.2.2.2.1	sorted(y&x&z&s") x ≤ z, x > y		IN2
	sorted(x&z&s") y ≤ x, x ≤ z, x > y		SO3

	sorted(x&z&s") x ≤ z, x > y	y ≤ x ⇐ x > y
	sorted(z&s") x ≤ z, x > y	SO3
	sorted(y&z&s") x ≤ z, x > y	sorted(z&s) ⇐ sorted(y&z&s) (L3)
	x ≤ z, x > y	premise elimination
	le(x,z) ≡ true le(x,y) ≡ false	LE1 LE2
2.2.2.2.2	sorted(y&z&insert(x,s")) x > z, x > y	IN3
	sorted(z&insert(x,s")) y ≤ z, x > z, x > y	SO3
	sorted(insert(x,z&s")) y ≤ z, x > z, x > y	IN3 (from right to left)
	sorted(z&s") z&s" >> s" y ≤ z, x > z, x > y	L1 as induction hypothesis
	sorted(z&s") y ≤ z, x > z, x > y	GR
	sorted(y&z&s") y ≤ z, x > z, x > y	sorted(z&s) ⇐ sorted(y&z&s) (L3)
	sorted(y&z&s") x > z, x > y	y ≤ z ⇐ sorted(y&z&s) (L4)
	x > z, x > y	premise elimination
	le(x,z) ≡ false le(x,y) ≡ false	LE2 LE2

The covering derived by these expansions is

s ≡ ε			(2.1)
s ≡ y&s'	le(x,y) ≡ true		(2.2.1)
s ≡ y&ε	le(x,y) ≡ false		(2.2.2.1)
s ≡ y&z&s"	le(x,y) ≡ false	le(x,z) ≡ true	(2.2.2.2.1)
s ≡ y&z&s"	le(x,y) ≡ false	le(x,z) ≡ false	(2.2.2.2.2)

It is extendable to a case matrix by replacing s' with ε and z&s", respectively. Note that lemmas L3 and L4 constitute the inverse of SO3. They are inductive theorems, but do not hold in all term-generated models of SORT. As inverses of an axiom, L3 and L4 are consequences of the *closed world assumption* (cf. Sect. 1).

Finally, L2 is proved by inductive {{L2}}-expansions:

3	seqToBag(insert(x,s)) ≡ seqToBag(x&s) (L2)		
3.1	seqToBag(insert(x,ε)) ≡ seqToBag(x&ε)	ε/s	
	seqToBag(x&ε) ≡ seqToBag(x&ε)		IN1
	∅		unification
3.2	seqToBag(insert(x,y&s')) ≡ seqToBag(x&y&s')	y&s'/s	
3.2.1	seqToBag(x&y&s') ≡ seqToBag(x&y&s') x ≤ y		IN2
	x ≤ y		unification
	le(x,y) ≡ true		LE1
3.2.2	seqToBag(y&insert(x,s')) ≡ seqToBag(x&y&s') x > y		IN3
	add(y,seqToBag(insert(x,s'))) ≡ add(x,seqToBag(y&s')) x > y		BA2
	y&s' >> s' add(y,seqToBag(x&s')) ≡ add(x,seqToBag(y&s')) x > y		L2 as induction hypothesis
	add(y,add(x,seqToBag(s'))) ≡ add(x,add(y,seqToBag(s'))) x > y		GR, BA2
	x > y		BA3
	le(x,y) ≡ false		LE2

The covering derived by these expansions is a case matrix, namely:

s ≡ ε		(3.1)
s ≡ y&s'	le(x,y) ≡ true	(3.2.1)
s ≡ y&s'	le(x,y) ≡ false	(3.2.2) ■

5. How to Prove the Ground Completeness of Term Sets

Theorem 4.7 provides a proof method for inductive theorems where case matrices are presupposed both at the beginning and at the end of the proof; at the beginning because we have to start out from a constructor-based clause set the important property of which is that its premises constitute a (constructor-based) case matrix; at the end because the final goal-substitution pairs must correspond to a (not necessarily constructor-based) case matrix.

Apart from syntactic conditions, a set of goals is a case matrix if it is built up from ground complete sets of output terms. As we mentioned in Section 3, the constructor condition can be derived immediately from the Church-Rosser property of AX. Ground completeness, however, is a condition that needs its own proof methods.

Theorem provers devote a considerable amount of work to checking that the functions used have been defined completely (cf. [BM79], [Hut86]). At first sight, this does not seem to be necessary for proving theorems. But most proofs are carried out by case reasoning and thus the question arises whether a case distinction is complete. When it is presented as a case matrix CM, the question just amounts to whether the right-hand sides of CM-equations "cover" the left-hand sides. This leads to a new verification problem where induction is needed again. However, one may run into a cycle when *this* proof is also based on a case distinction. The problem can be overcome by expressing *these* cases on a "lower level", in terms of a particular set of *base* terms. In consequence, ground terms must be *base-representable,* which is indeed a sort of functional completeness. (For dealing with partial functions, non-base-representable terms are admitted, too. However, for simplifying the presentation, we do not consider such cases here.)

Moreover, ground completeness is an *existential* statement.

Both deviations from the kind of theorems considered in previous sections call for a particular method for proving ground completeness.

Definition Let *BOP* $\subseteq$ OP be a set of *base operations. GBT* denotes the set of ground *base* terms, i.e. ground terms over BOP. We assume that for all s $\in$ S, GBT_s is nonempty. A ground term Ft is *innermost* if F $\notin$ BOP and t is a base term (tuple).

A term t is *base-representable* if for all f $\in$ GBT^X there is a base term that is AX-equivalent to t[f]. *BR* denotes the set of base-representable terms. Subsets of BR are called base-representable sets. A substitution is base-representable if f(X) is base-representable. ∎

Base-representability can be expressed in terms of a base existential theorem, i.e. a goal with existentially quantified variables:

Definition A goal ψ is a *base existential theorem* if for all f $\in$ GBT^X there is g $\in$ GBT^X such that $AX \vdash_C \psi[f|IN+g]$. ∎

Proposition 5.1 *Let GEN be a set of input terms such that each innermost term is subsumed by some t* $\in$ *GEN. GBT is ground complete (cf. Sect. 3) or, equivalently, GEN is base-representable, if for some output variables* $x_1,...,x_n$, $\psi(GEN) = \{t_i \equiv x_i \mid 1 \le i \le n\}$ *is a base existential theorem.* ∎

The analogue of Lemma 4.6 for proving base existential theorems reads as follows.

Lemma 5.2 *Let* ψ *be a goal. If for all f* $\in$ GBT^X *there are g* $\in$ BR^X *and a closed inductive M-expansion* $\langle\psi,id\rangle \vdash_{EX(\psi)} \langle\emptyset,g\rangle$ *such that f|IN and g|IN are AX-equivalent, then* ψ *is a base existential theorem.* ∎

While Lemma 4.6 establishes the correctness of inductive expansion w.r.t. *universally* quantified clauses, the previous lemma deals with existential theorems and thus requires *closed* expansions where the existentially quantified output variables of induction hypotheses are not instantiated. This is necessary because an induction hypothesis assures a validating instantiation, but in general not the one *non*-closed expansions would generate.

A second deviation of Lemma 5.2 from Lemma 4.6 concerns the range of substitutions. Since 5.2 deals with *base* theorems and with expansions into a *base* case matrix (see below), the given substitution f is a base substitution and the derived substitution g must be base-representable.

Finite coverings of GBT^X should be given as base case matrices:

Definition A set C of terms is *ground base complete* if each ground base term is subsumed by some c $\in$ C. A finite set CM of finite sets of equations is a *base case matrix with input* IN_0 $\subseteq$ X if either CM = $\{\emptyset\}$ or CM = $\{\{t \equiv c_i\} \cup \gamma \mid 1 \le i \le n, \gamma \in CM_i\}$ for a base-representable term t, a ground base complete set $\{c_1,...,c_n\}$ of output base terms, and base case matrices $CM_1,...,CM_n$ with input $IN_0 \cup var(c_1)$, ..., $IN_0 \cup var(c_n)$, respectively, such that var(t) $\subseteq$ IN_0 and for all $1 \le i \le n$, $var(c_i) \cap IN_0 = \emptyset$. ∎

The following result is concluded from Lemma 5.2 just as Theorem 4.7 is derived from Lemma 4.6.

Theorem 5.3 *Let GEN be a set of input terms and* $\psi = \psi(GEN)$ *(cf. Prop. 5.1) such that each innermost term is subsumed by some t* $\in$ *GEN. GBT is ground complete or, equivalently, GEN is base-representable, if there is a finite set of closed inductive* ψ*-expansions*

$$\langle\psi,id\rangle \vdash_{EX(\psi)} \langle\delta_1,g_1\rangle,$$
$$\cdots$$
$$\langle\psi,id\rangle \vdash_{EI(\psi)} \langle\delta_n,g_n\rangle$$

such that $g_1,...,g_n \in BR^X$ *and*

$$\delta_1 \cup \{x{\equiv}g_1x \mid x \in IN\}$$
$$\cdots$$
$$\delta_n \cup \{x{\equiv}g_nx \mid x \in IN\}$$

is a base case matrix, called the derived covering. ∎

Example 5.4 When claiming that the coverings derived in Example 4.8 are case matrices we have assumed that the sets C1 = {ε,y&s'}, C2 = {ε,x&ε,x&y&s'} and C3 = {true,false} are ground complete for the terms s and le(x,y), respectively. For justifying this statement with the help of Theorem 5.3 we choose {true, false, 0, _+1, ε, _&_, ∅, add(_,_)} as the set BOP of base operations. Innermost terms are, for instance, given by sorted(0&ε) and insert(0+1,ε).

Of course, if GBT is ground complete, then, in particular, C1, C2 and C3 are ground complete for s and le(x,y), respectively, as required. Suppose that the base-representability of GEN0 = {le(x,y), seqToBag(s)} has already been shown. Let GEN1 = {insert(x,s), sort(s)}. Since each innermost term is subsumed by some t ∈ GEN0 ∪ GEN1, GBT is ground complete if and only if GEN1 is base-representable. Hence by Thm. 5.3, it is sufficient to construct closed inductive expansions of T1 = {insert(x,s)≡s_o} and T2 = {sort(s)≡s_o}.

	goal	*substitution*	*axioms and lemmas applied*
1	insert(x,s)≡s_o (T1)		
1.1	∅	ε/s, x&ε/s_o	IN1
1.2	x ≤ y	y&s'/s, x&y&s'/s_o	IN2
	le(x,y) ≡ true		LE1
1.3	y&insert(x,s') ≡ s_o x > y	y&s'/s	IN3
	y&s_1 ≡ s_o y&s' >> s' le(x,y) ≡ false		T1 as induction hypothesis LE2
	le(x,y) ≡ false	y&s_1/s_o	unification, GR

The derived covering is the base case matrix {{s≡ε}, {s≡y&s', le(x,y)≡true}, {s≡y&s', le(x,y)≡false}}. (By assumption, le(x,y) ∈ GEN0 is base-representable.) By Thm. 5.3, expansions 1.1-1.3 imply that insert(x,s) is base-representable. Note that the induction step in expansion 1.3 is closed because the (output) variable s_1 is not replaced later on.

2	sort(s)≡s_o (T2)		
2.1	∅	ε/s, ε/s_o	IS1
2.2	insert(x,sort(s')) ≡ s_o	x&s'/s	IS3

insert(x,s$_1$) ≡ s$_0$		T2 as induction hypothesis
∅	insert(x,s$_1$)/s$_0$	unification, GR

The derived covering is the base case matrix is {(s≡ε), (s≡x&s')}. Since insert(x,s$_1$) is base-representable, we conclude from Thm. 5.3 that sort(s) is base-representable, too. ∎

6. Conclusion

We have presented a calculus for proving inductive theorems by resolving and paramodulating upon axioms, lemmas and induction hypotheses. The theorems must be given as constructor-based sets of Horn clauses. The derivations end up with a ground complete set of goal-substitution pairs. As a criterion for ground completeness, we introduced the notion of a case matrix, which reduces the completeness requirement from goal-substitution pairs to term sets. Constructors and ground complete term sets are the only non-syntactical notions associated with inductive expansion. As to ground completeness, we have shown in Section 5 how this property can be proved with the help of closed inductive expansions. As to constructors, one may refer to *goal reduction,* which extends term rewriting to a rule for transforming goals:

> *Reduction Rule* Let δ be a goal, x ∈ single(δ), u≡u'⇐ϑ ∈ AX and f be a substitution. Then
> δ[u[f]/x] ⊢ δ[u'[f]/x]∪ϑ[f].

A goal reduction stops successfully if a goal consisting of reflexive equations has been obtained:

> *Success Rule* Let γ be a goal consisting of equations of the form t≡t such that the Reduction Rule is not applicable to γ. Then γ ⊢ ∅.

$\vdash_R$ denotes the corresponding inference relation. AX is called ***Church-Rosser*** if all proofs using the cut calculus have a "reduction counterpart", i.e., AX $\vdash_C$ γ implies γ $\vdash_R$ ∅. The literature is full of criteria for the Church-Rosser property (cf., e.g., [Pad88a]). It yields the following constructor criterion: If AX is Church-Rosser on ground goals and no left-hand side of a conditional equation of AX "overlaps" a term of a term set T, then T is a set of constructors.

Another consequence of the Church-Rosser property is the possibility of restricting paramodulation to the more effective rule of ***narrowing***, invented by [Lan75]. Indeed, the crucial Lemma 4.6 could also be based upon rules different from resolution and paramodulation, provided that they are complete in the sense of Thm. 4.1 (perhaps only for a particular class of specifications, like Church-Rosser ones.) More detailed suggestions concerning this line of developing inductive proof methods are given in [Pad88e]. It must be noted, however, that every restriction of the inference rules might prevent induction hypotheses from being generated. For instance, narrowing does not admit applying an equation from right to left as we did in expansion 2.2.2.2.2 of Example 4.8. But this application was necessary for proceeding with an L1-induction step.

Inductive completion, the current alternative for proving inductive theorems (cf. Sect. 1), is also based upon the Church-Rosser property. In spite of the resemblance between narrowing steps and the basic steps of inductive completion, i.e., the construction of ***critical pairs*** (pointed out in [Der87b], Sect. 4.2), there is an important difference between "inductive narrowing" and inductive completion. In the latter case, the Church-Rosser property is extended from the axioms to the conjecture that is to be proved. In fact, sophisticated Church-Rosser criteria take into account the special role of the conjecture (cf., e.g., [Fri86], [Küc87], [HK88], [Pad88d]). Nevertheless, many examples have shown that the remaining conditions are more difficult to establish than the constructor-based clause set requirement .

References

[BM79] R.S. Boyer, J.S. Moore, A Computational Logic, Academic Press (1979)

[Bur69] R.M. Burstall, Proving Properties of Programs by Structural Induction, Comp. J. 12 (1969) 41-48

[Der87a] N. Dershowitz, Termination of Rewriting, J. Symbolic Comp. 3 (1987) 69-115

[Der87b] N. Dershowitz, Completion and its Applications, Report (1987)

[Fri 86] L. Fribourg, A Strong Restriction of the Inductive Completion Procedure, Proc. ICALP '86, Springer LNCS 226 (1986) 105-115

[GG88] S.J. Garland, J.V. Guttag, Inductive Methods for Reasoning about Abstract Data Types, Proc. POPL '88 (1988) 219-228

[GHM88] A. Geser, H. Hußmann, A. Mück, A Compiler for a Class of Conditional Term Rewriting Systems, Proc. Conditional Term Rewriting Systems '87, Springer LNCS 308 (1988) 84-90

[HH82] G. Huet, J.M. Hullot, Proofs by Induction in Equational Theories with Constructors, J. Comp. and Syst. Sci. 25 (1982) 239-266

[HK88] D. Hofbauer, R. Kutsche, Proving Inductive Theorems Based on Term Rewriting Systems, Proc. Algebraic and Logic Programming, Math. Research 49, Akademie-Verlag Berlin (1988) 180-190

[HR87] J. Hsiang, M. Rusinowitch, On Word Problems in Equational Theories, Proc. ICALP '87, Springer LNCS 267 (1987) 54-71

[Hut86] D. Hutter, Using Resolution and Paramodulation for Induction Proofs, Proc. 10th GWAI, Springer Informatik-Fachberichte 124 (1986) 265-276

[JK86] J.-P. Jouannaud, E. Kounalis, Automatic Proofs by Induction in Equational Theories without Constructors, IEEE Symp. Logic in Comp. Sci. (1986) 358-366

[KM87] D. Kapur, D.R. Musser, Proof by Consistency, Artificial Intelligence 31 (1987) 125-157

[Küc87] W. Küchlin, Inductive Completion by Ground Proof Transformation, Proc. Resolution of Equations in Algebraic Structures, Austin (1987)

[Lan75] D.S. Lankford, Canonical Inference, Report ATP-32, Univ. of Texas at Austin (1975)

[MW80] Z. Manna, R. Waldinger, A Deductive Approach to Program Synthesis, ACM TOPLAS 2 (1980) 90-121

[MW87] Z. Manna, R. Waldinger, How to Clear a Block: A Theory of Plans, J. Automated Reasoning 3 (1987) 343-377

[Pad87] P. Padawitz, ECDS - A Rewrite Rule Based Interpreter for a Programming Language with Abstraction and Communication, Report MIP-8703, Univ. Passau (1987)

[Pad88a] P. Padawitz, Computing in Horn Clause Theories, EATCS Monographs on Theor. Comp. Sci. 16, Springer (1988)

[Pad88b] P. Padawitz, Can Inductive Proofs be Automated? EATCS Bulletin 35 (1988) 163-170

[Pad88c] P. Padawitz, Program Verification Revisited (1988), submitted

[Pad88d] P. Padawitz, Proof by Consistency of Conditional Equations (1988), submitted

[Pad88e] P. Padawitz, Reduction and Narrowing for Horn Clause Theories (1988), submitted

[Rei78] R. Reiter, On Closed World Data Bases, in: H. Gallaire, J. Minker, eds., Logic and Data Bases, Plenum Press, New York (1978) 55-76

[Rob65] J.A. Robinson, A Machine-Oriented Logic Based on the Resolution Principle, J. ACM 12 (1965) 23-41

[RW69] G. Robinson, L. Wos, Paramodulation and Theorem-Proving on First-Order Theories with Equality, in: Machine Intelligence 4, Edinburgh Univ. Press (1969) 135-150

Local Model Checking in the Modal Mu-Calculus

Colin Stirling and David Walker
Department of Computer Science
University of Edinburgh
Edinburgh EH9 3JZ, U.K.

1 Motivation

The modal mu-calculus, due to Pratt and Kozen [Pr, Ko], is a natural extension of dynamic logic. It is also one method of obtaining a branching time temporal logic from a modal logic [EL]. Furthermore, it extends Hennessy-Milner logic, thereby offering a natural temporal logic for Milner's CCS, and process systems in general. (Discussion of the uses of the mu-calculus for CCS can be found in [GS,Ho,La,St,Sti2].) Within this context we are especially interested in whether or not a particular state, or process, in a finite model satisfies a mu-calculus formula. This is a different enterprise from that addressed by Emerson and Lei [EL] who ask if a given formula is satisfiable in a given finite model. Their model checker appeals to standard approximation techniques for computing the set of states which satisfy a fixpoint formula. But then one has to compute *all* the states or processes in the model which satisfy that formula.

In this paper we present a local model checker for the mu-calculus, as a tableau system. It checks whether or not a particular state satisfies a formula. Instead of using approximation techniques there is an implicit use of fixpoint induction (inspired by [La]). A maximal fixpoint formula, in effect, expresses a safety property. One shows that the assumption that a state has such a property leads to no unforeseen consequences. In contrast, a minimal fixpoint formula expresses a liveness property. Therefore one has to establish that the property holds of a particular state. Formulae involving alternating fixpoints [EL] introduce subtleties. However the resulting tableau system is natural and an equivalent version of it has been implemented by Rance Cleaveland [Cl].

In section 2 we describe the syntax and semantics of the modal mu-calculus. A small extension to the calculus, the addition of propositional constants, is detailed in section 3. The model checker, presented as a tableau system, is given in section 4, while the proofs of its soundness, completeness and decidability are the topic of section 6. Finally, in section 5 we use the model checker to analyse a mutual exclusion algorithm when translated into CCS.

2 The modal mu-calculus

The set of formulae of the modal mu-calculus is defined by:

$$A ::= Z \mid Q \mid \neg A \mid A \wedge A \mid [a]A \mid \nu Z.\, A$$

where Z ranges over propositional variables, Q over atomic propositions, and a over a set of (action) labels. One restriction on $\nu Z.\, A$ is that each free occurrence of Z in A lies within the scope of an even number of negations. Derived operators are defined in the familiar way: $A \vee B$ is $\neg(\neg A \wedge \neg B)$; $\langle a \rangle A$ is $\neg [a] \neg A$; and $\mu Z.\, A$ is $\neg \nu Z.\, \neg A[Z := \neg Z]$, where $A[Z := \neg Z]$ is the result of substituting $\neg Z$ for each free occurrence of Z in A.

The mu-calculus, with action labels drawn from a set *Act*, is interpreted on labelled transition systems $\mathcal{T}$ which are pairs of the form $\mathcal{T} = (S, \{\xrightarrow{a} \mid a \in Act\})$. S (or $S_{\mathcal{T}}$) is a nonempty set of states, and for each $a \in Act$, $\xrightarrow{a}$ is a transition relation on states. We write $s \xrightarrow{a} s'$ instead of $(s, s') \in \xrightarrow{a}$. Labelled transition systems are popular structures for modelling concurrent systems, [Mi, Pn], including process algebras such as CCS. S is then a set (or algebra) of processes and $s \xrightarrow{a} s'$ means that process s may become s' by preforming the action a. In this context the mu-calculus can be viewed as a branching time temporal logic for CCS, a natural extension of the modal logic in [HM].

A model $\mathcal{M}$ for the mu-calculus is a pair $\mathcal{M} = (\mathcal{T}, V)$ where $\mathcal{T}$ (or $\mathcal{T}_{\mathcal{M}}$) is a transition system and V (or $V_{\mathcal{M}}$) is a valuation assigning sets of states to atomic propositions and variables: $V(Q) \subseteq S_{\mathcal{T}}$ and $V(Z) \subseteq S_{\mathcal{T}}$. We assume the customary updating notation: $V[S'/Z]$ is the valuation V' which agrees with V except that $V'(Z) = S'$. Finally the set of states satisfying A in a model $\mathcal{M} = (\mathcal{T}, V)$ is inductively defined as $\| A \|_V^{\mathcal{T}}$ (where for ease of notation we drop the index $\mathcal{T}$ which is assumed to be fixed):

$$\begin{aligned}
\| Z \|_V &= V(Z) \\
\| Q \|_V &= V(Q) \\
\| \neg A \|_V &= S_{\mathcal{T}} - \| A \|_V \\
\| A \wedge B \|_V &= \| A \|_V \cap \| B \|_V \\
\| [a]A \|_V &= \{ s \in S_{\mathcal{T}} \mid \forall s'.\ \text{if } s \xrightarrow{a} s' \text{ then } s' \in \| A \|_V \} \\
\| \nu Z.A \|_V &= \bigcup \{ S' \subseteq S_{\mathcal{T}} \mid S' \subseteq \| A \|_{V[S'/Z]} \}
\end{aligned}$$

The expected clause for the derived operator $\mu Z.$ is:

$$\| \mu Z.A \|_V = \bigcap \{ S' \subseteq S_{\mathcal{T}} \mid \| A \|_{V[S'/Z]} \subseteq S' \}$$

A simple example is the model $\mathcal{M} = (\mathcal{T}, V)$ where $\mathcal{T}$ is

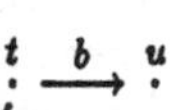

and $V(Q) = \emptyset$ for all atomic Q. Let R be the formula $\langle b \rangle$**true**. Let A and B be the formulae

$$\begin{aligned}
A &\equiv \nu Z.\mu Y.\langle a \rangle((R \wedge Z) \vee Y) \\
B &\equiv \mu Y.\nu Z.\langle a \rangle((R \vee Y) \wedge Z)
\end{aligned}$$

Now

$$\begin{aligned}
\| A \|_V^{\mathcal{T}} &= \{s, t\} \\
\| B \|_V^{\mathcal{T}} &= \emptyset
\end{aligned}$$

The formula A expresses that on some a^ω path R holds infinitely often, while B expresses that on some a^ω path R holds almost always. In CCS, where states are processes, u represents the process $\mathbf{0}$ (*Nil*) which can preform no actions, while s and t are the processes

$$\begin{aligned}
s &= \text{fix}Z.\,a.\,(b.\mathbf{0} + a.\,Z) \\
t &= \text{fix}Z.\,b.\mathbf{0} + a.\,a.\,Z
\end{aligned}$$

Hence both processes s and t have the property expressed by A.

A model is *finite* if its set of states is finite. Our interest is in the particular question: does state, or process, s have the property expressed by the formula A in the finite model $\mathcal{M} = (\mathcal{T}, V)$, i.e. is $s \in \| A \|_V^{\mathcal{T}}$? A natural technique is to compute the set $\| A \|_V$, [EL], using approximation techniques when A contains fixpoint subformulae. For instance, using semantic approximants, if V is a valuation let $V_0 = V[S_{\mathcal{T}}/Z]$ and $V_{i+1} = V_i[\| A \|_{V_i} /Z]$. Then because the model is finite we know that

$$\| \nu Z. A \|_V = \bigcap_{i \geq 0} V_i(Z)$$

Also by finiteness we know that there is $i \geq 0$ such that $V_i(Z) = V_{i+1}(Z)$, and for such an i, $V_i(Z) = \| \nu Z. A \|_V$. Finally one just needs to check whether or not the required state s is in this set. (A minimal fixpoint formula $\neg \nu Z. A$ can be dealt with by computing either $S_{\mathcal{T}} - \| \nu Z. A \|_V$ or $\bigcup_{i \geq 0} V_i(Z)$ where $V_0 = V[\emptyset/Z]$ and $V_{i+1} = V_i[\| \neg A[Z := \neg Z] \|_{V_i} /Z]$.) But this technique is not intended to be sensitive to the fact that we are interested only in whether or not the particular state s lies in $\| A \|_V$.

An apparent localisation is to appeal, instead, to syntactic approximants. Let $(\nu Z. A)^0 = \mathbf{true}$ and $(\nu Z. A)^{i+1} = A[Z := (\nu Z. A)^i]$. Then again because of finiteness we know that

$$s \in \| \nu Z. A \|_V \text{ iff } \forall i \geq 0.\, s \in \| (\nu Z. A)^i \|_V$$

But again it is necessary to compute the complete fixpoint set, i.e. the set $S' = \| (\nu Z. A)^i \|_V$ where $\| (\nu Z. A)^i \|_V = \| (\nu Z. A)^{i+1} \|_V$. For there is no guarantee that if for some j, $s \in \| (\nu Z. A)^j \|_V \cap \| (\nu Z. A)^{j+1} \|_V$ then also $s \in \| \nu Z. A \|_V$.

An alternative, more local, approach to model checking (which does not depend on computing complete fixpoint sets) is to appeal to fixpoint induction. The idea is that $s \in \| \nu Z. A \|_V$ if the assumption that $s \in \| \nu Z. A \|_V$ *implies* $s \in \| A[Z := \nu Z. A] \|_V$; and in the case of a minimal fixpoint formula, $s \in \| \mu Y. A \|_V$ if the assumption that $s \notin \| \mu Y. A \|_V$ implies $s \in \| A[Y := \mu Y. A] \|_V$. This technique is used by Larsen [La] for a logic which disallows alternating fixpoints: each formula contains only maximal fixpoints or only minimal fixpoints. The major problem here, especially in the presence of formulae containing alternating fixpoints, is that of logically understanding assumptions of the form $s \in \| \nu Z. A \|_V$ and $s \notin \| \mu Y. A \|_V$ as well as the notion of implication. The simple local tableau technique which we offer below not only caters for the full modal mu-calculus but also has a natural logical interpretation. There is, however, a small cost: a need to extend the mu-calculus to include propositional constants and definition lists.

3 Adding constants and definition lists

The syntax of the mu-calculus is extended to embrace a family of propositional constant symbols. Associated with a constant U is a declaration of the form $U = A$ where A is a closed formula, possibly containing previously declared constant symbols. A *definition list* is a sequence Δ of declarations $U_1 = A_1, \ldots, U_n = A_n$ such that $U_i \neq U_j$ whenever $i \neq j$ and such that each constant occurring in A_i is one of $U_1, \ldots, U_{i-1}$. This means that a prefix of a definition list is itself a definition list. When Δ as above is such a list we let $dom(\Delta) = \{U_1, \ldots, U_n\}$ and $\Delta(U_i) = A_i$. Moreover, if Δ is a definition list, $U \notin dom(\Delta)$ and each constant occurring in A is in $dom(\Delta)$, then $\Delta \cdot U = A$ is the definition list which is the result of appending $U = A$ to Δ. A definition list Δ is *admissible for B* if every constant occurring in B is declared in Δ. In this circumstance we let B_Δ be the formula B in the 'environment' Δ (see Definition 1). The interpretation of formulae is now extended to formulae relative to admissible definition lists by, in effect, treating constants as variables.

Definition 1 If $\Delta : U_1 = A_1, \ldots, U_n = A_n$ is admissible for B then $\| B_\Delta \|_V =_{df} \| B \|_{V_n}$ where $V_0 = V$ and $V_{i+1} = V_i[\| A_{i+1} \|_{V_i} / U_{i+1}]$.

This interpretation accords with the expected meaning of B_Δ in terms of syntactic substitution.

Lemma 2 $\| B_{\Delta \cdot U=A} \|_V = \| (B[U := A])_\Delta \|_V$.

Proof: By induction on the structure of B. □

A corollary, invoked later, is that if U does not occur in B then $B_{\Delta \cdot U=A}$ has the same meaning as B_Δ.

4 The model checker

The model checker is a tableau system for testing whether or not a state s has the property expressed by a closed formula A in a finite model $\mathcal{M}$. As is common in tableau systems, the rules are inverse natural deduction type rules. Here they are built from 'sequents' of the form $s \vdash^{\mathcal{M}}_{\Delta} A$, proof-theoretic analogues of $s \in \| A_\Delta \|^{\mathcal{T}}_V$. Each rule is of the form

$$\frac{s \vdash^{\mathcal{M}}_{\Delta} A}{s_1 \vdash^{\mathcal{M}}_{\Delta_1} A_1 \ldots s_k \vdash^{\mathcal{M}}_{\Delta_k} A_k}$$

where $k > 0$, possibly with side conditions. The premise sequent $s \vdash^{\mathcal{M}}_{\Delta} A$ is the goal to be achieved while the consequents are the subgoals, which are determined by the structure of the model 'near s,' the definition list Δ and the structure of A. Often, in the sequel, the index $\mathcal{M}$ is dropped from the sequents. The intermediate use of definition lists is essential, as they keep track of the 'dynamically changing' subformulae as fixpoints are unrolled. This is the key to the technique. Condition $\mathcal{C}$, the side-condition on the constant rules, is explained later as it is a condition on proof trees, rather than on the particular sequents of the premises.

$$\frac{s \vdash_\Delta \neg\neg A}{s \vdash_\Delta A} \qquad \frac{s \vdash_\Delta A \wedge B}{s \vdash_\Delta A \quad s \vdash_\Delta B}$$

$$\frac{s \vdash_\Delta \neg(A \wedge B)}{s \vdash_\Delta \neg A} \qquad \frac{s \vdash_\Delta \neg(A \wedge B)}{s \vdash_\Delta \neg B}$$

$$\frac{s \vdash_\Delta [a]A}{s_1 \vdash_\Delta A \ldots s_n \vdash_\Delta A} \quad \{s_1, \ldots, s_n\} = \{s' \mid s \xrightarrow{a} s'\}$$

$$\frac{s \vdash_\Delta \neg[a]A}{s' \vdash_\Delta \neg A} \quad s \xrightarrow{a} s'$$

$$\frac{s \vdash_\Delta \nu Z.\, A}{s \vdash_{\Delta'} U} \quad \Delta' \text{ is } \Delta \cdot U = \nu Z.\, A$$

$$\frac{s \vdash_\Delta \neg\nu Z.\, A}{s \vdash_{\Delta'} U} \quad \Delta' \text{ is } \Delta \cdot U = \neg\nu Z.\, A$$

$$\frac{s \vdash_\Delta U}{s \vdash_\Delta A[Z := U]} \quad \mathcal{C} \text{ and } \Delta(U) = \nu Z.\, A$$

$$\frac{s \vdash_{\Delta} U}{s \vdash_{\Delta} \neg A[Z := \neg U]} \quad \mathcal{C} \text{ and } \Delta(U) = \neg \nu Z.\, A$$

A *tableau* for $s \vdash^{\mathcal{M}} A$ is a maximal proof tree whose root is labelled with the sequent $s \vdash^{\mathcal{M}} A$ (where we omit the definition list when, as here, it is empty). The sequents labelling the immediate successors of a node labelled $s \vdash^{\mathcal{M}}_{\Delta} A$ are determined by an application of one of the rules, dependent on the structure of A. For simplicity we have allowed non-determinism in the result sequents in the cases of $\neg(A \wedge B)$ and $\neg[a]A$, rather than entangling proof trees with or-branching as well as and-branching. Maximality means that no rule applies to a sequent labelling a leaf of a tableau. The rules for booleans and modal operators are straightforward. New constants are introduced in the case of fixpoint formulae, while the rules for constants unroll the fixpoints they abbreviate when condition $\mathcal{C}$ holds. This condition is just that no node above the current premise, $s \vdash^{\mathcal{M}}_{\Delta} U$, in the proof tree is labelled $s \vdash^{\mathcal{M}}_{\Delta'} U$ for some Δ'. So failure of the condition, when there is a sequent $s \vdash^{\mathcal{M}}_{\Delta'} U$ above $s \vdash^{\mathcal{M}}_{\Delta} U$, enforces termination. In fact the presence of condition $\mathcal{C}$ guarantees that when $\mathcal{M}$ is finite any tableau for $s \vdash^{\mathcal{M}} A$ is of finite depth. Notice that all the rules are backwards sound. For example, in the case of the rule for maximal fixpoints, if Δ' is $\Delta \cdot U = \nu Z.\, A$ and $s \in \| U_{\Delta'} \|_V$, then by Lemma 2, $s \in \| \nu Z.\, A_{\Delta} \|_V$. Hence if the leaves of a (finite) tableau are *true*, i.e. if whenever $s \vdash_{\Delta} A$ labels a leaf, $s \in \| A_{\Delta} \|_V$, then so is the root.

A *successful* tableau for $s \vdash^{\mathcal{M}} A$ is a finite tableau in which every leaf is labelled by a sequent $t \vdash^{\mathcal{M}}_{\Delta} B$ fulfilling one of the following requirements:

(i)	$B = Q$ and $t \in V_{\mathcal{M}}(Q)$
(ii)	$B = \neg Q$ and $t \notin V_{\mathcal{M}}(Q)$
(iii)	$B = [a]C$
(iv)	$B = U$ and $\Delta(U) = \nu Z.\, C$

A successful tableau contains only true leaves. This is clear for leaves fulfilling (i) and (ii). Maximality of a tableau guarantees it for leaves satisfying (iii), because then $\{t' \mid t \xrightarrow{a} t'\} = \emptyset$. Of more interest is (iv): if $t \vdash^{\mathcal{M}}_{\Delta'} U$ labels a node in a tableau above a node labelled $t \vdash^{\mathcal{M}}_{\Delta} U$ where $\Delta(U) = \nu Z.\, A$, then indeed $t \in \| U_{\Delta} \|_{V_{\mathcal{M}}}$ (provided that the other leaves beneath $t \vdash^{\mathcal{M}}_{\Delta'} U$ are also true). An unsuccessful tableau has at least one false leaf, such as a leaf labelled $t \vdash^{\mathcal{M}}_{\Delta} Q$ where $t \notin V_{\mathcal{M}}(Q)$. Again, the most interesting failure is when a leaf is labelled $t \vdash^{\mathcal{M}}_{\Delta} U$ where $\Delta(U) = \neg \nu Z.\, A$ and above it is a node labelled $t \vdash^{\mathcal{M}}_{\Delta'} U$.

Tableau rules for the derived operators are just reformulations of some of the negation rules:

$$\frac{s \vdash_{\Delta} A \vee B}{s \vdash_{\Delta} A} \qquad \frac{s \vdash_{\Delta} A \vee B}{s \vdash_{\Delta} B}$$

$$\frac{s \vdash_{\Delta} \langle a \rangle A}{s' \vdash_{\Delta} A} \quad s \xrightarrow{a} s'$$

$$\frac{s \vdash_{\Delta} \mu Z.\, A}{s \vdash_{\Delta'} U} \quad \Delta' \text{ is } \Delta \cdot U = \mu Z.\, A$$

$$\frac{s \vdash_{\Delta} U}{s \vdash_{\Delta} A[Z := U]} \quad \mathcal{C} \text{ and } \Delta(U) = \mu Z.\, A$$

If these operators were also taken as primitive (as in the case of normal forms) then the definition of successful tableau would remain unchanged.

The two important theorems follow. Their proofs are given in section 6 below. For both we assume that $\mathcal{M}$ is finite. Theorem 4 affirms soundness and completeness, while Theorem 3 amounts to decidability (since there can be only a finite number of tableaux for $s \vdash^{\mathcal{M}} A$, up to renaming of constants). Discussion of the complexity of the model checker will be contained in the full version of the paper.

Theorem 3 Every tableau for $s \vdash^{\mathcal{M}} A$ is finite.

Theorem 4 $s \vdash^{\mathcal{M}} A$ has a successful tableau if and only if $s \in \| A \|_{V_{\mathcal{M}}}$.

By employing more complex sequents the side condition C on the two constant rules can be replaced with a condition on sequents. Let an *extended sequent* have the form

$$\alpha \longrightarrow s \vdash_{\Delta} A$$

where α is a finite set of sequents, each of which is of the form $t \vdash_{\Delta'} U$: the idea is that α contains all sequents above $s \vdash_{\Delta} A$ whose formula is a constant. The rules earlier can be trivially expanded to extended sequents. Two sample examples are:

$$\frac{\alpha \longrightarrow s \vdash_{\Delta} A \wedge B}{\alpha \longrightarrow s \vdash_{\Delta} A \qquad \alpha \longrightarrow s \vdash_{\Delta} B}$$

$$\frac{\alpha \longrightarrow s \vdash_{\Delta} U}{\alpha, s \vdash_{\Delta} U \longrightarrow s \vdash_{\Delta} A[Z := U]} \quad s \vdash_{\Delta'} U \notin \alpha \text{ for any } \Delta' \text{ and } \Delta(U) = \nu Z.\, A$$

Now the side condition C is replaced by: $s \vdash_{\Delta'} U \notin \alpha$ for any Δ'. This simple reformulation of the rules is akin to the formalisation of sequent calculi from natural deduction systems. It is also possible to dispense with the use of constants but at the expense of a complex subformula test [Cl].

5 Applications

We begin with two examples to illustrate the tableau method. Suppose $\mathcal{M} = (\mathcal{T}, V)$ is the model where $\mathcal{T}$ may be pictured as

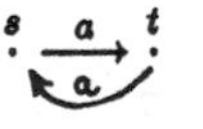

and $V(Q) = \{t\}$. Consider the formulae

$$\begin{aligned} A &\equiv \nu Z.\, \mu Y.\, [a]((Q \wedge Z) \vee Y) \\ B &\equiv \mu Y.\, \nu Z.\, [a]((Q \vee Y) \wedge Z) \end{aligned}$$

which in $\mathcal{M}$ express, respectively, that on all paths Q holds infinitely often, and that on all paths Q holds almost always. We present a successful tableau for $s \vdash^{\mathcal{M}} A$ and show that every tableau for $t \vdash^{\mathcal{M}} B$ is unsuccessful.

In the following successful tableau for $s \vdash^{\mathcal{M}} A$,

$$\begin{aligned} \Delta_1 &= (U_1 = A) \\ \Delta_2 &= \Delta_1 \cdot (U_2 = A_1) \\ \Delta_3 &= \Delta_2 \cdot (U_3 = A_1) \end{aligned}$$

where $A_1 = \mu Y. [a]((Q \wedge U_1) \vee Y)$.

$$s \vdash A$$

$$s \vdash_{\Delta_1} U_1$$

$$s \vdash_{\Delta_1} A_1$$

$$s \vdash_{\Delta_2} U_2$$

$$s \vdash_{\Delta_2} [a]((Q \wedge U_1) \vee U_2)$$

$$t \vdash_{\Delta_2} (Q \wedge U_1) \vee U_2$$

$$t \vdash_{\Delta_2} Q \wedge U_1$$

$$t \vdash_{\Delta_2} Q \qquad\qquad t \vdash_{\Delta_2} U_1$$

$$t \vdash_{\Delta_2} A_1$$

$$t \vdash_{\Delta_3} U_3$$

$$t \vdash_{\Delta_3} [a]((Q \wedge U_1) \vee U_3$$

$$s \vdash_{\Delta_3} (Q \wedge U_1) \vee U_3$$

$$s \vdash_{\Delta_3} U_3$$

$$s \vdash_{\Delta_3} [a]((Q \wedge U_1) \vee U_3)$$

$$t \vdash_{\Delta_3} (Q \wedge U_1) \vee U_3$$

$$t \vdash_{\Delta_3} Q \wedge U_1$$

$$t \vdash_{\Delta_3} Q \qquad\qquad\qquad\qquad t \vdash_{\Delta_3} U_1$$

In the following unsuccessful tableau for $s \vdash^M B$,

$$\Delta_1 = (U_1 = B)$$
$$\Delta_2 = \Delta_1 \cdot (U_2 = B_1)$$
$$\Delta_3 = \Delta_2 \cdot (U_3 = B_1)$$

where $B_1 = \nu Z. [a]((Q \vee U_1) \wedge Z)$.

$$t \vdash B$$
$$\overline{t \vdash_{\Delta_1} U_1}$$
$$\overline{t \vdash_{\Delta_1} B_1}$$
$$\overline{t \vdash_{\Delta_2} U_2}$$
$$\overline{t \vdash_{\Delta_2} [a]((Q \vee U_1) \wedge U_2)}$$
$$\overline{s \vdash_{\Delta_2} (Q \vee U_1) \wedge U_2}$$
$$\overline{s \vdash_{\Delta_2} Q \vee U_1 \qquad\qquad s \vdash_{\Delta_2} U_2}$$
$$\overline{s \vdash_{\Delta_2} U_1} \qquad\qquad \vdots$$
$$\overline{s \vdash_{\Delta_2} B_1}$$
$$\overline{s \vdash_{\Delta_3} U_3}$$
$$\overline{s \vdash_{\Delta_3} [a]((Q \vee U_1) \wedge U_3)}$$
$$\overline{t \vdash_{\Delta_3} (Q \vee U_1) \wedge U_3}$$
$$\overline{t \vdash_{\Delta_3} Q \vee U_1 \qquad\qquad t \vdash_{\Delta_3} U_3}$$
$$\overline{t \vdash_{\Delta_3} Q} \qquad\qquad \overline{t \vdash_{\Delta_3} [a]((Q \vee U_1) \wedge U_3)}$$
$$\overline{s \vdash_{\Delta_3} (Q \vee U_1) \wedge U_3}$$
$$\overline{s \vdash_{\Delta_3} Q \vee U_1 \qquad\qquad s \vdash_{\Delta_3} U_3}$$
$$\overline{s \vdash_{\Delta_3} U_1}$$

An important area of application of the model checker is to Milner's CCS [Mi]. An equivalent version of the checker has been implemented by Rance Cleaveland [Cl] in the Concurrency Workbench (a joint UK SERC venture between Sussex and Edinburgh Universities [CPS]). The operational semantics of CCS is given in terms of labelled transition systems. However, there is more than one transition system associated with CCS according to whether or not the τ action is observable. This distinction is marked by the differing transition relations $\xrightarrow{a}$ and $\xRightarrow{a}$ for $a \in Act$. In fact, the action sets differ too: there is the relation $\xrightarrow{\tau}$ but not $\xRightarrow{\tau}$; and there is the relation $\xRightarrow{\varepsilon}$, meaning zero or more silent moves, but not $\xrightarrow{\varepsilon}$. Thus, there are two different Hennessy-Milner logics for CCS [HM], each characterising the appropriate (strong or weak) bisimulation equivalence. Their extension to include fixpoints preserves this characterisation [Sti2]. These are sublanguages of the modal mu-calculus—for their sole atomic sentence is the constant `true`.

We now offer a more substantial example: an analysis of Knuth's mutual exclusion algorithm [Kn] when translated into CCS. Knuth's algorithm is given by the concurrent composition of the two programs when $i = 1$ and $i = 2$, and where j is the index of the other program:

while true **do**
begin

⟨ noncritical section ⟩ ;
L_0: $c_i := 1$;
L_1: **if** $k = i$ **then goto** L_2 ;
if $c_j \neq 0$ **then goto** L_1 ;
L_2: $c_i := 2$;
if $c_j = 2$ **then goto** L_0 ;
$k := i$;
⟨ critical section ⟩ ;
$k := j$;
$c_i := 0$;

end ;

The variable c_1 (c_2) of program one (two) may take the values 0, 1 or 2; initially its value is 0. When translated into CCS [Mi,Wa], the algorithm, assuming the initial value of k to be 1, becomes the agent *Knuth* below. For the example we let capital letters range over CCS processes (states of the CCS transition system). Here we are assuming that τ is not observable (so the transition relations are of the form $\stackrel{a}{\Longrightarrow}$). Each program variable is represented a family of agents. Thus the variable k with current value 1 is represented as an agent $K1$ which may perform actions corresponding to the reading of the value 1 and the writing of the values 1 and 2 by the two programs. The agents are:

$$Knuth =_{df} (P_1 \mid P_2 \mid K1 \mid C_1 0 \mid C_2 0)\backslash L$$

where L is the union of the sorts of the variables and

$$\begin{array}{lcl}
K1 & =_{df} & kw1.\,K1 + kw2.\,K2 + \overline{kw1}.\,K1 \\
K2 & =_{df} & kw1.\,K1 + kw2.\,K2 + \overline{kw2}.\,K2
\end{array}$$

$$\begin{array}{lcl}
C_1 0 & =_{df} & c_1w0.\,C_1 0 + c_1w1.\,C_1 1 + c_1w2.\,C_1 2 + \overline{c_1r0}.\,C_1 0 \\
C_1 1 & =_{df} & c_1w0.\,C_1 0 + c_1w1.\,C_1 1 + c_1w2.\,C_1 2 + \overline{c_1r1}.\,C_1 1 \\
C_1 2 & =_{df} & c_1w0.\,C_1 0 + c_1w1.\,C_1 1 + c_1w2.\,C_1 2 + \overline{c_1r2}.\,C_1 2
\end{array}$$

$$\begin{array}{lcl}
C_2 0 & =_{df} & c_2w0.\,C_2 0 + c_2w1.\,C_2 1 + c_2w2.\,C_2 2 + \overline{c_2r0}.\,C_2 0 \\
C_2 1 & =_{df} & c_2w0.\,C_2 0 + c_2w1.\,C_2 1 + c_2w2.\,C_2 2 + \overline{c_2r1}.\,C_2 1 \\
C_2 2 & =_{df} & c_2w0.\,C_2 0 + c_2w1.\,C_2 1 + c_2w2.\,C_2 2 + \overline{c_2r2}.\,C_2 2
\end{array}$$

$$\begin{array}{lcl}
P_1 & =_{df} & \tau.\,P_{11} + \tau.\,0 \\
P_{11} & =_{df} & \overline{c_1w1}.\,req_1.\,P_{12} \\
P_{12} & =_{df} & kr1.\,P_{14} + kr2.\,P_{13} \\
P_{13} & =_{df} & c_2r0.\,P_{14} + c_2r1.\,P_{12} + c_2r2.\,P_{12} \\
P_{14} & =_{df} & \overline{c_1w2}.\,P_{15} \\
P_{15} & =_{df} & c_2r0.\,P_{16} + c_2r1.\,P_{16} + c_2r2.\,P_{17} \\
P_{16} & =_{df} & \overline{kw1}.\,enter_1.\,exit_1.\,\overline{kw2}.\,\overline{c_1w0}.\,P_1 \\
P_{17} & =_{df} & \overline{c_1w1}.\,P_{12}
\end{array}$$

$$
\begin{array}{lll}
P_2 & =_{df} & \tau.\,P_{21} + \tau.\,0 \\
P_{21} & =_{df} & \overline{c_2w1}.\,req_2.\,P_{22} \\
P_{22} & =_{df} & kr2.\,P_{24} + kr1.\,P_{23} \\
P_{23} & =_{df} & c_1r0.\,P_{24} + c_1r1.\,P_{22} + c_1r2.\,P_{22} \\
P_{24} & =_{df} & \overline{c_2w2}.\,P_{25} \\
P_{25} & =_{df} & c_1r0.\,P_{26} + c_1r1.\,P_{26} + c_1r2.\,P_{27} \\
P_{26} & =_{df} & \overline{kw2}.\,enter_2.\,exit_2.\,\overline{kw1}.\,\overline{c_2w0}.\,P_2 \\
P_{27} & =_{df} & \overline{c_2w1}.\,P_{22}
\end{array}
$$

Some remarks on this representation may be helpful. The critical section of process P_i, where $i = 1$ or 2, is modelled as a pair of actions $enter_i$ and $exit_i$ representing, respectively, entry to and exit from the critical section. The noncritical section of each process is modelled as a summation, one summand of which represents the possibility that the process may halt, the other that it may proceed to request execution of its critical section. An action req_i appears in the definition of P_i. Its occurrence indicates that process P_i has 'just' indicated that it wishes to execute its critical section (by setting c_i to true). The reason for including these 'probes' will become clear below. Note also the presence of the agents P_{i7} and the way in which the statement **goto** L_0 is represented. The reason for this choice is that only the first $\overline{c_iw1}$ action (setting c_i to 1) is considered as signifying the initiation of an attempt by process i to execute its critical section.

The agent *Knuth* has sort $K = \{enter_i, exit_i, req_i \mid i = 1, 2\}$. We introduce two derived modal operators:

$$
\begin{array}{lll}
[K]A & \equiv & \bigwedge_{a\in K}[a]A \\
\langle K\rangle A & \equiv & \bigvee_{a\in K}\langle a\rangle A
\end{array}
$$

We consider two questions. Firstly, does the algorithm preserve mutual exclusion? And secondly, is the algorithm live (in the sense that if a process requests execution of its critical section it will eventually enter its critical section)? We express these questions as follows.

1. We say that Knuth's algorithm *preserves mutual exclusion* iff

$$Knuth \models \texttt{PME}$$

where PME ('preserves mutual exclusion') is the following formula:

$$\nu Z.\,((\neg(\langle exit_1\rangle\texttt{true} \wedge \langle exit_2\rangle\texttt{true})) \wedge [K]Z)$$

2. We say that Knuth's algorithm *is live* iff

$$Knuth \models \texttt{IL}$$

where IL is the formula

$$\nu Z.\,([req_1]\texttt{EICS1} \wedge [req_2]\texttt{EICS2}) \wedge [K]Z$$

where for $i = 1, 2$, EICSi ('eventually in critical section i') is the formula

$$\mu Y.\,[\varepsilon](((\langle exit_i\rangle\texttt{true} \vee ([K]Y \wedge \langle K\rangle\texttt{true}))$$

Some clarifying remarks may be helpful.

(i) Process i is 'in its critical section' if P_i reaches a state in which it may perform the action $exit_i$. The formula PME is satisfied by an agent P of sort K iff for any $s \in K^*$ and agent P', if $P \stackrel{s}{\Longrightarrow} P'$ then $P' \not\models \langle exit_1\rangle \mathtt{true} \wedge \langle exit_2\rangle \mathtt{true}$. Thus $Knuth \models$ PME iff it never reaches a state with both P_1 and P_2 in their critical sections.

(ii) $P \models$ EICSi iff there are no sequence $\langle a_j \mid j < \omega\rangle \in K^\omega$ and no sequence $\langle Q_j \mid j < \omega\rangle$ of agents such that $Q_0 = Q$ and for all j, $Q_j \stackrel{a_j}{\Longrightarrow} Q_{j+1}$ and $Q_j \not\models \langle exit_i\rangle\mathtt{true}$. Thus $Knuth \models$ IL iff for $i = 1, 2$, there is no path on which occur infinitely-many visible actions and on which there is a 'probe' req_i (indicating that P_i has requested execution of its critical section) which is not followed by a corresponding action $enter_i$.

Using the Concurrency Workbench we have verified that Knuth's algorithm preserves mutual exclusion and is live (for more details see [Wa]). The process $Knuth$ consists of a number of agents in parallel. A more enterprising model checker would try to verify liveness and safety properties of $Knuth$ by verifying appropriate subproperties of its components. Proof rules for structured model checking for the modal sublanguage of the mu-calculus are presented in [Stil]. We hope that these rules can be extended to the full mu-calculus.

6 Proofs of termination, soundness and completeness

We now prove the main results, theorems 3 and 4. First a little notation.

If B is a formula then $C(B)$ is the set of constants occurring in B. Recall from section 4 that a tableau is a maximal proof tree with root labelled $s \vdash^{\mathcal{M}} A$. Given two nodes n and n' in a tableau with n' an immediate successor of n, we say that the sequent $s' \vdash_{\Delta'} B'$ labelling n' *succeeds* the sequent $s \vdash_\Delta B$ labelling n. Also, given two nodes n and n' in a tableau labelled $s \vdash_\Delta U$ and $s' \vdash_{\Delta'} U'$ respectively, we say that $s' \vdash_{\Delta'} U'$ *C-succeeds* $s \vdash_\Delta U$ iff there is a sequence $\langle n_1, \ldots, n_k\rangle$ of nodes such that $n_1 = n$, $n_k = n'$, for $1 \le i < k$, n_{i+1} is an immediate successor of n_i, and for $1 < i < k$, the formula of the sequent labelling n_i is not a constant.

Next we define a useful nonnegative integer measure, the *degree*, $d(B)$, of a closed formula B:

$$d(Q) = 0 \qquad d(\neg Q) = 0$$

$$\begin{aligned} d(U) &= 0 \\ d(\neg\neg B) &= 1 + d(B) \end{aligned}$$

$$\begin{aligned} d(B \wedge C) &= 1 + \max\{d(B), d(C)\} & \qquad d(\neg(B \wedge C) &= 1 + \max\{d(\neg B), d(\neg C)\} \\ d([a]B) &= 1 + d(B) & d(\neg([a]B) &= 1 + d(\neg B) \\ d(\nu Z.\, B) &= 1 + d(B[Z := U]) & d(\neg\nu Z.\, B) &= 1 + d(\neg B[Z := \neg U]) \end{aligned}$$

We extend this definition to sequents as follows:

$$d(s \vdash_\Delta B) = \begin{cases} d(B) & \text{if } B \text{ is not a constant} \\ d(\Delta(B)) & \text{otherwise} \end{cases}$$

Lemma 3.1 (i) If $s' \vdash_{\Delta'} B'$ succeeds $s \vdash_\Delta B$ and B' is not a constant, then $d(s' \vdash_{\Delta'} B') < d(s \vdash_\Delta B)$.

(ii) If $s' \vdash_{\Delta'} U'$ $\mathcal{C}$-succeeds $s \vdash_\Delta U$, then either $U' \in \mathcal{C}(\Delta(U)) \cup \{U\}$, or $d(s' \vdash_{\Delta'} U') < d(s \vdash_\Delta U)$ and $\mathcal{C}(\Delta(U')) \subseteq \mathcal{C}(\Delta(U)) \cup \{U\}$.

(iii) Suppose Δ is a prefix of Δ' and $U \in dom(\Delta)$. Then for any s, s' $d(s \vdash_\Delta U) = d(s' \vdash_{\Delta'} U)$.

Proof: (i) By inspection of the the tableau rules and the definition of degree.

(ii) Suppose $\Delta(U) = \nu Z.B$. Then either U' is a subformula of $B[Z := U]$, when $U' \in \mathcal{C}(\Delta(U)) \cup \{U\}$, or U' is introduced as $\nu Z'.C$ $(\neg\nu Z'.C)$ which is a subformula of $B[Z := U]$, in which case $d(\nu Z'.C) < d(s \vdash_\Delta U)$ and $\mathcal{C}(\nu Z'.C) \subseteq \mathcal{C}(\Delta(U)) \cup \{U\}$ (and similarly for $\neg\nu Z'.C$).

(iii) Immediate from the definition. □

We now prove the termination theorem.

Theorem 3 Every tableau for $s \vdash^{\mathcal{M}} A$ is finite.

Proof: We omit the index $\mathcal{M}$.

Suppose there is an infinite tableau τ for $s \vdash A$. Since τ is finite-branching, there is an infinite path π through τ. Let $\sigma = \langle s_i \vdash_{\Delta_i} A_i \mid i < \omega\rangle$ be the sequence of sequents labelling the nodes of π. Since for each i, $s_{i+1} \vdash_{\Delta_{i+1}} A_{i+1}$ succeeds $s_i \vdash_{\Delta_i} A_i$, from Lemma 3.1(i) it follows that for infinitely many i, A_i is a constant. Also, since $\mathcal{M}$ is finite, no one constant appears infinitely often on π.

Consider the subsequence $\sigma' = \langle s'_i \vdash_{\Delta'_i} U_i \mid i < \omega\rangle$ of σ consisting of those sequents whose formulae are constants. Note that for each i, $s'_{i+1} \vdash_{\Delta'_{i+1}} U_{i+1}$ $\mathcal{C}$-succeeds $s'_i \vdash_{\Delta'_i} U_i$. Suppose i_0 is the largest i with $U_i = U_0$. Then since $\mathcal{C}(\Delta'_0(U_0)) = \emptyset$, by Lemma 3.1(ii), $d(s'_{i_0+1} \vdash_{\Delta'_{i_0+1}} U_{i_0+1}) < d(s'_{i_0} \vdash_{\Delta'_{i_0}} U_0)$ and $\mathcal{C}(\Delta'_{i_0+1}(U_{i_0+1})) \subseteq \{U_0\}$.

Now suppose i_1 is the largest i with $U_i = U_{i_0+1}$. Then again by Lemma 3.1(ii), $d(s'_{i_1+1} \vdash_{\Delta'_{i_1+1}} U_{i_1+1}) < d(s'_{i_1} \vdash_{\Delta'_{i_1}} U_{i_0+1})$ and $\mathcal{C}(\Delta'_{i_1+1}(U_{i_1+1})) \subseteq \{U_0, U_{i_0+1}\}$. By Lemma 3.1(iii), $d(s'_{i_1+1} \vdash_{\Delta'_{i_1+1}} U_{i_1+1}) < d(s'_{i_0+1} \vdash_{\Delta'_{i_0+1}} U_{i_0+1}) < d(s'_0 \vdash_{\Delta'_0} U_0)$.

By repeating this argument sufficiently often we obtain a contradiction since d is a nonnegative integer measure. □

Now we come to the proofs of soundness and completeness.

Theorem 4 $s \vdash^{\mathcal{M}} A$ has a successful tableau if and only if $s \in \| A \|_{\mathcal{V}_{\mathcal{M}}}$.

Proof: First some notation and a standard lemma.

If $B = \nu Z.D$ then $B^0 = \mathbf{true}$ and $B^{i+1} = D[Z := B^i]$.

If $B = \neg\nu Z.D$ then $B^0 = \mathbf{false}$ and $B^{i+1} = \neg D[Z := \neg B^i]$.

Lemma 4.1 ($\mathcal{M}$ finite)

(i) If $B = \nu Z.D$ and $s \notin \| B_\Delta \|_\mathcal{V}$, then there is $n < \omega$ such that $s \in \| (B^n)_\Delta \|_\mathcal{V} - \| (B^{n+1})_\Delta \|_\mathcal{V}$.

(ii) If $C = \neg\nu Z.D$ and $s \in \| C_\Delta \|_\mathcal{V}$, then there is $n < \omega$ such that $s \in \| (C^{n+1})_\Delta \|_\mathcal{V} - \| (C^n)_\Delta \|_\mathcal{V}$. □

We omit the indices $\mathcal{M}$ and $V_{\mathcal{M}}$.
($\Longrightarrow$) Suppose $s \vdash A$ has a successful tableau τ. If all the leaves of τ are true (i.e. if whenever $t \vdash_\Delta B$ labels a leaf then $t \in \| B_\Delta \|$), then all the nodes of τ are true: for, as we noted earlier, the rules are backwards sound. So it suffices to show that all the leaves of τ are true.

If a leaf is labelled $t \vdash_\Delta B$ with $B = Q$, $\neg Q$ or $[a]C$, then it is certainly true. Hence any false leaf must be labelled $t \vdash_\Delta U$ with $\Delta(U) = \nu Z.B$. Suppose there is a false leaf. From amongst all false leaves choose one, labelled $t \vdash_\Sigma U$ say, such that there is no constant U' introduced before U in τ for which there is a false leaf labelled $t' \vdash_{\Sigma'} U'$ for some t', Σ'. Consider the subtableau τ_1 of τ whose root is the node, labelled $s \vdash_\Delta U$ say, at which U is introduced in τ. For each of the false leaves of τ labelled $t \vdash_\Sigma U$ for some t, Σ, by Lemma 4.1(i) there is $n < \omega$ such that $t \in \| (\nu Z.B)^n_\Sigma \| - \| (\nu Z.B)^{n+1}_\Sigma \|$ where $\Delta(U) = \nu Z.B$. Choose such a leaf l, labelled $t \vdash_\Sigma U$ say, such that the corresponding n is as small as possible. Note that since l is a leaf, there is above l in τ_1 a node k, the *companion node* of l, labelled $t \vdash_{\Sigma'} U$ for some Σ'.

Now transform the tableau τ_1 into a new tableau τ_1^* by replacing each definition list Δ' in a sequent of τ_1 by $\Delta'[(\nu Z.B)^n/U]$. An examination of the rules shows that if the leaves of τ_1^* are true then all the nodes of τ_1^* are true: the only rule which could prevent this, namely

$$\frac{s' \vdash_{\Delta'} \nu Z.B}{s' \vdash_{\Delta''} U} \quad \Delta'' \text{ is } \Delta' \cdot U = (\nu Z.B)^n$$

is not applied in τ_1^* since the root of τ_1^* is labelled $s \vdash_{\Delta[(\nu Z.B)^n/U]} U$. But the image of the successor of the companion node k of l under the transformation is false since it is labelled $t \vdash_{\Sigma'[(\nu Z.B)^n/U]} B[Z := U]$ and $t \notin \| (\nu Z.B)^{n+1}_{\Sigma'} \|$. Therefore some leaf of τ_1^* is false.

Suppose $t' \vdash_{\Delta''} U'$ labels such a false leaf where the corresponding leaf of τ_1 is labelled $t' \vdash_{\Delta'} U'$ so that $\Delta'' = \Delta'[(\nu Z.B)^n/U]$. Then by the choice of n we have that $U' \neq U$. Moreover, U' is not introduced before U in τ, since otherwise, by the observation immediately following Lemma 2, the leaf of τ labelled $t' \vdash_{\Delta'} U'$ would be false, contradicting the choice of U. Hence U' is introduced after U in τ.

But now we may apply the entire argument above to the tableau τ_1^*. And so on. But this contradicts Theorem 3, that every tableau is finite.

($\Longleftarrow$) We build a *pseudo-tableau* with root $s \vdash A$. The rules for pseudo-tableaux differ from those for tableaux in just one case: the rule for constants defined as minimal fixpoints. The pseudo-tableau rule is

$$\frac{t \vdash_\Delta U}{t \vdash_{\Delta'} \neg B[Z := \neg U]} \quad C, \text{ and } \Delta(U) = \neg\nu Z.B \text{ or } (\neg\nu Z.B)^n$$

where $\Delta' = \Delta[(\neg\nu Z.B)^k/U]$ with k such that $s \in \| (\neg\nu Z.B)^{k+1}_\Delta \| - \| (\neg\nu Z.B)^k_\Delta \|$. Note that by Lemma 4.1(ii), if $t \in \| U_\Delta \|$ then this rule is applicable (provided C holds), and in such a case, if $\Delta(U) = (\neg\nu Z.B)^n$ and $\Delta'(U) = (\neg\nu Z.B)^k$, then $k < n$. We assume the same termination conditions for pseudo-tableaux as for tableaux. Moreover, defining the degree function as in the proof of Theorem 3 with $d(\Delta(U)) = d(\neg\nu Z.B)$ when $\Delta(U) = (\neg\nu Z.B)^n$, then by an argument similar to that in the proof of Theorem 3 we have that every pseudo-tableau for $s \vdash A$ is finite (provided $\mathcal{M}$ is finite). Finally we define the notion of a *successful* pseudo-tableau as for tableaux with the requirement that no leaf is labelled $t \vdash_\Delta U$ where $\Delta(U) = (\neg\nu Z.B)^n$.

A successful pseudo-tableau can be transformed into a successful tableau simply by updating the definition lists, changing $\Delta(U)$ from $(\neg\nu Z.B)^n$ to $\neg\nu Z.B$ as necessary. Hence it suffices to show that there is a successful pseudo-tableau for $s \vdash A$. Such a pseudo-tableau may be constructed as follows.

Its root is labelled $s \vdash A$ and is true. Suppose $t \vdash_\Delta B$ labels a leaf of the partial pseudo-tableau and $t \in \| B_\Delta \|$. We define the successors of this node in the pseudo-tableau as follows depending on the structure of B.

(1) $B = Q$ or $\neg Q$: the node has no successors.

(2) $B = \neg\neg C$: the node has single true successor labelled $t \vdash_\Delta C$.

(3) $B = C \wedge D$ or $\neg(C \wedge D)$: if $B = C \wedge D$ then the node has two successors, one labelled $t \vdash_\Delta C$, the other $t \vdash_\Delta D$. Since $t \in \| B_\Delta \|$, the successors are true. If $B = \neg(B \wedge C)$ there is one true successor labelled $t \vdash_\Delta \neg C$ or $t \vdash_\Delta \neg D$.

(4) $B = [a]C$ or $\neg[a]C$: similar to (2) with the extra possibility that $\{t' \mid t \xrightarrow{a} t'\} = \emptyset$ in which case the node has no successors.

(5) $B = \nu Z. C$ or $\neg \nu Z. C$: if $B = \nu Z. C$ then since $t \in \| B_\Delta \|$, $t \in \| U_{\Delta'} \|$ where Δ' is $\Delta \cdot U = \nu Z. C$. Similarly for $\neg \nu Z. C$.

(6) $B = U$: if C holds and $\Delta(U) = \neg \nu Z. C$ or $(\neg \nu Z. C)^n$ then by Lemma 4.1 there is k with $t \in \| (\neg \nu Z. C)^{k+1}_\Delta \| - \| (\neg \nu Z. C)^k_\Delta \|$, when $t \in \| \neg C[Z := \neg U]_{\Delta'} \|$ where $\Delta' = \Delta[(\neg \nu Z. C)^k / U]$. The case $\Delta(U) = \nu Z. C$ is simpler.

By the remarks above we thus obtain a pseudo-tableau in which all the nodes are true. The only possible impediment to its success could be that $t \vdash_\Delta U$ labels a leaf where $\Delta(U) = (\neg \nu Z. B)^k$. But by the choices of k in the construction this is impossible. □

Acknowledgments

We would like to thank Rance Cleaveland, Kim Larsen and Bernhard Steffen for comments and discussions about model checking. The second author was supported by a grant from the Venture Research Unit of BP.

References

[Cl] R. Cleaveland, *Tableau-Based Model Checking in the Propositional Mu-Calculus*, typewritten paper 1988.

[CPS] R. Cleaveland, J. Parrow and B. Steffen, *The Concurrency Workbench*, to appear.

[EL] E. Emerson and C. Lei, *Efficient model checking in fragments of the propositional mu-calculus*, Proc. Symposium on Logic in Computer Science, Cambridge, Mass., 267–278, 1986.

[GS] S. Graf and J. Sifakis, *A modal characterization of observational congruence of finite terms of CCS*, Information and Control 68, 125–145, 1986.

[HM] M. Hennessy and R. Milner, *Algebraic laws for nondeterminism and concurrency*, JACM 32, 137–161, 1985.

[Ho] S. Holmström, *Hennessy-Milner Logic with Recursion as a Specification Language, and a Refinement Calculus based on it*, Report 44 Programming Methodology Group, University of Göteborg, 1988.

[Kn] D. Knuth, *Additional Comments on a Problem in Concurrent Programming Control*, Comm. ACM 9/5, 1966.

[Ko] D. Kozen, *Results on the propositional mu-calculus*, Theoretical Computer Science 27, 333-354, 1983.

[La] K. Larsen, *Proof systems for Hennessy-Milner logic with recursion*, Proc. CAAP 1988.

[Mi] R. Milner, *A Calculus of Communicating Systems*, Springer Lecture Notes ion Computer Science, vol. 92, 1980.

[Pn] A. Pnueli, *Specification and development of reactive systems*, Information Processing 86, North-Holland, 854–858, 1986.
[Pr] V. Pratt, *A decidable μ-calculus*, Proc. 22nd. FOCS, 421-27, 1981.
[St] B. Steffen, *Characteristic formulae*, University of Edinburgh report, 1988.
[Sti1] C. Stirling, *Modal Logics for Communicating Systems*, Theoretical Computer Science 49 311–347, 1987.
[Sti2] C. Stirling, *Temporal Logics for CCS*, to appear in Proc. of REX Workshop, 1988.
[Wa] D. Walker, *Automated Analysis of Mutual Exclusion Algorithms Using CCS*, submitted for publication, 1988.

Vol. 296: R. Janßen (Ed.), Trends in Computer Algebra. Proceedings, 1987. V, 197 pages. 1988.

Vol. 297: E.N. Houstis, T.S. Papatheodorou, C.D. Polychronopoulos (Eds.), Supercomputing. Proceedings, 1987. X, 1093 pages. 1988.

Vol. 298: M. Main, A. Melton, M. Mislove, D. Schmidt (Eds.), Mathematical Foundations of Programming Language Semantics. Proceedings, 1987. VIII, 637 pages. 1988.

Vol. 299: M. Dauchet, M. Nivat (Eds.), CAAP '88. Proceedings, 1988. VI, 304 pages. 1988.

Vol. 300: H. Ganzinger (Ed.), ESOP '88. Proceedings, 1988. VI, 381 pages. 1988.

Vol. 301: J. Kittler (Ed.), Pattern Recognition. Proceedings, 1988. VII, 668 pages. 1988.

Vol. 302: D.M. Yellin, Attribute Grammar Inversion and Source-to-source Translation. VIII, 176 pages. 1988.

Vol. 303: J.W. Schmidt, S. Ceri, M. Missikoff (Eds.), Advances in Database Technology – EDBT '88. X, 620 pages. 1988.

Vol. 304: W.L. Price, D. Chaum (Eds.), Advances in Cryptology – EUROCRYPT '87. Proceedings, 1987. VII, 314 pages. 1988.

Vol. 305: J. Biskup, J. Demetrovics, J. Paredaens, B. Thalheim (Eds.), MFDBS 87. Proceedings, 1987. V, 247 pages. 1988.

Vol. 306: M. Boscarol, L. Carlucci Aiello, G. Levi (Eds.), Foundations of Logic and Functional Programming. Proceedings, 1986. V, 218 pages. 1988.

Vol. 307: Th. Beth, M. Clausen (Eds.), Applicable Algebra, Error-Correcting Codes, Combinatorics and Computer Algebra. Proceedings, 1986. VI, 215 pages. 1988.

Vol. 308: S. Kaplan, J.-P. Jouannaud (Eds.), Conditional Term Rewriting Systems. Proceedings, 1987. VI, 278 pages. 1988.

Vol. 309: J. Nehmer (Ed.), Experiences with Distributed Systems. Proceedings, 1987. VI, 292 pages. 1988.

Vol. 310: E. Lusk, R. Overbeek (Eds.), 9th International Conference on Automated Deduction. Proceedings, 1988. X, 775 pages. 1988.

Vol. 311: G. Cohen, P. Godlewski (Eds.), Coding Theory and Applications 1986. Proceedings, 1986. XIV, 196 pages. 1988.

Vol. 312: J. van Leeuwen (Ed.), Distributed Algorithms 1987. Proceedings, 1987. VII, 430 pages. 1988.

Vol. 313: B. Bouchon, L. Saitta, R.R. Yager (Eds.), Uncertainty and Intelligent Systems. IPMU '88. Proceedings, 1988. VIII, 408 pages. 1988.

Vol. 314: H. Göttler, H.J. Schneider (Eds.), Graph-Theoretic Concepts in Computer Science. Proceedings, 1987. VI, 254 pages. 1988.

Vol. 315: K. Furukawa, H. Tanaka, T. Fujisaki (Eds.), Logic Programming '87. Proceedings, 1987. VI, 327 pages. 1988.

Vol. 316: C. Choffrut (Ed.), Automata Networks. Proceedings, 1986. VII, 125 pages. 1988.

Vol. 317: T. Lepistö, A. Salomaa (Eds.), Automata, Languages and Programming. Proceedings, 1988. XI, 741 pages. 1988.

Vol. 318: R. Karlsson, A. Lingas (Eds.), SWAT 88. Proceedings, 1988. VI, 262 pages. 1988.

Vol. 319: J.H. Reif (Ed.), VLSI Algorithms and Architectures – AWOC 88. Proceedings, 1988. X, 476 pages. 1988.

Vol. 320: A. Blaser (Ed.), Natural Language at the Computer. Proceedings, 1988. III, 176 pages. 1988.

Vol. 321: J. Zwiers, Compositionality, Concurrency and Partial Correctness. VI, 272 pages. 1989.

Vol. 322: S. Gjessing, K. Nygaard (Eds.), ECOOP '88. European Conference on Object-Oriented Programming. Proceedings, 1988. VI, 410 pages. 1988.

Vol. 323: P. Deransart, M. Jourdan, B. Lorho, Attribute Grammars. IX, 232 pages. 1988.

Vol. 324: M.P. Chytil, L. Janiga, V. Koubek (Eds.), Mathematical Foundations of Computer Science 1988. Proceedings. IX, 562 pages. 1988.

Vol. 325: G. Brassard, Modern Cryptology. VI, 107 pages. 1988.

Vol. 326: M. Gyssens, J. Paredaens, D. Van Gucht (Eds.), ICDT '88. 2nd International Conference on Database Theory. Proceedings, 1988. VI, 409 pages. 1988.

Vol. 327: G.A. Ford (Ed.), Software Engineering Education. Proceedings, 1988. V, 207 pages. 1988.

Vol. 328: R. Bloomfield, L. Marshall, R. Jones (Eds.), VDM '88. VDM – The Way Ahead. Proceedings, 1988. IX, 499 pages. 1988.

Vol. 329: E. Börger, H. Kleine Büning, M.M. Richter (Eds.), CSL '87. 1st Workshop on Computer Science Logic. Proceedings, 1987. VI, 346 pages. 1988.

Vol. 330: C.G. Günther (Ed.), Advances in Cryptology – EUROCRYPT '88. Proceedings, 1988. XI, 473 pages. 1988.

Vol. 331: M. Joseph (Ed.), Formal Techniques in Real-Time and Fault-Tolerant Systems. Proceedings, 1988. VI, 229 pages. 1988.

Vol. 332: D. Sannella, A. Tarlecki (Eds.), Recent Trends in Data Type Specification. V, 259 pages. 1988.

Vol. 333: H. Noltemeier (Ed.), Computational Geometry and its Applications. Proceedings, 1988. VI, 252 pages. 1988.

Vol. 334: K.R. Dittrich (Ed.), Advances in Object-Oriented Database Systems. Proceedings, 1988. VII, 373 pages. 1988.

Vol. 335: F.A. Vogt (Ed.), CONCURRENCY 88. Proceedings, 1988. VI, 401 pages. 1988.

Vol. 336: B.R. Donald, Error Detection and Recovery in Robotics. XXIV, 314 pages. 1989.

Vol. 337: O. Günther, Efficient Structures for Geometric Data Management. XI, 135 pages. 1988.

Vol. 338: K.V. Nori, S. Kumar (Eds.), Foundations of Software Technology and Theoretical Computer Science. Proceedings, 1988. IX, 520 pages. 1988.

Vol. 339: M. Rafanelli, J.C. Klensin, P. Svensson (Eds.), Statistical and Scientific Database Management. Proceedings, 1988. IX, 454 pages. 1989.

Vol. 340: G. Rozenberg (Ed.), Advances in Petri Nets 1988. VI, 439 pages. 1988.

Vol. 341: S. Bittanti (Ed.), Software Reliability Modelling and Identification. VII, 209 pages. 1988.

Vol. 342: G. Wolf, T. Legendi, U. Schendel (Eds.), Parcella '88. Proceedings, 1988. 380 pages. 1989.

Vol. 343: J. Grabowski, P. Lescanne, W. Wechler (Eds.), Algebraic and Logic Programming. Proceedings, 1988. 278 pages. 1988.

Vol. 344: J. van Leeuwen, Graph-Theoretic Concepts in Computer Science. Proceedings, 1988. VII, 459 pages. 1989.

Vol. 345: R.T. Nossum (Ed.), Advanced Topics in Artificial Intelligence. VII, 233 pages. 1988 (Subseries LNAI).

Vol. 346: M. Reinfrank, J. de Kleer, M.L. Ginsberg, E. Sandewall (Eds.), Non-Monotonic Reasoning. Proceedings, 1988. XIV, 237 pages. 1989 (Subseries LNAI).

Vol. 347: K. Morik (Ed.), Knowledge Representation and Organization in Machine Learning. XV, 319 pages. 1989 (Subseries LNAI).

Vol. 348: P. Deransart, B. Lorho, J. Małuszyński (Eds.), Programming Language Implementation and Logic Programming. Proceedings, 1988. VI, 299 pages. 1989.

Vol. 349: B. Monien, R. Cori (Eds.), STACS 89. Proceedings, 1989. VIII, 544 pages. 1989.

Vol. 350: A. Törn, A. Žilinskas, Global Optimization. X, 255 pages. 1989.

Vol. 351: J. Díaz, F. Orejas (Eds.), TAPSOFT '89. Volume 1. Proceedings, 1989. X, 383 pages. 1989.

Vol. 352: J. Díaz, F. Orejas (Eds.), TAPSOFT '89. Volume 2. Proceedings, 1989. X, 389 pages. 1989.

This series reports new developments in computer science research and teaching – quickly, informally and at a high level. The type of material considered for publication includes preliminary drafts of original papers and monographs, technical reports of high quality and broad interest, advanced level lectures, reports of meetings, provided they are of exceptional interest and focused on a single topic. The timeliness of a manuscript is more important than its form which may be unfinished or tentative. If possible, a subject index should be included. Publication of Lecture Notes is intended as a service to the international computer science community, in that a commercial publisher, Springer-Verlag, can offer a wide distribution of documents which would otherwise have a restricted readership. Once published and copyrighted, they can be documented in the scientific literature.

Manuscripts

Manuscripts should be no less than 100 and preferably no more than 500 pages in length.

They are reproduced by a photographic process and therefore must be typed with extreme care. Symbols not on the typewriter should be inserted by hand in indelible black ink. Corrections to the typescript should be made by pasting in the new text or painting out errors with white correction fluid. Authors receive 75 free copies and are free to use the material in other publications. The typescript is reduced slightly in size during reproduction; best results will not be obtained unless the text on any one page is kept within the overall limit of 18 x 26.5 cm (7 x 10½ inches). On request, the publisher will supply special paper with the typing area outlined.

Manuscripts should be sent to Prof. G. Goos, GMD Forschungsstelle an der Universität Karlsruhe, Haid- und Neu-Str. 7, 7500 Karlsruhe 1, Germany, Prof. J. Hartmanis, Cornell University, Dept. of Computer Science, Ithaca, NY/USA 14850, or directly to Springer-Verlag Heidelberg.

Springer-Verlag, Heidelberger Platz 3, D-1000 Berlin 33
Springer-Verlag, Tiergartenstraße 17, D-6900 Heidelberg 1
Springer-Verlag, 175 Fifth Avenue, New York, NY 10010/USA
Springer-Verlag, 37-3, Hongo 3-chome, Bunkyo-ku, Tokyo 113, Japan

ISBN 3-540-50939-9
ISBN 0-387-50939-9